T4-AJS-574

GEOGRAPHIC INFORMATION SYSTEMS IN TRANSPORTATION RESEARCH

Related books and journals

Books

BELL (ed.)	Transportation Networks
ETTEMA & TIMMERMANS (eds.)	Activity-Based Approaches to Travel Analysis
GÄRLING ET AL (eds.)	Theoretical Foundations of Travel Choice Modelling
STOPHER & LEE-GOSSELIN (eds.)	Understanding Travel Behaviour In An Era of Change
TILANUS (ed.)	Information Systems in Logistics and Transportation

Journals

Accident Analysis and Prevention
Editor: Frank Haight

Transport Policy
Editors: Moshe Ben-Akiva, Yoshitsugu Hayashi and John Preston

Transportation Research Parts A-F

GEOGRAPHIC INFORMATION SYSTEMS IN TRANSPORTATION RESEARCH

Edited by

JEAN-CLAUDE THILL

Department of Geography and National Center for Geographic Information and Analysis, State University of New York at Buffalo

2000

PERGAMON

An imprint of Elsevier Science

Amsterdam – London – Oxford – New York – Paris – Shannon – Tokyo

ELSEVIER SCIENCE Ltd
The Boulevard, Langford Lane
Kidlington, Oxford OX5 1GB, UK

First edition 2000

Library of Congress Cataloging in Publication Data
A catalog record from the Library of Congress has been applied for.

British Library Cataloguing in Publication Data
A catalogue record from the British Library has been applied for.

ISBN: 0-08-043630-7

∞ The paper used in this publication meets the requirements of ANSI/NISO Z39.48-1992 (Permanence of Paper).
Printed in The Netherlands.

CONTENTS

VOLUME I

PERGAMON

Transportation Research Part C 8 (2000) 1

TRANSPORTATION RESEARCH PART C

www.elsevier.com/locate/trc

Preface

The idea for this book grew out of the difficulties encountered by graduate students enrolled in my *Transportation Modeling and GIS* class in grasping the depth and breadth of the geographic information systems (GIS) technology in transportation research. The long separate evolution of research in the fields of transportation and GIS has finally reached the point where paths overlap and continue in great unison. GIS has proven to be an approach to spatial research that dramatically enhances efficiency and effectiveness. Sorely missing in this rapidly growing field of application is a proper framing of the wealth of knowledge accumulated so far in apply GIS to transportation research questions. This volume is precisely intended to fill this void, with the ultimate goal of supporting the dissemination of recent research into graduate education and of hastening technology adoption in the field of transportation. It appears to be the first original book devoted exclusively to this topic.

This volume consists of 22 original papers solicited to represent the broad base of contemporary research themes in transportation GIS. Forty-five scholars from America, Europe and Asia have contributed their knowledge to produce a unique compilation of recent developments in the field. The quality of this volume has been greatly enhanced by the unselfish professionalism and dedication of 75 colleagues who advised me in the form of anonymous referee reports on individual manuscripts.

Special arrangements with Elsevier Science Ltd. made possible the dual publication of all the contributions in the form of the present book and as volume 8 of the periodical *Transportation Research Part C: Emerging Technologies*. In spite of the stringent schedule imposed by this arrangement, contributing authors showed great diligence throughout the entire process. Professor Stephen Ritchie, Editor of *Transportation Research Part C: Emerging Technologies* provided valuable oversight and guidance in meeting the editorial standards of the journal. His support during this effort has been remarkable. Finally, I am thankful for the patience, availability, and dedication of the editorial staff at Elsevier Science Ltd., particularly Chris Pringle and Leighton Chipperfield.

Jean-Claude Thill

Department of Geography and National Center for Geog. Info & Analysis,
State University of New York at Buffalo,
Buffalo, NY 14261, USA
E-mail address: jcthill@acsu.buffalo.edu

PII: S0968-090X(00)00030-9

PERGAMON

Transportation Research Part C 8 (2000) 3–12

TRANSPORTATION RESEARCH PART C

www.elsevier.com/locate/trc

Geographic information systems for transportation in perspective

Jean-Claude Thill *

Department of Geography and National Center for Geographic Information and Analysis, State University of New York at Buffalo, Buffalo, NY 14261, USA

Abstract

The late 1980s saw the first widespread use of Geographic Information Systems (GIS) in transportation research and management. Due to the specific requirements of transportation applications and of the rather late adoption of this information technology in transportation, research has been directed toward enhancing existing GIS approaches to enable the full range of capabilities needed in transportation research and management. This paper places the concept of transportation GIS in the broader perspective of research in GIS and Geographic Information Science. The emphasis is placed on the requirements specific of the transportation domain of application of this emerging information technology as well as on core research challenges. © 2000 Elsevier Science Ltd. All rights reserved.

Keywords: GIS-T; Geographic information systems

1. Introduction

It is quite a paradox that the field of transportation has at last come to embrace Geographic Information Systems (GISs) as a key technology to support its research and operational needs, while some of the early pioneers of GIS at the University of Washington and Northwestern University were in fact transportation scientists. Over three decades have passed since these missed opportunities for cross-fertilization. GIS has since evolved and matured from a tool, to a technology, and finally to a legitimate domain of scientific inquiry called Geographic Information Science (Goodchild, 1992a). In the meantime, transportation has distanced itself from its historical roots in geographic and spatial sciences, but has also become increasingly multi-

* Tel.: +1 716 645 2722; fax: +1 716 645 2329.
E-mail address: jcthill@acsu.buffalo.edu (J.-C. Thill).

0968-090X/00/$ - see front matter
PII: S0968-090X(00)00029-2

disciplinary, thus reflecting the multi-faceted reality of transportation infrastructure and flows and movements of passengers and freight.

In the United States, the multi-disciplinary outlook of transportation has been fostered by several key pieces of federal legislation passed in the 1990s (Meyer, 1999), including the Clean Air Act Amendments, the Intermodal Surface Transportation Efficiency Act, the American with Disabilities Act, and the Transportation Equity Act for the 21st century. All four acts contained explicit requirements for local and state governments to consider transportation systems through their interdependence with other natural, social, or economic systems. From the new integrative mission of transportation studies grew the need for enhanced approaches to store, manipulate, and analyze data spanning multiple themes: for instance, highway infrastructure, peak-time traffic flow, transit offering (fare, frequency, reliability), population ethnicity, work force participation, and air quality. Offering a data management and modeling platform capable of integrating a vast array of data from various sources, captured at different resolutions (street segments, census tracts, traffic analysis zones, street corner, etc.), and on seemingly unrelated themes, GIS has positioned itself as the ultimate information integration technology. In a GIS, integration proceeds by referencing all objects to some common locational framework. With proper conversion rules and algorithms, data stored in different scales, projections and data models can be registered to the same underlying referencing framework. It can be argued that the present adoption of GIS in transportation brings the field to a full circle as it is rediscovering the primacy of space and place, two concepts that launched the systematic study of transportation in Geography and Regional Science in the 1950s. The acronym GIS-T is often employed to refer to the application and adaptation of GIS to research, planning, and management in transportation.

In this paper, I give an overview of the nature of GIS and place its evolution in context. I also discuss the specificity and requirements of GIS in transportation. Finally, some of the core research themes in transportation GIS are highlighted.

2. The nature of geographic information systems

GISs are computer-based systems for the capture, storage, manipulation, display, and analysis of geographic information. The multiple functionality afforded by GIS distinguishes it from older technologies. The integration of multiple functionalities within one rather seamless environment dispenses users from mastering a collection of disparate, and specialized technologies. As it turns out, this aspect is often held by organizations as one of the decisive criteria in their decision to adopt GIS technology because of its efficiency benefits.

The functional complexity of GIS is what makes it a system different from any other. Without geo-visualization capability, the GIS is merely a database management engine endowed with some power to extract meaningful relationship between data entities. Without analytical capability, GIS would be reduced to an automated mapping application. Without database management features, GIS would be unable to capture spatial and topological relationships between geo-referenced entities if these relationships were not pre-defined.

What sets GIS apart from other database management systems (DBMS) is not the nature of the information handled. Indeed GIS and DBMS may contain exactly the same information, say fatal accidents occurring on New York state highways during a given year. The difference between the two systems is "under the hood", namely in the way information is referenced. A DBMS references accidents by some unique index or combination of indexes, such as the date of occurrence, the vehicle make, or the weather conditions. By contrast, information is all about a geographic description of the surface of the Earth in a GIS. Each accident record is a geographic event in the sense that it is tied to a unique location defined in a given referencing framework (global, national or local datum). With the spatial referencing of objects, topology of the data can be defined, which in turn enables a host of spatial query operations of objects and set of objects. For instance, the task "identify all accidents that occurred within 100 meters of any intersection on urban arterials" requires little effort because of the spatial indexing of all accident and roadway link objects in the GIS databases.

The concept of GIS traces its roots back to a handful of research initiatives in the US, Canada and Europe during the late 1950s. It is widely acknowledged that the first real GIS was the Canada Geographic Information System set up for the Canada Land Inventory. The reader will find complete histories of GIS in Coppock and Rhind (1991) and Foresman (1998). It suffices to say here that the development of the concept and its implementations is closely associated with the requirements of land information systems. Early transportation applications of GIS were few and unable to create a momentum in GIS research sufficient to remediate the known limitations of the technology in handling transportation data, in interfacing with complex analytical network-based models, and in fitting in existing enterprise-wide business model. Even the US Bureau of the Census's Dual Independent Map Encoding (DIME) system – precursor of the Topologically Integrated Geographic Encoding and Referencing (TIGER) system – does not qualify as a proper effort to enhance transportation GIS because of its crude topology.

A GIS is a spatial representation, or model, of the data used to depict a portion of the surface of the earth (Frank, 1992). In the transportation context, three classes of GIS models are relevant (Goodchild, 1992b, 1998).

- Field models, or representation of the continuous variation of a phenomenon over space. Terrain elevation uses this model.
- Discrete models, according to which discrete entities (points, lines or polygons) populate space. Highway rest areas, toll barriers, urbanized areas may use this model.
- Network models to represent topologically connected linear entities (such as roads, rail lines, or airlines) that are fixed in the continuous reference surface.

While all three models may be useful in transportation, the network model built around the concepts of arc and node plays the most prominent role in this application domain because single- and multi-modal infrastructure networks are vital in enabling and supporting passenger and freight movement. In fact, many transportation applications only require a network model to represent data. Examples of such applications include:

- pavement and other facility management systems;
- real-time and off-line routing procedures, including emergency vehicle dispatching and traffic assignment in the four-step urban transportation planning process;
- web-based traffic information systems and trip planning engines;
- in-vehicle navigation systems;
- real-time congestion management and accident detection.

3. What is special about GIS-T?

The previous section has identified the main data models of GIS. It also stressed that a common trait of research on, and with, GIS-T is its reliance on the network data model, at times at the exclusion of any other data model. This is not to say that other domains of application have no use for network representations, but when used at all, networks play a rather peripheral role. The network model is elegantly simple, yet functional. With its arc and node structure, it represents the one-dimensional network object in reference to the two- (or three-) dimensional surface of the earth. Arcs and nodes themselves are primitives of the discrete entity model. Their locational referencing is absolute, usually two-dimensional: x and y coordinates, longitude/latitude, for example.

Once proper topology has been defined on the network, the network model supports basic as well as advanced forms of network analysis (Waters, 1999; Souleyrette and Strauss, 2000), from location–allocation modeling to vehicle routing and scheduling and traffic assignment, and finally to network connectivity optimization and design. Network-based GIS enables thus the study of flows and movements, which lies at the core of transportation research.

Land information systems and early population census information systems have not conceived of roadways, railways and other transportation infrastructure "as features for analysis in and of themselves" (O'Neill and Harper, 2000) because transportation lines, like other linear features – most singularly streams – primarily serve to delineate polygons. In transportation research, however, there is a compelling need for attributing infrastructure lines. This is in line with the primary mission of transportation agencies to be custodian of the transportation infrastructure in their jurisdiction, and maintain it in good operating condition (Petzold and Freund, 1990). Furthermore, most models of network analysis mentioned above incorporate some measure of travel impedance on each link of the network, while some of them also use link-specific traffic capacity attributes. It is also well known that the external validity of many models is greatly enhanced by a better representation of traffic conditions at nodes on the network (at grade intersections, freeway entrances or exits). Nodes are remarkable locations on the network where various restrictions to movement may exist and delay often develops in relation to the mixing of traffic streams. Node attributes may entail a rather elaborate description of an intersection by traffic priorities, the presence of traffic signals, their timing and phase, among others.

So far, the implicit assumption has been that network links are homogeneous. This may hold true in some systems, but not in others. Number of lanes, pavement width, pavement condition, posted speed are all but a few attributes that cannot be constrained to be constant between terminal nodes of a link. Similarly, on the National Highway System, traffic parameters of speed, flow and capacity cannot be expected to be constant between widely spaced junctions. The dynamic nature of these distributed attributes of the network precludes that the network be permanently edited to maintain the homogeneity of each link on each attribute. Instead an attribute can be viewed as a spatial (linear) event occurring on the network. The variation of the attribute can be referenced to discrete locations measured by relative positions on a linear feature belonging to the network. In this approach, attributes are linearly referenced and linked dynamically to the entities forming the network (Scarponcini, 1999). Early research in GIS in transportation led Dueker (1987), Fletcher (1987), and Vonderohe et al. (1993) to identify the critical need for this capability in GIS-T. Traffic accidents, bridges, traffic signs, and other zero-dimensional events can also take advantage of linear referencing systems for linking to the one-dimensional transportation infrastructure.

Though the basic network data model is already a domain-specific departure from conventional GIS data modeling, it does not suffice to handle the complexity embedded in transportation network data. As pointed by Goodchild (1998), extensions are needed to handle particular structures. The following three meaningful extensions were recognized by Goodchild (1998).

- Planar versus non-planar model, wherein topological representation differs from cartographic representation by not forcing nodes at cartographic intersections. Non-planarity allows the representation of freeway overpasses as well as of turn prohibitions. Navigational databases must conform to a non-planar model.
- Turn tables contain properties of the turn between any pair of links connected on the network. Properties can be binary (allowed, disallowed), or cardinal measurements (for instance, expected delay through an intersection).
- Links are objects formed of traffic lanes. A structure allowing for this object-oriented view of the infrastructure needs to define topology between lanes. It may store attributes for individual lanes.

Certainly the need for these and other extensions to the base network model is not universal. It can be motivated by the resolution and the geographic scale of the representation. Ultimately, need is dictated by the specific GIS-T application. The representation afforded by extensions of this kind is essential for the development of navigational databases. On the other hand, it is most likely to be superfluous in a task of transportation planning for an entire metropolitan area.

To sum up, GIS in transportation is more than just one more domain of application of generic GIS functionality. GIS-T has several data modeling, data manipulation, and data analysis requirements that are not fulfilled by conventional GIS. The final report of NCHRP Project 20-27 (Vonderohe et al., 1993) theorizes GIS-T as the product of the cross-fertilization of an enhanced GIS and an enhanced Transportation Information System (TIS). See Fig. 1. To quote Vonderohe et al. (1993), "the necessary enhancement to existing TISs is the structuring of the attribute

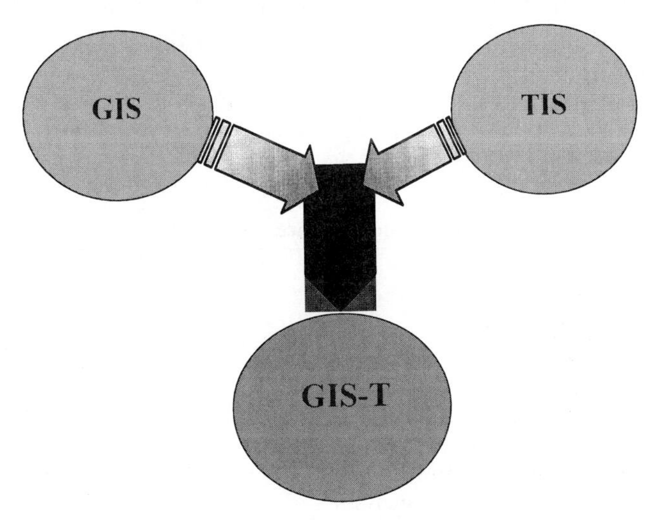

Fig. 1. GIS-T, product of an enhanced GIS and an enhanced TIS. After Vonderohe et al. (1993).

databases to provide consistent location reference data in a form compatible with the GIS, which in turn has been enhanced to represent and process geographic data in the forms required for transportation applications" (p. 11).

4. The challenges brought by suffix "T"

The current flurry of research activity in and around GIS-T is a clear sign of the interest of transportation researchers and professionals for this still emerging technology. Some of the new trends discernible in state-of-the-art research in GIS-T merely echo the transformations of GIS per se. Some of these transformations are technologically motivated (Fletcher, 2000). Others are part of an agenda set forth by the University Consortium for Geographic Information Science (UCGIS) to strengthen the scientific basis of this emerging discipline born to the GIS technology. In 1997, UCGIS outlined a research agenda composed of research priorities in ten areas. These priorities are listed in Table 1. The state of research in GIS-T with respect to each of the UCGIS priorities has recently been compiled by Wiggins et al. (2000). This section points to selected research themes and challenges that are particularly salient in contemporary GIS-T research.

Research themes and challenges may manifest themselves differently on different functional aspects of GIS. It is imperative therefore to discuss them in a functional framework. For the sake of the exposition, GIS functionality is here organized in relation to the level of intensity of data processing involved. A commonplace framework derived from this line of thought identifies three functional groups: data management, which concerns storage and retrieval of data; data manipulation, which refers to the creation of new data out of raw data; and data analysis or analytical modeling. See McCormack and Nyerges (1997) for a similar framework in the context of GIS-T. Interestingly, requirements associated to each group are not independent. As data manipulation requires data storage, and modeling is built on the latter two, requirements and challenges are cumulative. This hierarchical view of functionality is depicted in Fig. 2. This logic is also followed in the organization of the contributions included in this volume dedicated to GIS in transportation research. Many of the themes introduced in the rest of this paper are further developed in these contributions.

Table 1
UCGIS Research Priorities for Geographic Information Science

(1) Spatial data acquisition and integration
(2) Distributed computing
(3) Extensions to geographic representation
(4) Cognition of geographic information
(5) Interoperability of geographic information
(6) Scale
(7) Spatial analysis in a GIS environment
(8) The future of the spatial information infrastructure
(9) Uncertainty in spatial data and GIS-based analyses
(10) GIS and society

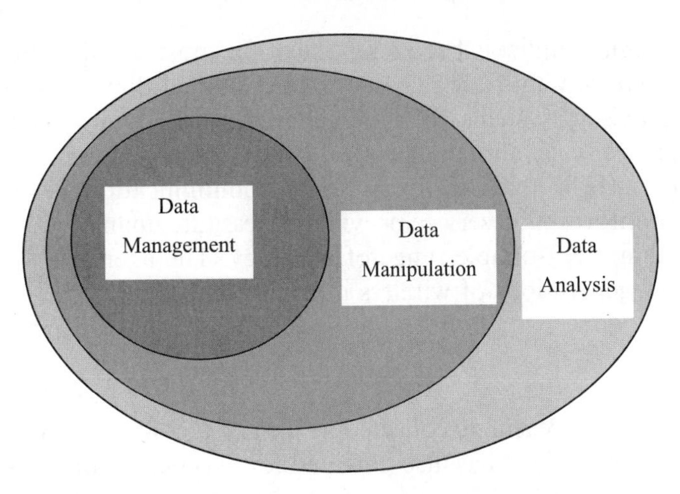

Fig. 2. Hierarchical model of data management, data manipulation, and data analysis functional groups.

4.1. Legacy data management system

Transportation agencies have a long tradition of maintaining comprehensive inventories of the transportation infrastructure, of its condition and usage by the public. As a norm, multiple legacy TIS co-exist within an agency. It is not uncommon that each TIS handles a single type of information (bridge inventory, highway planning network, pavement management system, accident inventory, etc.) with its own data, and is runs on its own hardware and software platform. A serious challenge of GIS-T is to transfer disparate data into a unified data management system that preserves access to legacy data, and allows for their integration to meet the demands of multi-thematic analysis. GIS-T can play this integrator role if, and only if, frameworks are established for the communication and exchange of data among disparate data models so as to enable multivariate analysis and modeling and support decision making in transportation policy and management. Some of the options available include generic relational data models, new dynamic segmentation data standards, and object-oriented data models.

4.2. Data interoperability

Transportation data are typically maintained by a variety of agencies and private data providers. Each data source may have its own data model, and because of different data capture techniques and standards, accuracy may be quite heterogeneous across data sets. Diversity of data models and approaches available to circumvent this problem are mentioned in Section 4.1. Errors in data position, topology, classification and inclusion, naming and attributing, linear measurement render conflation of data of various sources an onerous task with haphazard outcomes. Advances in this domain will only be possible if the GIS-T research agenda follows a three-prong approach: algorithms for map matching, models of error and error propagation in transportation data (in particular in the context of one-dimensional data model), data quality standards and data exchange standards.

The interoperability issue is quickly becoming one of the most pressing themes in GIS-T as geo-referenced data find their way into the market place. Detailed digital street databases populate and routing and dispatching systems for emergency services, and vehicle navigation systems available to the general public and to fleets of commercial vehicles. Elements of Intelligent Transportation Systems (ITS) that involve wireless communication between motorists and a traffic control center or information service provider necessitate unambiguous identification of the motorist locations within a reasonable range of accuracy. The agenda set forth above will contribute to enabling new generations of wireless information services.

4.3. Real-time GIS-T

Geo-referenced data are increasingly collected as part of a continuous process rather than at a few pre-set moments in time. Need has also emerged for accessing data on a real-time basis. For instance, continuous streams of traffic data from vehicles carrying toll transponders on parts of New York state's freeway system are fed into computational algorithms for early accident detection. In other metropolitan regions, probe vehicles equipped with a Global Positioning System (GPS) device provide speed data to the Traffic Management Center, which in turn disseminates congestion information and forecasts to wireless information service providers, thus fitting in the area's Congestion Management System. This (quasi) real-time traffic data is also a primary input of world-wide-web applications discussed below. Real-time data storage, retrieval, processing, and analysis are presently not meeting the needs of society when it comes to geo-referenced data. Quicker access data models, and more powerful spatial data fusion techniques and dynamic routing algorithms are needed to take advantage of real-time traffic information.

4.4. Large data sets

Real-world transportation problems tend to involve large amounts of geo-referenced data and large networks. Visualization techniques on which GIS mapping is based are inherited from an age where data was not abundant. GIS-T will benefit from an evolution in Geographic Information Science research towards close integration of geo-visualization principles and computational methods of knowledge discovery and data mining. As the latter are still very much in their infancy, no tangible outcome should be expected in the near future. With GIS-T, the complexity is compounded by the difficulty to visualize information on the single dimension of a network.

The sheer size of transportation data sets often require innovative system designs that manage both to optimize speed and accuracy of the display of information and to optimize the run time of algorithms and analytical tools of flow and network analysis.

4.5. Distributed computing

The connectivity offered by the Internet technology has transformed the relationship between the computer, the software application, the data, and the user. Computing has emerged as a mobile, distributed, and ubiquitous reality. Web-based GIS applications have become commonplace, including in the domain of transportation. Real-time transit route and schedule information, road construction, traffic information are examples of applications than are currently available. Remaining challenges revolve around bringing to the Internet client-server environment the power of desktop GIS-T. This entails the development of more powerful and robust analytical tools to fit the limited distributed computing resources and limited bandwidth on communication networks. Also, system architectures will need to be judiciously designed to make efficient use of local and remote computing resources.

The future – no longer distant – of mobile computing is with Internet-enabled Personal Digital Assistants (PDA), Personal Navigation Assistants (PNA), and other on-board computing devices. All the issues brought up earlier in this section are considerably magnified in this setting due to the more severe constraints on bandwidth and local computing resources. An issue re-emerging in this context is that of geo-referencing of remote service users and tracking of their movement in real-time.

5. Conclusions

The late 1980s saw the first widespread use of GIS in transportation research and management. Due to the specific requirements of transportation applications and of the rather late adoption of this information technology in transportation, research has been directed toward enhancing existing GIS approaches to enable the full range of capabilities needed in transportation research and management.

This paper placed the concept of transportation GIS in the broader perspective of research in geographic information systems and Geographic Information Science. The emphasis was placed on the requirements specific of the transportation domain of application of this emerging information technology. The paper concluded with a synopsis of dominant themes in current research in GIS for transportation. Successful pursuit of this agenda should solidify the position of GIS as an integrative system for transportation research and management.

References

Coppock, J.T., Rhind, D.W., 1991. The history of GIS. In: Maguire, D.J., Goodchild, M.F., Rhind, D.W. (Eds.), Geographical Information Systems: Principle and Applications, vol. 1. Longman, Harlow, UK, pp. 21–43.

Dueker, K.J., 1987. Geographic information systems and computer-aided mapping. Journal of the American Planning Association 53, 383–390.

Fletcher, D.R., 1987. Modelling GIS transportation networks. In: Proceedings of the 25th Annual Conference of the Urban and regional Information Systems Association, Fort Lauderdale, pp. 84–92.

Fletcher, D.R., 2000. GIS-T in the new millenium – A look forward. In: Transportation in the New Millenium, Transportation Research Board, Washington, DC (CD-Rom).

Foresman, T.W., 1998. The History of Geographic Information Systems: Perspectives from the Pioneers. Prentice-Hall, Upper Saddle River, NJ.

Frank, A.U., 1992. Spatial concept, geometric data models, and geometric data structures. Computer and Geosciences 18, 409–417.

Goodchild, M.F., 1992a. Geographic information science. International Journal of Geographic Information Systems 6, 31–45.

Goodchild, M.F., 1992b. Geographical data modeling. Computers and Geosciences 18, 401–408.

Goodchild, M.F., 1998. Geographic information systems and disaggregate transportation planning. Geographical Systems 5, 19–44.

McCormack, E., Nyerges, T., 1997. What transportation modeling needs from a GIS: a conceptual framework. Transportation Planning and technology 21, 5–23.

Meyer, M.D., 1999. Transportation planning in the 21st Century. TR News 204, 15–22.

O'Neill, W., Harper, E.A., 2000. Implementation of linear referencing systems in GIS. In: Easa, S., Chan, Y. (Eds.), Urban Planning and Development Applications of GIS. American Society of Civil Engineers, Reston, VA, pp. 79–98.

Petzold, R.G., Freund, D.M., 1990. Potential for geographic information systems in transportation planning and highway infrastructure management. Transportation Research Record 1261, 1–9.

Scarponcini, P., 1999. Generalized model for linear referencing. In: Advances in Geographic Information Systems, Proceeding of the Seventh International Symposium, ACM GIS'99, ACM, pp. 53–59.

Souleyrette, R.R., Strauss, T.R., 2000. Transportation. In: Easa, S., Chan, Y. (Eds.), Urban Planning and Development Applications of GIS. American Society of Civil Engineers, Reston, VA, pp. 117–132.

Vonderohe, A.P., Travis, L., Smith, R.L., Tsai, V., 1993. Adaptation of geographic information systems for transportation. National Cooperative Highway Research Program Report 359, Transportation Research Board, Washington, DC.

Waters, N.M., 1999. Transportation GIS: GIS-T. In: Longley, P.A., Goodchild, M.F., Maguire, D.J., Rhind, D.W. (Eds.), Geographic information systems, vol. 2: Management Issues and Applications. Wiley, New York, pp. 827–844.

Wiggins, L., Dueker, K., Ferreira, J., Merry, C., Peng, Z.-R., Spear, B., 2000. Application challenges for geographic information science: Implications for research, education and policy for transportation planning and management. Journal of the Urban and Regional Information Systems Association 12 (2), 51–59.

PERGAMON

Transportation Research Part C 8 (2000) 13–36

TRANSPORTATION RESEARCH PART C

www.elsevier.com/locate/trc

A geographic information system framework for transportation data sharing

Kenneth J. Dueker [a,*,1], J. Allison Butler [b]

[a] *Center for Urban Studies, Portland State University, Portland, OR 97207-0751, USA*
[b] *Department of GIS, Hamilton County TN, USA*

Abstract

This paper develops a framework and principles for sharing transportation data. The framework is intended to clarify roles among participants, data producers, data integrators, and data users. The principles are intended to provide guidance for the participants. Both the framework and the principles are based on an enterprise geographic information systems-transportation (GIS-T) data model that defines relations among transportation data elements. The data model guards against ambiguities and provides a basis for the development of the framework and principles for sharing transportation data.

There are two central principles. First is the uncoupling of graphics, topology, position, and characteristics. Second is the establishment of a schema for transportation features and their identifiers. An underlying principle is the need for a common data model that holds transportation features, not their graphical representations, as the objects of interest. Attributes of transportation features are represented as linear and point events. These are located along the feature using linear referencing. Sharing of transportation data involves exchange of relevant transportation features and events, not links and nodes of application-specific databases. Strategies for sharing transportation features follow from this approach. The key strategy is to identify features in the database to facilitate a transactional update system, one that does not require rebuilding the entire database anew. This feature-oriented enterprise GIS-T database becomes the basis for building separate application-specific network databases.

Keywords: GIS-T; Data sharing; Data models; Linear locational referencing systems

* Corresponding author. Tel.: +1-503-725-8480; fax: +1-503-725-8480.
E-mail address: duekerk@aol.com (K.J. Dueker).
[1] www.upa.pdx.edu/cus/

PII: S0968-090X(00)00006-1

1. Introduction

The problem with sharing and maintaining geographic information systems-transportation data (GIS-T data) among applications is the diversity of formats that lead to inconsistencies, inaccuracies, and duplication. This diversity is due to differences among data models that make it difficult to achieve consistent representations of the transportation system. Yet there are legitimate differences in requirements that lead to application-specific definitions and representations of transportation objects and their geometry. This has resulted in multiple and inconsistent digital representations of various parts of the transportation system.

Currently, we have different networks to support applications, such as vehicle navigation systems, overweight truck routing, facility location and address geocoding, emergency management, or, for a reference layer of roads, resource management applications. These applications define roads differently. Some include paths and trails, private streets, alleys, and resource roads and some do not. In addition, the level of detail and spatial accuracy differs. The challenge is to establish means of data exchange among these disparate representations that lead to improvements in accuracy, consistency, and completeness. What is more important in the long run is the need to share updates to maintain currency in representing the transportation system. At the same time, it is important to recognize the legitimacy of specialized representations needed to support different applications.

The following metaphor is offered to make the elusive concepts in this paper more understandable:

> Transportation features (TFs) are like strings of pasta. A bowl of spaghetti contains TFs of all types: freeways, arterials, local roads, alleys, paths and trails, airports, pipelines, and railroads. Users identify the types of TFs they want and select from the bowl those pieces needed for vehicle navigation, emergency dispatch, pizza delivery, walking, or bicycling. Then, using a "clean and build" procedure, users can construct application-specific networks.

This paper develops a framework and principles for sharing transportation data to achieve more accurate representations of transportation data in an intermediate form, from which application-specific databases can be generated. The framework is intended to clarify roles among participants, data producers, data integrators, and data users. The principles are intended to provide guidance for the participants. Both the framework and the principles are based on an enterprise GIS-T data model that defines relations among transportation data elements. The data model guards against ambiguities and provides a basis for the development of the framework and principles for sharing transportation data.

This paper assesses the problems of current approaches to sharing transportation data and of standards and data models for transportation data. The paper then describes an enterprise GIS-T data model that is designed for data sharing, from which end users of transportation data can generate application-specific databases. This is followed by a discussion of the requirements of transportation data participants. This leads to a formulation of principles for sharing transportation feature data and discussion of ways to identify transportation features in databases. This is followed by a discussion of issues to foster data sharing to improve accuracy, completeness, and currency of databases used for GIS-T applications.

2. Current approaches to sharing GIS-T data

Data sharing is more than simply having the ability to occasionally import information from someone else's database. The business needs driving dynamic data sharing include those for multi-agency local government GIS infrastructures, where E-911 (emergency dispatch) public works, and property assessment organizations need to utilize a common database. The property assessor gets a new subdivision plat on which recently constructed streets are shown. The E-911 center needs to know where the new streets are located; the building inspector needs to know all the new lot addresses; and the public works department will need to establish new trash pick-up routes and pavement management segments. Somehow, the information on the plat has to be communicated accurately, efficiently and quickly to these many users, each with a unique linear and/or non-linear location referencing system.

The reality in most cases is that the enterprise approach to GIS is late on the scene, with the different agency-specific GIS applications having been developed in isolation. Until such a time when these systems come to use a single enterprise-wide data infrastructure, these agencies will need to frequently exchange data sets. This paper attempts to set the stage for establishing comprehensive transportation data exchange mechanisms by describing a way to integrate them in an enterprise GIS-T data model.

Sharing GIS-T data is both an important issue and a difficult one. It is important because there are many organizations that produce or use GIS-T data; it is difficult because there are many ways to segment and cartographically represent transportation system elements. There is a lack of agreement among transportation organizations in defining transportation objects and in the spatial accuracy with which they are represented cartographically. This lack of agreement leads to difficulty in conflating or integrating two views of the same or adjacent linear objects.[2]

There are two problems in defining transportation objects: different definitions of roads and different criteria with which to break roads into logical segments. The logical segments become objects in the database that we will refer to as "transportation features". We have selected this term in order to include more than just roads. Roadways, railroads, transit systems, shipping lanes, and air routes are all linear features that utilize the same basic network data model, which utilizes linear travel paths between points of intersection. Since they all use the same basic data model, we will generally restrict our discussion to roadways for simplicity.

Transportation features become the building blocks for specific applications. Persons building vehicle navigation databases need to include private roads that are open for public use. "Paper streets", those which are not yet constructed and that cannot be navigated, should be omitted. Yet public organizations responsible for road maintenance follow different rules. They omit private roads and include planned public roads on their maps. Similarly, two organizations responsible for roads on resource lands, the Forest Service and the Bureau of Land Management have quite different definitions of roads.[3] Most organizations that maintain databases of roads

[2] Sperling and Sharp (1999) describe conflation of US Bureau of the Census TIGER files with local street centerline files as the "automatic matching and transfer of features and attributes from one geo-spatial database into another".

[3] The Forest Service defines roads as any visible track, while the BLM limits roads to tracks that can be traversed by a normal vehicle.

break them into logical segments to create discrete transportation features according to some business interests, such as a change of pavement type, jurisdiction, functional type, or at all intersections.

These differences in original purpose for transportation databases create a difficult arena for sharing data with others. The data sharing arena includes data producers, data users, and, increasingly, data integrators who collect data from the field, legacy databases, or other data producers or users and reorganize it for new uses and/or to maintain currency. A healthy data sharing environment suggests that data producers embed registration points and feature identifiers in their original data to facilitate importing and registration of foreign or legacy cartography and attribute data.

The need is for standards for data sharing among organizations, both public and private. However, standards are difficult to develop because system requirements of advanced applications of GIS-T technology differ in spatial and temporal accuracy and details of such features as ramps and lanes. Further, differing levels of real-time use of systems dictate response requirements of databases and interfaces.

A more systematic approach to GIS-T data sharing of GIS data calls for relief from the conflation technique that requires the matching of spatial objects of separate data sets. It is a problematic process due to the need to simultaneously match both topological and geometric properties. This is exemplified by efforts of transferring TIGER attributes to more accurate vector files (Brown et al., 1995; Tomaselli, 1994). This approach to matching spatial data of similar scale works well, if the data were captured using the same data model or criteria by which to define and segment roads. If roads are not segmented into similar spatial objects, then conflation is not a satisfactory way to share transportation data. Some (Sester et al., 1998; Walter and Fritsch, 1999) work toward automating the conflation process, while Devogele et al. (1998) call for the need to develop an integrated schema from data models to facilitate data sharing and/or interoperability.

In addition, sharing of transportation data is not a one-time issue. In a larger context, it is a means of disseminating data about changes in the transportation system. Management of updates in order to maintain current representations of transportation systems is a growing concern. Until the present time, the effort has been on building the initial database. Attention is now turning to maintaining currency of databases of increasing detail and complexity. The GIS-T community is in need of guidance on this issue. We neither see sufficient progress on the conflation approach, nor do we see adequate consensus to support schema integration to support the wide range of applications of GIS-T data.

Sester et al. (1998) identify an alternative to the bottom-up approach of conflation: a top-down approach known as semantic data integration. While maintaining the richness of attribute detail along features by linear referencing, our application of this top-down approach to GIS-T requires the aggregation of spatial object primitives into larger (longer) transportation features that can and need be uniquely identified. Similarly, the transportation features in our enterprise GIS-T data model are not topological spatial objects. This enables the building of application-specific networks from selections of a single and consistent set of underlying data. Our enterprise GIS-T data model also allows for multiple spatial representations to accommodate the need for both abstract arterial-level data and detailed representations of roads, including freeway ramps and local streets.

Consequently, our enterprise GIS-T data model falls somewhere between one that specifies an end product like TIGER or GDF, and a data exchange standard like SDTS or DIGEST. "Between" implies it is more than a neutral standard, and it is more than a format for a database to support a specific application. We call it an enterprise GIS-T data model to convey the notion of a standard approach to maintaining business data in a transportation organization about the transportation system for which they are responsible. One master set of transportation features enables maintenance of current knowledge about the system. Selection of transportation features by type and from the available multiple geometric representations enables building of a number of application-specific networks for functions, such as vehicle navigation and emergency, pavement, bridge, and congestion management.

3. Assessment of data models used in GIS-T

There are several GIS data models used for transportation applications. GDF (Geographic Data File Standard, 1995 and 1999), NCHRP 20–27 (Vonderohe et al., 1998), and our enterprise GIS-T data model (Dueker and Butler, 1998) have been developed specifically for transportation, while others, such as Chan (1998) and TIGER, have broader applications. They differ in their robustness in representing transportation systems and in handling updates.

Current data models used to represent streets and roads (e.g., TIGER line and ArcInfo georelational data model) integrate the cartography, the network link, and attributes of the link into a single linear spatial object. This analysis questions the "integrated data model" and calls for an unbundled approach to facilitate data sharing and maintenance of GIS-T databases.

The tighter the integration of topology, geometry, and attributes, the more information has to be input before the update is usable. The loosely coupled approach of our enterprise GIS-T data model allows each piece of the bundle to stand on its own and be entered separately (perhaps by different organizations). Each piece of the bundle is thus available for use as soon as it is in the database. In this way, the loosely coupled approach requires less business process revisions; it better fits the way people work.

The GDF aggregated-way feature may be the closest one that matches our transportation feature. GDF, for all intents, is a fully topological data model that requires full specification of cartography, topology, and attributes for any useful data sharing to occur, although pure cartography can be exchanged. There is no explicit support for multi-link objects inside the transmission protocol. The use of "aggregated way" appears to be external, a way to allow agencies to connect their data sets to the GDF objects using a one-to-many relationship. Use of GDF appears to require prior agreement on roadway segmentation.

GDF is designed to operate at two levels. Level 1 consists of simple features, road elements and junctions. Level 2 consists of complex features, roads, and intersections. Road is a topological entity representing a complex feature composed of road elements joining two intersections. A freeway interchange is represented as an intersection node, a complex feature at Level 2, and as road elements and junctions to represent ramps and ramp connections at Level 1. Adding intersections in GDF requires the splitting of roads.

This limitation that road entities are bounded by intersections is identical to the topological structure of TIGER. The addition of alleys, private roads, pedestrian paths, and statistical or political boundaries will require a splitting of topological edges. This increases the difficulty of sharing data among networks that are separately maintained. Alternatively, newer GIS software, such as ArcInfo 8 is based on data models that break this limitation by allowing vector data to be non-topological line features or topological features. Our enterprise GIS-T data model builds on that principle and separates transportation features from topological links. It is an intermediate form from which databases to support applications can be generated.

4. GIS-T data modeling issues

A National Cooperative Highway Research Program project, 20-27(3), is in the process of trying to bring consensus in the area of GIS-T data models (Adams et al., 1998). Meanwhile, there are two different approaches. Our enterprise GIS-T data model represents a feature (object) database approach. It is best suited for a federated systems environment with legacy data of varying spatial accuracy supplied and used by a wide variety of agencies acting in concert as a single enterprise. This approach relies on the traditional model of linear location referencing systems (linear LRS), which utilize roadway identifiers and linear offset measurements to locate attributes of a highway. The field rules for a linear LRS form its location referencing method (LRM).

An alternative approach can be characterized as a location (geometry) approach (Sutton, 1999). This approach embraces the use of earth-based, two- and three-dimension positions, such as those that might be derived from a series of GPS-derived coordinates while traveling down a highway. Thus, location becomes the way in which various data about roads are integrated.

This geometry approach would work well in a strong centralized environment, in which the location of transportation features would be redigitized with high precision using GPS. This would be needed to enable linking by coordinate snapping of spatially accurate tracking or events to a spatially accurate map base. However, this approach has not been formally stated or tested. Issues, such as repeatability of GPS positions, how to abstract networks, how to relate to other location referencing systems, and representation at smaller scales have not been adequately addressed.

Our GIS-T data model supports both approaches by employing the anchor point object proposed by Vonderohe and Hepworth (1998), a feature that can be described in both location referencing systems. We endorse the use of these objects as registration points to interface between the two systems. But the fact remains that both approaches must have a way of defining a transportation feature. A series of GPS coordinates does not make a line to describe a roadway centerline unless someone connects the dots and calls it by a common name. Defining a common name for a roadway is also a requirement of the linear LRS approach. Therefore, our foundation business rules address how to combine the roadway pieces into features and attach the attributes that relate to them.

The NCHRP consensus approach will have to resolve differences in the definitions of basic transportation entities and primitives that exist among the various data models. For example, a road is a topological entity in GDF, but the Transportation Feature in our enterprise GIS-T data model is not a topological entity. We directly connect the cartography to the transportation feature, not to the linear datum as in the NCHRP 20-27 data model. Making the linear datum the principal entity requires transferring topology to the datum and thence to the cartography. Using the NCHRP 20-27 data model approach, a dynamic segmentation process to display transportation feature attributes requires conversion from network measures to node offsets then to anchor point offsets.

There are missing elements in both approaches that are being addressed in the NCHRP consensus data modeling effort. One is the problem of treating time and the third spatial dimension more explicitly in the data model. The other problem is with GIS data models in general, which become paramount in GIS-T. In traditional GIS, a spatial object is defined by its location. Consequently there is no suitable way to represent a moving object, like a vehicle, package shipment, or storm in such a GIS. There needs to be a new dynamic tracking or moving object class in GIS, especially in a GIS-T. There are three approaches. One is a static object with frequently changing positions. Another is a new object class with location as an attribute rather than part of the definition. Yet another is a moving object construct with starting location and attributes of direction, speed, and destination to define a moving object. Emerging object-oriented GIS platforms offer a way to do this by treating location as an attribute of an object.

Section 5 presents the enterprise GIS-T data model. The model brings consistency to the representation of transportation data and provides guidance to the data sharing participants. The model also provides the basis for the subsequent development of data sharing principles.

5. The enterprise GIS-T data model

In previous works, Dueker and Butler provide a data model well suited for GIS-T data sharing. It unbundles the geometry, topology, and attributes to facilitate separate maintenance and enables extraction of data for different uses. The original paper provided a general introduction to information system design by offering tutorial appendices on such topics as user requirements analysis, data modeling, and business rule construction. The tutorial information was offered to enhance the reader's understanding of the transportation data model described in the main body of the paper. That model was proposed as a comprehensive description of the entire scope of transportation data that might be housed in a State Department of Transportation. While business rules on which the model was founded were presented, the accommodation of implementation details was the main thrust of the work.

One unfortunate outcome of that paper was the general difficulty readers had at taking in the big picture of a comprehensive data model and the myriad of database tables needed to implement it. Reader comments on that paper have motivated us to create a new version, contained herein, that is based on the formal business rules upon which our enterprise GIS-T data model is founded. These business rules provide the basis for the ensuing discussion of GIS-T data sharing issues.

The simplified version of the enterprise GIS-T data model is shown in Fig. 1.

Entities, the things about which we wish to store information, are shown in boxes, while the relationships between entities are shown using lines. Each entity type has been identified by a

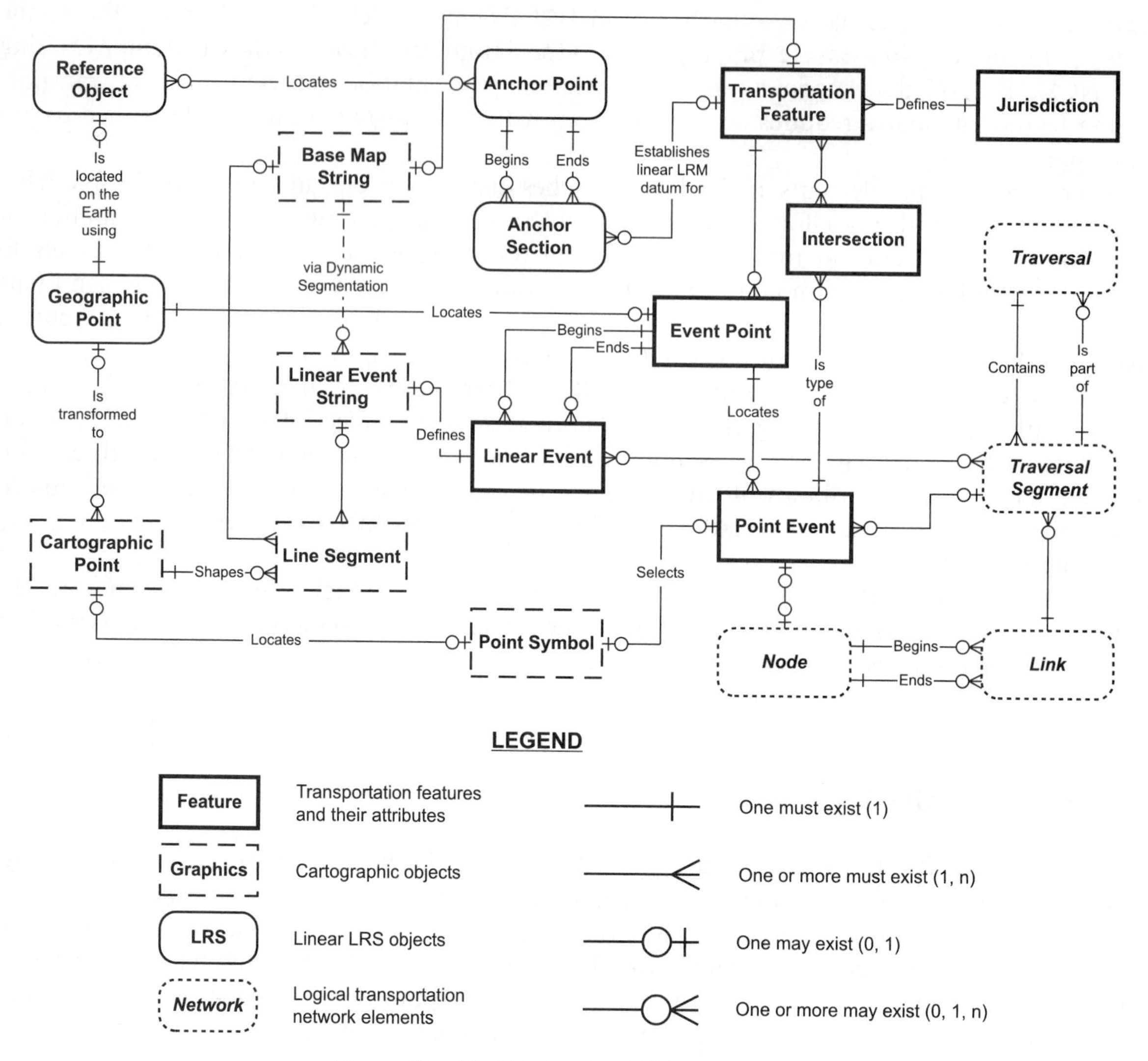

Fig. 1. Simplified Enterprise GIS-T data model.

special style. Each relationship type has been shown using descriptive text and connection symbols. Each group of entities for a single type can be treated as a stand-alone data set.

An entity is a discrete part of our world, one for which we want to store information. An entity is a basic building block of data models. Entities are related to each other through verb-oriented statements, such as, "A jurisdiction defines one or more transportation features", and its corollary, "A transportation feature must be defined in the context of a single jurisdiction". Relationships are the other building blocks of data models. A relationship without explicitly stated verbs generally can be read as an ownership relationship, such as, "A base map string has one or more line segments". Attributes are part of the entities, not separate entities. For example, a line segment may have width and color attributes, which are aspects of the line segment entity, not separate entities owned by line segment.

5.1. Transportation features and their attributes

The first group of entities we need to examine is the one that contains our basic transportation elements. The group consists of six entities: Jurisdiction, Transportation feature, Event point, Linear event, Point event, and Intersection. Definitions of these entities and their relationships are as follows:

• *Jurisdiction: The political or other context for designating transportation features and their names, which may be merely numerical references unique within the jurisdiction.* A jurisdiction sets the context for defining the extent and name for each transportation feature and is primarily geographic. Jurisdiction carries no other burden; i.e., it does not mean which agency has maintenance responsibility. A State Department of Transportation may choose to subdivide the State highway system on a county basis, with each county-specific portion of a roadway having its own identifier. In this instance, "county" would be the value of Jurisdiction. The maintenance jurisdiction in which a transportation feature may be located would be an attribute of the feature, which could be stored as part of the transportation feature entity if it applied to the entire feature, or a separate linear event. Jurisdiction need not be the same for all transportation feature types. Airports can be named on a national basis, interstate freeways named on state basis, and local streets named on a zip code basis. *Relationships:* Jurisdiction defines one or more transportation feature. Transportation feature must be defined within the context of Jurisdiction.

• *Transportation feature: An identifiable element of the transportation system.* A transportation feature can be like a point (interchange or bridge), a line (road or railroad), or an area (rail yard or airport). Some transportation features can consist of other features. This may be the case with bridges and roadways. In a roadway inventory, the bridge is an event that occurs on a specific roadway. The location of the bridge would typically be defined by a linear LRS location description. In a bridge inventory, the bridge is a transportation feature in its own right, and its location may be defined by a set of earth coordinates. *Relationships:* Transportation Feature may have one or more event point. Event point must be defined on a single transportation feature. Transportation Feature may contain one or more intersection. Intersection must be owned by one or more Transportation Feature.

• *Event: An attribute, occurrence, or physical component of a transportation feature.* Attributes include functional class, speed limit, pavement type, and state road number. These things are not tangible but describe a tangible element, such as a road. Occurrences include traffic crashes and projects. Physical components include the number of lanes, guardrails, signs, bridges, intersections, and other tangible things that are field-identifiable elements. There are three event subtypes. A given event instance may be expressed as more than one subtype:

Point event: A component or attribute that is found at a single location (one event point). Point events may occur independently or on transportation features of the linear or area form.

Linear event: A component or attribute that is found along a segment of a linear transportation feature. Linear events are defined by two event points (beginning and ending). Linear events may occur only on linear transportation features.

Area event: A transportation feature component or a non-transportation entity that affects a transportation feature. Areas can be explicitly represented as polygons or implicitly represented as to where they intersect transportation features. The implicit option is called an area event and is represented through related linear and point events. For example, an area event could be a city. The city could be expressed by creating a linear event for the portion of a transportation feature located within it, or as point events, where the city limits cross a transportation feature. Another example could be a park-and-ride lot, which would be stored as a point event located where the driveway to the lot intersects the adjacent road (transportation feature). Area events may be applicable for any kind of transportation feature. Area event as a discrete entity is omitted from the simplified model since such events are almost always expressed in a transportation database using linear and point events. The omission of Area Event also lets us drop polygon, its corresponding cartographic entity.

• *Event point*: *The location where an event occurs on a transportation feature.* Event points are located on a transportation feature as an offset distance measure from the beginning of the transportation feature. Most transportation databases use event point measures made in units of 0.01 miles. The smaller the measurement unit, the higher the resolution of the database, i.e., the closer two events can be and still be stored as separate events. A resolution of 0.01 miles means that two events within 52.8 feet of each other may be stored as being located at the same position along the transportation feature. Event point locations are stored using field measures in real-world units, not cartographic units. Event points locate events on the transportation system while cartographic points locate representative graphical elements on a map. In addition to locating Event Points by a linear measure along transportation features, they can be located by direct coordinate measurement. Digitizing maps or field measurements from surveying or GPS does this. Just as linear measures along transportation features can be converted to coordinates by interpolation along shape files representing the transportation feature using dynamic segmentation, coordinates can be snapped to transportation features and converted to linear measures by a reverse of dynamic segmentation. *Relationships*: Transportation Feature may possess one or more Event Points. Event Point must be defined in the context of one Transportation Feature. Event Point may locate one or more Point Event. Point Event must be located (on Transportation Feature) by one Event Point. Event Point may locate the beginning or end of one or more Linear Event. Linear Event must be located by a beginning Event Point and an ending Event Point. Event Point represents Geographic Point. Geographic Point locates Event Point.

• *Intersection*: *A special type of Point Event which may be owned by more than one transportation feature.* From the perspective of a single transportation feature, an intersection has but one owner; i.e., that transportation feature. Although in actuality, an intersection may appear as a point event on more than one transportation feature. Relying solely on the Transportation Feature owns Point Event relationship will result in redundant data storage in that each transportation feature would be required to record the same intersection data. The use of an Intersection entity allows the location of the intersection to be stored as a point event on each intersecting transportation feature but stores the intersection characteristics as part of intersection entity attributes. It may be advantageous to treat any type of transportation system junction or crossing as an intersection. For example, treat all bridges as intersections in order to link bridge height and weight data to the roads going over and under the bridge. *Relationships*: Intersection must correspond to one or more Point Event. Point Event may represent one Intersection. Intersection must involve one or more Transportation Feature. Transportation Feature may possess one or more Intersection.

These entity and relationship definitions are based on the following business rules:

1. Transportation features are contained completely within a single jurisdiction. A transportation feature that exists physically in more than one jurisdiction may be administratively subdivided at the jurisdiction boundary as a data management approach analogous to the tiling of GIS maps.
2. The physical path of a linear transportation feature can be arbitrarily defined, although specific users may adopt rules for defining a transportation feature. The model is independent of these rules.
3. A transportation feature's unique identifier (name) in the database must be independent of its real-world name, which is an attribute of the feature, even if the adopted rule is to follow a named path to define a transportation feature.
4. An external public key identifier may be constructed using a widely accepted naming convention. The public key is independent of the primary (internal) key and the real-world name.
5. Point and area features must be expressed relatively to linear highway features. For example, consider that a city is an area feature. In a roadway database, the extent of a city is defined by the portion of each roadway within the city. To store the relevant information about the city, define the portion of each roadway inside the city by stating the beginning and ending points (city limits) using linear LRS measures.
6. The relationship between roadways at intersections must be stored and data about intersections must be recorded efficiently without duplication.

5.2. Network features and their attributes

The second group of entities consists of four entities that contain information about the connectivity (topology) of the transportation network: Node, Link, Traversal, and Traversal Segment. Networks are subdivided into segments called links, which begin and end at nodes. A traversal is a path through the network that is composed of traversal segments, each of which may be formed from one or more links and their attributes.

• *Node*: *A zero-dimension object that represents the topological junction between two or more links, or the endpoint of a link*. In their most common form, nodes will correspond to intersections and other point events. Not every intersection need be represented as a node. It is sufficient for a routing application, for example, to include only those decision points (i.e., where the course may be altered) that have been deemed suitable. *Relationships*: Node may represent one Point Event. Node may begin one or more Link. Node may end one or more Link. In most cases, Node must begin or end at least one Link (may not be true if centroid nodes are included in data set).

• *Link*: *A one-dimension object that represents the logical connections between nodes*. Links, as used here, are dimensionless in that they only specify the possible connections and not the actual distance between nodes. We could include an attribute for valid directions, but we have chosen to recommend two alternative implementation strategies. One is to order nodes; the other is to only include valid nodes that may be reached from a given node. In the ordered node approach, if one could travel from Node A to Node B or from B to A, as would be the case with a two-way road, then a link table would include the entries A,B and B,A. A corresponding one-way road would omit one of the node pair entries. In the valid destination node approach, a two-node entry for Node A would be B and one for B would be A. If, however, the link A–B were a one-way road with the direction of travel being from B to A, then there would not be an entry for Node B in the node table record for Node A. *Relationships*: Link must begin at one Node. Link must end at one Node. Link may be part of one or more Traversal Segment.

- *Traversal*: *A path or route through a portion of a transportation network consisting of one or more segments.* We have chosen to make a distinction in application between static traversals, such as that formed by a state road crossing many counties, and a dynamic traversal, such as that created to route an overweight truck from origin to destination. These are functionally the same things, however, their duration of use is different. Both are constructed by selecting traversal segments with a specified set of attributes. For the state road traversal, this would be those traversal segments with the same road number. For the overweight truck route traversal, this would be those that could support the weight of the truck and possibly match other criteria, such as low traffic volumes. *Relationships*: Traversal contains one or more Traversal Segment. Traversal Segment may be part of one or more Traversal.
- *Traversal segment*: *A link and its relevant attributes.* A traversal segment is both topological and physical. It combines the connectivity of the link with the attributes of the transportation feature segment it represents. We have elected to utilize this separate entity to store link attributes for two reasons. First, we wanted to keep the purely topological information separate in order to facilitate data sharing and our central concept of unbundling. Second, we wanted to be able to select the attribute set to match the needs of the application, while preserving a normalized data structure for transportation features and their attributes. Traversal Segments may be viewed as a special linear event that corresponds to the path of the related link. Traversal Segment attributes may be derived by finding the desired linear and/or point event records that fall within the same roadway segment. The actual selected values may be minimums (e.g., bridge with the lowest weight bearing capacity, lowest overhead clearance, etc.) or maximums (e.g., highest traffic volume, greatest population, etc.). *Relationships*: Traversal Segment may be part of one or more Traversal. Traversal Segment must represent one Link. Traversal Segment may include attributes of one or more Point Event. Traversal Segment may include attributes of one or more Linear Event. Linear Event may be applicable to one or more Traversal Segment. Point Event may be applicable to one or more Traversal Segment.

These entity and relationship definitions are based on the following business rules:

1. Link-node data structures may be used to express certain attributes that conform to that structure, such as the distances or travel times between intersections (block is a link, intersection is a node).
2. Static and dynamic paths (traversals) must be defined through the highway network. A static path is one that is defined by a slowly changing characteristic, such as state road number or a bus route. A dynamic path is one created to serve an immediate and frequently changing need, such as to route an overweight vehicle to its destination or to direct a rider to the nearest bus stop.

5.3. Linear LRS objects and their attributes

We have provided four entities to tie transportation features to the surface of the earth in 2- or 3-coordinate spaces. These entities are derived from those proposed by Vonderohe and Hepworth (1998).

• *Anchor point*: *A zero-dimension object representing the end or beginning of an anchor section and serving to relate that terminus to an earth-based location. Position on earth and on any related transportation feature are mandatory attributes.* By storing the earth-based and linear LRS coordinates for an anchor point, the database model provides a registration mechanism to tie the transportation features to the ground. This facilitates cartographic conflation. (We actually recommend storing the linear LRS data in the anchor section records since the position is specific to each terminating anchor section.) *Relationships*: Anchor Point may be located by one or more Reference Object. Anchor Point may begin one or more anchor section. Anchor Point may end one or more Anchor Section. From a practical perspective, Anchor Point must begin or end at least one Anchor Section.

• *Anchor section*: *A one-dimension object providing a logical representation of all or part of a transportation feature. Length is a mandatory attribute.* An anchor section begins and ends at an ordered pair of anchor points. This ordering provides an indication of direction of increasing linear LRS measures. Anchor sections and points differ from links and nodes, by including more than topological information and an incomplete specification of topology. For example, anchor point locations are defined in possibly several systems; node locations are not defined at all. Ordered anchor point pairs for defining anchor section direction fail to indicate whether traffic can flow in the opposite direction. *Relationships*: Anchor Section may establish the linear LRS datum for one Transportation Feature. Transportation Feature may be defined by one or more Anchor Section. Anchor Section must begin at one Anchor Point. Anchor Section must end at one Anchor Point.

• *Reference object*: *A physical object or recoverable location to which the position of anchor points may be conveniently related.* Many anchor point locations are likely to be conceptually easy to define but physically hard to find in the field. An example is the middle of an interchange or intersection. It is easy to recognize but hard to precisely and consistently locate. Tying the location to a physical object, such as a traffic signal pole, allows the precise location of the anchor section to be readily found using direction and distance from the reference object. Anchor Point and Reference Object may be merged to form a single entity. *Relationships*: Reference Object is located on the surface of the earth using one Geographic Point (which may have many coordinate descriptions). Reference Object may be used to locate one or more Anchor Point.

• *Geographic point*: *A zero-dimension object carrying the real-world (earth-based) location of a reference object.* We have been deliberately restrictive in our definition of Geographic Point as a means of simplifying the model. In reality, every other entity in the real world could be described using a geographic point. What we want to emphasize here is the translation of geographic points, defined minimally for reference objects, to cartographic points as a means of facilitating conflation. *Relationships*: Geographic Point may locate one Reference Object. Geographic Point may be transformed to one or more Cartographic Point (each with its own cartographic datum and coordinate system). Event Point represents Geographic Point. Geographic Point locates Event Point.

The business rules that produced this portion of the data model are:

1. Many transportation features may be efficiently located on the surface of the earth using non-linear location referencing systems based on GPS and other earth geode methods. Most typically, these features are of the area and point geometric forms. Point and linear attributes on linear features may be secondarily defined using earth coordinates. However, this approach will

not unambiguously define a point on the linear feature *from a graphic perspective* due to differences (errors) between the map and the real world, the measured position on the earth and its equivalent on the map, and between the measured position and the true position.
2. The model must provide the data entities needed to store a linear datum developed to meet positional accuracy and data sharing needs, but do not require that such a datum exist.

5.4. Cartographic objects

We use the term 'cartography' to refer to the map or picture produced by a GIS application. There are five cartographic entities that may be used to visually express the location and shape of transportation features, linear events, and point events. The following entity definitions are independent of proprietary software terms:

- *Cartographic point*: *The internal address reference for placing a single point on the surface of a two- or three-dimension map (manifold)*. Each mapping environment has its own way of defining single locations, i.e., its cartographic datum. *Relationships*: Cartographic Point may represent one Geographic Point. Cartographic Point may shape one or more Line Segment. Cartographic Point may locate one Point Symbol.
- *Line segment*: *A straight connection or otherwise defined mathematical path between two cartographic locations*. All line segments must be defined using beginning and ending cartographic locations (points). *Relationships*: Line Segment is shaped by one or more Cartographic Point. Line Segment may be part of base Map String. Line Segment may be part of Linear Event String.
- *Base map string*: *A connected non-branching sequence of line segments, usually specified as an ordered sequence of vertices, which cartographically defines the shape of a transportation feature.* Our basic implementation suggestion for mapping transportation features is to start with an equivalency between each entire transportation feature and a single line object. This line object will be composed of one or more line segments. Multiple spatial representations are enabled. A transportation feature can be represented by more than one base map string. *Relationships*: Base Map String may be used to describe the shape and position of one Transportation Feature. Base Map String must consist of one or more Line Segment.
- *Linear event string*: *A connected non-branching sequence of line segments, usually specified as an ordered sequence of vertices, which cartographically defines the shape of a linear event occurring on a transportation feature*. Our expectation here is that most linear event strings will be created as needed by dynamic segmentation, which uses straight-line interpolation to extract the portion of a base map string that corresponds to the location of a linear event using linear LRS measures. Line style, width, and color may all be used to distinguish the attribute being described. *Relationships*: Linear Event String must represent one Linear Event. Linear Event String must consist of one or more Line Segment.
- *Point symbol*: *A cartographic object that is used to show the position and nature of a point-like real-world feature*. Just as linear features can be shown using a line, point features can be shown using a point symbol. *Relationships*: Point Symbol may be located on a map using one cartographic Point. Point Event may be illustrated by point symbol, which may also select the correct point symbol instance to display.

The business rules that produced this portion of the data model are:
1. Separate graphic representations from attribute data to provide scale independence; provides "one database-many maps" functionality.
2. Utilize dynamic segmentation and field-based linear LRS measures to create cartographic objects that correspond to linear events and to position point symbols.
3. Provide a means of relating real-world measures to map locations.

5.5. Applications of enterprise GIS-T data model

More details of the model and examples of implementation are contained in the original presentation (Dueker and Butler, 1998). For example, attributes that are offsets from the centerline, such as signs, guardrails, or number of lanes, are handled by means of point or linear events. Linear events, such as lanes can be reformulated in several ways. A linear event could be represented as a recursive one-to-many relationship for Transportation Feature entity, if we want to look at the lanes as transportation features owned by a larger one. Or each lane could be a standalone feature. The data model itself is substantially indifferent to the details of transportation feature construction. We choose not to specify one or the other as the application may drive one over the other.

The data model supports non-planar graphs although we encourage the use of intersections for over- or underpasses to store clearances and restrictions. Although our enterprise GIS-T data model does not support temporal data explicitly, time is an attribute of transportation features or of the point and linear events associated with them.

Whether this data model or another, a common understanding of the transportation system is needed in order to share data effectively. The data model facilitates selection of appropriate transportation features from different databases and spatially registering them and creating new application-specific networks. The GIS open systems concept applied to interoperability of transportation data requires a common feature schema, which is consistent with our transportation features. In addition, the data model separates the update and maintenance of transportation features from the networks used in specific applications. Different networks can be generated from a common set of transportation features and one update process can support a set of applications.

5.6. Comparison of enterprise GIS-T and NCHRP 20-27 data models

The connection between the NCHRP 20-27 data model and the Dueker–Butler data model is considerable and implicitly apparent from the common use of several terms and methods. Recognizing the considerable contribution of 20-27 towards consensus building in the field, we sought to adopt, wherever possible, the conventions, approach, and definitions of the 20-27 data model. A basic component of our model is the concept of transportation features. Transportation features have many of the characteristics of the 20-27 model's traversals. For example, a linear transportation feature can have points located along its length by the use of a linear LRS. However, the concept of transportation feature includes area and point forms, not just linear ones, where the term 'traversal' does not intuitively apply. We have retained the traversal concept for linear transportation features, even to the point of using the same definition as in 20-27, but the primary linear LRS is tied to the transportation feature. (Positions on traversals may be defined using a subordinate linear LRS, including one based on time, e.g., a transit route running on all or part of several transportation features.)

However, there are fundamental differences. First, we do not require transportation features to be composed of one or more links. In other words, topology is not a requirement for transportation features. Second, we relate the cartography to the transportation feature, not to the linear datum. To some degree we also adopt a third difference by embedding the linear LRS in the transportation feature; however, the implementation of this approach and that of the 20-27 data model appear to be identical. A fourth difference is the expansion of datum entities to include reference objects that are more readily recoverable in the field than anchor points. To some degree, our inclusion of reference objects is intended to accommodate the proposed national intelligent transportation system (ITS) Datum, which consists of a group of reference objects and the rules to define them.

One of the biggest differences is the fifth one: incorporation of cartography. The 20-27 data model provides support for cartographic conflation in that it can supply position information in the form of earth coordinates for anchor points. These coordinates can be mapped in the cartographic environment and be used to relate graphical objects to the datum and improve the quality of maps. This approach leaves out many graphical objects and provides little roadway shape information. Cartographic conflation remains a substantially manual process, but one that is a necessary pre-requisite for data sharing. The requirement for transferring topology, imposed by the use of links and nodes to connect traversals (a.k.a., linear transportation features) to the datum and thence to the cartography places a still greater burden on data sharing.

6. Transportation data participants

The enterprise GIS-T data model provides participants in data sharing with a framework for clarifying roles. Departments of transportation or public works are organizations that have ownership and maintenance responsibilities for transportation infrastructure. They are data producers, data integrators, and data users both for internal uses and for other organizations. Internal uses vary considerably. Planning, design, construction, and maintenance divisions are substantially independent of one another. The solution is not a single GIS-T but an enterprise approach to GIS-T data.

Motorists and the general public are primarily data users. Organizations that use the transportation system in their business, the police, delivery services, etc., often rely on data integrators to provide transportation data in the form of maps and networks for location, path finding, and routing. Increasingly, users are demanding current, logically correct, and spatially accurate transportation data in interoperable digital form for large regions that span many jurisdictions. Currently there are neither technical and institutional processes to achieve a single integrated database to handle those diverse needs, nor is it likely that such processes will be developed and sustained. Rather, principles need to be established to guide data sharing and the development of application-specific transportation databases that can be assembled without costly redundant recollection of source data and updates from the field each time.

There are two participants whose accuracy requirements drive the data sharing process:

- Emergency management, E-9-1-1 and computer-aided (emergency) dispatch (CAD) have the most demanding need for currency and completeness.
- Vehicle navigation applications, which may include CAD, have the most demanding need for spatial accuracy of street centerline files. This is sometimes referred to as "map matching" of GPS-derived location of vehicles to the correct street or road in the road database. Identifying the correct ramp of a complex freeway interchange on which a disabled vehicle is located is a particularly demanding task. Similarly, electronic (ITS) toll collection applications may require tracking vehicles by lane of multiple lane facilities.

Others have less demanding needs for temporal accuracy (currency), completeness, and spatial accuracy.

7. Transportation data sharing principles

Successful data sharing requires a common schema or data model that is flexible enough to handle the needs of diverse participants. The flow of data from providers to users calls for capturing data once, and delivery to users in various forms, time frames, and spatial scales.

Principles for successful sharing of transportation data among participants must address a variety of issues: definition and identification of transportation features, cartography and spatial accuracy, generation of application-specific network representations, and interoperability.

Two important principles follow from the GIS-T data model:

- Transportation features are bounded by jurisdictions, not intersections. This is not to say that the underlying cartography could not have other forms, only that the link between attribute data and cartography must occur through transportation features rather than spatial primitives or network topology.
- Attributes of transportation features are represented as linear or point events and are located along the feature using linear referencing.

These two principles enable longer transportation features than is the case in link-based networks and reduces the number of transportation features that must be maintained to represent the system. Adding network detail or additional attributes does not necessarily increase the number of features. Additional detail can be added by linearly referenced event tables and analyzed and visualized using dynamic segmentation.

Butler and Dueker (1998) also identified important data sharing principles:

- Transportation features must be uniquely identified to facilitate sharing of data among participants. Participants need to identify common features in sharing data.
- Transportation data producers need to include a standardized unique identifier with each transportation feature.

The latter principle leads to subsidiary principles for assignment of identifiers:

- Segment major arterial facilities at jurisdictional boundaries or major intersections if consistency with traffic assignment networks is desired.

- Collect minor road facilities by street name or route number to minimize the number of unique identifiers. But do not use names or routes as identifiers. They change.

There are several other principles that are offered to reduce the amount of manual coding and conflation, and thereby ease compliance with the data sharing principles. These are offered to avoid the need for simultaneous conflation of cartography and topology with a process to resolve inconsistent segments:

- Exchange attribute data as event tables for logical transportation features, i.e., without shape points.
- Exchange cartography without topology.
- There is no need to code topology, let the GIS generate application-specific networks from a selection of appropriate transportation features.
- Minimize manual coding of transportation feature identifiers by embedding existing identifiers into more global identifiers and using scripts to bulk-assign state and county codes.

Fig. 2 shows how using a transportation feature-based identifier to exchange data avoids most of the problems of conflating cartographic and topological objects. In this example, the heavy line represents the common transportation feature in all three schemas. The top and middle versions use topological (link/node) internal data structures but with different sets of links and nodes (the top version has more). The bottom version uses simple line strings. From an implementation perspective, the top two might be ArcInfo-based coverages, while the bottom is representative of line strings used in GeoMedia, ArcSDE, and ArcView. Dependence on conflation to make the cartography and topology the same must proceed the practical sharing of data. However, if all three had previously combined their primitive objects to create the higher-level transportation feature and given it a common identifier, then there would be no need to make the maps and

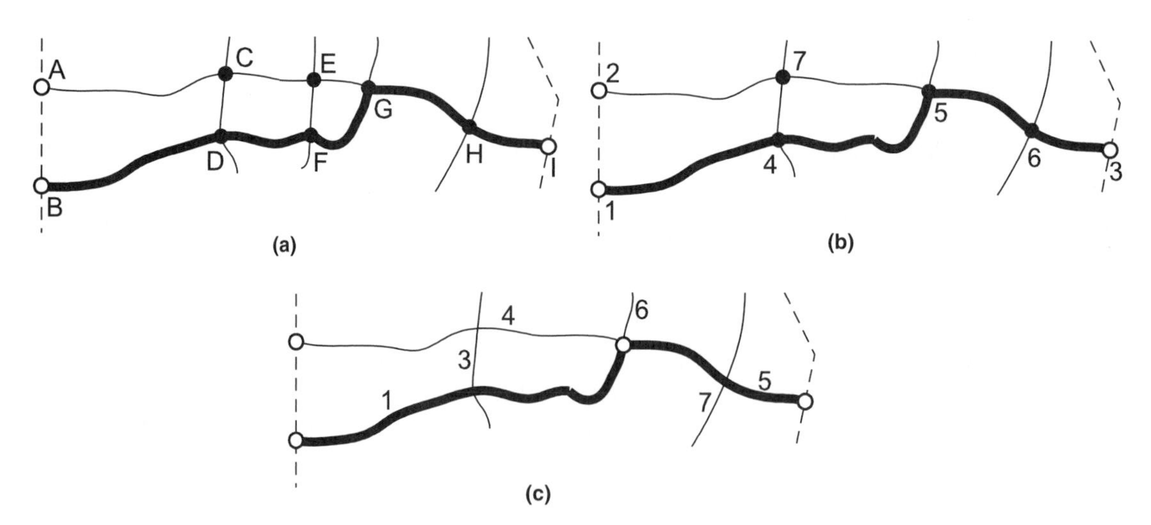

Fig. 2. Comparison of cartographic and topological model approaches: (a) Transportation feature defined by links B–D, D–F, F–G, G–H, and H–I; (b) Transportation feature defined by links 1–4, 4–5, 5–6, and 6–3; (c) Transportation feature defined by lines 1 and 5.

internal structures the same. The shape of the line, the scale of the map, the number of line segments, the topological structure (if any), and other implementation-specific issues would disappear. Data sharing would be direct and immediate.

7.1. Issues of definition and identification of transportation features

A transportation feature can be like a point (interchange or bridge), a line (road or railroad), or an area (rail yard or airport). Nevertheless, some applications may restrict their databases to roads, pedestrian paths, or waterways. The important point is to code the type of transportation feature so that the type can be used to select those features of common interest for sharing of data.

Butler and Dueker (1998) proposed an internet-like address identifier for transportation features. Similarly, the Oregon road base information team subcommittee (ORBITS) (Bosworth et al., 1998) and the NSDI framework transportation identification standard (FGDC, 1999) have proposed a roadway identifier schema. The NSDI proposal will likely prevail when it is complete.

The purpose of assigning a stable and unique identifier to each transportation feature is to eliminate or reduce reliance on traditional conflation processes to reconcile different transportation databases. Unique identifiers are used to match transportation features between databases without relying on matching coordinates and links.

A case study was conducted to test methods of assigning both the ORBITS and NSDI identification codes. The ORBITS approach collects or divides roadway features and identifies them with a unique code. In the context of the case study, decision rules for breaking or collecting roadway sections and procedures for bulk assignment of higher level codes to sequenced numbered roadway features were developed. The assignment of NSDI codes to roadway features was similar, except that point identifiers were also assigned to beginning and ending points of roadway features. The ORBITS team chose not to code the topology, leaving that to be generated if needed, as it is too application-specific to be of general use.

The ORBITS case study developed different decision rules for assignment of transportation feature identifiers to arterial roads and local roads. Urban arterial roads are segmented at intersections of major arterials. Arterial road identifiers in urban counties are a concatenation of state and county FIPS codes and a concatenation of i and j traffic assignment network node numbers. In rural counties, we would recommend arterial roads be assigned a code that is the concatenation state and county FIPS codes and the State Department of Transportation or county Department of Transportation road identifier. Portland Metro desired a finer breakdown of uniquely identified major roads than the rural rule would have accomplished.

The decision rule for assigning codes to local roads is to collect connected TIGER lines that have the same name and assign them sequence numbers concatenated with state and county FIPS codes. Some judgment has to be applied to deal with interruptions in connectedness of local streets. When there are minor interruptions, the same code is assigned to the local road with the common name. When interruptions are more than minor, separate identifiers are assigned. Also, there are situations where name changes occur arbitrarily, such as at municipal boundaries where a different identifier may not be needed. Bender et al. (1999) provide a description of a case study of assigning NSDI identifiers.

Even with the collection of links of legacy databases into larger transportation features, the assignment of identifiers is an onerous task, especially if mandated without assistance.

Transportation organizations may not be willing to undertake such an effort in the absence of a payoff or need. A multi-participant project that requires integration of data from various sources may be needed to provide the reason and incentive to assign transportation feature identifiers. The usual approach relies on a single transportation organization to conflate linework from various sources. This may be an adequate solution to build a spatial database, but not for maintenance over the long term.

In the absence of identifiers, projects that involve massive data integration are very difficult to update and maintain. Thus, the identifier approach is particularly crucial to apply to roadway additions, deletions, and modifications. The key to database maintenance is to manage the incorporation of updated transportation features. The fundamental strategy is to identify features in the database to facilitate a transactional update system, one that does not require rebuilding the entire database anew.

7.2. Implications for the NSDI proposal

The present draft NSDI framework transportation identification standard (FGDC, 1999) satisfies many of the stated business rules of our enterprise GIS-T data model. But the proposed standard is not based on a formal data model and as a result suffers from ambiguities. Framework transportation segments (FTSegs) have a unique identifier and like anchor sections, FTSegs are defined by beginning and ending points (framework transportation reference points, or FTRPs). Without a formal data model, the NSDI proposal lacks guidance on whether length of FTSegs is mandatory or optional, or how to code an interchange.

The implicit topology of FTSegs, created by stating the terminal FTRPs, is equivalent to that in our anchor point/anchor section structure. The NSDI proposal requires an intermediate FTRP to be defined if there are multiple paths between the two FTRPs defining a given FTSeg. The explicit topology of FTSegs and intermediate FTRPs – which do not create new segments but are located using a distance offset along the segment – is essentially the same as our point event. Both intermediate FTRPs and explicit topology can be expressed using our point event entity.

A continuing problem with the NSDI proposal is the need to serve both logical and physical descriptions of the transportation network. Some users may need to use a single FTRP to represent an entire interchange that for another user may require a dozen of more FTRPs. We have proposed the use of an intersection entity to serve as a general, or logical, object for what may be one or more physical position descriptions for all or part of a complex structure.

The NSDI proposal meets the transportation data sharing needs for unique and public identifiers. It almost meets the requirements of a national transportation datum, such as that which has been listed as a pre-requisite for many ITS deployments, needing only to make length a mandatory attribute. It almost provides a clear statement of network topology, falling short only in its ambiguity of connection specification.

And yet, we are not satisfied. Data users need to exchange not only attributes relevant to an entire transportation feature, but also those that apply to a portion of a feature. A proposed standard structure for constructing a universal transportation data exchange format seems necessary for people to be able to exchange entire data sets, not simply identify which road they are talking about. The absence of linework to represent the FTSegs is a serious omission in a "spatial" standard, in the eyes of many potential users. For it is the need for more accurate linework that drives most spatial data sharing needs. But more accurate linework is accompanied by a more detailed representation of transportation features. Yet the NSDI Framework Transportation Identification Standard is a major effort in the process of forging a true data exchange mechanism.

7.3. Cartography and spatial accuracy issues

Issues which states face in constructing a road layer for a statewide GIS illustrate the problem of sharing transportation data. The problem is stitching together data from various sources and vintages.

Typically, State Departments of Transportation have a roadway database for highway inventory. Attributes of roads are recorded by milepoint and associated with a straight-line chart for visualization. Some states have incorporated the linearly referenced data into a GIS and use dynamic segmentation for analysis and visualization. The spatial scale of the cartography ranges from1:24 000 to 1:100 000.

Similarly, the spatial scale of the road layer used by natural resource agencies is from 1:24 000 or 1:100 000 USGS sources but with very little attribution and with uneven currency. However, the USGS digital orthophotography quarter-quadrangle program offers the opportunity for states to update their road vectors and register them using 1:12 000 imagery.[4] This ought to provide sufficient spatial and temporal accuracy for E-911 and for vehicle navigation (snapping vehicle GPS tracking data to road vectors). However, these sources may not be sufficiently accurate to distinguish road lanes, which are needed in urban areas for dynamic vehicle navigation and road pricing (e.g., snapping vehicle GPS tracking data to lanes and ramps to relate to lane-specific volumes and speeds from loops, imaging and vehicle probes).

7.4. Spatial database completeness and currency issues

Vehicle tracking will require proper positioning, both in terms of which road the vehicle is on and where it is on that road. The ITS community has proposed a "cross-streets" profile that provides this information in a message format that includes the name of the current street and the location as the terminal cross-streets for the present block on which the vehicle is located. One reason for this approach is to avoid the need to have precise GPS-map concurrence regarding spatial position. Researchers have discovered, however, that current maps do not have sufficiently reliable and complete street name attributes for this schema to be routinely implemented. (Noronha, 1999) Currently, the ITS database community is proposing to use the coordinates of intersections as well as street addresses and may add street type for locational referencing.

If work must be done to populate databases with sufficient information to identify which street and block a vehicle traversing, then it seems appropriate to develop a more complete approach that avoids the remaining problems, such as differences in spelling that may arise for street names in different databases. The purpose of a transportation feature identification schema (perhaps with a redundant cross-street index) is to insure completeness and currency of databases.

---

[4] Some regions have more accurate orthophotography.

Development of an identification schema requires guidelines or standards for segmenting of arterial roads, insuring completeness of local road segments, and collecting them into larger chunks for assignment of an identifier. Similarly, development of standards for coding ramps and lanes is needed. But most importantly, a typology of transportation features needs to be developed to accommodate different definitions of roads and non-road features for different applications. This is particularly important where databases are developed by means of the vertical integration of databases from different organizations for the same geographic area or jurisdiction, say integrating state, county, and city data.

Identifiers for transportation features facilitate transactional maintenance, additions, deletions, and changes to transportation features with periodic issuance of new editions for less time-sensitive applications. Real-time users will require dynamic changes issued as linear and point events, or modifications to links and nodes to reflect lane/street closures or construction detours.

7.5. Network issues

In-vehicle navigation systems will provide the greatest challenge in terms of spatial and temporal accuracy for road map databases. Current technology supports generalized client-based networks for minimum path routing (based on typical speeds or impedances) that produces instructions in terms of street names and turns. This is based on a road map base that snaps GPS vehicle-tracking data to road vectors.

In the near future, we are likely to see detailed server-based dynamic routing based on current network traffic conditions with instructions including ramp and signage details and snapping of GPS vehicle tracking data to lanes and ramps. The coding of topology using formal and widely recognized transportation feature identifiers will allow vehicle routing to be done without reliance on maps.

The chief problem with transportation networks is the perception that there is one base network that will satisfy all applications. We contend this is a false premise, as someone will always want to add or delete links. Networks ought to be application-specific, and consequently, the network cannot be the building block of sharable or interoperable transportation data.

7.6. Interoperability issues

Transportation feature identifiers provide a common key by which to relate data to achieve interoperability among transportation databases. Nevertheless, relating databases on the fly may not perform well for real-time applications. Performance may be a problem in relating to a highway inventory database to check underpass clearance for a dynamic pathfinding application for rerouting traffic due to emergency incidents. Instead, clearances may need to be pre-computed and stored as link attributes in the dynamic pathfinding application.

Full interoperability suggests "plug and play", meaning Vendor A's data can be read by Vendor B's system and vice versa. In transportation, this will be difficult to achieve because of the varied nature of applications that require data in specific forms, and the size of typical regions for which consistent data would be needed.

It is difficult to achieve consistent and accurate data for a region as large as the nation, considering the temporal data streams that will be created by vehicle tracking systems and by video cameras used to estimate vehicle flow rates. This stream of data will require format standards. Dailey et al. (1999) provides a self-describing method for transfer of data in real-time for ITS applications.

8. Conclusion

Sharing of GIS-T data poses challenges. This paper identified the issues and developed a framework and principles to address them. The central principle is the establishment of a schema for transportation features and their identifiers. An underlying principle is the need for a common data model that holds transportation features as the object of interest, and that attributes of transportation features are represented as linear and point events that are located along the feature using linear referencing. Until there is agreement on these principles, data sharing and interoperability will not progress well. This lack of agreement stems from the current state of flux with respect to GIS-T data models. This problem should diminish with the completion of the NCHRP 20-27(3) project and final adoption of the NSDI framework transportation identification standard.

In this context, sharing of transportation data involves exchange of relevant transportation features and events, not links and nodes of application-specific databases. This is a major departure from the existing process of conflation. The exchange of more fundamental features is encouraged in recognition that each application has quite specific requirements for their end-use database, but all users have need for basic transportation features. The major contribution of our enterprise GIS-T data model is this separation of the maintenance database from many application databases. The term enterprise takes on a broader meaning of a shared stewardship of data about the larger transportation system.

Strategies for the sharing of transportation features follow from this approach. The first is to enlist state and local cooperators to construct transportation features by registering existing transportation vector data from TIGER and local sources to digital orthophotography, such as the USGS orthophotography quarter-quadrangles. A second stage would be to update the vectors using replications of vehicle tracking data. The fundamental strategy is to identify features in the database to facilitate a transactional update system, one that does not require rebuilding the entire database anew.

Although this approach to the sharing of transportation data needs to be refined, it provides a better framework than that which currently exists. There is no common approach among the communities of ITS, vehicle navigation database vendors, NSDI, state and local transportation organizations, and E-911. However, we are encouraged by the continuing efforts of these groups to reach a more consistent approach that will facilitate data sharing by reducing the number of inconsistent and duplicative representations.

Transportation features are the fundamental objects in the database and must be uniquely and permanently identified. Similarly, additions, deletions, and modifications to transportation features must be identified to facilitate database updates and maintenance. This approach, or data model, separates the update and maintenance of transportation features from the links of networks used in specific applications. Thus, different networks can be generated from a common set of transportation features and one update process can support a set of applications.

References

Adams, T.M., Vonderohe, A.P., Butler, J.A., 1998. Multimodal, multidimensional location referencing system modeling issues. Prepared for NCHRP 20–27(3). Workshop on Functional Specifications for Multimodal, Multidimensional Transportation Location Referencing Systems, Washington, DC, 3–5 December 1998.

Bender, P., Bosworth, M., Dueker, K., 1999. ORBIT: The Oregon Road Base Information Team: the Canby case study report. Center for Urban Studies, Portland State University, Catalog Number PR111. http://www.upa.pdx.edu/CUS/PUBS/contents.html.

Bosworth, M., Dueker, K., Wuest, P., 1998. ORBIT: The Oregon Road Base Information Team: a draft summary report. Center for Urban Studies, Portland State University, Catalog Number PR106, 19 p. http://www.upa.pdx.edu/CUS/PUBS/contents.html.

Brown, J., Rao, A., Baran, J., 1995. Are you conflated? Integrating TIGER and other data sets through automated network conflation. In: GIS-T Symposium Proceedings of the American Association of State Highway and Transportation Officials, 220–229.

Butler, J.A., Dueker, K., 1998. A proposed method of transportation feature identification. Center for Urban Studies, Portland State University, Catalog Number DP97-8, 18 p. http://www.upa.pdx.edu/CUS/PUBS/contents.html.

Chan, K., 1998. DIGEST: A primer for the international GIS standard. CRC Press, Boca Raton, FL.

Dailey, D, Meyers, D., Friedland, N., 1999. A self describing data transfer methodology for ITS applications. Preprint CD-ROM. 78th Annual Meeting, January 1999, Transportation Research Board.

Dueker, K., Butler, J.A., 1998. GIS-T enterprise data model with suggested implementation choices. Journal of the Urban and Regional Information Systems Association 10 (1), 12–36.

Devogele, T., Parent, C., Spaccapietra, S., 1998. On spatial database integration. International Journal of Geographic Information Sciences 12 (4), 335–352.

FGDC: Federal Geographic Data Committee (Ground Transportation Subcommittee), 1999. NSDI Framework Transportation Identification Standard – Working Draft, 1999 May. http://www.bts.gov/gis/fgdc/web_intr.html.

GDF: Geographic Data Files Standard, version 3.0, 1995. Working Group 7.2 of CEN Technical Committee 278, European Committee for Standardization.

GDF: Geographic Data Files Standard, version 5.0, 1999. ISO/TC 204/SC/WG3. ISO/WD 19990722-2.

Noronha, V., 1999. Spatial data interoperability for ITS. Presentation at Annual Meeting of the Transportation Research Board, Washington, DC.

Sester, M., Anders, K., Walker, V., 1998. Linking objects of different spatial data sets by integration and aggregation. GeoInformatica 2 (4), 335–358.

Sperling, J., Sharp, S., 1999. A prototype cooperative effort to enhance TIGER. Journal of the Urban and Regional Information Systems Association 11 (2), 35–42.

Sutton, J., 1999. Object-oriented network data structures: the road to interoperability. Presentation at GIS-T Symposium, April 1999, San Diego.

Tomaselli, L., 1994. Topological transfer: evolving linear GIS accuracy. URISA Proceedings, 245–259.

Vonderohe, A., Chou, C., Sun, F., Adams, T., 1998. A generic data model for linear referencing systems. Research Results Digest Number 218. National Cooperative Highway Research Program, Transportation Research Board.

Vonderohe, A., Hepworth, T., 1998. A methodology for design of measurement systems for linear referencing. Journal of the Urban and Regional Information Systems Association 10 (1), 48–56.

Walter, V., Fritsch, D., 1999. Matching spatial data sets: a statistical approach. International Journal of Geographic Information Sciences 13 (5), 445–473.

PERGAMON

Transportation Research Part C 8 (2000) 37–52

TRANSPORTATION RESEARCH PART C

www.elsevier.com/locate/trc

Dynamic location: an iconic model to synchronize temporal and spatial transportation data

John C. Sutton [a,*,1], Max M. Wyman [a,b,2]

[a] *GIS/Trans Ltd., Suite 700, 8555 16th Street, Silver Spring, MD 20910, USA*

[b] *Terra Genesis, Earth Modeling and Systems Restructuring Group, USA*

Abstract

This paper describes a model to synchronize the management and query of temporal and spatially referenced transportation data in geographic information systems (GIS). The model employs a method referred to as dynamic location, which facilitates spatial intersect queries from geographic shapes without the use of topological relationships. This is the inverse of how dynamic segmentation works in GIS. In contrast to dynamic segmentation, dynamic location stores geometry as an object within a single database field. This is an efficient, precise iconic model superseding the need for data decomposition into a complex set of tables. As an object model, the dynamic location process lends itself to high performance in an Internet, data-intensive, enterprise environment. Linear events are stored as $\{x, y\}$ features, and not referenced to any route system. Route systems are built from $\{x, y, m\}$ values (m for measure) and serve as number lines for mathematical operations. Any $\{x, y\}$ object can then be referenced to either the Cartesian grid or any selected number line. This method offers the benefits of linear referencing, while making full use of a stable geodetic datum. Combinations of any $\{x, y\}$ events may be placed over any $\{x, y, m\}$ number line (route) and an intersect determined by looking through stacked $\{x, y, m\}$ vertices of the coincident shapes. Since both geometry and shape reside in the same record, the use of "begin" and "end" dates facilitates full spatial and temporal version control. From a business process perspective, this creates a spatially enabled database, pulling GIS business functions back into the information technology mainstream. © 2000 Elsevier Science Ltd. All rights reserved.

Keywords: Dynamic location; Temporal version control; Combined linear and geodetic referencing

* Corresponding author. Tel.: +1-301-495-0217.

E-mail addresses: jsutton@gistrans.com (J.C. Sutton), max.wyman@TerraGenesis.com (M.M. Wyman).

[1] www.gistrans.com.

[2] Tel.: +1-480-345-0447; www.TerraGenesis.com.

PII: S0968-090X(00)00028-0

1. Introduction

The purpose of this paper is to review some of the problems of integrating transportation data in geographic information systems (GIS) and to present a new data model that provides a robust method for referencing and managing temporal and spatial changes. The model has been developed from field practice and experience, and utilizes off-the-shelf database and GIS software components. The model has been tested on road network files only, as described below, but should be applicable to transit networks also. The model, referred to as dynamic location, has been developed in response to some of the limitations in GIS, with respect to the management of networks and linearly referenced data. It also takes advantage of some of the technology changes affecting the use of geospatial technologies, especially the use of object-oriented programming techniques and object relational database management systems (OBDBMS). The tool is a synergistic counterpart to dynamic segmentation.

Linear referencing is a core transportation method because many transportation features are linear in nature. This referencing process is easily implemented in a tabular environment, but complicated by limitations within GIS. GIS has been an important information technology tool in transportation because of its ability to perform a spatial intersect on network overlays, an analysis structured query language (SQL) cannot support because data order (location) must be considered. There are two ways to conduct a spatial intersect: the intersection of topologically related tables, or, the mathematical graphing and subsequent measuring of graphed spatial relationships. Topological methods separate attribute data from geometry and are a requirement when greater accuracy is afforded by physically measuring the earth, say with a calibrated odometer, and storing the values. These accurate values can then be mapped for display using dynamic segmentation. Although this process preserves field-measured accuracy, it separates data from its GIS geometry (Dueker and Vrana, 1992; Adams and Vonderohe, 1998; Bespalko et al., 1998), and this prevents the use of common IT tools such as temporal and spatial version control and database rollback. Four problems result.

First, there is not a one-to-one relationship between data records and spatial references. This is an interesting point, because tools such as dynamic segmentation consider the ability to map many-to-one relationships an asset. In many cases it is. However, much more real-world infrastructure and events are discrete, and when treated as an object, a one-to-one relationship is created. This opens the door for a new tool that maximizes the one-to-one object relationship to synchronize data and geometry, and thus regain spatial and temporal control.

Second, additional problems are encountered when poor map scale or generalization cannot elucidate modern transportation infrastructure (Sutton, 1997; Bespalko et al., 1998, 1999). Such situations are called network pathologies (Sutton, 1995, 1996a), such as ramps and connectors, frontage and service roads, interchanges, stacked decks, high-occupancy lanes and contra-flow lanes whose directionality may vary with time. Similar difficulties arise when integrating and managing multiple network topologies of the real-world network feature, for example, integrating street centerline files and model networks for highway and transit demand modeling (Mainguenaud, 1995; Kwan et al., 1996; Horowitz, 1997; Kwan and Golledge, 1997; Sutton, 1996b, 1997).

Since small-scale generalized maps could not mathematically manage complicated geometry, topological intersect tools emphasized many-to-one relationships and not graphical intersect means. In the case of transportation infrastructure management, it was easier to physically measure the distances between field infrastructures using a calibrated odometer, build many-to-one relational tables, and live with the synchronization problems of topology. However, accurate technologies such as GPS, meter-level imagery, and light distance and ranging (LIDAR) can create mapping grids that surpass the accuracy that can be gained from traditional field measurement techniques (Burkholder, 1997). There is opportunity to utilize GIS as a Cartesian model of those data. Dynamic segmentation relates tabular data to the prevailing GIS Cartesian grid, while dynamic location assumes a GIS that is sufficient to preserve and mathematically manipulate data without tables.

Third, because relative locations are a function of a route and distance from origin (DFO), data management is complicated when DFO points are moved, routes are renamed, or roadbeds are realigned. Experience has shown that there are many difficulties in integrating relative locations (route and DFO) used in the transportation industry and the geodetic coordinates (latitude and longitude, state plane coordinates (SPC), or universal transverse mercator (UTM)) used by most other mapping disciplines. Attempts by researchers to develop a standard methodology have complex implementations (Fletcher et al., 1995, 1996; Vonderohe and Hepworth, 1996; Dueker and Butler, 1998a). Further, data are neither readily exchanged between different linear referencing systems (LRS) nor with the three-dimensional coordinate systems used by GPS. There is incompatibility between relative and absolute locations. For example, a hazardous waste spill of Route 66, milepost 23 is not easy to pass on to others who are not using this method and without the geographic coordinates.

Last, the topology necessary to accomplish a spatial intersect (network overlay) by dynamic segmentation is complex, computationally intensive, and fails to supply query and analysis results in a reasonable time period within an enterprise environment. Object models streamline this process and consume less bandwidth; however, a linear spatial intersect process has been unavailable, preventing the adoption of object models within transportation GIS. Dynamic location provides this tool.

2. Approaches to the problem

As indicated above, attempts have been made to resolve these problems. These implementations fall into three broad categories.

2.1. Dynamic segmentation data models

These are proprietary data models produced by the GIS vendors. The two most prominent models are Environmental Systems Research Institute's (ESRI's) Arc/Info dynamic segmentation module and Intergraph's MGE segment manager method. Both software products use different methods to manage and query linearly referenced data, which reflect their underlying data models for managing spatial data and topology. More recently, Bentley (GeoDynSeg) and Oracle (Oracle Spatial) have developed their own methods of linear referencing that attempt to incorporate some of the elements recommended in the National Cooperative Highway Research Program (NCHRP) 20-27 research program (see below). These latter programs are built with object-oriented methods. ESRI and Intergraph are also developing new versions of their dynamic segmentation programs that are similarly object based.

2.2. Generic road data models

Since 1994, GIS-T researchers and practitioners have been attempting to develop a generic data model for linear referencing that would be compatible with existing field practices and GIS software. Beginning with the initial NCHRP 20-27 project on adaptation of GIS for transportation (Vonderohe et al., 1995), the research evolved into the GIS-T Linear Referencing Pooled-Fund Study (Fletcher et al., 1995), and then to the current NCHRP 20-27 (3) project on guidelines for the implementation of multimodal transportation location referencing systems (Adams and Vonderohe, 1998). At present, two states, Iowa and Minnesota, are implementing these models using Intergraph and ESRI software, respectively.

2.3. Framework road data models

Unlike the NCHRP approach, the framework data models are less prescriptive and less complex. They are developed around procedures rather than complex data models, and emphasize organizational cooperation (Westcott, 1997). The National Spatial Data Infrastructure (NSDI) Framework model provides guidelines on road classifications and definition of road segments (FGDC Ground Transportation Subcommittee, 1998). These include the anchor point–anchor section elements proposed by the NCHRP, which act as datum for the referencing of network elements, routes and event features. Unlike the NCHRP 20-27 model, which is primarily focused on integrating linearly referenced data, the NSDI framework is intended to enable integration of disparate data sets. For example, by identifying anchor points that can act as connectors between different network files, such as state roads and city street maps. A more formal model for classifying road features has been put forward by Dueker and Butler (1998b). Both of these models rely upon the classification of road features that agencies will adopt. However, if an agency does not accept the classification or the use of shared anchor points as the datum, the implementation is moot.

It should be noted that these models are not mutually exclusive and attempts are being made to integrate components of each in different case studies. The applications in Iowa and Minnesota DOTs' are examples of this. And the NSDI framework developers have also attempted to build upon some of the results of the NCHRP projects (Westcott, 1997). However, implementation experience is still sparse and no agreement exists as to what is the best approach.

One of the major problems with implementing the generic or framework road data models is the compatibility with off-the-shelf GIS software. In almost every case, the dynamic segmentation tools in GIS have to be customized to the application. This has significant performance implications, and can create complex data models that are difficult to operationalize. Another constraint is that the dynamic segmentation tools are not available on desktop GIS. Most users with ArcView, Geomedia and other desktop GIS software do not have access to the full set of dynamic segmentation tools. As the use of GIS grows in DOT's and other transportation agencies, the ability to access and use tools that can manage and query linearly referenced data in GIS becomes critical.

As introduced above, GIS can be one of two things: a map linked to data (a relational model), or data stored as a mathematically tractable model (an object or iconic model). Since mapping technologies could not support sufficiently reproducible mathematical models, both research and application have stressed GIS as a linkage to accurate tabular data. Tools such as dynamic segmentation assumed it was best to keep data separate from maps, and live with the problems of synchronizing data and geometry.

Dynamic location takes the second approach. If sufficient data are available to build an iconic GIS model, what new tools are required and what new benefits result? GIS software has already shifted into the object model world. It creates a one-to-one relationship between real-world and database features thus providing temporal and spatial synchronization. What has not been provided is a means of performing linear spatial intersect.

3. Dynamic location

Dynamic location is a graphical method for spatial intersect in a linear referencing environment. It is an iconic model, where real-world objects are modeled at their true Cartesian locations. A bus stop would thus have an $\{x, y\}$ pair and not a route DFO. A section of concrete would be modeled as a line attributed as concrete, and not a table entry with "from" and "to" DFO measures. A complete data set of pavement type would be a complete network representation. Since these are discrete objects, the temporal and spatial version control tools available to database management systems can be used.

All transportation events are thus carried as spatial objects with $\{x, y\}$ coordinates just like any other GIS object. Signs, bus shelters, accidents, mile markers are $\{x, y\}$ points. Bus routes and guard rails are $\{x, y\}$ lines with no reference to any route or linear referencing system. Ride quality, pavement, and speed limits are modeled as complete networks of $\{x, y\}$ locations, again with no reference to any route or linear referencing system. These discrete objects are called event dynamic location objects (EDLO). GIS software already provides this ability.

A second object class is defined, the route dynamic location object (RDLO). While a Cartesian grid provides a two-dimensional measurement framework, the RDLO is a one-dimensional number line built from $\{x, y, m\}$ coordinate tuples ("m" for measure, see Fig. 1). This RDLO is then laid out across the Cartesian grid creating a one-to-one correspondence between relative and absolute location (see Fig. 2). Geometric shape is stored as an object within a field of each

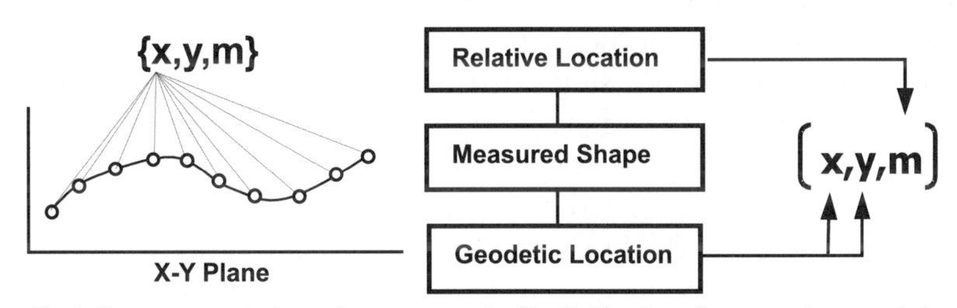

Fig. 1. RDLO is modeled with $\{x, y, m\}$ coordinate tuples at each vertex. EDLO only require $\{x, y\}$ pairs.

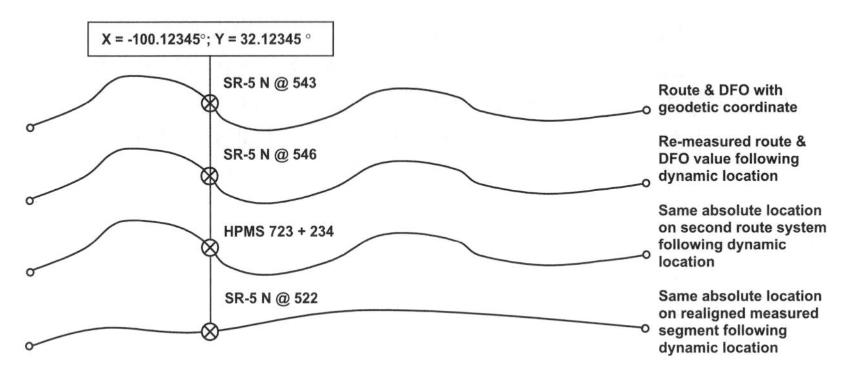

Fig. 2. Using dynamic location to move location between relative spaces.

dynamic location objects (DLO) database record, and facilitates spatial intersect between EDLO and RDLO. From and to distance fields are not required because they may be derived by looking at the EDLO end point locations relative to the $\{x, y, m\}$ coordinate of any desired RDLO. Any combination of EDLO may be intersected with any other event or route DLO in this manner. Old views of separate geometry with detached tables of explicit *from*, *to*, and *route* values are discarded. Another way of conceptualizing the EDLO and RDLO is as logical networks (EDLO) referencing the underlying geometric network (e.g., RDLO street centerline file). The logical EDLO (e.g., pavement section or bus route) correspond to the geometric network (e.g., highway 270) by location (linear and absolute), through spatial coincidence. No complex linkage of tables to topology is required.

This builds a stable geodetic data model with the functional benefits of linear referencing. Many RDLO can follow the same ground track, but calibrated to any datum or linear reference system. Any EDLO can then be referenced to any location referencing system, linear or Cartesian. Since each record stores both the spatial object and attribute data, any table or set of values may be developed on the fly by placing the desired EDLO against any RDLO and an intersect table built by reading down the number line (RDLO). If a roadbed is renamed, realigned, or recalibrated, EDLO do not change, because they are $\{x, y\}$ absolute locations not assigned to any route. Bring in a new RDLO and develop DFO values as desired.

The dynamic location process is illustrated in Fig. 3. The desired intersect product is a table with the segmentation of surface type and condition in terms of a stated route system. The process begins by spatially selecting the surface type and surface condition EDLO that meet query constraints, as well as the desired RDLO to view the solution. In this example, the three-component cursor set contains nine records (4 surface type, 3 surface condition and 2 route sections), each equipped with geographic shape and a single attribute. Since the cursor is both small and independent of topological constraint, data may be rapidly transmitted from the server to the client.

Once records have been selected by the server, the client plots the three DLO sets. Next, the *from* and *to* EDLO points, and associated single attribute values, are defined and linked to the RDLO. Last, the points along the RDLO are read in increasing order to build the query table at the bottom half of Fig. 3. The tabular data are not used to generate a geographic picture as in dynamic segmentation, but rather geographic shapes are used to create ordered tabular data in the output query table.

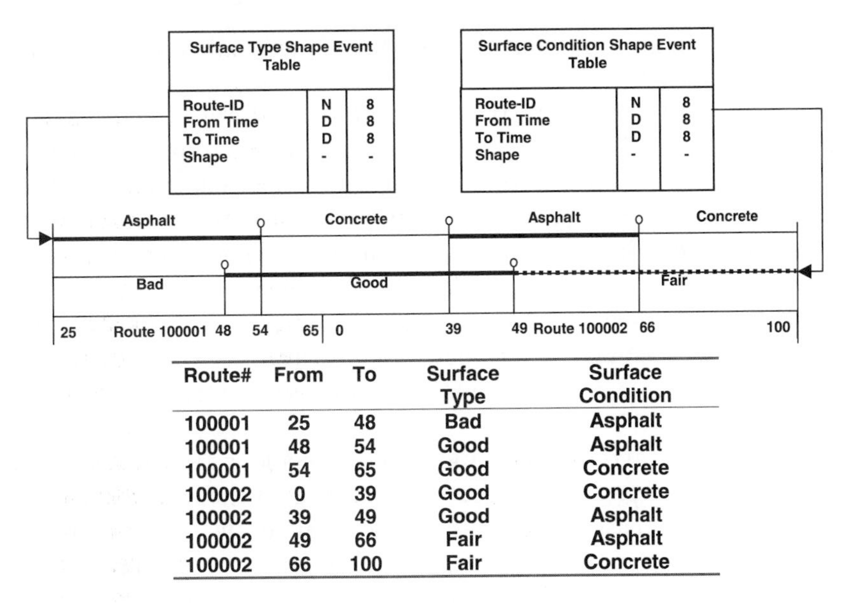

Route#	From	To	Surface Type	Surface Condition
100001	25	48	Bad	Asphalt
100001	48	54	Good	Asphalt
100001	54	65	Good	Concrete
100002	0	39	Good	Concrete
100002	39	49	Good	Asphalt
100002	49	66	Fair	Asphalt
100002	66	100	Fair	Concrete

Fig. 3. Building spatial intersect tables using dynamic location.

This same method can also transfer data between any set of referencing systems in terms of multiple attributes (intersect), different linear measurement systems, and linear and geodetic referencing systems. Using snapping procedures, it is not necessary to use systems of precisely coincident shapes; however, loss of precision is encountered as scales decrease and data become spatially inconsistent. It is also important to note that only precise, that is repeatable, data should be used (e.g., replication of an agreed upon set of roadbed centerlines). Inaccurate data, such as smaller scale maps, can also be used without loss of information. The key is the replication of the same consistent shape.

As shown in Fig. 4, RDLO and EDLO have equal status, and have no relational linkages. Relationships between selected RDLO and EDLO are developed on demand through geodetic stacking or proximity. It is important to note that by definition, the object model provides version control because spatial and attribute data may be stored in the same record. This is by design, and the fundamental benefit of the object model. However, the object model cannot be readily adapted to GIS-T business practice because no network overlay tools are available outside the dynamic

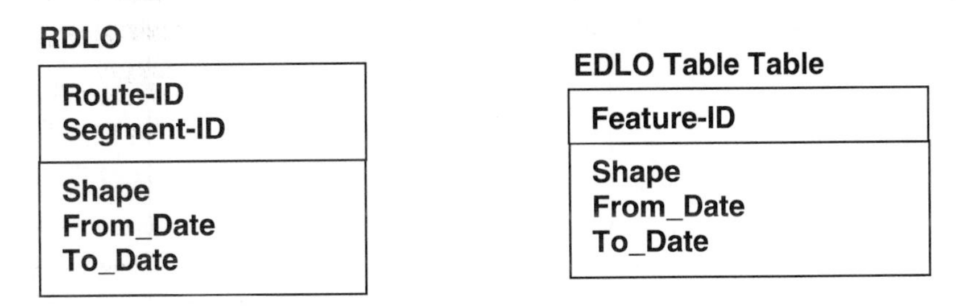

Fig. 4. There are no relationships between $\{x, y\}$ EDLO event and $\{x, y, m\}$ RDLO.

segmentation. Dynamic location provides this intersect functionality, and thus the full benefit of the spatial object model may be exploited.

One of the advantages of this model is that different types of objects can be referenced to more complex, larger scale geometry. This is illustrated in Fig. 5, where four different types of objects or transportation elements can be referenced to the base map. In each case, the absolute location is retained even when the object may be referenced linearly for network modeling purposes.

The DLO object model can integrate four types of spatial objects (see Fig. 5):

1. *Route objects* that form the navigable network on the centerlines. Dynamic segmentation and dynamic location can reference and develop data using these networks. The set of all route objects contains calibrated roadbed centerlines, modal travelway centrelines (e.g., HOV lanes), transport route events, such as transit stops, and other logical networks, such as a pedestrian network.
2. *File objects* group non-spatial data by location. File object icons are located at various physical network locations, for example, an intersection. Clicking on the icon will present the user with a pick list of file information associated with that icon. Example files would include road infrastructure drawings, traffic signal timing diagrams, and traffic count data.

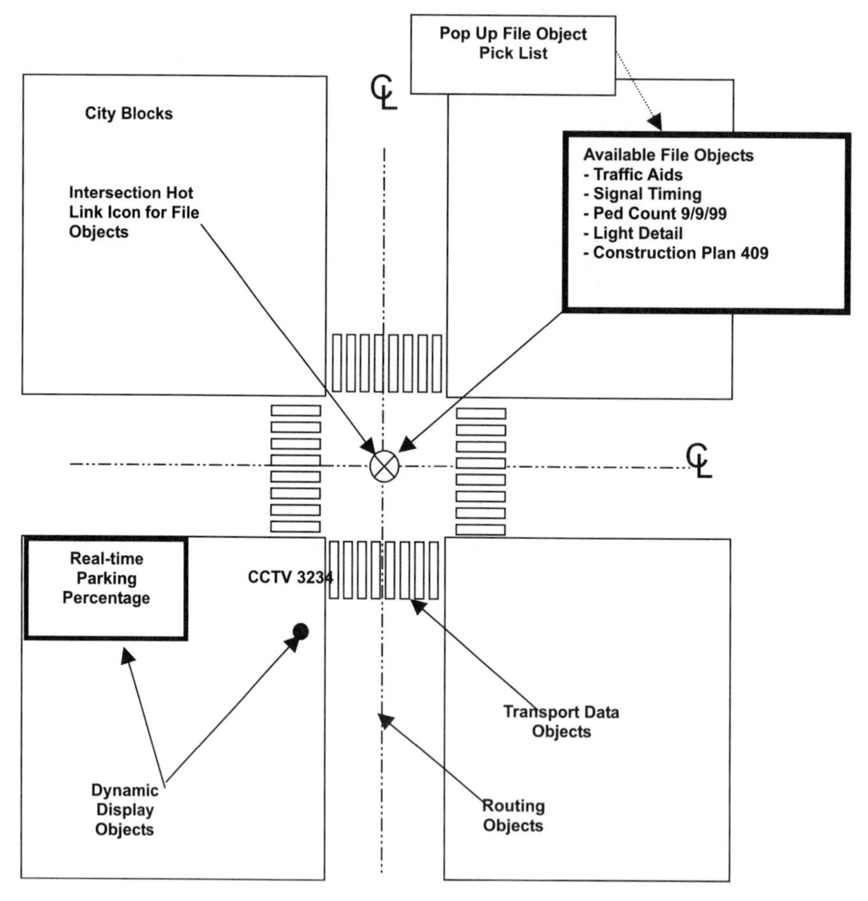

Fig. 5. Attaching data to logical and physical networks.

3. *Transport data objects* are iconic models of transport infrastructure and are located at their real world map locations using drawn shapes of real-world shape. Attribute data are then attached via the feature ID to the object. Examples include road signs or crosswalks for pedestrians.
4. *Dynamic link objects* are icons that represent the real-time state of bit streams. Icons are placed at their real-world map locations and can display traffic signal status, loop detector status, CCTV images, and other real-time data.

4. DLO creation

DLO creation, either RDLO or EDLO, begins with the adoption of a base network. Once the base network has been adopted, it must be maintained, as EDLO and RDLO are replicated from it (thus precise, not necessarily accurate). Since dynamic location supports an object model, it is important to adopt a network of sufficient scale to avoid network pathologies. Roadbed based models have proven the most successful mapping strategy for dynamic location, as described in the case study later.

The adopted network should be viewed as raw material in pure link-node form. That is, one continuous line network between intersects and cul-de-sacs. From this raw material, RDLO and EDLO may be marked out. For the creation of RDLO, select appropriate DFO start points and use these points to create the desired route system from the base network, then calibrate the $\{x, y, m\}$ tuples accordingly. To create a second RDLO, a different set of DFO points is applied to the original unsplit base network, which is then calibrated. Likewise, pavement attribute change points may be collected as GPS points and tagged with the attributes leading into that point. These points are then used to mark out the EDLO, and the attributes from the points are snapped onto the EDLO line segments. Note how this system is designed to collect field data as absolute points, and not measured linear segments. Mathematical operation is carried out on the GIS Cartesian grid, and not in tables of calibrated odometer distances. This marks a fundamental shift in GIS, from a map linked to data, to a geometric and mathematically tractable model of data.

Changes in EDLO are simple because geometry does not change. Old data are closed out with from and to dates, with new concurrent geometry opened. When the base geometry changes, there is a point where the new network departs the old roadbase and a point where the new geometry rejoins the old network. These two points are used to cut the old section out of all stacked geometry sets. The retired pieces are closed out with from and to dates, and new sections with the new geometry replicated and placed in the database. For example, if there were a mile stretch of concrete roadbed whose last half-mile was realigned, the section would be cut at the half-mile point. The first half-mile would keep its old attribute set and start date, while the second section after the half-mile point would receive the new geometry and attributes. The process has been proven easy to set-up, and effective, using standard GIS programming languages.

5. Temporal control and archive

Version control tracks the history of a network, referencing system, and associated data over time, such that data from any point in time may be recalled and presented over a network representation of the correct time period. This procedure has not been well implemented within GIS-T because the practice of dynamic segmentation separates events from their respective geometric locations. Traditionally, geometric changes to the roadbed have been calibrated with the routes and events by remeasurement using the linear referencing system. This led to inaccuracies in data location on those sections unaffected by the road realignment. The NCHRP and other data models deal with these anomalies through the use of reference points to provide greater control in the remeasurement process. Retired sections of road with the old ID can be queried as can new sections with the new ID, but they cannot be queried together. This is because they are referenced to separate route systems, control sections, and reference points. Even where the same datum is used (anchor sections and anchor points in the NCHRP model) the location of event data still relies on linear measurement, and without absolute location to correspond to the underlying cartographic representation, versioning is problematic. In practice, events, routes and route systems are duplicated for each time period that a geometric change occurs, which introduces a lot of redundancy and overhead into the data management process. In contrast, though object models provide this capability, they have not been adopted by GIS-T because spatio-temporal intersect capabilities are not supported in GIS software.

Since dynamic location provides spatial intersect capabilities, the object model may be used to manage geometry and attribute information in a single monomial record as temporally bound and controlled with *begin* and *end* dates. Following any query, spatial and/or temporal, selected attributes will already contain the applicable network for the time period in question. Point events will be a set of points, linear events a set of lines, and continuous events as a representation of the complete network at a given point in time. Route and EDLO will be added to the database, opened and closed with begin and end dates, but never removed.

To answer the query, “What was the pavement composition and alignment as they appeared in 1997”, a spatial index query is conducted first to extract all pavement records within the geographic area of interest, independent of date (see Fig. 6). A second SQL passes through the cursor set selects those spatial records meeting temporal constraints. Note how expired DLO resides within the active database. Since time-expired event and RDLO contain their own geometry, placing historical events over the correct historical network is simple.

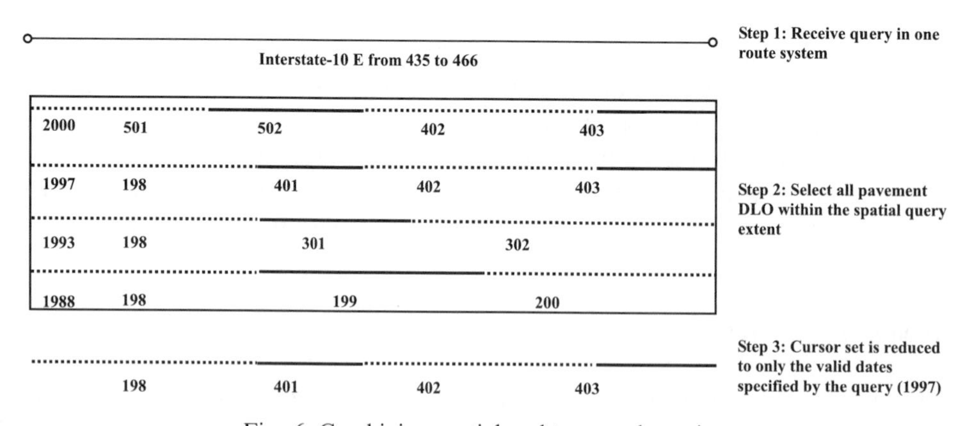

Fig. 6. Combining spatial and temporal queries.

As new transportation features are added or subtracted, new shapes are added to the database, while old shapes are closed with an "end date" entry and maintained in the same database. Route renaming and realignments are both addressed in this manner. Since dynamic location is essentially a linearly enabled geodetic model, recalibration of measures need not be accomplished.

6. Case study: Texas linear management system

The location object concept is being applied in the design of the transportation data layer for Texas Strategic Mapping (StratMap) project, sponsored by the Texas Water Development Board but involving several state agencies and the USGS. StratMap is designed to integrate disparate networks from Texas DOT, counties, cities and other sources in a unified data layer. In addition, the network will have multiple attributes associated with it. Thus, multiple spatial and attribute data are being conflated into a single network file. It was apparent that the traditional methods of integrating the data in GIS whether by network conflation or linear referencing would not be robust enough to accomplish the task. There were also issues of tracking data changes over time as well as being able to integrate spatial data at different levels of resolution and detail, from 1:1000 to 1:24 000 scale. The data model also had to accommodate use of GPS and ortho-rectified images that may be used to update data in future, at great accuracy (~1 m precision). It was decided to use the DLO data model, that had previously been developed for the Texas Department of Transportation to enable TxDOT to integrate the Texas Reference Marker (TRM) System (the state LRS), with a new GPS based referencing schema. By adopting the Texas Linear Measurement System (TLMS), StratMap will be consistent with TxDOT, and using the same methodology for managing and updating the transportation data layer.

The DLO model as outlined above was programmed in the GIS environment in use at TxDOT, namely Arc/Info and ArcView. One of the reasons why TxDOT were open to the new TLMS data model is because the integration of the TRM in Arc/Info relied upon dynamic segmentation, and only those users trained in this method and with access to Arc/Info could perform the spatial queries required. TxDOT were therefore looking for a desktop GIS solution, and one that was easier to learn and quicker to execute. The DLO method accomplishes both these objectives. For example, tests conducted on a sample data set in Arc/Info using dynamic segmentation versus the DLO query method showed dramatic performance enhancement. The same result in the DLO was accomplished in a fraction of the time it takes dynamic segmentation to compute the spatial intersect. This is because most of the data processing is performed in the database rather than the GIS, without the added overhead of managing network topology.

Fig. 7 illustrates the format of the temporally controlled DLO table used in TLMS, while Fig. 8 illustrates the entities of the entire model. Note how events and geometry are controlled by the dates they are opened and closed to traffic. A second set of dates is used to mark when the objects were entered and retired from the database. The prior date pair offers version control, and the latter database rollback functionality. The lock feature is provided to prevent any attribute changes, should a piece of geometry enter an edit mode or transition stage. Surface length is carried as a separate field because TLMS was designed as a surface model, where route segments are always measured against a digital elevation model. As configured in Fig. 8, the model accommodates multiple LRS, geodetic coordinates (GPS), and linear, continuous, and point EDLO.

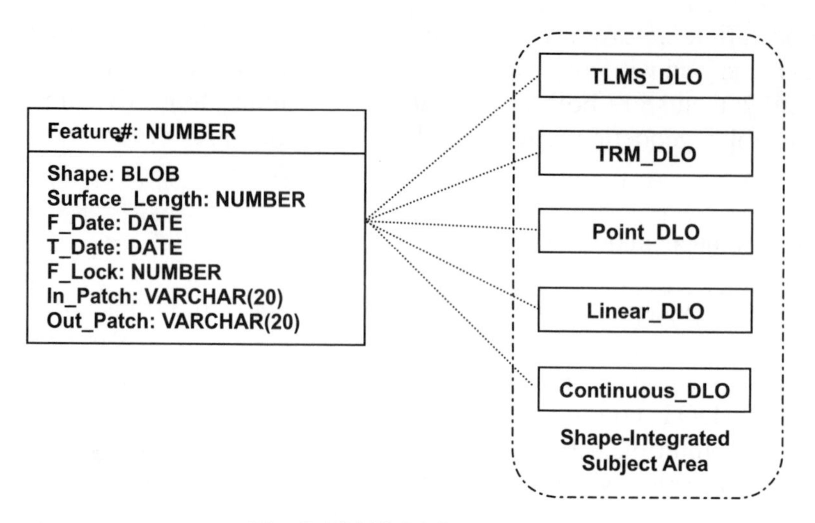

Fig. 7. TLMS DLO structure.

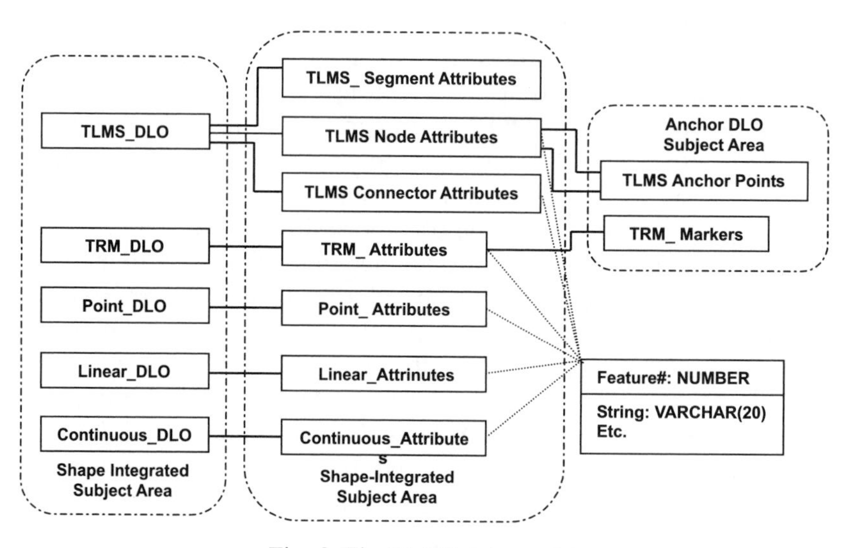

Fig. 8. The TLMS data model.

Anchor points are not required by the model, but are used here to carry field descriptors for begin and end points.

The DLO data model prototype was developed using ArcView 3.1 and Access database management system. ArcView's measured shape file format, polyLineM, was utilized to link features referenced by absolute $(x, y, \{z\}$ coordinates) and linear measurement (m value). The polyLineM data type joins the functionality of both geodetic and LRS. The $\{x, y, m\}$ coordinate tuple creates a one-to-one correspondence between relative and absolute locations. This builds a stable geodetic data model with the functional benefits of linear referencing. Roads and transportation features are modeled as separate and independent geometric artifacts, i.e.,

DLO, creating a system of single attribute, static-length segments for all linear features. As illustrated in Fig. 9, feature and road DLO are developed on demand through geodetic stacking or proximity. Geometric shape is stored as a binary large object (BLOB) within a field of each DLO database record, and facilitates spatial intersect between feature and road DLO. From and to distance fields are not required because they may be derived by looking at the feature DLO end point locations relative to the $\{x, y, m\}$ coordinate of any desired road DLO. Any combination of feature DLO may be intersected with any other feature of road DLO in this manner. The DLO is time-stamped, so temporal queries for any feature are feasible.

By treating roads and features as logical networks, with absolute and relative location, the model overcomes one of the traditional weaknesses in linear referencing, namely calibrating logical network position with physical (geometric) network location. All networks, geometric and logical, are treated as separate DLO's, therefore calibration issues do not arise. This does not mean that problems of feature location accuracy do not occur, but it provides a more robust model for managing data errors or inconsistencies because DLO do not demand that linearly referenced feature data be tied to physical road network. The linkage is established by proximity. If features are offset from the road, they can be integrated by snapping to the network, but the absolute location value is retained even though the relative location has changed. Thus, enhancements to the location accuracy of the network following GPS survey, for example, can be accommodated. In the StratMap project, the database is being enhanced through a variety of methods including use of aerial photography. Fig. 10 shows an example of the use of aerial

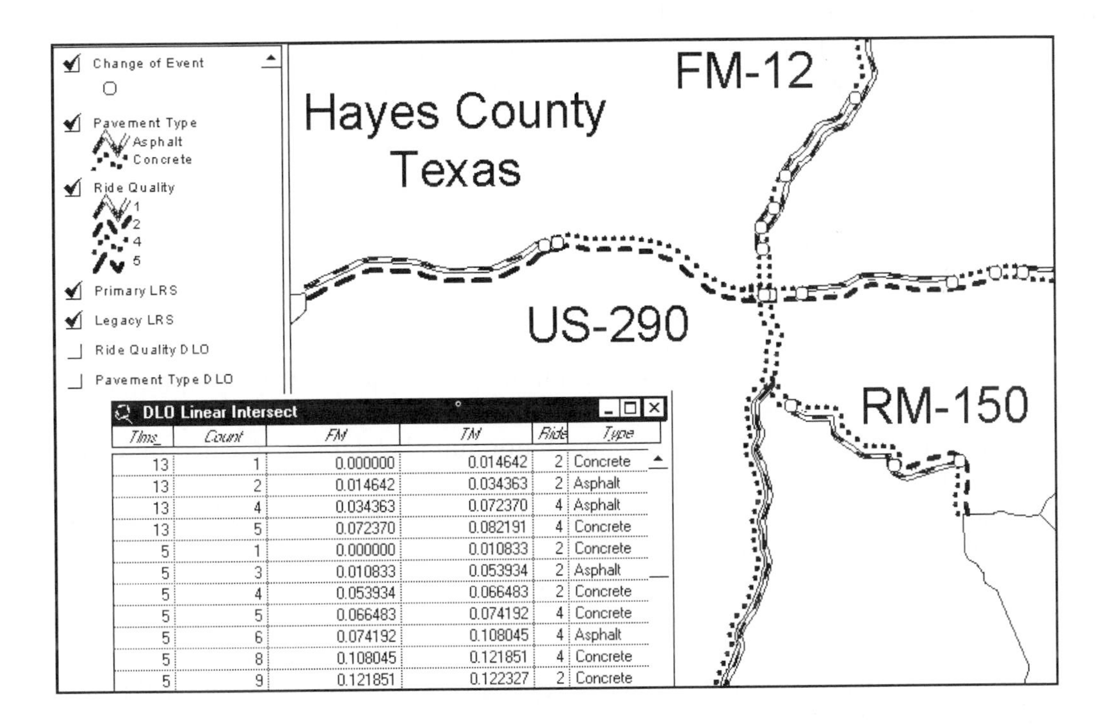

Time	Count	FM	TM	Ride	Type
13	1	0.000000	0.014642	2	Concrete
13	2	0.014642	0.034363	2	Asphalt
13	4	0.034363	0.072370	4	Asphalt
13	5	0.072370	0.082191	4	Concrete
5	1	0.000000	0.010833	2	Concrete
5	3	0.010833	0.053934	2	Asphalt
5	4	0.053934	0.066483	2	Concrete
5	5	0.066483	0.074192	4	Concrete
5	6	0.074192	0.108045	4	Asphalt
5	8	0.108045	0.121851	4	Concrete
5	9	0.121851	0.122327	2	Concrete

Fig. 9. Texas TLMS interface showing results of a dynamic location query.

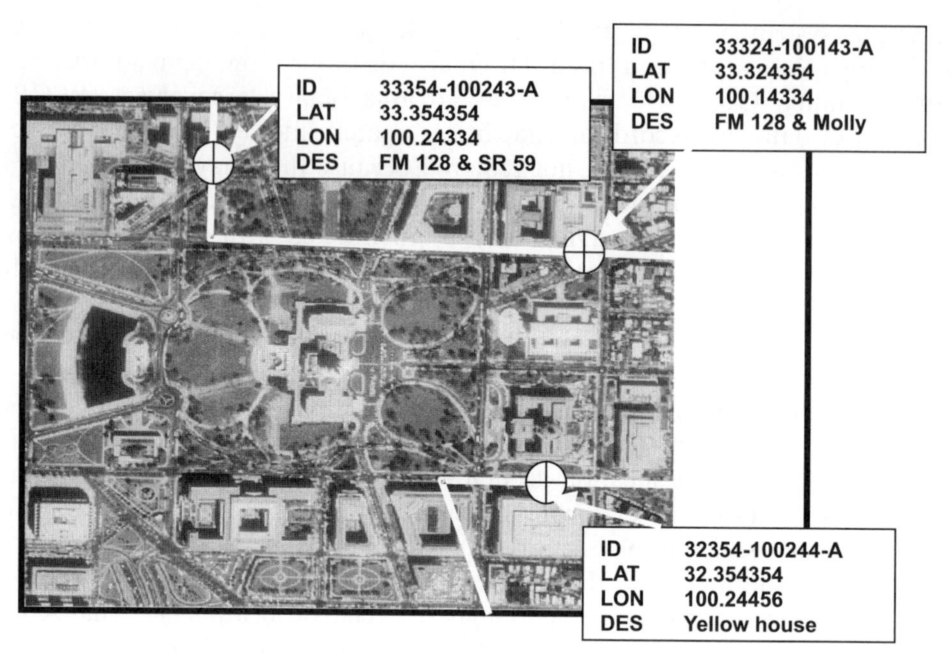

Fig. 10. Aerial photography used to identify features on road DLO.

photograph to locate features on the highway that may serve as anchor points for linear referencing of state highway features or intersection points with other network files. Note how the Feature ID is a concatenation of the latitude of longitude thus establishing the absolute location for the designated feature.

7. Summary and conclusions

Dynamic segmentation builds geographic displays from tabular data on demand, and suffers the pitfalls of storing tabular data separate from its geometry. Dynamic location is an object-oriented method of encapsulating transportation attributes with their geographic shape and location in a single database record. This facilitates full spatial and temporal query, storage, and version control. In contrast to dynamic segmentation, tabular data are created on demand from geographic shape.

Spatial intersect between any number of EDLO is accomplished by stacking DLO end points over any RDLO and extracting the underlying measure from the $\{x, y, m\}$ coordinate tuple. Data may be swiftly exchanged between referencing systems through a stable geodetic datum, while enjoying the benefits of linear referencing.

The dynamic location model changes how linear data are stored in an automated environment. Geographic shapes are easily stored as binary large objects in a single database field, and become a precise iconic model that supersedes the need for data decomposition into tabular form. Relationships between separate iconic objects are inherently bound to the geodetic datum, using geodetic location as the only data integrator. By using iconic objects rather than relationship tables, dynamic location is a paradigmatic shift in data management process, and thus the way transportation data are collected, stored, and used.

While dynamic location is a different model to dynamic segmentation, the two are not mutually incompatible. It is still possible to use dynamic segmentation with the DLO data or create DLO's from dynamic segmentation routes and events. In this way, the DLO extends the capabilities of GIS-T and puts network overlay queries on the users' desktops.

Finally, the DLO model has been developed and tested on a specific GIS platform. However, the data model is robust enough to work with other GIS software that are capable of storing and managing measurements in their network data structure, and can import/export data from database management systems. The BLOB format is supported by Access and Oracle, for example, and is employed to minimize storage and transmission – a critical factor in Internet access to spatial data and operations on the server. The DLO model has proved to be a robust method that meets these requirements using industry standard technology components.

References

Adams, T., Vonderohe, A.J., 1998. Guidelines for the implementation of multimodal transportation location referencing systems, project design. National Cooperative Highway Research Project 20-27(3), Transportation Research Board, Washington, DC.

Bespalko, S.J., Sutton, J.C., Wyman, M.M., Vander Veer, J.A., Sindt, A.D., 1998. Linear referencing systems and three dimensional GIS. Paper presented at the 77th Annual Meeting of the Transportation Research Board, Washington, DC.

Bespalko, S.J., Sulsky, D.L., Dison, H.L., Wyman, M.M., 1999. A nonlinear method for correcting GPS data. Paper presented at the 78th Annual Meeting of the Transportation Research Board, Washington, DC.

Burkholder, E.F., 1997. The global spatial data model: a tool designed for surveyors. Professional Surveyor Magazine 17 (7), 21–24.

Dueker, K.J., Vrana, 1992. Dynamic segmentation revisited: a milepoint linear data model. URISA Journal 4 (2), 94–105.

Dueker, K.J., Butler, A.J., 1998a. GIS-T Enterprise Data Model with Suggested Implementation Choices. Center for Urban Studies, School of Urban and Public Affairs, Portland State University, Portland, OR.

Dueker, K.J., Butler, A.J., 1998b. A Proposed Method for Transportation Feature Identification. Center for Urban Studies, School of Urban and Public Affairs, Portland State University, Portland, OR.

FGDC Ground Transportation Subcommittee, 1998. NSDI Framework Road Data Model Standard (Proposal). 31 March.

Fletcher, D., Henderson, T., Espinoza, J., 1995. GIS-T/ISTEA Management Systems Server–Net Prototype Pooled Fund Study Systems Analysis and Preliminary Design. Albuquerque, NM.

Fletcher, D., Espinoza, R., Gordon, S., Spear, B., Vonderohe, A., 26–27 July 1996. The case for a unified linear referencing system. Presented at a workshop on Enterprise Location Referencing Systems: Policies, Procedures and Standards for Implementation, Department of Geography, University of Utah.

Horowitz, A.J., 1997. Integrating GIS concepts into transportation network data structures. Transportation Planning and Technology 21, 139–153.

Kwan, M., Golledge, R., Speigle, J., 1996. A review of object oriented approaches in geographic information systems for transportation modeling. Paper 412, The University of California Transportation Center.

Kwan, M., Golledge, R., 1997. Developing and object oriented testbed for modeling transportation networks. Paper 406, The University of California Transportation Center.

Mainguenaud, M., 1995. Modelling the network component of geographic information systems. International Journal of Geographical Information Systems 9 (6), 575–593.

Sutton, J., 1995. Network Pathologies Phase 1 Report. Sandia National Laboratories, Project AH-2266, November.
Sutton, J., 1996a. Network Pathologies Phase 2 Report. Sandia National Laboratories, Project AH-2266, March.
Sutton, J., 1996b. The role of GIS in regional transportation planning. Transportation Research Record 1518, 25–31.
Sutton, J., 1997. Data attribution and network representation issues in GIS and transportation. Transportation Planning and Technology 21, 25–44.
Vonderohe, A.P., Chou, C.L., Sun, F., Adams, T., 1995. Results of a workshop on a generic data model for linear referencing systems. In: Alan Vonderohe (Ed.), Summary of a workshop sponsored by the National Cooperative Highway Research Program (NCHRP), Project No. 20-27. University of Wisconsin, Madison, WI.
Vonderohe, A.P., Hepworth, T.D., 1996. A Methodology for Design of a Linear Referencing System for Surface Transportation, Final Report. Project AT-4567, Sandia National Laboratories.
Westcott, B. (Ed.), 1997. NSDI Framework Road Data Modeling Workshop. Summary of a workshop held in Wrightsville Beach, NC, 3–5 December.

PERGAMON

Transportation Research Part C 8 (2000) 53–69

TRANSPORTATION RESEARCH PART C

www.elsevier.com/locate/trc

Map accuracy and location expression in transportation – reality and prospects

Val Noronha *, Michael F. Goodchild [1]

Vehicle Intelligence and Transportation Analysis Laboratory, NCGIA, University of California, Santa Barbara, CA 93106-4060, USA

Abstract

We were contracted to test a suite of proposed location messaging standards for the intelligent transportation systems (ITS) industry. We studied six different databases for the County of Santa Barbara, documented types and magnitudes of error, and examined the likely success of the proposed standards. This paper synthesizes the test results and identifies caveats for the user community as well as challenges to academia. We conclude that, first, current messaging proposals are inadequate, and superior methods are required to convey both location and a measure of confidence to the recipient. Second, there is a need to develop methods to correct map data geometrically, so that location is more accurately captured, stored and communicated, particularly in mission critical applications such as emergency servicing. To address this, we have developed methods for comparing maps and adjusting them in real time. Third, there must be standards for centerline map accuracy, that reflect the data models and functions associated with transportation. © 2000 Elsevier Science Ltd. All rights reserved.

Keywords: Location referencing; Intelligent transportation systems; Geographic information systems; Street network databases; Map database interoperability; Transportation datums

1. Introduction

Although it is several years, even decades since the advent of digital maps, global positioning systems (GPS) technology and machine-searchable street names and coordinates, it can be surprisingly difficult for the average person to describe a location, even one limited to the discrete

* Corresponding author. Tel.: +1-805-893-8992.
E-mail addresses: noronha@ncgia.ucsb.edu (V. Noronha), good@ncgia.ucsb.edu (M.F. Goodchild).
[1] Tel.: +1-805-893-8049.

PII: S0968-090X(00)00020-6

confines of a street network. This problem is faced daily by people reporting accidents and vehicle breakdowns, and by the emergency service personnel receiving and servicing those calls.

Recent technological development in intelligent transportation systems (ITS) allows vehicles and pedestrians to be tracked in real time, using GPS, cellular phone tower triangulation, inductive loop detectors embedded in highway pavement, or closed circuit television (CCTV) cameras; even satellite imagery is being proposed as a tracking option. The methods have applications in emergency servicing, real-time highway information provision (e.g. congestion, construction, fog) and traffic management, hazardous material management, travel demand studies, law enforcement and criminology. They are certain to generate a large volume of geographically referenced data in the coming years. But given the inherent difficulty in describing location, the positional references in such data could be ambiguous or erroneous, and the errors could propagate through subsequent processing of the data.

In the United States, ITS is a major area of technology research, its development abetted by two transportation bills, the 1991 Intermodal Surface Transportation Efficiency Act (ISTEA), and the 1998 Transportation Equity Act for the 21st century (TEA-21). ISTEA spurred the initial development of ITS concepts, notably the national ITS architecture (USDOT, 1997), but it became clear by the mid-1990s that significant operational problems remained to be resolved. One was the need for standards for interoperability, particularly in location referencing. TEA-21 explicitly addressed this, and in 1998 the US Department of Transportation released a list of "critical standards" (USDOT, 1998) for immediate research, development and testing. Interoperability in general, and location referencing in particular, featured prominently in this list.

Emergency management services (EMS) is an important component of ITS, because they present challenges similar to incident management (IM) services in ITS, and because the road system is the medium of delivery of service to most non-road emergencies. The problem of determining the precise location of an incident, identification of appropriate resources, facilitation of service delivery by signal pre-emption, and real-time traffic management or evacuation, are issues that cut across the boundaries of EMS and ITS.

Another aspect of transportation research, with potentially large impacts on infrastructure, is the periodic systematic study of travel behavior, best exemplified by the Federal Highway Administration's Nationwide Personal Transportation Survey (NPTS). Individuals are asked to maintain diaries over a period of days or weeks, recording each trip they make for personal or business reasons, listing time of day, origin and destination, intervening stops and route. Battelle (1997) conducted an experimental variant on the standard methodology in Lexington, Kentucky, using GPS loggers to supplement the respondents' answers; similar methods have since been explored in Georgia and Texas. A study by the California Air Resources Board tracked trucks with GPS to study their traffic patterns, and their implications in air quality. Such new technology solutions are obviously data rich, but there is the potential for locations – origins, destinations and routes – referenced in the data to be incorrectly interpreted due to GPS and map data errors. Clearly the need for accuracy is greater in some applications than in others.

Problems with location reporting have been observed anecdotally for a number of years, but there has been relatively little scientific documentation of the scale of the problem and its solution. Some researchers have focused on the problem of address matching and geocoding (e.g. Kim and Nitz, 1994), and a number of studies have examined GPS error independently (e.g. Quiroga and

Bullock, 1998). Neither of these addresses the broader issues of location reporting in a transportation context.

This paper examines these issues analytically, synthesizing results of formal tests conducted by the Vehicle Intelligence and Transportation Analysis Laboratory (VITAL) in response to the need for standardization in the ITS industry. It expands on ideas by Noronha (1999). Section 2 presents the need for location expression and exchange with particular reference to ITS scenarios. Section 3 examines the types of error commonly found in contemporary digital maps, and their impact on location referencing. Section 4 discusses short- and long-term measures to improve the accuracy of location reporting. The paper concludes with speculations on future requirements, and the academic, technical and managerial challenges that lie ahead.

2. Location expression and the interoperability problem

For transportation purposes, location referencing is the description of location of a static or dynamic object, in one, two or three dimensions. In the case of dynamic objects there may be reference to a fourth dimension. The object may be a disabled vehicle, which is static and has a relatively easily defined location. Alternately, it may be an incidence of congestion, the boundaries of which are not easily defined, though if modeled appropriately, congestion can be sampled using loop detectors and CCTV. A toxic plume is an example of an object that is dynamic, that may have poorly defined boundaries, and whose location is not easily measured. This paper does not attempt to deal with these difficult cases; it addresses the relatively simple instances of disabled vehicles and similar point incidents.

ITS scenarios require cooperative action between parties, each operating in a free market with respect to proprietary databases and systems. Take an example of a vehicle collision that results in casualties, on an urban freeway interchange during rush hour. Numerous agencies, both public and private, are involved in the response – police, fire engines, ambulances and tow trucks. Traffic management centers (TMCs) and private information service providers (ISPs) monitor the incident, advise motorists and divert traffic. Each agency employs an internal communications code that is meaningful in the context of its operations. GPS users quote coordinates, tow truck operators use intersecting street names, highway maintenance crews operate with respect to linear measures from a defined starting point, some may apply an undocumented nickname such as "Suicide Bend" to an accident-prone section. No matter what the language, agencies must understand the precise location of the incident, its location with respect to other ramps and roads, and the dynamics of its impact on traffic conditions. The purpose of standardization is to ensure that parties communicate unambiguously and error-free with respect to mission-critical information items.

We use the term location expression (LX) to mean the description of location by coordinates, street names or other means. The process of communicating an LX is location expression exchange, or LXX. An object that changes location over time, such as a moving vehicle, is described by multiple pairings of an LX with a time expression, (TX) (note that time is also subject to uncertainty and variation in expression, e.g. "1430 h PST", "yesterday" or "every Tuesday in the summer"). The LX/LXX nomenclature isolates the components of the location reporting problem. Our concern with LXX is specifically in the context of machine-to-machine communication, where human intervention and interpretation are not possible. ITS scenarios of vehicles making emergency calls on behalf of their unconscious drivers rely on such exclusively machine-based interaction.

At first sight, the transformation between LXs, from coordinates to street names, linear or grid references, is a simple geometric process; even if agencies work with different coordinate systems, such as Universal Transverse Mercator (UTM) and State Plane projections, translations are well defined mathematically and relatively easily achieved. In reality, there are two problems that conspire to make LXX an error-prone process. The first is the accuracy of the initial position capture and expression. The highway incident above is typically reported to emergency services by several cellular phone callers, their descriptions perhaps pointing to different locations. Police often record accident data at the nearest intersection to an incident, because they do not have the means to measure the distance from the intersection. Even if the incident location were reported by GPS coordinates, the 10–100 m GPS error could position the incident on the wrong highway ramp or on an adjacent service road. The second problem is the error in the reference map with which the location is interpreted. This is a complex problem, and is discussed in detail in the next section.

3. Error in map databases

To understand the extent of the spatial interoperability problem, one must appreciate the types and degree of error in digital map databases. In general there are errors in position, classification and inclusion, names and other descriptive attributes, linear measurement and topological relationships. The following discussion is based on tests conducted in the County of Santa Barbara, California, using six databases from public and private sources, representing the principal street map vendors for ITS in the US. Although we use the term "error" liberally in this discussion, bear in mind that these databases were developed for different purposes, from different scales of survey and mapping, and that their sale prices vary from $1500 to $45 000. We do not identify vendors or compare products, and our comments are intended not to be critical of vendors, but to illustrate the kinds of operational hurdles that exist and must be overcome by users.

Testing involves sampling a set of point locations, either lab-generated or selected in the field (field locations are associated with landmarks such as utility poles, and documented by notes and photographs). In some cases, sample points are generated at regular or random intervals along network polylines; in other tests, they are required to be at commonly identifiable locations such as intersections. Major streets are examined separately, on the grounds that initial ITS deployment would be on highways and major arterials only. The sample points are transferred to each database in turn by the LX method under test, and an appropriate measure of success applied to examine the accuracy of the transfer. The test results are obviously specific to the County of Santa Barbara and the time of acquisition of the data sets (May, 1997). The county does present a representative and average cross-section of street styles, both urban freeways and sinuous mountain roads, but it is reasonable to believe that results would be better with databases focused on large urban centers, and conversely poorer in more remote locations.

3.1. Position

Although digital maps are sometimes thought to be exceptionally precise, there is no reason why they should be better than hardcopy products. Whether paper or digital, a map is an iconic representation of a reality that can be interpreted in different ways. That reality itself may not be constant. For example, street centerlines and intersections are defined precisely at the time of construction planning, but once the survey stakes are removed and the asphalt laid, the centerline is represented by a stripe of paint, which wanders slightly each time there is construction activity or re-striping. Consequently there is about ±0.5 m of lateral uncertainty in a centerline, and ±2–3 m in the definition of a large intersection. A gore point (the "V" at a highway exit, where the right shoulder stripe of the highway and the left shoulder stripe of the exit meet) may wander about 10 m due to striping imprecision – this is not immediately evident from the vantage point of a moving vehicle. These are examples of uncertainty in a real-world object; in addition there is error in the measurement and representation of that object. The earth's frame of global reference is itself subject to some uncertainty due to wobbling, tectonic movement and tides, and true sub-metre accuracy is achievable only in relation to local monuments. Digitization of a centerline from aerial photographs or older maps inevitably introduces error and generalization, the amount depending on the source scale. Some vendors evolved into the digital era after long-established practices in the hardcopy business. For the sake of clarity or esthetics, they had deliberately distorted the alignment of some roads such as service roads running parallel to highways. Or they had taken the position – valid in the pre-GPS era – that absolute positional accuracy was unimportant as long as the map represented names, gross shapes and topological relationships accurately. The cumulative effect of these sources of uncertainty and error is that road alignments can differ as much as 100 m in urban areas, and 200 m on winding mountain roads. An incident at a freeway interchange can therefore appear to be on the wrong ramp (Fig. 1(a)), or in an adjacent neighborhood (Fig. 1(b)). It is unlikely that a third coordinate (elevation) can resolve this reliably,

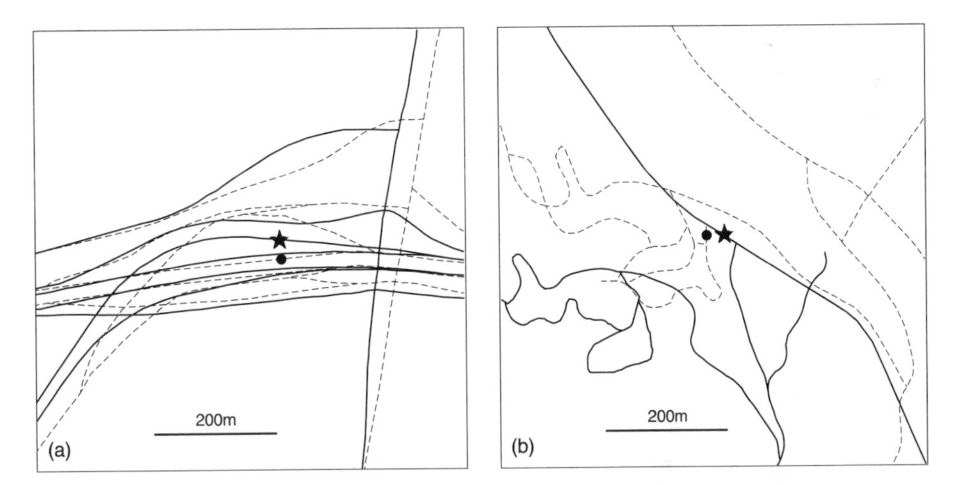

Fig. 1. Positional error. A coordinate from a vehicle on a ramp or highway in map A (star on solid line) snaps to the wrong ramp or street in map B (dot on broken line): (a) shows a freeway interchange in Santa Barbara; (b) is a highway and adjacent neighborhood on the outskirts of the city.

for two reasons: first, current databases do not contain the third dimension; secondly, error in GPS elevation readings is generally twice as high as horizontal error.

To correct error, one must first measure, understand and model it. For a single point, an obvious measure of error is the Euclidean distance between the point on the erroneous map, and

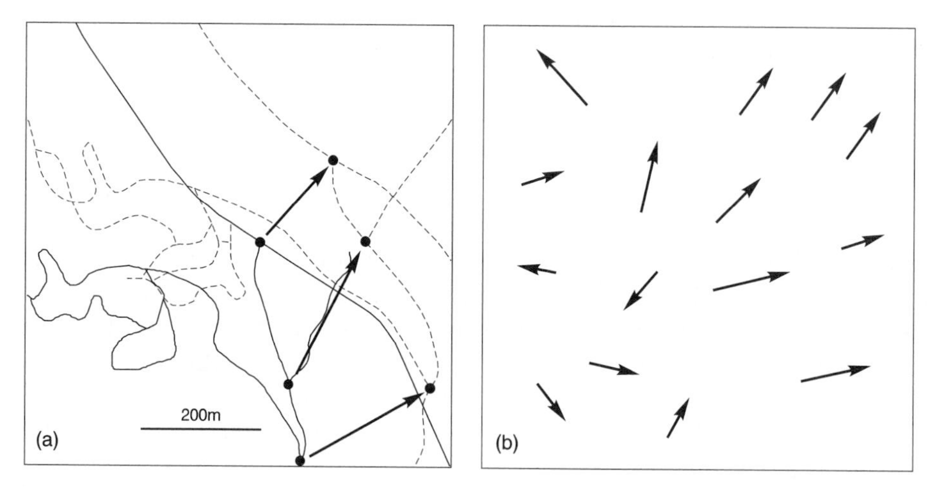

Fig. 2. (a) Error vectors constructed between points on map A and corresponding points on map B. (b) Vector field. Errors that are consistent in magnitude and direction (northeast of figure) are easy to model and to correct. When vectors are inconsistent (western portion of figure), extensive correction or complete re-survey may be required.

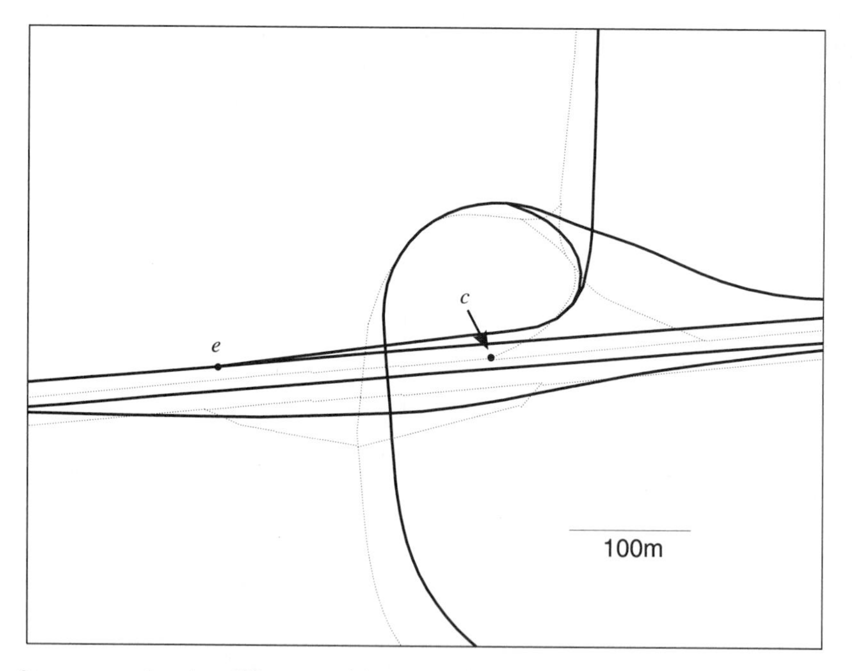

Fig. 3. Overlay of two maps showing different positions (c and e) where ramp meets freeway. When these are represented by planar-enforced models, topological relations clearly differ. The impact on linear measurements taken from this point is more than 200 m. Reproduced from Noronha (1999) with permission.

the true position of that point. The vectors between the points (Fig. 2(a)) show the magnitude as well as the direction of the error. Note that the term error assumes that one version is accurate and the other is not; this may not be easy to determine, considering the remarks in the last paragraph. Our analysis of error is based on an engineering scale map of the study area, and verification by differential GPS. Matching of points from one database to another is accomplished by a combination of automated and manual methods.

Extending the vector analysis to a larger area, one can develop a vector field (Goodchild and Hunter, 1997) that shows the variation in error over space (Fig. 2(b)), and suggests its origins. One could hypothesize that the field is continuous, in which case the error at a given point can be interpolated from that at surrounding points; our preliminary studies indicate that this is true at least to some degree (Church et al., 1998; Funk et al., 1998). Error direction and magnitude are often consistent over entire neighborhoods, but they change abruptly across some boundaries. This indicates that the map was developed piecemeal, at different times, or from different sources. Continuity of the error field also breaks down on freeways. Intersections of ramps with freeways are prone to high positional error that is difficult to model, because the paths intersect at a small angle, and a slight error in alignment translates into a large longitudinal displacement (Fig. 3).

3.2. Scale, classification, inclusion and topology

It was argued above that mapping necessarily entails an interpretation of reality. This is truest in the case of road inclusion and classification, which are closely connected with map scale and resolution. Dual carriageway freeways are single lines at one scale and double lines at larger map scales. Similarly there are differences in the treatment of traffic circles, median strips and channelized turn lanes. It is impossible to find two maps that agree on their categorization of "major" roads, let alone the fine distinctions between arterials and collectors. Discrepancies in classification lead to differences in inclusion. Driveways into condominiums and building complexes (e.g. hospitals) do not appear in all maps, because not all vendors consider these part of the street network. Inclusion is also impacted by the effectiveness of map update. The vendor that promptly includes new neighborhoods clearly shows several streets that another vendor does not.

A further consequence of discrepancies in inclusion is conflicts in topology (the term is used in the GIS sense, of connective relationships between points, lines and polygons). Clearly if one map shows a number of driveways off a principal road, while another map ignores them, the topological relations differ. Another aspect of topological inconsistency is the reduction of non-planar intersections (i.e. overpasses). GIS structures modeled on the TIGER database (Marx, 1986) impose planarity on all intersections due to the need to consider roads as potential polygon boundaries. Navigation turn tables are built into the database structure, with a large artificial impedance on turns that are not physically possible. Problems then arise when gross topological errors occur in the database, e.g. when a cloverleaf ramp meets a freeway east rather than west of an overpass (Fig. 3). Correction of the positional error requires extensive recoding of the topological relationships within the interchange.

It is tempting to study some of the above differences in inclusion and classification by characterizing each road section by some combination of name and position, and looking for a corresponding segment in the other database. This does not work because of (a) differences in naming, discussed in detail below, and (b) topological discrepancies that result in one map having long sections of road, while another fragments them into short segments. Nystuen et al. (1997) study both positional and inclusion differences between maps by buffering the centerlines in each map, and topologically intersecting the buffers. The degree of correspondence is analyzed spatially by gridding the area and measuring the intersections between buffers, cell by cell. This method is attractive because it mimics visual analysis and is intuitively valid; a drawback is that finer topological differences such as dual-line versus single-line highways do not affect the statistics appreciably, although these differences can have significant impacts on interoperability.

3.3. Names

In the light of the inadequacy of coordinates as an unambiguous LX, it has been suggested that street names, in the form of street addresses or intersections, would be more reliable. This is a reasonable argument up to a point – after all, the post office delivers mail based on street addresses where they exist, with little error. There are two caveats to this. First, mail delivery is facilitated by postal codes that enable sorting and reduce ambiguity, and also by intelligent human interpretation that easily substitutes an incorrect "Market Street" with the correct "Marquette Street". Secondly, transportation applications are not restricted to mail delivery routes. Many roads (notably freeway ramps) do not have names; for Santa Barbara County as a whole, 20–45% of records have blank name fields.

Further problems exist in the capture and storage of names. Abbreviations are inconsistent (e.g. "Av" versus "Ave"), parsing of fields varies (e.g. Spanish street types such as "Via del" should be separated as street type prefixes, but they are often incorporated into the proper name), and only one of our six databases provides for aliases that would equate "San Diego Freeway" with Interstate 5. There are consistency errors even within databases: one section of a road is correctly named "Winding Way" while another contains a typographic error, "Winting Way". Appropriate data structuring would avoid this.

Table 1
Four stretches of road, as named in five databases A–E[a]

Road	Database				
	A	B	C	D	E
1	E Camino Cielo	Camino Cielo	[blank]	[Street not present]	E Cm Cielo
2	San Marcos Pass Hwy 154	 Highway 154	[blank]	San Marcos Pass CA-154	San Marcos Pass [blank]
3	Mountain E Mountain	Mountain Park Bella Vista	[blank]	Mountain	W Mountain E Mountain Park Bella Vista
4	Foothill Cathedral Oaks	 Cathedral Oaks	Hwy 192	Foothill Cathedral Oaks	Foothill Cathedral Oaks

[a] A road may have several names over the sampled distance; databases may not agree on where changes occur. Only A provides for an alias. Reproduced from Noronha (1999) with permission.

Table 1 shows four randomly selected lengths of street, as named in five databases. The streets are selected visually, and some are so long as to have multiple names over their course. The variation in names offers a quick insight into the matching problem.

For the reasons outlined above, success rates in matching names across databases are lower in ITS applications than even the 60–75% often reported for address matching as used with street addresses. Intelligent software does exist to deal with some of the problems, e.g. to equate Winding with Winting, Market with Marquette; and even CA-101 with HWY-101 (Fellegi and Sunter, 1969; Knuth, 1973; Jaro, 1989); but software alone cannot resolve the more difficult cases, particularly where blank names are encountered.

3.4. Linear measurement

Prior to the advent of GPS, the only practical way to describe a location on a remote section of road was by linear measurement from a defined starting point. Transportation engineers continue to use linear references today, sometimes to the exclusion of maps and two-dimensional referencing. While motorists' experience of linear referencing is limited to the 100 m resolution of vehicle odometers, professionals employ distance measuring instruments (DMI) for 1 m-resolution readings. DMI technology has evolved over the years, from mechanical or optical revolution counters attached to the wheels of a vehicle, to modern electronic pulse sensors that hook into the transmission.

There are three components to a linear reference: (a) the road on which the point lies, described by a road name or numeric identifier; (b) the origin and direction for the measurement; (c) the distance measurement. Error in (a) is eliminated if parties communicate using standard identifiers, else the comments on street name interoperability above apply. Specification of the origin (b) is similarly susceptible to identification error. The origin has to be described with precision equivalent to that of the measurement: if the distance is stated with 1 m precision, it is not sufficient to specify the origin simply as the intersection of two streets, when the intersection is 30 m wide. Finally the measurement (c) is subject to errors, depending on curvature and other physical characteristics of the road. Throughout this paper we assume offsets to be measured along the road (in the case of the DMI) or along its polyline representation.

The linear LXX question is two-fold: (i) how well a location expressed by other means (e.g. coordinates or street names) translates to a linear reference, and (ii) how well a linear reference measured by a DMI, or with respect to one map database, describes that location with reference to a different map database. Regarding (i), in the Santa Barbara databases, there is on an average a 40% chance that a GPS point (with 100 m selective availability error) snaps to the wrong road, and a further 10% chance that it snaps to the wrong topological section of the correct road. The average linear reference error from such a point is about 90 m, assuming no recalibration points (VITAL, 1999). If the point is specified directly from DMI measurement, there are errors due to instrument limitations, calibration and operation, usually the greater of 1% or 2–10 m. On the matter of (ii), the relationship between polyline geometry, generalization and linear measurement has been studied from a number of standpoints (Mandelbrot, 1967; Douglas and Poiker, 1973; Buttenfield, 1985). It is generally well known that as a polyline is subjected to increasing generalization, the digitized length decreases. The Santa Barbara data set shows average differences of 8% in digitized length (longest versus shortest version of the same road) on a variety of roads, with a range of 1–16%. The deviation of digitized length from DMI length for a given data set over all sample roads ranges from 0.5% to 4% with an average of 3% and a worst case of 16%. Compensation for this error can be applied by normalizing a linear offset, i.e. expressing the offset as the proportion of its distance from the origin, to the total length of the road section (clearly this adds to the cost of specifying the location, because it requires two measurements to be made). Absolute linear transfer errors are about 50 m on average, and normalized transfer errors are about 25 m; errors up to nearly 1 km are encountered in exceptional cases, due to longitudinal errors in defining the intersections of ramps with highways (VITAL, 1999).

4. Error remediation

Given the immediate need for incident reporting and amelioration of transportation problems, it is necessary to examine strategies that enable the deployment of services using current databases. Taking an ITS example, a motorist may buy an in-vehicle database from vendor A, receive data from a TMC that operates with vendor B, and pass messages to an emergency center that uses vendor C. Regardless of which pair of databases is involved in a transaction, the message must be received unambiguously and error-free.

In addition to ITS concerns, there is the general problem of backward compatibility of databases. A wealth of legacy data is attached to databases of differing positional quality. For example, an emergency center may attach a note to a 1980s-vintage database, that 872 Whistler Highway is the farm of John Brown, with a large red barn near the driveway. It is important that when the emergency center upgrades to a more current database, the details of the Brown farm be preserved. This section examines how this might be done.

4.1. Interoperability standards

Fletcher (1999) defines four levels of interaction, the "Four I's": (a) integration, in which parties share hardware and software from a common vendor, or use a single database in which internal identifiers are universally understood, e.g. within a small office; (b) interoperability, where vendors may differ, but the semantics of objects and processes are standardized, and systems differ only in the internal details of implementation; (c) interfacing, where there are differences in semantics, traditions, systems and vendors, and communication is achieved by means of third-party translators that attempt to harmonize semantics to the extent possible; and (d) independence, in which parties fail to communicate because their systems are radically different. Interfacing is generally considered less elegant than is interoperation, and attracts irreverent terms such as "duct tape". But given the considerable problems with map databases, we may not have the luxury of restricting ourselves to any one of these I's, hence the word interoperability is used in the broadest sense below.

4.1.1. Messaging

One obvious candidate for standardization is the message used to communicate a location. In the language of the Open Systems Interconnect (OSI) model (Day and Zimmerman, 1983), this discussion of message standardization pertains to the upper level layers, application and presentation. In the United States the Society of Automotive Engineers (SAE) publishes J2374, the national Location Referencing Message Specification (SAE, 1998), which currently has the status of an "information report" rather than a standard. J2374 is a proposed set of seven message profiles to communicate a location, each profile generally being a form of LX. A user group chooses a profile to suit its mode of operations, for example a location derived from GPS is usually transmitted using a coordinate profile, whereas a linear GIS-T application employs a linear referencing profile. J2374 does not specify how the message is to be interpreted at the receiving end; it does not contain metadata or any explicit measure of message reliability; and it provides only for one-way communication – there is no provision for the originator to know whether the message has been successfully received and correctly interpreted.

The cross-streets profile (XSP) is the J2374 profile proposed for widest use in communicating between TMCs, and from TMCs or ISPs to vehicles; it is also the basis of SAE's J1746, the ISP-to-vehicle standard. The original XSP described a location in terms of: (a) the name of the principal street on which the location lies; (b) the names of the two cross-streets that bound the segment of the principal street; and (c) the linear offset along the principal street, measured from one of the cross-streets (Fig. 4). At first sight this appears to be an adequate way to represent location, but on close examination, and bearing in mind the context of machine-to-machine communications, there are several problems. First, there is no provision in the XSP for geographic region, so a matching triad of street names anywhere in the world is a potential candidate. Second, due to the high incidence of blank name fields in current databases, a XSP message can be composed successfully (i.e. with all three street names non-blank) in only 33% of test cases. Third, when certain street configurations are encountered (e.g. crescents, intertwining roads with multiple crossings, closed loops) the logic of the XSP breaks down. In sum, due to these and other problems with name matching outlined above, the original XSP is successful (i.e. accurate and unambiguous) only in about 25% of all cases – this applies even to major roads, because highway ramps are unnamed. Following our original suite of tests (VITAL, 1999) we recommended that the XSP be strengthened by including a pair of coordinates in the profile, to establish general location. The measure of success is now complex, because the enhanced XSP contains two entirely independent ways of expressing location, which could be in conflict; or there could be several partial but different fuzzy lexical matches with the cross-street triad, which would have to be ranked by some rule based on the strength of lexical match and distance from the coordinates. In general, the success rate for the enhanced XSP is 50–80%. This still leaves a large proportion of cases that fail, or are ambiguous. Details on the tests are reported by VITAL (1999) and Noronha et al. (1999).

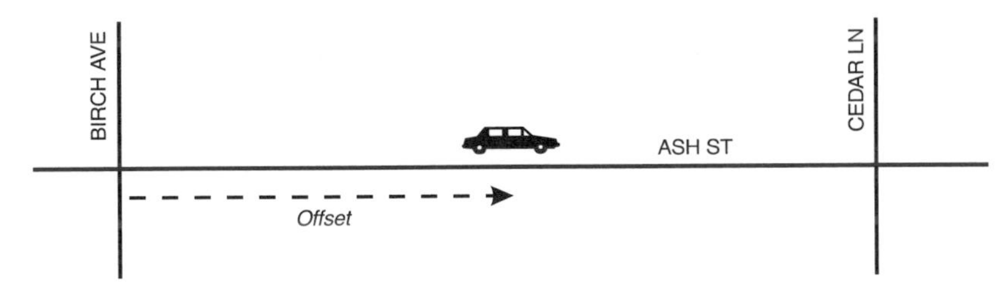

Fig. 4. The J2374 cross-street profile expresses location in terms of a principal street (Ash), two cross-streets (Birch and Cedar) and the offset distance from the first cross-street.

An obvious need in messaging is for the user to have a measure of information reliability, and it may be important for the sender to know whether the message has been satisfactorily received. VITAL has therefore proposed a robust, theoretically 100% successful protocol termed "LX-100", for use in mission-critical LXX. LX-100 mimics the negotiation process that takes place in verbal communication when one describes an emergency or any other location. Given the luxury of time, say to describe the venue of an office party, one employs a combination of street names, landmarks, absolute and relative navigation directives, which are together redundant; the receiver processes the information, detects success or failure based on redundant clues, and requests additional information in the event of failure. Similar principles are used for error checking in digital packet data transfer, around which the Internet Protocol is built; the European standards development and testing initiative, Extensive Validation of Identification Concepts in Europe (EVIDENCE), also uses redundant LX in its concept of the intersection location (ILOC).

LX-100 is still under development, and many issues and algorithms remain to be resolved. If the process is to approach 100% success, it must make assumptions about the quality and fitness for use of reference databases.

4.1.2. Database quality

The issue of database quality standards is sensitive, because it directly impacts the operations and financial viability of data vendors – and standards can be successful only if players subscribe to them. It takes large investments to build high-quality network databases, and if the benefits are expected to accrue only at some undefined time in the future, data vendors understandably channel resources only into areas of unquestionable benefit. Therefore a plea for standards must be cast in an incremental and economically feasible framework.

One obvious and realizable area for standardization is street naming, especially abbreviation, parsing of fields, and treatment of shared names (aliases such as San Diego Freeway and I-5, as well as cases where highways bear more than one designation when they merge temporarily). It is essential that standards govern the naming of highways and ramps, particularly because ramps form the topological intersections that define highway segments. There is also a need, though not as urgent, to employ classification standards based on road width and traffic volume, so that arterial roads are defined consistently across databases. All these are relatively easy to achieve, and would greatly improve interoperability in the short term.

The more challenging issues for standardization are those related to map scale, resolution, inclusion, topology, currency, coordinate and attribute accuracy. Compliance with some of these would require extensive field work and re-survey, and perhaps entirely new data models. Resolution standards have to specify that divided highways, traffic circles and channelized lanes should be explicitly represented if they meet certain criteria of size, separation and accessibility. Topological and data structuring standards must ensure that non-planar intersections are appropriately represented. Coordinate accuracy is a function of map scale, and a vendor could argue that a map meets standards for 1:100 000 mapping, although the corresponding ± 50 m accuracy may not be sufficient for some applications. There may be a need to certify a product's fitness for use in an application.

4.1.3. Datums

The more ambitious aspects of standardization, such as coordinate accuracy, may not be realizable by private vendors alone, and government invention may be required. In the 1980s and

1990s, the ready availability of DIME and TIGER files in the US spawned a large number of value-added products; similarly it may be that a high-level effort to standardize coordinates may be an appropriate government initiative in the 21st Century. This does not necessarily require re-survey and development of a new high-quality data base: a skeleton of well defined points will suffice in most areas. The ITS Datum (Siegel et al., 1996) is a proposed set of monuments designed to serve map accuracy needs in ITS, in both the short and long terms. In the short term, it provides an interim vehicle for vendors to improve the accuracy and interoperability of their products; in the long term it offers a unifying set of reference points for data exchange, the benefits of which need not be limited to ITS.

Preliminary study at VITAL indicates that the accuracy of ITS Datum points would vary with the needs of applications (Table 2), but would generally be defined to ±3–5m. Density would be higher in areas of greater road density, for example, each ramp in a freeway interchange would require at least one Datum point, probably at the intersection of the ramp with the freeway. A test database of about 5000 points has been compiled, and a rubber-streeting™ algorithm developed to match one data set to another geometrically, in real time (Fig. 5). The algorithm works satisfactorily on the test area shown, but it is now clear that mere specification of some points with accurate coordinates, identifiers and segment lengths will not suffice, and further work remains to be done on the development of appropriate data models. If the Datum is properly designed and implemented, there will be no need for manual identification of points of correspondence between databases, as was required to derive the experimental results reported above.

Two other national datums are currently being discussed: a linear datum (Vonderohe et al., 1995) to improve accuracy of linear measurement in GIS-T, and a road identification standard (FGDC, 1998), to enable interoperation of public sector databases as part of the National Spatial Database Infrastructure (NSDI). The work on the linear datum is described elsewhere in this book. It proposes a conceptual network of anchor sections, segments of road with precisely defined lengths. Anchor sections terminate at anchor points, which are defined by recoverable field locations (not necessarily intersections) that are selected solely to achieve the objective of optimizing linear accuracy. The NSDI proposes a set of points and line segments that are readily identifiable in the field and in databases (e.g. intersections), to which standard identifiers are applied, that serves as public keys to enable data exchange. The design enables data sharing between dual-line and single-line representations of the same road, and largely considers two-or three-dimensional positional accuracy to be irrelevant because interoperability is achieved by the use of standard identifiers, not positional similarity between objects. Clearly there are parallels

Table 2
Error tolerance in ITS applications, present and future[a]

Period	Locale	Lateral tolerance (m)	Longitudinal tolerance (m)
Present (2000)	Open highway	25	50–100
	Urban/interchange	5	50
Future (2015)	Open highway	1	10–20
	Urban/interchange	1	5

[a] The numbers are coarse estimates; the absolute values are less important than the relative values and the structure of the table.

Fig. 5. VITAL's rubber-streeting algorithm applies a geometric adjustment to one map to make it agree with another more accurately surveyed map or datum. Reproduced from Noronha (1999) with permission.

between these concepts and the ITS Datum, which incorporates all the requirements of the other standards – linear accuracy, standard identifiers – and three-dimensional accuracy.

4.2. Other solutions to the interoperability challenge

VITAL (1999) has shown that while some of the interoperability test results with current databases are poor, results with differential GPS and better quality databases are excellent, at least for current application requirements. One might legitimately argue that new surveys will soon brighten ITS interoperability prospects. While this is true of the future, there are two causes for caution. First, GPS-based re-survey is technologically possible today, but remains largely undone because of cost. An alternate way to improve public sector databases is by integration of construction and engineering data into transportation GIS; but for technical and management reasons, there are few jurisdictions where this has been done effectively. Second, improved coordinate accuracy is only one aspect of quality. Legacy attribute data have to be merged with new coordinates and perhaps revised topology – that can be best achieved by a datum that links old databases with new by means of common identifiers.

Another argument is that the interoperability problem can be avoided by employing just one database across applications. There are several reasons why this is not practical. First, the functional requirements of a database are not universal. Data needs vary considerably between users – for example, there are situations where single-line representation of freeways is appropriate and preferable to dual-lines. Second, in a market-driven economy that is stimulated by competition, there is a danger in allowing any vendor, private or public, a monopoly over a critical information resource such as road centerlines. However, in the short term, the single-vendor approach may be most expedient.

5. Conclusions and prospects

The preceding sections have documented the types and magnitude of interoperability problems in location reporting. The data errors are presented not as a case of vendor malpractice, but as a consequence of an evolving state of art that is currently being overtaken by increasingly ambitious user requirements. As recently as 1980, development of street network databases was driven by demographic applications that required accuracy at no better than the general neighborhood or block face. Now just 20 years later, users are equipped with inexpensive GPS and portable computers; they seek navigation solutions, they demand carriageway resolution, 1–20 m accuracy and current information on attributes that change frequently (e.g. one-ways and turn restrictions, even construction and congestion). Emerging methodologies for emergency management and other applications are built around these heightened expectations. Inevitably, in the short term some of these requirements will not be met.

There are a number of initiatives under way, e.g. intelligent messaging and datums, to address these problems. One could argue that as the quality of databases evolves, the need for these solutions will diminish. On the other hand, it is reasonable to speculate that requirements will be even more stringent in the future, and that solutions will have to keep pace. One obvious area of development is improved resolution, from carriageway to lane. It is likely that navigational instructions will direct drivers to move into a certain lane in preparation for an approaching maneuver. The California Department of Transportation (Caltrans) is planning for a traffic management future where diversion of highway traffic is initiated several kilometres ahead of an incident, one lane at a time, to prevent concentration of such traffic into surrounding arterials.

These developments raise academic challenges. Data models must evolve to accommodate lanes (e.g. Fohl et al., 1996), and public initiatives to improve coordinate accuracy, such as datums, must anticipate the evolution of user requirements. New messaging methods have to be developed for traffic information scenarios that not only report current status, but also describe previous or future events, such as the anticipated course of a toxic plume – some work on multi-dimensional data modeling is already under way, sponsored by the National Cooperative Highway Research Program (NCHRP) project 20–27(3). Based on the recent pace of development it is reasonable to believe that despite GPS, or perhaps because of it, the art and science of location reporting will remain an important area of endeavor over the coming decades.

Acknowledgements

This article is based on research supported by the California Department of Transportation, Interagency Agreement 65V250, and the US Department of Transportation, Federal Highway Administration, ITS Joint Program Office, Contract DTFH61-91-Y-30066.

References

Battelle Memorial Institute, 1997. Global positioning systems for personal travel surveys: Lexington area travel data collection test. Final Report. United States department of transportation, Federal Highway Administration, Office of Highway Management, Washington, DC.

Buttenfield, B.P., 1985. Treatment of the cartographic line. Cartographica 22 (2), 1–26.

Church, R., Curtin, K., Fohl, P., Funk, C., Goodchild, M., Kyriakidis, P., Noronha, V., 1998. Positional distortion in geographic data sets as a barrier to interoperation. Proceedings, ACSM Baltimore [document available at www.ncgia.ucsb.edu/vital].

Day, J.D., Zimmerman, H., 1983. The OSI reference model. Proceedings of the IEEE 71, 1334–1340.

Douglas, D.H., Poiker (formerly Peucker), T.K., 1973. Algorithms for the reduction of the number of points required to represent a digitized line or its caricature. Canadian Cartographer 10, 112–122.

Fellegi, I.P., Sunter, A., 1969. A theory for record linkage. Journal of the American Statistical Association 64 (328), 1183–1210.

FGDC, 1998. Road Data Model – Content Standard and Implementation Guide (Working Draft). Federal Geographic Data Committee, Ground Transportation Subcommittee, Washington, DC.

Fletcher, D., 1999. Road data model workshop, Washington, DC. Meeting Notes.

Fohl, P., Curtin, K., Goodchild, M.F., Church, R.L., 1996. A non-planar, lane-based navigable data model for ITS. In: Kraak, M.J., Molenaar M. (Eds.), Proceedings of the Spatial Data Handling, Delft, 12–16 August, pp. 7B/17–29.

Funk, C., Curtin, K., Goodchild, M., Montello, D., Noronha, V., 1998. Formulation and test of a model of positional distortion fields. In: Proceedings of the Third International Symposium on Spatial Accuracy Assessment in Natural Resources and Environmental Sciences, Quebec City [URL www.ncgia.ucsb.edu/vital].

Goodchild, M., Hunter, G., 1997. A simple positional accuracy measure for linear features. International Journal of Geographical Information Systems 11 (3), 299–306.

Jaro, M., 1989. Advances in records linkage methodology as applied to matching the 1985 census of Tampa. Journal of the American Statistical Association 84 (406), 414–420.

Kim, K., Nitz, L., 1994. Application of automated records linkage software in traffic records analysis. Transportation Research Record 1467, 50–55.

Knuth, D.E., 1973. The art of computer programming. Sorting and Searching, vol. 3. Addison-Wesley, Reading, MA.

Mandelbrot, B.B., 1967. How long is the coast of Britain? Statistical self-similarity and fractional dimension. Science 156, 636–638.

Marx, R.W., 1986. The TIGER system: automating the geographic structure of the United States census. Government Publications Review 13, 181–201.

Noronha, V., 1999. Towards ITS map database interoperability – database error and rectification. Presented at International Workshop on GIS-T and ITS, Hong Kong. Forthcoming in GeoInformatica.

Noronha V., Goodchild M., Church R., Fohl P., 1999. Location expression standards for ITS – testing the LRMS cross streets profile (special issue on GIS Sharing and Standardization). Annals of Regional Science 33 (2) pp. 197–212.

Nystuen, J.D., Frank, A., Frank, L., 1997. Assessing topological similarity of spatial networks. In: Proceedings of the International Conference and Workshop on Interoperating Geographic Information Systems, Santa Barbara.

Quiroga, C.A., Bullock, D., 1998. Travel time studies with global positioning and geographic information systems: an integrated methodology. Transportation Research C 6, 101–127.

SAE, 1998. Surface Vehicle Information Report – Location Referencing Message Specification. Society of Automotive Engineers, Information Report J2374.

Siegel, D., Goodwin, C., Gordon, S.R., 1996. ITS Datum Final Design Report. United States Department of Transportation, FHWA Contract 61-94-Y-00001, Review Draft, 28 June 1996.

USDOT, 1997. The National ITS Architecture. A Framework for Integrated Transportation into the 21st Century. Federal Highway Administration, ITS Joint Program Office, Washington, DC. [URL www.odetics.com/itsarch].

USDOT, 1998. TEA-21 Critical Standards: Proposed Criteria and List of Critical Standards. United States Department of Transportation, ITS Joint Program Office, Washington, DC.

VITAL, 1999. The linear referencing profile – technical evaluation. United States Department of Transportation, FHWA Contract DTFH61-91-Y-30066, Final Report. [URL www.ncgia.ucsb.edu/vital].

Vonderohe, A.P., Chou, C.L., Sun, F., Adams, T., 1995. Results of a workshop on a generic data model for linear referencing systems. Proceedings, AASHTO Symposium on Geographic Information Systems in Transportation, Sparks NV.

PERGAMON

Transportation Research Part C 8 (2000) 71–89

TRANSPORTATION
RESEARCH
PART C

www.elsevier.com/locate/trc

A three-stage computational approach to network matching

Demin Xiong *

ITS Research Program, Oak Ridge National Laboratory, 1000 Bethel Valley Road, MS 6206, Oak Ridge, TN 37831-6206, USA

Abstract

Network matching is frequently needed for integrating data that come from different sources. Traditional ways of finding correspondences between networks are time-consuming and require considerable manual manipulation. This paper describes a three-stage matching algorithm (node matching, segment matching, and edge matching) that combines bottom-up and top-down procedures to carry out the matching computation. As it uses sensitive matching measures, the proposed algorithm promises good improvement to existing algorithms. An experiment of matching two waterway networks is reported in the paper. The results of this experiment demonstrate that a reasonable matching rate and good computational efficiency can be achieved with this algorithm. The paper also briefly discusses necessary improvements in areas of linear alignment, aspatial matching and higher-level matching.

Keywords: Network matching; Spatial data integration; Three-stage computation; GIS

1. Introduction

Matching road networks that come from different sources or have been created at different times is an important function for transportation data management and manipulation. Frequently, applications need to integrate data that are referenced by different networks for the same area (Rosen and Saalfeld, 1985; Saalfeld, 1988; Brown et al., 1995; Nystuen et al., 1997; Walter and Fritsch, 1999). A common procedure used to realize this integration is to establish correspondences among different networks; then data that are referenced on different networks are transferred to and integrated into one of the networks. The same procedure is necessary in situations where an existing database grows older and has to be updated with a newer version of the network. In this case, the newer version of the network functions as the control layer, and matches

* Tel.: +1-865-574-2696; fax: +1-865-574-3895.
E-mail address: xiongd@ornl.gov (D. Xiong).

0968-090X/00/$ - see front matter
PII: S0968-090X(00)00011-5

between this newer version and the earlier versions are established. Then data on the earlier versions of networks are conflated to the newer version.

Network matching also has important applications in image processing (Novak, 1992; Stilla, 1995; Wang, 1998). Raw images obtained from remote sensors usually contain various kinds of distortions. To geometrically rectify a distorted image, this image can be overlaid with a network map. By establishing correspondences between linear features on the image and on a network map, the image geometry can be transformed and rectified. Matching with a network map is also a good strategy for image recognition. When unknown linear features on an image can be matched with known features on a network map, the unknown features on the image can be recognized as the same features as the ones on the network map. After matching, the characteristics of the matched features can be extracted, and by referencing these characteristics, other features on the image that may not directly correspond to an existing map can be recognized.

Due to the importance of these applications, considerable effort has been devoted to studying and developing automated procedures for network matching. This research focuses on an investigation of an algorithm that consists of three types of matching: node matching, segment matching, and edge matching. Node matching establishes node correspondences between two networks using Euclidean distance and angle patterns formed by incident edges. Segment matching tracks segment pairs along potential matching edges. Edge matching takes input from segment matching and then derives matching measures at the edge level. Applying these matching measures allows identification of the best matching pairs.

In this research, a computational strategy that combines bottom-up and top-down computations is adopted. The bottom-up computation starts with node matching, then proceeds to segment matching, and finally ends up with edge matching. The top-down procedure is the reverse. At the initial stage of the top-down computation, potential edge-matching pairs are first hypothesized using screening criteria such as distance and angle difference. Then, segment matching proceeds. Through segment matching, matching measures for those hypothesized matches are computed, like matches are confirmed, and unlike matches are rejected. The purpose of combining the bottom-up and top-down computations is that the bottom-up computation will find matches, where node matches can be quickly established, while the top-down computation will find matches where node matches fail or network structures differ.

This three-stage matching procedure promises good improvements to previous matching procedures for two reasons. Firstly, previous research has frequently used node matching and edge matching, but not segment matching. Segment matching is important because it can be used to evaluate correspondences between each pair of segments on potentially matched edges. Through segment matching, matching measures between each pair of segments are first computed; then, overall matching measures at the edge level are obtained. Due to these detailed considerations, edge-matching measures derived from segment matching can be more sensitive in recognizing differences and similarities among different edge pairs. Using these measures, there is a better chance to find the best matches.

The other advantage of the current algorithm is in its overall computational procedure. Previous research has often used a bottom-up procedure as the main computational strategy. There are some exceptions (e.g., the two-stage matching method developed by Gabay and Doytsher (1994)). The current algorithm not only combines the bottom-up and top-down computations but also considers interaction and feedback between different stages of matching, especially in segment matching and edge matching, thereby providing some refinement to those procedures that already have a computational mechanism in place such as the one developed by Gabay and Doytsher (1994).

In the following discussions, Section 2 provides an overview of previous research on the subject of network matching; Section 3 describes the proposed matching algorithm; Section 4 reports on the matching experiment; and Section 5 provides conclusions and discussions of the current algorithm with respect to its overall performance and necessary improvements needed in the future.

2. Previous research

The problem of network matching has been studied in different disciplines, and related literature can be found in the areas of geographic information systems (GIS), cartography, transportation, and image processing. Due to diversified references, different terms have been used in the literature. These terms include *map matching*, *conflation*, *and linear alignment*. In this section, alternate terms will be used as well.

One of the earliest matching algorithms was developed by the US Census to integrate data from the US Geological Survey and the Census Bureau (Rosen and Saalfeld, 1985; Saalfeld, 1988). This algorithm was applied to match maps that have a link-node structure. A bottom-up computational procedure is utilized in this algorithm, which uses node matching as the starting point. After node matching, a rubber sheeting operation is implemented. In this operation, the geometry of one of the maps is adjusted to make it correlate better with the other map. Finally, matches of the corresponding links on the two maps are identified.

The method introduced by Gabay and Doytsher (1994) represents a major departure from previous matching algorithms. The major advantage of their method is that it is able to find both common elements on two maps and to reveal unique elements appearing on only one of the maps. This capability allows geometric inconsistency and differences of topological characteristics to be recognized and handled during the map matching process. The two-stage matching procedure proposed by Gabay and Doytsher (1994) is also a blueprint of the strategy that combines the bottom-up and top-down computations. In this two-stage matching procedure, lines that have matched end nodes are matched first, then unmatched lines are further evaluated, and final matches are obtained.

Brown et al. (1995) described a conflation system that was developed in a GIS environment. This system makes use of existing GIS functions such as rubber sheeting and dynamic segmentation to adjust network geometry and establish node and link correspondences of two matching networks. With the system, linear mappings along network edges are possible, and network attributes such as direction flags and distances can be assigned or computed automatically when these attributes are transferred from one network to another network.

Nystuen et al. (1997) studied a method that compares two networks for the purpose of evaluating the quality of digital network databases. In their research, they developed an automated procedure to associate network elements. Through nearest node analysis and buffering, corresponding nodes and edges on two participating networks are identified and statistics showing how well two networks correspond each other are generated. Their studies identified several problems arising from network comparison. These problems, including scale effects, definitional discrepancies, data model differences, and the need for pattern recognition, represent major research directions in network matching.

More recently, Walter and Fritsch (1999) developed a statistical matching algorithm that provides significant advancement to the state of the art. Walter and Fritsch's (1999) algorithm has two important properties. Firstly, their algorithm takes a relational matching approach that bases matching decisions not only on the characteristics of a matching pair, but also on whether their neighbors are matched or not. Secondly, information theory is utilized to derive sophisticated matching measures to facilitate matching decisions. By transforming the map-matching problem into a problem of searching for the best communication channel in a communication system, the algorithm obtains optimal matches by selecting matching pairs, whose mutual information is maximal. With a combination of these two properties and a bottom-up implementation, matches can be derived first at the local level, then for clusters, and eventually for the entire network.

Matching techniques have also been studied in image processing for purposes of image registration and recognition. Considerable literature can be found in the review article by Fonseca and Manjunath (1996). The method of relational matching or structural matching that was developed in computer vision (Shapiro, 1980; Shapiro and Haralick, 1981) is particularly relevant to network matching. Ventura et al. (1990) and Wang (1998) successfully applied this method in image registration. Walter and Fritsch (1999) incorporated the concept of relational matching in their network-matching algorithm.

Despite existing research, several issues remain to be addressed for automated network matching. In previous research, edge matching has usually been approached by matching end nodes of edges and by searching corresponding edges in a given buffer. In cases where multiple edges are close to each other, identifying best matches can be difficult. Segment matching is likely to be useful in dealing with this problem. By comparing edges at the segment level, one can measure similarity and differences between edges more accurately. Nevertheless, segment matching has not been adequately discussed in the literature. Integrating segment matching into the network-matching algorithm therefore represents a major contribution to the effort reported here.

As for the computation of network matching, a great deal of research has been done for the bottom-up procedure, but not much for the top-down procedure. The top-down procedure is generally needed in situations where a priori information about potential matches can be obtained. For instance, when matching candidates are identified or hypothesized on the basis of information about proximity or aspatial attributes, the top-down procedure can be conveniently utilized to verify these matches and find the best matches. Considering that the bottom-up procedure has its advantage in other matching tasks, this research takes an approach that combines both the bottom-up and top-down procedures. To do so, different procedures can be used to effectively handle different tasks.

3. The algorithm

This section introduces some notation and definitions and describes the overall procedure used to carry on the calculation of network matching and the individual algorithms that constitute the overall computational procedure.

3.1. Notation and definitions

For simplicity, the symbol $G = (N, E)$ is used to represent a directed network graph, planar or non-planar, with $N = \{n_1, \ldots, n_m\}$ representing the node set, and $E = \{e_1, \ldots, e_p\}$ representing the edge set. For matching purposes, each node will have a set of attributes, including an ID, the number of edges that are incident to the node, called degree $D(n)$, the incident edge set, $n(e) = \{e_1, \ldots, e_{D(n)}\}$, and a coordinate pair (x, y) representing the location of the node. An edge-attribute set will include an ID, from-node and to-node (e.g., n_i and n_j), and a set of coordinate pairs $\{(x_1, y_1), \ldots, (x_k, y_k)\}$ representing the shape points of the edge. The shape points on an edge also define a set of segments, and this segment set is denoted as $S = (s_1, \ldots, s_{k-1})$.

The matches of two network graphs $G_a(N^a, E^a)$ and $G_b(N^b, E^b)$ are defined with three sets of mappings: node mapping, $M_n = \{(n^a, n^b) | n^a \in N^a \text{ and } n^b \in N^b\}$; segment mapping, $M_{es} = \{(s_e^a, s_e^b) | s_e^a$ representing a segment on an edge, e^a, of the G_a, and s_e^b representing a segment on the edge e^b that corresponds to $e^a\}$; and edge mapping, $M_e = \{(e^a, e^b) | e^a \in E^a \text{ and } e^b \in E^b\}$. As precise correspondences between elements (nodes, segments, and edges) of two networks are rare, it is assumed that mappings between network matching pairs are based on edited network graphs rather than on the original ones. That is, network edges may be split or merged (or network nodes inserted or removed) when mappings between two networks are generated.

In this discussion, one of the networks, G_a, that participates in the matching is called the *reference network* (following the definition by Nystuen et al. (1997)), and the other participating network, G_b, is referred to as the *matching network*. In general, it is assumed that the reference network functions as a control or reference layer. When data from two networks need to be integrated, usually data from the matching network will be transferred to the reference network. In cases where networks need to be geometrically rectified, the reference network will be used as a control layer, and the geometry of the matching network will be transformed.

3.2. The computational strategy

The major objective of network matching is to generate three types of mappings: node mapping, segment mapping, and edge mapping. Naturally, calculations of these mappings involve three types of computations: node matching, segment matching, and edge matching. As network elements (nodes, segments, and edges) are intrinsically related to each other, it is not efficient to compute individual mappings separately. To integrate these three types of computations, we consider an overall computational strategy as shown in Fig. 1.

For this overall computational strategy, the matching computations are divided into two subprocesses, the bottom-up and the top-down. The bottom-up computation starts with matching network nodes using criteria of Euclidean distances between nodes and angle patterns of network edges that are incident to these nodes. After node correspondences are established, segment matching proceeds. Segment matching involves both tracking corresponding segments of a potential match and computing the likelihood of whether one segment indeed forms the counterpart of another segment. Finally, edge matching is carried out. In edge matching, matching measures obtained from segment matching are first aggregated at the edge level. Then these measures are applied to eliminate unlikely matches.

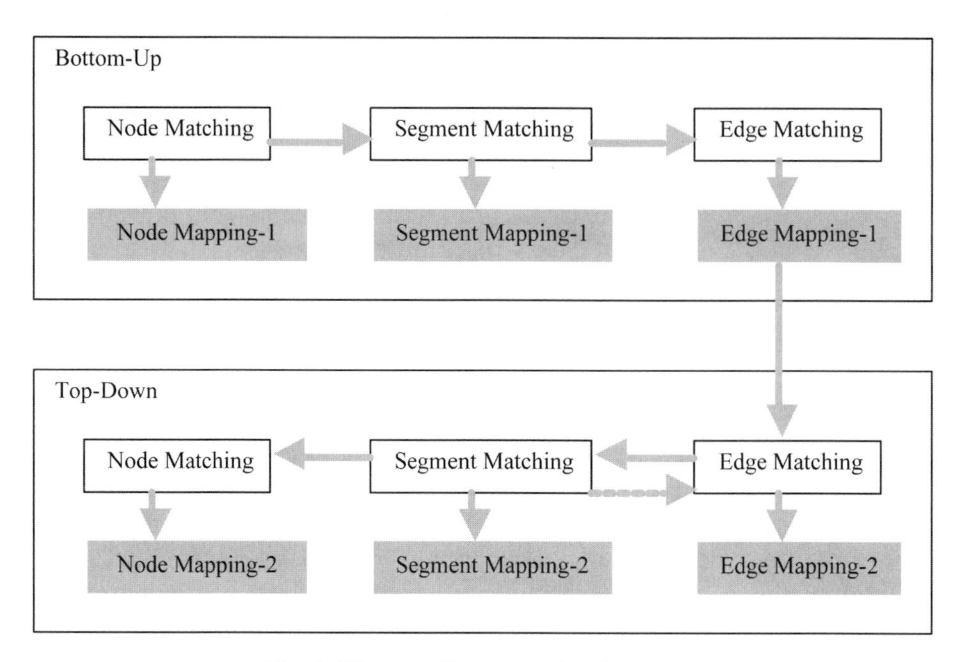

Fig. 1. The overall computational strategy.

By searching nearest neighbors and exploiting network topological relationships, the bottom-up computation can quickly find node and edge matches at locations where node correspondences can be reliably established, but this method may leave good matches unidentified at other locations. To deal with this problem, the top-down computation follows. The top-down computation assumes no node correspondences; instead, it proceeds with generating matches between network edges as the first step. With the bottom-up results, the top-down computation will need only to evaluate those edges that are not matched after the bottom-up matching. Note, however, that edge matching actually proceeds with two stages in the top-down computation. In the first stage, potential candidates on the matching network that may form a counterpart to a given edge on the reference network are first identified. After that, segment matching takes over, and for each pair of edge-matching candidates, segment mappings are generated. With these segment mappings, edge-matching procedure is reactivated to compute matching measures at the edge level (or sub-edge level if edges are split). And finally, decisions are made to find the best edge matches.

The top-down procedure can be independently implemented. In this case, the edge-mapping set from the bottom-up matching, as shown in Fig. 1, is assumed to be empty.

With the overall procedure described above, we can now treat each of the three matching procedures (node matching, segment matching, and edge matching) separately.

3.3. Node matching

Due to errors and distortions in network databases, the same nodes on different networks usually may not have an exact geometric match, e.g., $x_i^a \neq x_j^b$ and $y_i^a \neq y_j^b$. Instead, a gap exists between these two corresponding nodes

$$d_{ij}^{n} = \sqrt{(x_i^a - x_j^b)^2 + (y_i^a - y_j^b)^2}. \tag{1}$$

Assuming that geometric errors and distortions of the networks tend to be local and are bounded, the gap between a corresponding node pair d_{ij}^n can be used as a criterion to search for candidate node matches.

As each node has one or more edges incident to the node, if two nodes represent the counterparts of each other on two networks, the similarity of angle patterns formed between the nodes and edges can be used as another matching criterion. Assume the angle between two edges referenced to the same node is calculated by

$$\varphi_{kl}^{n} = \arccos\left(\frac{v_k^a \cdot v_l^b}{|v_k^a||v_l^b|}\right), \tag{2}$$

where v_k^a and v_l^b are vectors formed with e_k^a and e_l^b, and ($e_k^a \in E^a$; $e_l^b \in E^b$) and it is assumed that v_k^a and v_l^b always have a direction pointing away from the referenced node. Then, an angle-matching measure between two nodes can be evaluated with

$$d(\varphi)_{ij}^{n} = \min_{kl}(\varphi_{kl}^{n}). \tag{3}$$

The minimization sign used in Eq. (3) means that the minimum angle difference is used to determine whether at least one pair of edges is matched for these two corresponding nodes. With the distance and the angle difference criteria, node matching is computed with a decision function

$$F_{ij}^{n} = \begin{cases} 0 & \text{if } d_{ij}^{n} > \Delta, \\ 0 & \text{if } d(\varphi)_{ij}^{n} > \Phi, \\ 1 & \text{if } \mathrm{d}_{ij}^{n} = \text{min or 0 otherwise}, \end{cases} \tag{4}$$

where Δ and Φ are thresholds set for distance and angle difference, beyond which no matches will be allowed. In Eq. (4), 0 indicates no matches, and 1 indicates the two nodes are matched. The procedure to derive matches between nodes of two networks is straightforward. The inputs for this procedure include node and edge coordinates of the two networks, the maximum matching distance, and the maximum angle difference. The computation proceeds as follows:

Step 1. Initialize the node-mapping set (e.g., set $M_n = \{(n_1^a, \emptyset), \ldots, (n_m^a, \emptyset)\}$).

Step 2. For each node n^a on the reference network, find all the nodes on the matching network that are within the maximum distance Δ. Then, apply the matching criteria using Eq. (4), and find the best match.

Step 3. Output the mapping M_n, which now contains nodes on the matching network that correspond to the nodes on the reference network.

3.4. Segment matching

Segment matching establishes detailed correspondences at the segment level between edges. It can find corresponding locations of nodes that are represented on one network but not on the other network, and generate measures that can be aggregated at the edge level for edge-matching decisions. We utilize three criteria to evaluate segment matching: angle difference between segments, distance between segments, and matched segment length. The use of these three criteria is based upon the consideration that angle difference and distance are direct measures of how well two segments can correspond with each other, while the length can be used as a weighting factor for both the angle difference and the distance.

To effectively match segments, a configuration, as shown in Fig. 2, is used to determine relationships between segments of the two participating networks. In this configuration, s^a represents a segment on an edge of the reference network, and s^b represents a segment on an edge of the matching network. The angle difference between s^a and s^b can be computed using an equation similar to Eq. (2), the only difference is that the vectors used in the equation are now the two segments instead of the two edges.

Before the distance between the two segments is calculated, four distance measures are first derived: d^{AC}, d^{CE}, d^{BF}, and d^{DB}. As shown in Fig. 2, d^{AC} represents the distance from the starting point A of s^a to s^b, d^{CE} represents the distance from the starting point C of s^b to s^a, d^{BF} represents the distance from the ending point B of s^a to s^b, and d^{DB} represents the distance from the ending point D of s^b to s^a. The distance between the two segments then is computed with

$$d^s_{ab} = \frac{1}{2}\left(\min(d^{AC}, d^{CE}) + \min(d^{BF}, d^{DB})\right). \tag{5}$$

Notice that distance between the two segments can be defined in different ways – for instance, $\min(d^{AC}, d^{CE}, d^{BF}, d^{DB})$. Nevertheless, Eq. (5) provides an average distance measure that takes account of the distances at both ends of the segment pair. By contrast, measures such as $\min(d^{AC}, d^{CE}, d^{BF}, d^{DB})$ give distance measures only at certain points. When corresponding segments stretch into different directions, these measures may fail to provide sensitive information in a particular match. During the computation of the segment distance, the lengths of the effectively matched segments are also determined. In the case of Fig. 2, the matching lengths are d^{EB} on s^a and d^{CF} on s^b.

Segment matching in general involves intensive computations. To carry out these computations effectively, a tracking procedure that can coordinate with the node-matching and edge-matching computations is devised. The major steps for this tracking procedure are as follows:

Step 1. For a given edge pair (that may be derived from node matching or assumed as a potential matching candidate from edge matching), first compare their directions. If they are in opposite directions, then the order of the shape points for one of the two edges will be reversed.

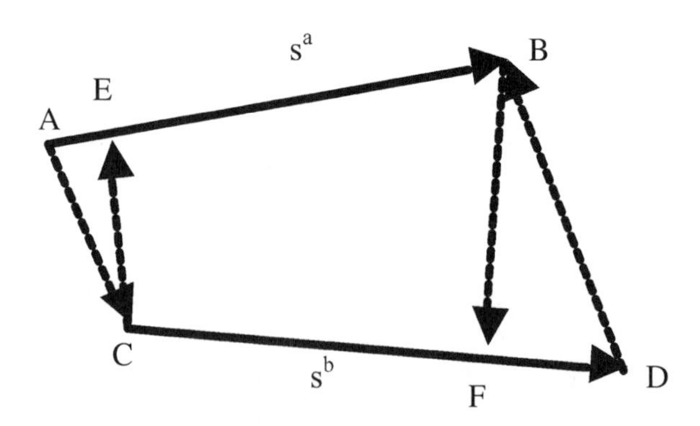

Fig. 2. Segment configuration.

Step 2. Search for the first pair of matching segments along the two edges. This matching pair must be within the allowed maximum distance and in the same direction.
Step 3. Compute the angle difference, distance, and matching lengths as described earlier for each segment pair. Continue the process until one or two of the edges end or break away. During the process, the decision on which segment is selected for the next computation is based on whether a segment is an overshot or an undershot (in Fig. 2, s^b is an overshot and s^a is an undershot). The segment that is next to a previously undershot segment will be selected for next computation (in Fig. 2, the segment next to s^a will be selected).

3.5. Edge matching

For edge matching, we use three matching measures: average angle difference, average distance, and total matched lengths. These three measures are derived from segment matching but aggregated at the edge level. The average angle difference is computed with the length-weighted average of segment angle differences. The average distance is the length-weighted average of segment distances. The total matched lengths are the total lengths of the edges on the reference network and on the matching network respectively. With these measures, the overall matching measure between two edges are calculated with

$$D^e_{ij} = \alpha d^e_{ij} + \beta d\phi^e_{ij} + \delta(L^e_i + L^e_j), \tag{6}$$

where i and j represent ith edge on the reference network and jth edge on the matching network; d^e_{ij}, $d\varphi^e_{ij}$, L^e_i, and L^e_j represent the average distance, average angle difference, total matched length on the edge of the reference network, and total matched length on the edge of the matching network separately; α, β, and δ are weighting factors used to balance the effect of different measures. In program implementation, a decision function is utilized to facilitate matching decisions:

$$F^e_{ij} = \begin{cases} 0 & \text{if } \mathrm{d}^e_{ij} > \Delta^e, \\ 0 & \text{if } (L^e_i < L^e \text{ and } L^{e'}_i > L^e) \text{ or } (L^e_j < L^e \text{ and } L^{e'}_j > L^e), \\ 1 & \text{if } D^e_{ij} = \min \text{ or } 0 \text{ otherwise,} \end{cases} \tag{7}$$

where 1 indicates a match, and 0 indicates no match; Δ^e represents the maximum distance allowed between two edges, and L^e represents the minimum length required for an edge match; for $D^e_{ij} = \min$, it is assumed that for all potential matches, the matching pair with the minimum overall measure will be identified to make the match. $L^{e'}_i$ and $L^{e'}_j$ are the total lengths of the edge pair, which may be different from the matched lengths, L^e_i and L^e_j. The reason to use $L^{e'}_i$ and $L^{e'}_j$ as constraints in Eq. (7) is to make sure that short edges are not eliminated from the matching process.

In designing procedures to carry out edge matching, two scenarios must be considered. The first scenario is when the segments between two edges have been matched (e.g., edge matching with the bottom-up computation). In this case, the task of edge matching is to apply the aggregated matching criteria derived from segment matching to validate a potential match. The second scenario is when edge matching starts without segment matching (e.g., edge matching with the top-down procedure). In this case, edge matching proceeds with selecting a set of potential matching candidates as the starting point, then segment matching follows. Through segment matching, edge-matching measures are obtained for each of the assumed matching pairs. With these measures, the best match will be finally identified. Since edge matching in the second scenario contains all the steps necessary for edge matching in the first scenario, the procedure of edge matching for the second scenario is outlined below:

Step 1: Initialize the edge-matching mapping, M_e, which starts as an empty set.

Step 2: For each edge on the reference network, find all the edges on the matching network that are located within the maximum matching distance. Before proceeding to segment matching, use a pre-processor to group segments to form sub-edges in cases where part of an edge is included within the maximum matching distance. Then, proceed with segment matching, and compute edge-matching measures for each potential edge pair. Finally, carry out the matching test (*substep 2.1*) to find the best match if there is one.

Substep 2.1: The matching test first checks whether matched edges meet the requirements of the minimum length, the maximum angle difference, and the maximum distance. Then, the matching measures of all the potential matching pairs that have met these minimum or maximum requirements are compared, and the pair with the maximum overall matching measure is identified as the best match. If no matching pairs meet the minimum and maximum requirements, then there will be no matches.

Step 3: During *Step 2*, the edge-mapping set, M_e, grows into a list that contains corresponding edges on the two networks. This list, however, may also contain entries where edges on the matching network match multiple edges on the reference network redundantly. Before the mapping set is finalized, use matching measures again to eliminate these overlapped mappings.

4. The experiment

To examine how well the proposed algorithm works, a computer program uses the above-described programming logic developed. This program was built in C and works in the Windows environment. Currently, the program is able to read and write networks in ESRI's shape file format and has the ability to perform fast searches for nearest points and segments. This section reports on the experiment of matching two versions of the US waterway networks using this program.

4.1. The data

The data used for the matching experiment are waterway networks for the eastern United States that are derived from the 1995 and 1996 versions of the national transportation atlas databases (NTAD) provided by the US Bureau of Transportation Statistics. These two networks were chosen because a major revision occurred between the 1995 and 1996 networks; it was therefore a good test case. Fig. 3 shows the 1995 network, and Fig. 4 shows the 1996 network. Both of the networks are shown on an Albers equal-area projection.

Given the map scales in Figs. 3 and 4, specific differences between these two networks are not obvious. Fig. 5 provides a detailed comparison of these two networks. As shown in Fig. 5, the 1996 network is represented with thick gray lines as the edges and big gray dots as the nodes. The 1995 network is represented with thin dark lines as the edges and small dark dots as the nodes.

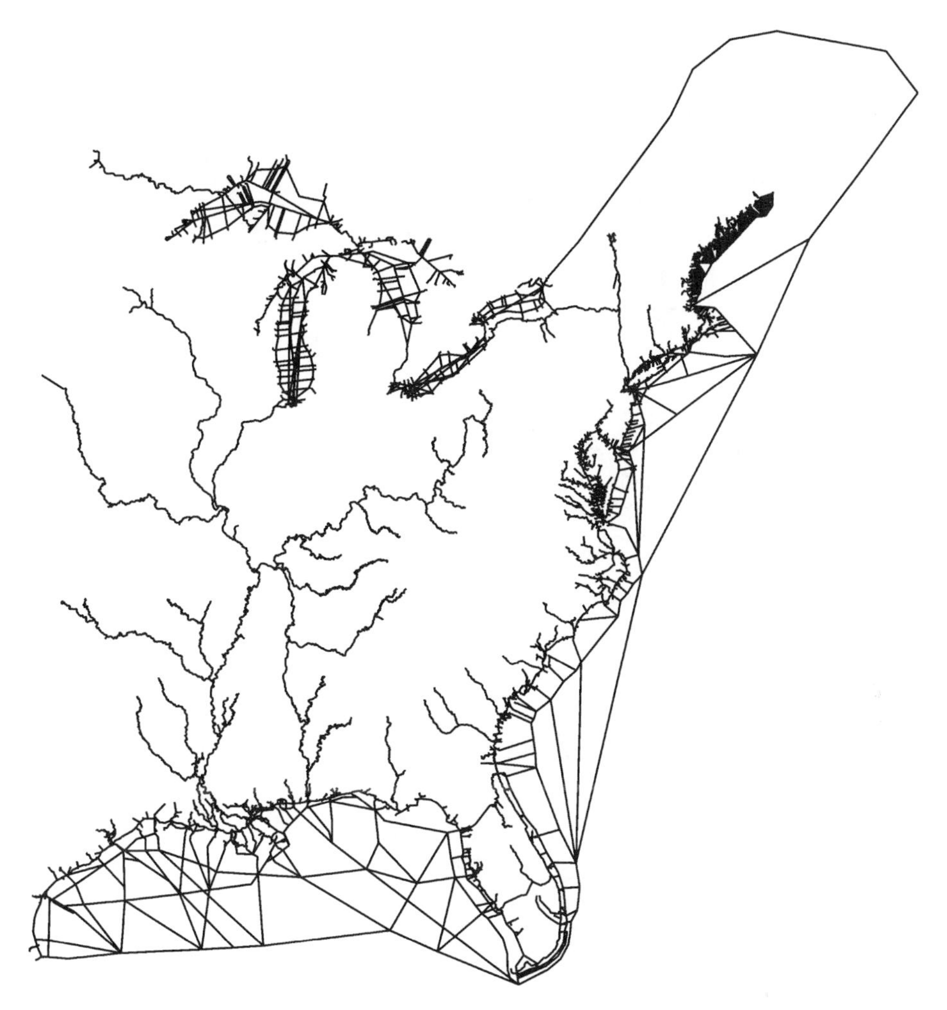

Fig. 3. The 1995 waterway network.

Three major differences between the 1996 network and the 1995 network can be identified. Firstly, some of the edges on the 1995 network have a coarse geometry, but these edges have been replaced with edges that have more naturally curved shapes on the 1996 network. Location shifts also occur on some of the edges of the 1996 network, which apparently is in an attempt to improve the accuracy of the network geometry. Secondly, in some cases, the 1996 network has shorter edges, which means that one edge on the 1995 network may correspond to several edges on the 1996 network. Thirdly, the 1996 network introduces some new edges that do not have correspondences on the 1995 network.

4.2. Bottom-up matching

In spite of differences, these two waterway networks have a very similar node-link structure and have limited coordinate discrepancies, which make them suitable for the use of the bottom-up

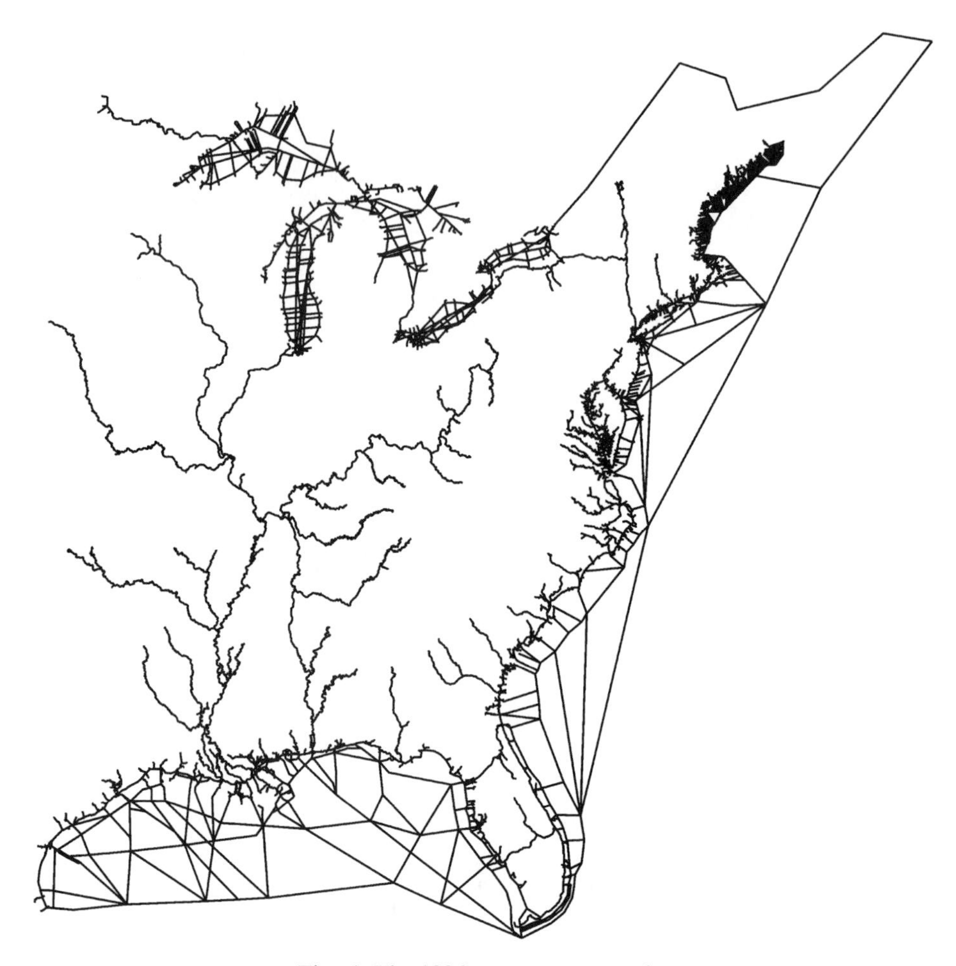

Fig. 4. The 1996 waterway network.

matching. To apply the bottom-up procedure, 10 km is used as the maximum distance criterion. Angle difference is not used in this particular matching because there are significant differences between the two versions of the networks for some waterway links, which result in large angle differences. In general, angle difference is a valuable piece of information for distinguishing nodes that are close to each other. To allow effective use of this information, yet to avoid mismatching at nodes with large angle differences, an ideal solution would be an adaptive procedure that is able to determine when the information should be turned on and when the information should be turned off. Such a procedure was not developed in this specific experiment.

Fig. 6 shows the distances between each of nearest node pairs. The vertical axis of Fig. 6 represents the distance. The horizontal axis represents the node IDs on the 1996 network. As Fig. 6 shows, a large number of nodes are perfectly matched (e.g., have 0 distance). Among the nodes that are not perfectly matched, most have the distance gap of less than 10 km. (As 10 km is used as the cut-off distance, distances greater than 10 km are shown as 10 km on the diagram.) Using the distance criterion, for all 4584 nodes on the 1996 network, 4551 nodes have a match on the 1995 network.

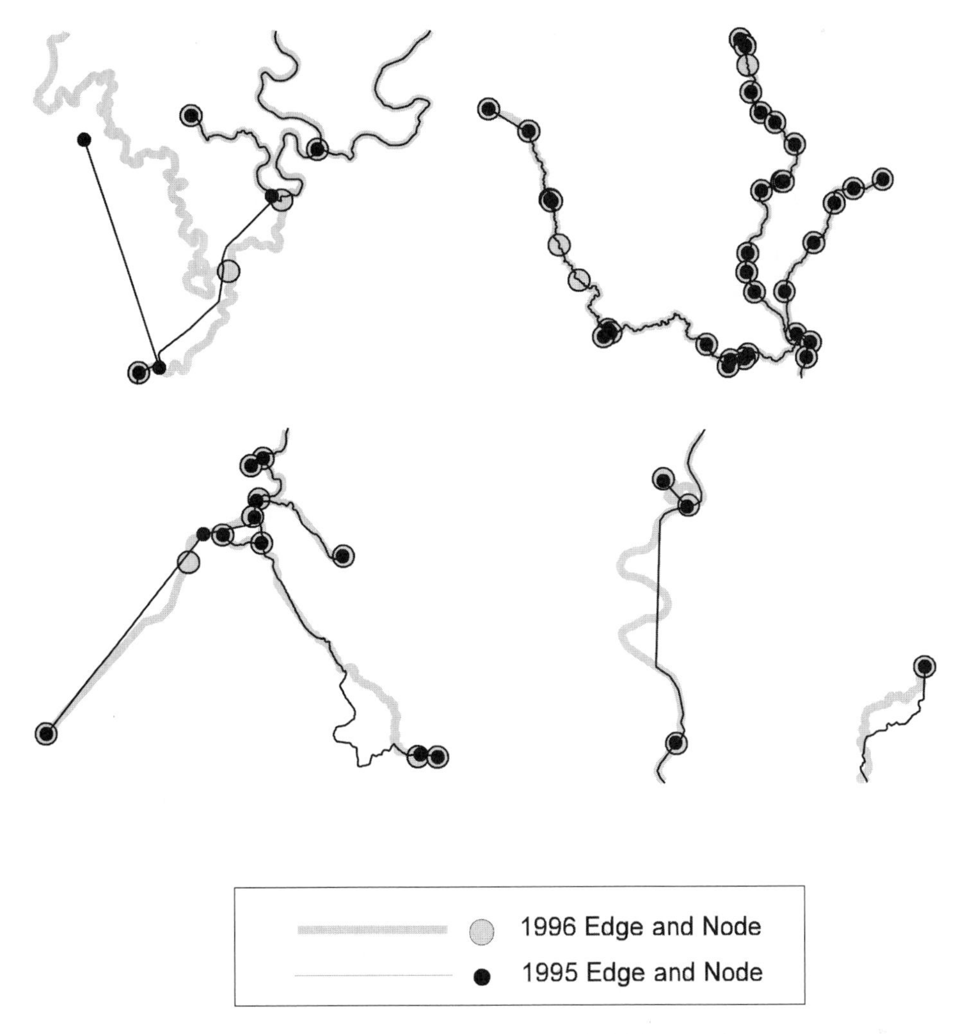

Fig. 5. Comparison of the 1995 and 1996 networks.

After node matching, edge correspondences are evaluated. This evaluation starts with segment matching, which generates matching measures for edges that have their end nodes matched. The computations of matching measures follow the general procedures outlined in Sections 3.4 and 3.5, but with some customization. In this specific application, each of the matching measures is transformed into a 0–6 point score, and the weighting factors α, β, and δ, as shown in Eq. (6), are set to 1.0. Finally, the three scores are added together to get the total score for each of the matching pairs.

Table 1 illustrates how the matching measures are transformed into the 0–6 point scores. The purpose of this transformation is to have a discretized score so that each of the measures used for the edge matching can be easily assessed. As is shown in Table 1, different intervals were chosen for the values of thresholds used for the discretization. The purpose was to provide a higher resolution for measures, where potential matches are more likely. As no evidence suggests that a

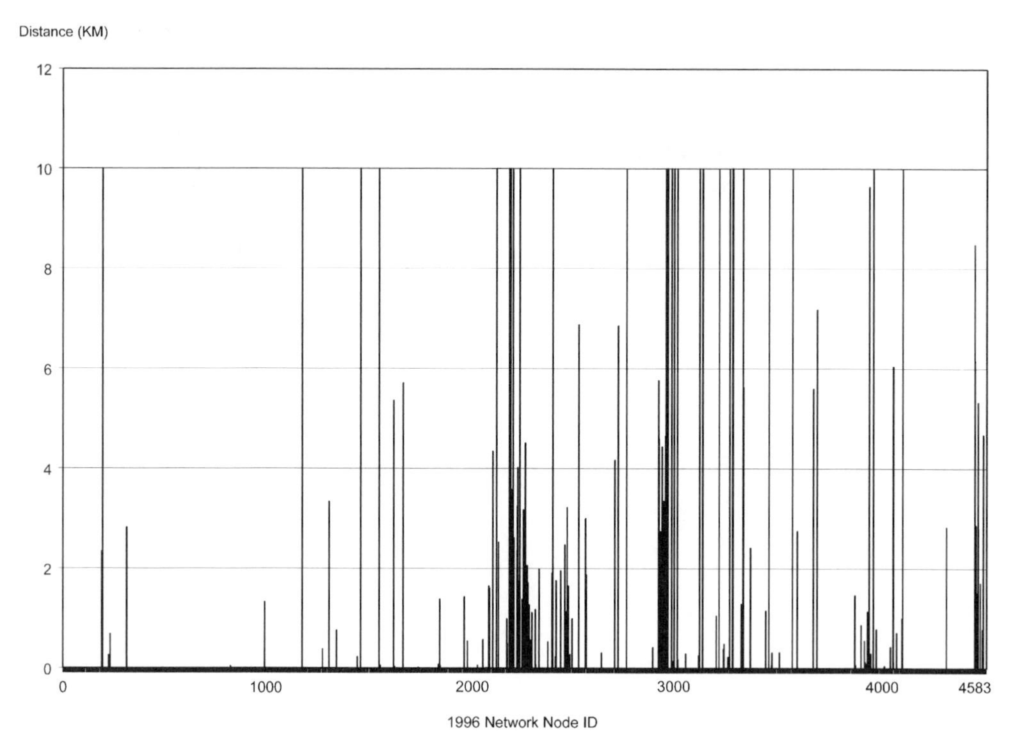

Fig. 6. Distances between nearest node pairs.

particular matching measure needs to be weighted more or less than other measures, 1.0 was chosen for each of the weighting factors to give equal weight to the measures.

When the total scores are computed for each of the edge pairs, those edge pairs that have a score less than 1 will be eliminated for a match. The matching result with these scores is that for all 5088 edges on the 1996 network, 4894 edges are matched to the 1995 network. A visual examination of these matched nodes and edges reveals no obvious mismatches.

4.3. Top-down matching

After the bottom-up matching, 194 edges on the 1996 network are left without a match. Some of these edges may not have a match, but some may have a counterpart on the 1995 network, but their end nodes are not exactly matched. To allow matching of additional edges, the top-down matching procedure is now utilized.

For the top-down matching, only those edges that are unmatched during the bottom-up matching are considered for further matching. The edge mapping set generated from the bottom-up matching is passed to the top-down matching procedure so that those unmatched edges can be identified. In the top-down matching, each edge on the 1996 network is evaluated separately. When a 1996 edge is selected, unmatched edges on the 1995 network that are close and parallel to the selected edge on the 1996 network are identified. Each of the candidate pairs is then matched at the segment level. After segment matching, matching scores between each edge pair are

Table 1
Transformation of matching measures to 0–6 point scores[a]

Point	Angle difference	Distance	Length
6	$d\varphi^e_{ij}/\Phi^e < 0.125$	$d^e_{ij}/\Delta^e < 0.125$	$L^e_i/L^e > 2.0$ and $L^e_j/L^e > 2.0$
5	$d\varphi^e_{ij}/\Phi^e >= 0.125$ and $d\varphi^e_{ij}/\Phi^e < 0.25$	$d^e_{ij}/\Delta^e >= 0.125$ and $d^e_{ij}/\Delta^e < 0.25$	$(L^e_i/L^e > 2.0$ and $L^e_j/L^e > 1.0)$ or $(L^e_i/L^e > 1.0$ and $L^e_j/L^e > 2.0)$
4	$d\varphi^e_{ij}/\Phi^e >= 0.25$ and $d\varphi^e_{ij}/\Phi^e < 0.325$	$d^e_{ij}/\Delta^e >= 0.25$ and $d^e_{ij}/\Delta^e < 0.325$	$L^e_i/L^e > 1.0$ and $L^e_j/L^e > 1.0$
3	$d\varphi^e_{ij}/\Phi^e >= 0.325$ and $d\varphi^e_{ij}/\Phi^e < 0.5$	$d^e_{ij}/\Delta^e >= 0.325$ and $d^e_{ij}/\Delta^e < 0.5$	$(L^e_i/L^e < 1.0$ and $L^{e'}_i < L^e)$ and $(L^e_i/L^e > 1.0$ or $L^e_j/L^e > 1.0)$
2	$d\varphi^e_{ij}/\Phi^e >= 0.5$ and $d\varphi^e_{ij}/\Phi^e < 0.75$	$d^e_{ij}/\Delta^e >= 0.5$ and $d^e_{ij}/\Delta^e < 0.75$	$(L^e_i/L^e < 1.0$ and $L^{e'}_i < L^e)$ and $(L^e_j/L^e < 1.0$ and $L^{e'}_j < L^e)$
1	$d\varphi^e_{ij}/\Phi^e >= 0.75$ and $d\varphi^e_{ij}/\Phi^e < 1.0$	$d^e_{ij}/\Delta^e >= 0.75$ and $d^e_{ij}/\Delta^e < 1.0$	N/A
0	$D\varphi^e_{ij}/\Phi^e <= 1.0$	$d^e_{ij}/\Delta^e >= 1.0$	$(L^e_i/L^e <= 1.0$ and $L^{e'}_i > L^e)$ or $(L^e_j/L^e <= 1.0$ and $L^{e'}_j > L^e)$

[a] Refer to Section 3.5 for symbol representations in the table.

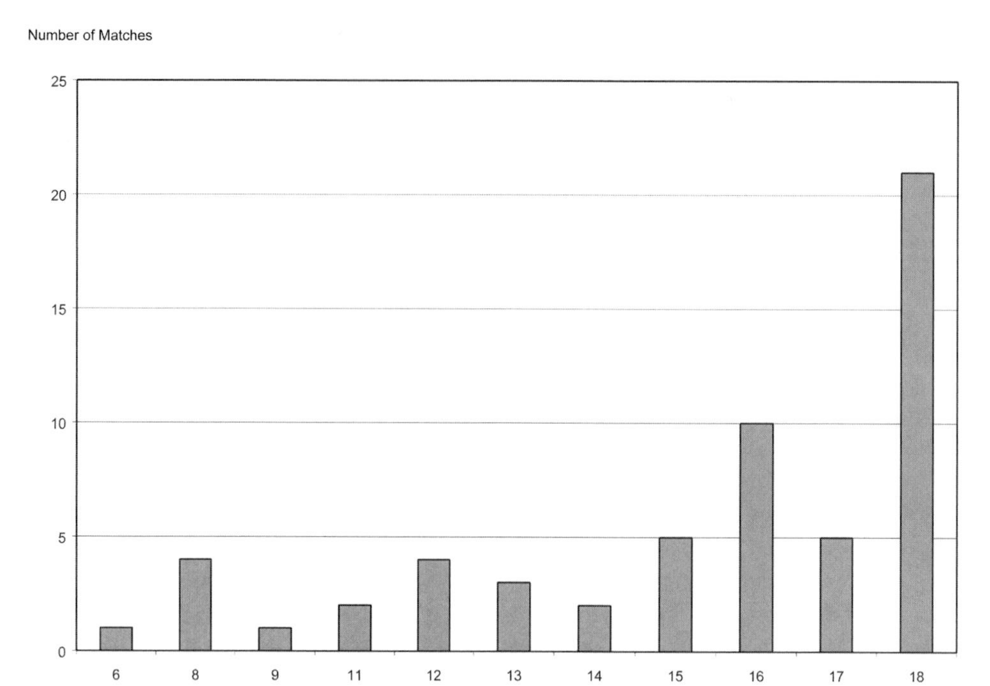

Fig. 7. Score distribution for matched edges.

computed, and used to determine the best match. For the 194 unmatched edges on the 1996 network, 58 edges then find their corresponding matches on the 1995 network.

Fig. 7 shows the frequency distribution of final matching scores for the 58 matched edges. As shown in Fig. 7, 21 out of 58 edges are matched at the highest score (18), and the lowest score where a match has been found is 6, which indicates that all matches are fairly certain and unambiguous. Typical cases for the 136 edges that remain unmatched include: (1) edges newly introduced on the 1996 network with no counterpart on the 1995 network; (2) edges whose geometry has changed significantly from the 1995 network to the 1996, making it difficult to determine whether they actually form a match; and (3) edges that are far apart and cannot be effectively matched with the use of the distance criterion.

5. Conclusions and discussions

The experiment results show that a reasonable match rate can be achieved with the proposed matching algorithm. The combined use of top-down and bottom-up computations is effective. As the waterway networks used in our experiment have a similar network structure, the bottom-up computation can quickly identify matches using node correspondences. In cases where node correspondences cannot be effectively identified, the top-down computation is able to find additional matches. The experiment also demonstrates that segment matching provides leverage when edges are compared in detail. With segment matching, finer resolution can be achieved when edges show subtle differences.

Currently, the program used to implement the proposed algorithm is able to read and write a ESRI's shape file format directly, which has proven to be very helpful. The data can be viewed and analyzed conveniently before the algorithm is applied. After matching computations, the results can be quickly evaluated. In addition, it is possible to take advantage of existing GIS functions for pre-processing and post-processing the network data, which is necessary in many real-world applications.

The computational performance of the proposed algorithm appears satisfactory based upon the current experiment. The entire matching process for the waterway networks took about 2 s on an Intel Pentium II processor. These 2 s included the time for reading the input files and writing the output files and the time spent on the actual matching computations. This computational effectiveness has a great deal to do with the use of a grid system for effectively indexing network nodes and edges. This grid system is very similar to the one introduced by Franklin et al. (1994). The major usefulness of this grid system is in searching for matching candidates during node and edge matching. Take node matching as an example, if indexes are not used, then $m * n$ calculations are necessary to find the nearest matches for two sets of nodes that have the node numbers m and n, respectively. If m and n are assumed to be the same size, then the computational complexity is on the order of n^2 or $O(n^2)$. When grid indexes are utilized, however, the computation number can be cut to $9 * m * n/g$, where g is the total number of cells. The factor of 9 is based on the assumption that the maximum search range is equal to or less than the edge length of a cell, so a maximum of nine cells will be searched in each of the searches. As the size of g is adjustable and can be made proportional to the size of n, the computational complexity is now actually on the order of n or $O(n)$.

With the use of the spatial indexes, as described above, an increase in network size will not significantly slow down the matching computation. Nevertheless, for extremely large networks such as those for cities like Los Angeles or New York, a large amount of data may overwhelm computer memories if they are loaded at once. In these cases, special care must be taken. For instance, these large networks can first be divided into several smaller networks; these smaller networks are then matched separately. Another factor that has to be considered in the analysis of computational performance is the similarity and difference of the participating networks. In general, networks with similar geometrical and topological characteristics can be more effectively matched. However, further study of this problem will be necessary in order to arrive at more substantive conclusions.

In spite of the preliminary success of this algorithm, additional improvement or extension of the algorithm is necessary to make it more effective and reliable. In the short term, it appears that improvements should be made in three areas. Firstly, the algorithm uses a segment-matching procedure that is based on distance and angle differences to track corresponding segments. The drawback of this procedure is that when network edges are distorted significantly, segment mappings derived from this procedure cannot faithfully reflect actual segment correspondences. If this occurs, matching measures derived from these segment mappings are not accurate, and node positions represented by the segment mappings cannot be determined precisely. Filin and Doytsher (1998) studied the problem and suggested that edges or curves should be divided into separate smooth sub-edges or sub-curves, and then mappings should be established between these sub-curves. Incorporation of this matching procedure into the current algorithm is likely to improve the accuracy of segment matching.

Secondly, the current research mainly deals with spatial characteristics of the participating networks. Yet, aspatial properties of these networks such as road names or mileposts also provide critical information for network matching. To make effective use of the aspatial properties, integration of aspatial matching with spatial matching represents an important strategy. In general, attributes of networks can be first evaluated to determine matches or potential matches (e.g., edges have the same road name or the same street address range). When uncertainties arise, spatial matching will be used as a tiebreaker. In this case, the top-down procedure can be called up to verify whether the matching candidates can be spatially matched. Although spatial matching and aspatial matching can be implemented independently, an integrated procedure will likely perform better.

The third area for improvement is in higher-level matching (e.g., matching at the cluster or sub-network level). The matching strategy proposed in this paper is focused on local matches, using Euclidean distance in searching for and matching individual matching pairs. In situations where counterparts are not immediate neighbors, correct matching will be difficult. If matching decisions can be evaluated at a higher level (e.g., clusters or sub-networks evaluated as a whole before individual matches are determined), there will be a better chance for higher matching accuracy. In this case, the relational matching approach adopted by Walter and Fritsch (1999) represents a good solution.

Acknowledgements

Sincere thanks are due to the three anonymous reviewers for their valuable comments.

References

Brown, J., Rao, A., Baran, J. 1995. Automated GIS conflation: coverage update problems and solutions. In: Proceedings of Geographic Information Systems for Transportation Symposium (GIS-T), Sparks, Nevada, USA, pp. 220–229.

Fonseca, L.M.G., Manjunath, B.S., 1996. Registration techniques for multisensor remotely sensed imagery. Photogrammetric Engineering and Remote Sensing 62, 1049–1056.

Filin, S., Doytsher, Y., 1998. Conflation – the linear matching issue. In: Proceedings of 1998 Annual Convention and Exhibition, American Congress on Surveying and Mapping, Baltimore, Maryland, USA.

Franklin, W.R., Sivaswami, V., Sun, D., Kankanhalli, M., Narayanaswami, C., 1994. Calculating the area of overlaid polygons without constructing the overlay. Cartography and Geographic Information Systems 21, 81–89.

Gabay, Y., Doytsher, Y., 1994. Automatic adjustment of line maps. In: Proceedings of the GIS/LIS'94 Annual Convention, Arizona, Phoenix, USA, pp. 333–341.

Novak, K., 1992. Rectification of digital imagery. Photogrammetric Engineering and Remote Sensing 58, 339–344.

Nystuen, J.D., Frank, A.I., Frank, Jr., L., 1997. Assessing topological similarity of spatial networks. In: Proceedings of International Conference on Interoperating Geographic Information Systems, Santa Barbara, California, USA.

Rosen, B., Saalfeld, A., 1985. Match criteria for automatic alignment. In: Proceedings of Auto-Carto VII, American Congress on Surveying and Mapping and American Society for Photogrammetry and Remote Sensing, pp. 1–20.

Saalfeld, A., 1988. Automated map compilation. International Journal of Geographical Information Systems 2, 217–218.

Shapiro, L.G., 1980. A structural model of shape. IEEE Transactions on Pattern Analysis and Machine Intelligence 2, 111–126.

Shapiro, L.G., Haralick, R.M., 1981. Structural description and inexact matching. IEEE Transactions on Pattern Analysis and Machine Intelligence 5, 504–519.

Stilla, U., 1995. Map-aided structural analysis of aerial images. ISPRS Journal of Photogrammetry and Remote Sensing 50, 3–10.

Walter, V., Fritsch, D., 1999. Matching spatial data sets: a statistical approach. International Journal of Geographic Information Science 13, 445–473.

Wang, Y., 1998. Principles and applications of structural image matching. ISPRS Journal of Photogrammetry and Remote Sensing 53, 154–165.

Ventura, A.D., Rampini, A., Schettini, R., 1990. Image registration by recognition of corresponding structures. IEEE Transactions on Geoscience and Remote Sensing 28, 305–314.

PERGAMON

Transportation Research Part C 8 (2000) 91–108

TRANSPORTATION
RESEARCH
PART C

www.elsevier.com/locate/trc

Some map matching algorithms for personal navigation assistants

Christopher E. White [a], David Bernstein [b,*], Alain L. Kornhauser [a]

[a] *Department of Operations Research and Financial Engineering, Princeton University, Princetown, NJ 08540, USA*
[b] *Department of Computer Science, James Madison University, MSC 143 Harrisonburg, VA 22807, USA*

Abstract

Third-generation personal navigation assistants (PNAs) (i.e., those that provide a map, the user's current location, and directions) must be able to reconcile the user's location with the underlying map. This process is known as *map matching*. Most existing research has focused on map matching when both the user's location and the map are known with a high degree of accuracy. However, there are many situations in which this is unlikely to be the case. Hence, this paper considers map matching algorithms that can be used to reconcile inaccurate locational data with an inaccurate map/network. © 2000 Published by Elsevier Science Ltd.

1. Introduction

There are three different types of personal navigation assistants (PNAs). First-generation PNAs simply provide the user with a map and the ability to search the map in a variety of ways (e.g., search for an address, search for a landmark, scroll, and pan). Second-generation PNAs provide both a map and the user's current location/position. Third-generation PNAs provide a map, the user's location, and directions of some kind.

It should be clear why we distinguish first-generation PNAs from second- and third-generation systems. Clearly, a system that provides the user's current location is much more complicated than one that does not, and generally requires both additional hardware and software. What may not be clear is why we distinguish between second- and third-generation PNAs.

The rationale for doing so is actually quite simple. In second-generation systems the location that is provided to the user need not coincide with the street system (or subway system, etc).

* Corresponding author. Tel.: +1-540-568-1671; fax: +1-540-568-2745.
E-mail address: bernstdh@jmu.edu (D. Bernstein).

PII: S0968-090X(00)00026-7

However, in order to provide directions, the user's location must coincide with a street (or subway line, etc.) when appropriate.

There are, in essence, three different ways to determine the user's location. The first is to use some form of dead reckoning (DR) in which the user's speed of movement, direction of movement, etc. is continuously used to update her/his location (Collier, 1990). The second is to use some form of ground-based *beacon* that broadcasts its location to nearby users (Iwaki et al., 1989). The third is to use some form of *radio/satellite positioning system* that transmits information that the PNA can use to determine the user's location. This last approach is by far the most popular, a great many PNAs use the global positioning system (GPS) to determine the user's location (Hofmann-Wellenhoff et al., 1994).

Given a GPS receiver, it is almost trivial to convert a first-generation PNA into a second-generation PNA (i.e., one that provides both a map and the user's location), and many people have done so. However, reconciling the user's location with the underlying map (or network) can be much more complicated. In other words, converting a second-generation PNA into a third-generation PNA can be quite difficult.

When both the user's location and the underlying network are very accurate, the reconciliation problem is thought to be straightforward – simply "snap" the location obtained from the GPS receiver to the nearest node or arc in the network. Hence, it is not surprising that a number of people are working on improving the accuracy of both the underlying network and the positioning system. In order to develop more accurate maps/networks, enormous "surveying" efforts are underway (Deretsky and Rodny, 1993; Schiff, 1993; Shibata, 1994). Some of these efforts are being undertaken by government agencies and others are being undertaken by private companies. In order to develop more accurate positioning systems, a great deal of attention is being given to combining data from multiple sources. Some systems combine GPS with dead reckoning systems (Degawa, 1992; Mattos, 1994; Kim, 1996), others use differential GPS (Blackwell, 1986), and others use multiple sources of data (sometimes including maps) and then filter or fuse the data in some way (Krakiwsky et al., 1988; Tanaka, 1990; Abousalem and Krakiwsky, 1993; Scott and Drane, 1994; Watanabe et al., 1994; Jo et al., 1996).

We are interested in situations in which it is not possible or desirable to improve the accuracy of the map/network and the user's location enough to make a simple "snapping" algorithm feasible. Such situations arise for many reasons. Firstly, not all PNAs are vehicle-based. Hence, it may not be possible to use DR or other data sources. Secondly, even if it is possible to develop a network/map that is accurate enough, such a network may not always be available. For example, the PNA may not have sufficient capacity to store the complete, accurate network at all times and hence, may need to either store inaccurate/incomplete networks or download less-detailed networks from either a local or central server. Thirdly, many facilities will probably never be available from map/network vendors and will need to be obtained on-the-fly from the facility, probably with limited accuracy. For example, vendors may not provide detailed networks/maps of airports, campuses (both corporate and university), large parking facilities, and shopping centers.

Hence, the purpose of this paper is to discuss some simple *map matching algorithms* that can be used to reconcile inaccurate locational data with an inaccurate map/network. We begin in the following section with a formal definition of the problem. We then discuss point-to-point, point-to-curve and curve-to-curve matching. In all three cases, we consider algorithms that only use geometric information and algorithms that also use topological information. Then, we consider the performance of each of the algorithms in practice (in an admittedly limited number of tests). Finally, we conclude with a discussion of possible future research directions.

Our objective in this paper is not to provide a definitive evaluation of different map matching algorithms. Rather, our objective is to describe some simple algorithms and to consider, both theoretically and in a small number of tests, why they might or might not work well in practice.

2. Problem statement

Our concern is with a person (or vehicle) moving along a finite system (or set) of streets, $\overline{\mathcal{N}}$. At a finite number, T, of points in time, denoted by $\{0, 1, \ldots, T\}$, we are provided with an estimate of this person's location. The person's actual location at time t is denoted by $\overline{P}^t$ and the estimate is denoted by P^t. Our goal is to determine the street in $\overline{\mathcal{N}}$ that contains $\overline{P}^t$. That is, we want to determine the street that the person is on at time, t.

Of course, we do not know the street system, $\overline{\mathcal{N}}$, exactly. Instead, as illustrated in Fig. 1, we have a *network representation*, $\mathcal{N}$, consisting of a set of curves in $\mathbb{R}^2$, each of which is called an *arc*. Each arc is assumed to be piece-wise linear. Hence, arc $A \in \mathcal{N}$ can be completely characterized by a finite sequence of points $(A^0, A^1, \ldots, A^{n_A})$ (i.e., the endpoints of the individual line segments that comprise A), each of which is in $\mathbb{R}^2$. The points A^0 and A^{n_A} are referred to as *nodes* while $(A^1, A^2, \ldots, A^{n_A-1})$ are referred to as *shape points*. A node is a point at which an arc terminates/begins (e.g., corresponding to a dead-end in the street system) or a point at which it is possible to move from one arc to another (e.g., corresponding to an intersection in the street system).

This problem is called a *map matching problem* because the goal is to match the estimated location, P^t, with an arc, A in the "map", $\mathcal{N}$, and then determine the street, $\overline{A} \in \overline{\mathcal{N}}$, that corresponds to the person's actual location, $\overline{P}^t$. A secondary goal is to determine the position on A that best corresponds to $\overline{P}^t$.[1]

In order to simplify the exposition, we assume that there is a one-to-one correspondence between the arcs in $\mathcal{N}$ and the streets in $\overline{\mathcal{N}}$. This assumption can easily be relaxed, however (and often does not hold in practice).

3. Map matching algorithms

There are a number of different ways to approach the map matching problem, each of which has advantages and disadvantages. We will briefly discuss several of them before moving on to a discussion of the specific algorithms that we considered.

[1] Not surprisingly, the problem considered here is similar to the map matching problem in mobile robotics. There, the problem is to establish a correspondence between a current local map and a stored global map.

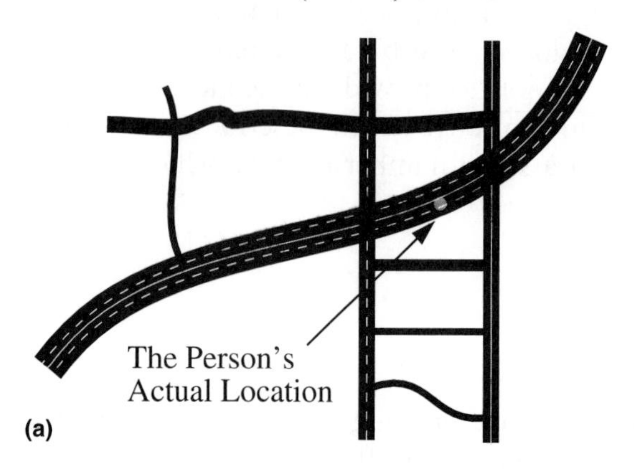

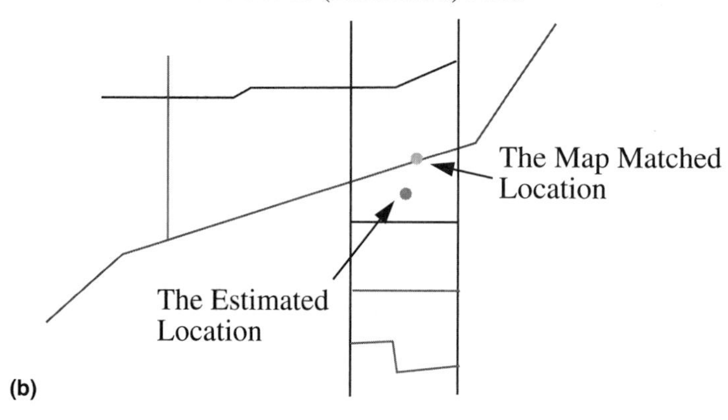

Fig. 1. The map matching problem.

3.1. Map matching as a search problem

One can view the map matching problem as a simple search problem. Then the problem is to match P^t to the "closest" node or shape point in the network.

A number of data structures and algorithms exist (see e.g., Bentley and Maurer, 1980; Fuchs et al., 1980) for identifying all of the points "near" a given point (often called a *range query*). It is then a simple matter to find the distance between P^t and every node and shape point that is within a "reasonable" distance of it (regardless of the metric used), and select the closest.

While this approach is both reasonably easy to implement and fast, it has many problems in practice. Perhaps most importantly, it depends critically on the way in which shape points are used in the network. To see this, consider the example shown in Fig. 2. Here, P^t is much closer to B^1 than it is to either A^0 or A^1, hence it will be matched to arc B even though it is intuitively clear that it should be matched to arc A. Hence, this kind of algorithm is very sensitive to the way in which the network was digitized. That is, other things being equal, arcs with more shape points are more likely to be matched to.

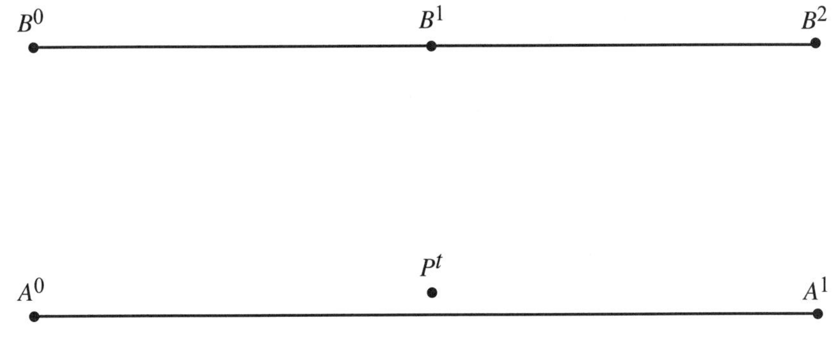

Fig. 2. One problem with point-to-point matching.

One might argue that this problem could be overcome simply by including more shape points for every arc. Unfortunately, this dramatically increases the size of the network and is not guaranteed to correct the problem.

3.2. Map matching as statistical estimation

One can also view map matching as a problem of statistical estimation. In this approach, one considers a sequence of points $(P^s, \ldots, P^t)$ and attempts to fit a curve to them. This curve is constrained to lie on the network.

This kind of approach has been explored in numerous papers (see e.g., Krakiwsky et al., 1988; Scott and Drane, 1994; Jo et al., 1996) and is quite appealing. It is particularly elegant when the model describing the "physics of motion" is simple (e.g., movement is only possible along a straight line). Unfortunately, in most practical applications, the physics of motion is dictated by (or constrained by) the network. This makes it quite difficult to model.

To understand why this is important, consider the network shown in Fig. 3. In this example, the positions $P^1 \ldots P^7$ have been recorded. Our objective is to fit a curve to these points, but the curve is constrained to lie on the network. In this case, there are two candidate curves, A and B (we ignore the rest of the network for simplicity).

In general (i.e., regardless of the metric), the curve P is closer to the curve B than it is to the curve A. Thus, if one uses a simple model of motion one will be led to match P to B rather than to A.

4. Algorithms used in this study

Our objective in this study was to combine the simplicity of the simple search approach with some of the ideas in the statistical approach. In the end, we implemented and tested four different algorithms.

Algorithm 1 is very simple. It finds nodes that are close to the GPS "tick" and finds the set of arcs that are incident to these nodes. It then finds the closest of these arcs and projects the point onto

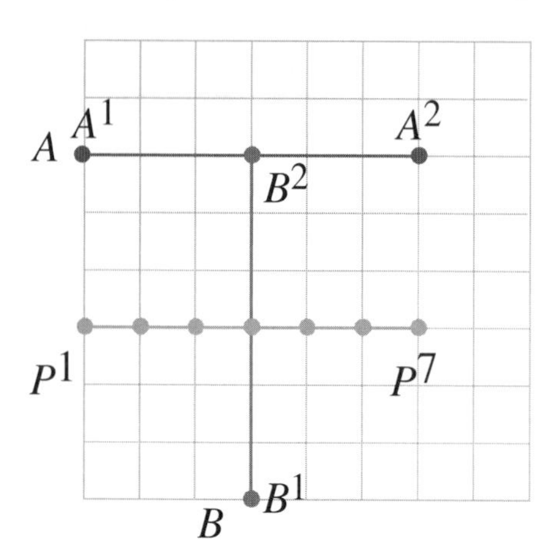

Fig. 3. Curve fitting.

that arc (using a minimum norm projection). As shown in Fig. 4, calculating the minimum distance between a point and a line segment is slightly more complicated than calculating the minimum distance between a point and a line. Calculating the minimum distance between p and the line segment between A^0 and A^1 is straightforward since it is the same as the minimum distance between p and the line through A^0 and A^1. However, when we calculate the distance between q and the line through A^0 and A^1, we see that the "perpendicular" intersects the line outside the line segment. Hence, we must also calculate the distance between q and both A^0 and A^1 and choose the smallest. Finally, since each arc is a piece-wise linear curve, we must find the minimum distance from the point of interest to each of the line segments that comprise A and select the smallest. Thus, calculating the minimum distance between a point, P^t, and an arc A, involves finding the minimum distance between P^t and the line segments $\{\lambda A^0 + (1-\lambda)A^1, \lambda \in [0,1]\}, \{\lambda A^1 + (1-\lambda)A^2, \lambda \in [0,1]\}, \ldots, \{\lambda A^{n_A - 1} + (1-\lambda)A^{n_A}, \lambda \in [0,1]\}$ and choosing the smallest.

Obviously, this algorithm has many shortcomings. Firstly, it does not make use of "historical" information and this can cause problems of the kind illustrated in Fig. 5. The estimated position P^2 is equally close to arcs A and B. However, given P^0 and P^1 it seems clear that P^2 should be matched to arc A.

Another problem with this algorithm is that it can be quite "unstable". This is illustrated in Fig. 6. The points P^0, P^1, and P^2 are all equidistant from arcs A and B. But, it turns out that P^0 and P^2 are slightly closer to A and P^1 is slightly closer to B. Hence, the matching oscillates back and forth between the two.

Hence, Algorithm 1 will play the role of a "straw man". It is fast, easy to implement, and should be easy to beat.

Algorithm 2 is identical to Algorithm 1 except that it makes use of "heading" information.[2] If the heading of the PNA is not comparable to the heading of the arc, then the arc is discarded. So,

[2] As discussed below, this calculation is actually performed by the GPS receiver.

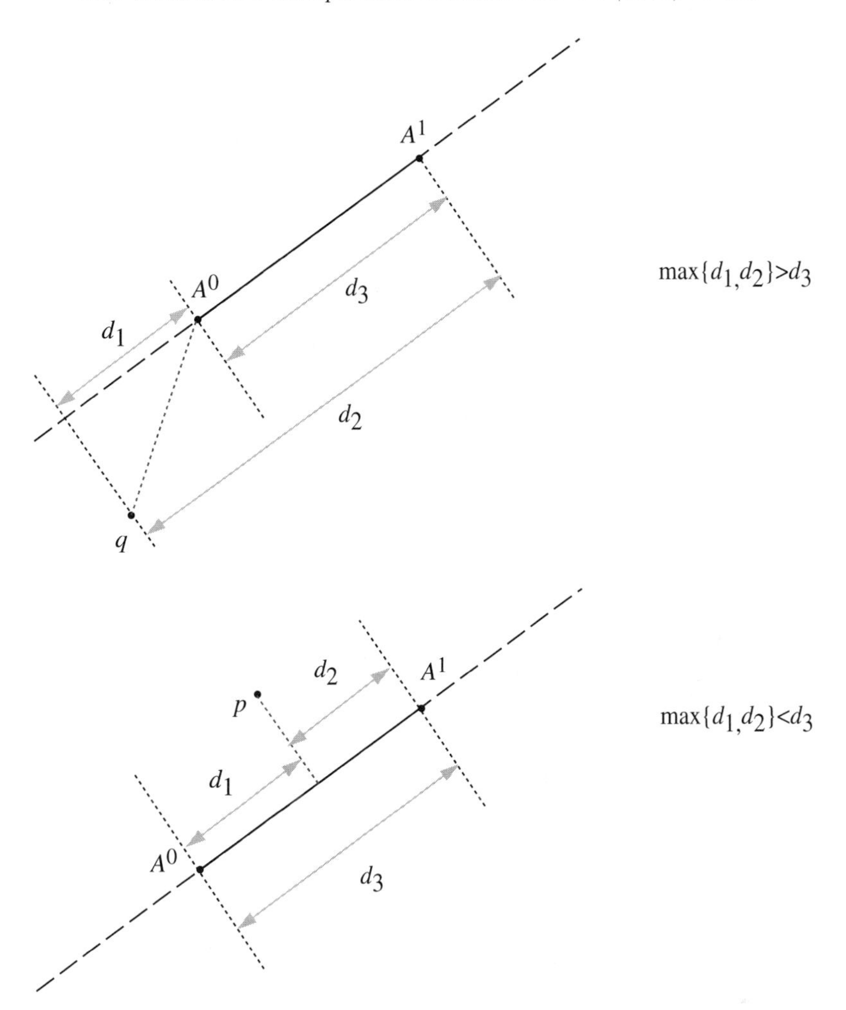

Fig. 4. The distance between a point and a segment.

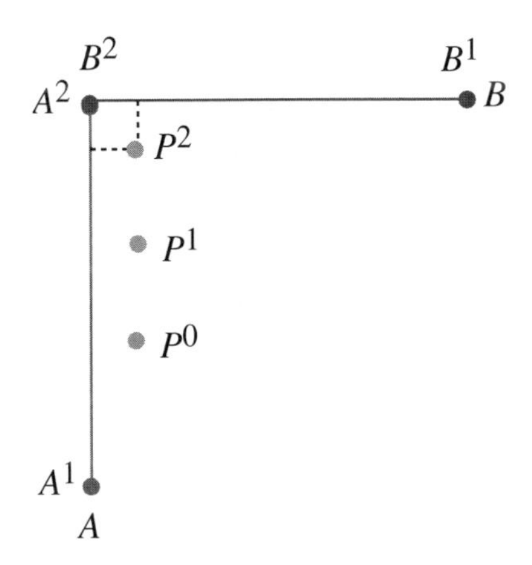

Fig. 5. One problem with point-to-curve matching.

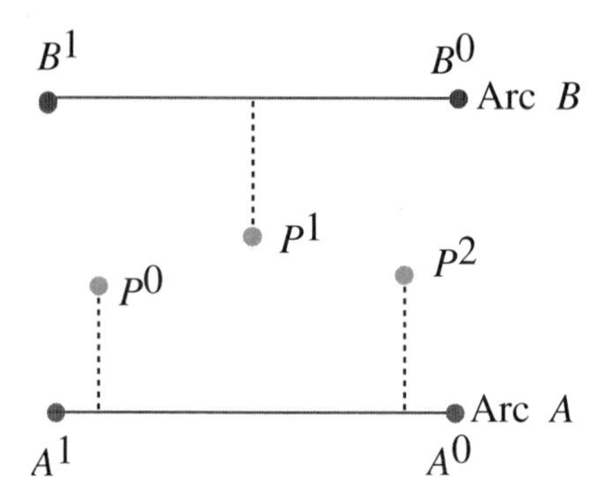

Fig. 6. Another problem with point-to-curve matching.

for example, a GPS tick will not be matched to an arc that is perpendicular to the current direction of travel (as occurred in the example in Fig. 3). Again, this algorithm is included primarily as a straw man.

Algorithm 3 is a variant of Algorithm 2 that uses topological information. In particular, whereas Algorithms 1 and 2 only use a range query to locate candidate nodes (and, hence, candidate arcs), Algorithm 3 also uses connectivity information. Specifically, if the algorithm has confidence in the previous match, it will use the topology of the network to locate candidate nodes for the next match. That is, it will only consider arcs that are reachable (in a topological sense) from the "current" arc. On the other hand, if the algorithm does not have confidence in the previous match, it will use a range query. A match is considered "good" if the error is less than the minimum of 0.15 km and twice the average error in the matches obtained thus far.

Perhaps the easiest way to understand how topological information can be used is to consider the example shown in Fig. 7. Suppose we know that the person was initially at P^0. We then know

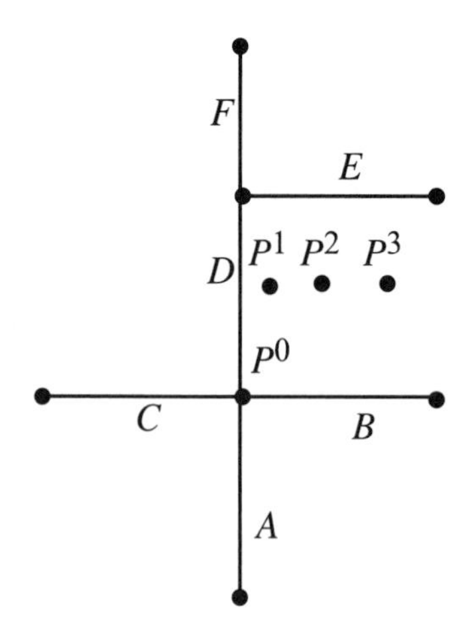

Fig. 7. Using topological information.

that P^1 can only be on A, B, C, or D. In fact, given a sufficiently small amount of time between measurements, we might also know that P^3 can only be matched to A, B, C, or D. This kind of information could prevent us from mistakenly matching P^3 to, say E.

This is, in our opinion, the easiest way to incorporate the physics of motion into a search-based algorithm. While it is not nearly as sophisticated as the filtering algorithms discussed earlier, it does have the potential to preclude many kinds of errors. Of course, it can be quite sensitive to the threshold that is used. That is, one bad match that you have confidence in can lead to a sequence of bad matches. On the other hand, if you make it difficult to call a match good, then it will behave much like Algorithm 2. The threshold we chose (i.e., 0.15 km) was based on an "expected" GPS error of 0.10 m.

Algorithm 4 uses curve-to-curve matching. Firstly, it locates candidate nodes using the same techniques as in Algorithm 3. Then, given a candidate node, it constructs piece-wise linear curves from the set of paths that originate from that node. Secondly, it constructs a piece-wise linear curve using the points $(P^s, \ldots, P^t)$ and calculates the distance between this curve and the curves corresponding to the network. Finally, it selects the closest curve and projects the point onto that curve.

The details of this algorithm are illustrated in Fig. 8. P^1 and P^2, represent the previous and current points, respectively. The point P^0 is a candidate node. The dotted line is the piece-wise linear curve, P, constructed from P^0, P^1, and P^2. The two segments of this piece-wise linear curve are 8 and 5 units in length, respectively.

There are many ways to calculate the distance between two curves. Of particular importance in this context is the way in which curves of different lengths are handled. As shown in Fig. 9, one can either calculate the distance between the two curves using their actual lengths, or calculate the distance between "subcurves" of equal length. We use the latter approach.

Specifically, returning to Fig. 8, to calculate the distance between the curve and A we use the two points a and b. a is the point that is 8 units along arc A starting from P^0 and b is the point that

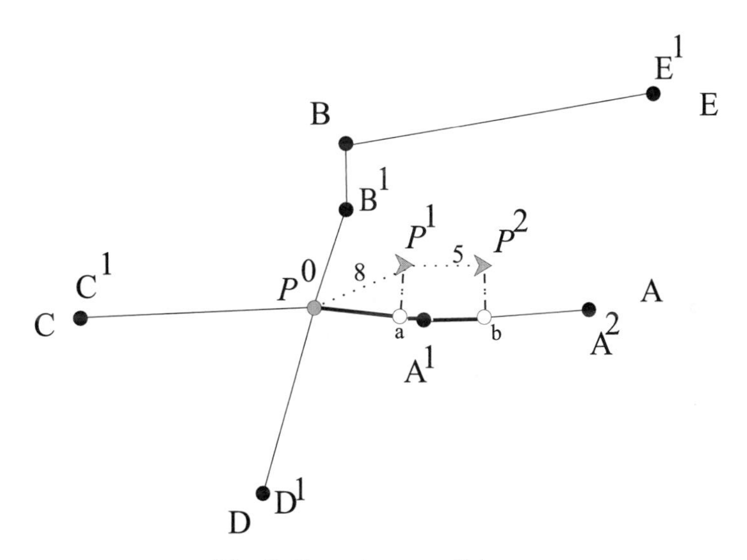

Fig. 8. Curve-to-curve distances.

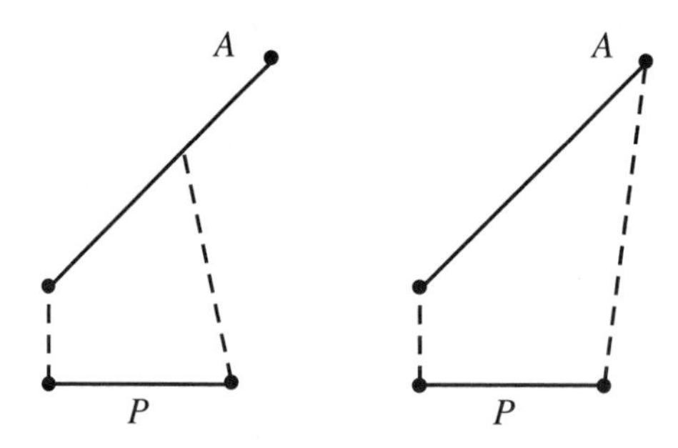

Fig. 9. The distance between curves of different length.

is of 5 units from a along arc A. The distance between P and A is then calculated as $\|P^1 - a\|_2 + \|P^2 - b\|_2$.

A similar process is used for all C and D, and for the other candidate nodes. P^2 is then projected onto the closest.

5. Performance of the algorithms

Unfortunately, there are no standard data sets that can be used to evaluate map matching algorithms. Indeed, very few evaluations have even been reported. This is in part because of the proprietary nature of much of this research and in part because of the difficulty of conducting this kind of evaluation. Hence, we constructed our own test bed of four routes.

Obviously, no conclusions can be drawn from this admittedly limited evaluation. We think the results are interesting nonetheless.

5.1. Data collection

The street network we used for this study was taken from the 1997 TIGER/Line files for Mercer County, New Jersey. We collected data while traveling on four pre-determined routes. All of the routes were "in-town". That is, they did not involve highways or arterials.

We determined our actual location by manually instructing the PNA to record the system time as we entered each link. We determined our estimated location using a GPS receiver that was manufactured by TravRoute using a chipset from Rockwell (12 channels, differential correction was turned off). We ran the receiver in NMEA-0183 mode and used the GPRMC sentences (which includes information about the time, latitude and longitude, speed, heading, and satellite count). We recorded one GPS tick per second, along with the system time when it was generated.

Note that, given this sampling scheme, the data will be spatially biased. Unfortunately, given current technology, it is not possible to obtain a spatially unbiased sample for two reasons. Firstly, the satellites themselves broadcast messages on a pre-determined (temporal) schedule. Secondly, the GPS receiver cannot know when it has traveled a given distance, and, even if it

Table 1
General route information

Route	No. of arcs	Avg. arc length (km)
1	12	0.1706
2	12	0.2249
3	14	0.1928
4	16	0.6083

could, it would not necessarily be able to determine its location at that time. Thus, an algorithm that cannot perform well in the presence of the spatial bias is not a good algorithm for our purposes.

Note also that the GPS receiver did not provide us with raw satellite information; we let the GPS receiver do as much processing as it could in the time available to it. In fact, the GPS ticks we used were actually "smoothed" by the receiver. Of course, this filtering does introduce error that we would prefer to avoid. Unfortunately, it is not clear how to avoid this kind of error. In particular, the system is almost always under-identified (i.e., fewer than four satellites are visible) or over-identified (i.e., more than four satellites are visible). We cannot ignore all such observations because, in practice, they occur frequently (indeed, sometimes for an entire trip). Since every method of handling the identification problem introduces error, we chose to use the filtering scheme that Rockwell built into their chip set.

Table 1 provides some information about the routes that were used. The four routes had between 12 and 16 arcs, most of which had a speed limit of 25 mph. The speed limit was never exceeded while the data were being collected, but it was frequently not realized because of other vehicles on the road. Routes 1, 2, and 3 have similar mean arc lengths. Route 4 has a considerably longer mean arc length because it included a few "long" links.

5.2. The evaluation metric

To evaluate the algorithms we calculated the percentage of correct matches on a given route. That is, each estimated location was matched to an arc in the network. The match was deemed to be correct if that arc was actually being traversed at that time and it was deemed to be incorrect otherwise. While this metric is clearly not ideal, it is difficult to develop a better alternative because it is very difficult to measure the PNA's "true" location at each point in time.[3]

5.3. Results

Unfortunately, space prevents us from presenting maps of all of the results. We will, instead, try to summarize them. Given the small number and limited variability of routes in the field test, we

[3] Of course, it is possible to record the PNA's true location on a closed test track. Unfortunately, the only test tracks available to us were small and had simple topologies. Almost all algorithms work well on test tracks.

have chosen not to present a statistical analysis of the data. Indeed, we caution against drawing any strong conclusions from these results.

Table 2 displays, for each of the four pre-planned routes used in the study, the percentage of correct matches attained by each algorithm.

Overall, the best algorithm only correctly matches between 66% and 86% of the GPS ticks, however, this is not as bad as it might sound. Recall that one GPS tick was generated each second. This means that relatively more ticks are generated near intersections (because speeds are always lower near intersections and are sometimes zero) and the map matching problem is much more difficult near intersections since several arcs are very close to each intersection. In addition, the GPS receiver itself performs much more poorly near intersections because the speed of the vehicle is lower. Hence, the error in the GPS ticks is much larger at intersections.

This is relatively easy to see on a map. Fig. 10 contains a portion of the arcs that compose Route 1. In the figure, there are two sets of arrow-heads. The lighter arrow-heads represent the GPS ticks and the darker arrow-heads represent the map matched locations that were produced by Algorithm 1. As you can see, most of the problems occur at intersections.

Table 2
Matching rates for route/algorithm pairs

Algorithm	Route 1	Route 2	Route 3	Route 4
1	0.534	0.677	0.618	0.608
2	**0.663**	**0.736**	0.855	0.681
3	0.661	0.707	**0.858**	0.664
4	0.617	0.726	0.771	**0.687**

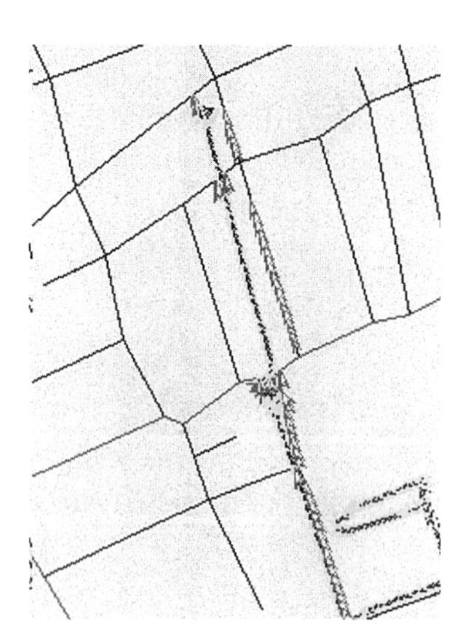

Fig. 10. Matching results of Algorithm 1 on a portion of Route 1.

As expected, Algorithm 1, has the worst performance. In contrast, Algorithm 2, performs reasonably well. Oddly, Algorithm 3 does not out-perform Algorithm 2. Referring to the figures, Algorithm 3 achieves the best performance on Route 3 (slightly outperforming Algorithm 2), the second best performance on Route 1 (slightly less than Algorithm 2), and the third best on Routes 2 and 4. The most complex algorithm, Algorithm 4, does not consistently out-perform either Algorithm 2 or Algorithm 3.

Part of the difference in performance is clearly due to performance at intersections. Fig. 11 displays the same "ticks" on the same portion of Route 1 as in Fig. 10 except that the darker arrow-heads in this figure were produced by Algorithm 2. As you can see, these two algorithms had comparable performance away from the intersections. However, because Algorithm 2 uses heading information, it performs much better near intersections, even though the heading information is sometimes inaccurate at lower speeds. This result holds in general.

We were, to say the least, disappointed by the relatively poor performance of Algorithms 3 and 4. In spite of our expectations, they did not out-perform Algorithm 2, which is considerably simpler.

One is next led to ask when the algorithms performed well and when they performed poorly. We considered six factors: the length of the matched arc, the speed of the PNA, the distance from

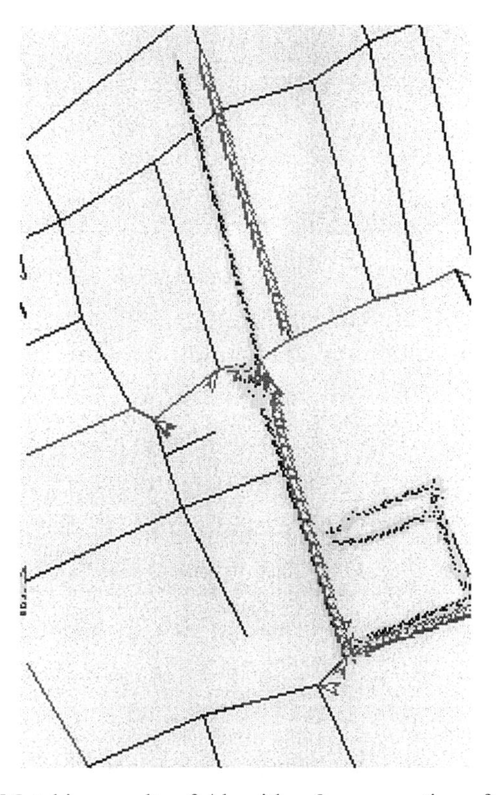

Fig. 11. Matching results of Algorithm 2 on a portion of Route 1.

Table 3
Detailed performance of Algorithm 1 (point-to-curve matching)

Algorithm 1						
Attribute	Match		Route 1	Route 2	Route 3	Route 4
Arc length (km)	Correct	Mean	0.246	0.290	0.268	1.12
		S.D.	0.074	0.089	0.072	1.12
	Incorrect	Mean	0.146	0.239	0.191	0.345
		S.D.	0.098	0.116	0.115	0.375
Speed (mph)	Correct	Mean	21.8	21.5	22.2	32.8
		S.D.	5.43	5.92	4.69	14.4
	Incorrect	Mean	15.8	17.8	15.2	32.4
		S.D.	6.16	7.12	6.16	38.1
Closest arc (km)	Correct	Mean	0.025	0.020	0.021	0.025
		S.D.	0.019	0.023	0.019	0.022
	Incorrect	Mean	0.020	0.023	0.020	0.079
		S.D.	0.013	0.022	0.015	0.096
Next closest arc (km)	Correct	Mean	0.065	0.061	0.069	0.144
		S.D.	0.027	0.032	0.032	0.087
	Incorrect	Mean	0.040	0.037	0.039	0.122
		S.D.	0.019	0.024	0.019	0.107
Serial correlation	Correct	Mean	10.3	18.6	12.8	16.7
		S.D.	2.13	2.02	1.92	4.33
	Incorrect	Mean	8.18	8.88	7.91	11.5
		S.D.	3.43	3.12	2.49	3.55
Closest intersection (km)	Correct	Mean	0.073	0.076	0.079	0.173
		S.D.	0.042	0.047	0.047	0.127
	Incorrect	Mean	0.040	0.035	0.038	0.063
		S.D.	0.022	0.025	0.024	0.060

the GPS tick to the matched arc, the distance from the GPS tick to the next best arc, the serial correlation in the GPS ticks, and the distance from the GPS tick to the nearest intersection. This information is summarized in Tables 3–6.

As you can see, all algorithms were more likely to produce an incorrect match when the PNA is traversing a relatively short road segment. This is almost certainly because the PNA is never far from an intersection.

As one might expect, all of the algorithms worked better with "better" GPS ticks (i.e., when the distance between the location approximation and the closest arc is relatively small). Note, however, that because of errors in the map database, this does not necessarily mean that a more accurate GPS receiver (e.g., a DGPS receiver) will yield better results.

Speed can also play a role and can be associated with arc length. Indeed, correct matches tend to occur at greater speeds than incorrect matches. This may simply be because the mean speed of travel is higher on longer arcs (and, hence, the GPS readings tend to be better). However, in the case of Algorithm 4, it may also be a direct result of the inclusion of topological information. Oddly, however, Algorithm 4 seems to perform badly at high speeds on Route 4.

Table 4
Detailed performance of Algorithm 2 (point-to-curve matching supplemented with heading information)

Algorithm 2						
Attribute	Match		Route 1	Route 2	Route 3	Route 4
Arc length (km)	Correct	Mean	0.223	0.284	0.246	0.098
		S.D.	0.084	0.094	0.097	1.043
	Incorrect	Mean	0.153	0.246	0.192	0.466
		S.D.	0.111	0.115	0.094	0.706
Speed (mph)	Correct	Mean	21.2	21.8	20.4	30.9
		S.D.	5.70	5.62	5.85	11.0
	Incorrect	Mean	14.7	16.0	14.3	36.5
		S.D.	5.79	7.08	6.21	43.6
Closest arc (km)	Correct	Mean	0.030	0.023	0.028	0.032
		S.D.	0.020	0.024	0.022	0.032
	Incorrect	Mean	0.048	0.042	0.050	0.175
		S.D.	0.073	0.061	0.072	0.197
Next closest arc (km)	Correct	Mean	0.073	0.066	0.072	0.186
		S.D.	0.031	0.035	0.028	0.143
	Incorrect	Mean	0.093	0.066	0.070	0.259
		S.D.	0.10	0.081	0.099	0.23
Serial correlation	Correct	Mean	10.7	12.5	16.3	20.0
		S.D.	1.91	2.86	3.52	3.90
	Incorrect	Mean	5.00	4.46	2.75	9.36
		S.D.	3.04	1.65	0.94	3.48
Closest intersection (km)	Correct	Mean	0.065	0.076	0.063	0.145
		S.D.	0.044	0.048	0.049	0.116
	Incorrect	Mean	0.054	0.047	0.038	0.094
		S.D.	0.076	0.064	0.072	0.147

6. Conclusions and future research

In this paper, we have described several algorithms (or parts of algorithms) for matching an estimated position to a network representation of the street system and attempted to evaluate four of them. Obviously, more work needs to be done.

Firstly, more attention needs to be given to how different algorithms can be compared empirically. This is a particularly thorny problem because it is quite difficult to measure the "true position" outside of a laboratory or test track. In addition, it is not immediately clear what measures of performance are most appropriate or what scenarios should be evaluated. In an abstract sense, it is clear that we would like the algorithm to perform perfectly when the errors go to zero, but it is not entirely clear what that means in practice.

Secondly, a wider variety of algorithms need to be evaluated. Particular attention needs to be given to the problems that arise at intersections. While there are many things that can be tried (including turning the algorithm off when speeds are very low), it is not immediately obvious what approach will work best. In addition, one can argue that intersections are the most important portion of the network since most route changes occur at intersections. We are currently in the

Table 5
Detailed performance of Algorithm 3 (point-to-curve matching supplemented with heading information and connectivity)

Algorithm 3						
Attribute	Match		Route 1	Route 2	Route 3	Route 4
Arc length (km)	Correct	Mean	0.219	0.279	0.250	0.942
		S.D.	0.082	0.092	0.098	1.02
	Incorrect	Mean	0.163	0.267	0.177	0.569
		S.D.	0.117	0.117	0.073	0.846
Speed (mph)	Correct	Mean	21.2	21.3	20.1	31.1
		S.D.	5.78	5.84	5.97	11.1
	Incorrect	Mean	15.6	18.4	16.6	35.7
		S.D.	5.78	7.60	7.00	42.6
Closest arc (km)	Correct	Mean	0.029	0.019	0.029	0.033
		S.D.	0.020	0.021	0.023	0.033
	Incorrect	Mean	0.035	0.058	0.051	0.158
		S.D.	0.032	0.062	0.066	0.178
Next closest arc (km)	Correct	Mean	0.106	0.116	0.119	0.639
		S.D.	0.059	0.084	0.059	0.633
	Incorrect	Mean	0.079	0.105	0.098	0.258
		S.D.	0.053	0.104	0.105	0.217
Serial correlation	Correct	Mean	10.3	16.9	19.3	22.8
		S.D.	1.81	2.69	3.35	3.33
	Incorrect	Mean	4.85	7.00	3.20	12.5
		S.D.	3.01	2.75	2.46	3.45
Closest intersection (km)	Correct	Mean	0.061	0.071	0.059	0.288
		S.D.	0.04	0.046	0.045	0.385
	Incorrect	Mean	0.033	0.044	0.019	0.084
		S.D.	0.039	0.059	0.056	0.140

process of developing and testing both statistical algorithms and search-based algorithms that make use of additional information (e.g., speeds and speed limits, preferences for turns of different types, and road classification).

Thirdly, we do not yet know whether these results will hold for other filtering schemes. It may be that, in practice, different map matching algorithms will need to be used with different kinds of filtering schemes (and, hence, hardware). This requires further study.

Fourthly, the algorithms need to be evaluated on a wider array of routes. Our discussion here focused on in-town routes rather than highway routes because most algorithms work well on highways. However, more in-town routes should be considered. In addition, it is important to consider truly urban routes because of the problems that arise in "urban canyons". It is also important to consider the impact of weather conditions, season (i.e., the amount of foliage that blocks the view of the sky), time of day, and other factors. The limited number and variety of routes considered here obviously prevent us from drawing any real conclusions.

Finally, more work needs to be done on the ways in which mistakes in map matching influence the overall performance of the PNA. In some situations, directions do not change much as a result

Table 6
Detailed performance of Algorithm 4 (curve-to-curve matching)

Algorithm 4						
Attribute	Match		Route 1	Route 2	Route 3	Route 4
Arc length (km)	Correct	Mean	0.233	0.288	0.250	1.04
		S.D.	0.081	0.087	0.094	1.08
	Incorrect	Mean	0.147	0.238	0.199	0.347
		S.D.	0.106	0.124	0.104	0.416
Speed (mph)	Correct	Mean	21.4	21.2	21.0	31.5
		S.D.	5.30	6.12	5.64	10.7
	Incorrect	Mean	16.0	19.2	15.8	36.0
		S.D.	6.19	6.51	5.78	44.3
Closest arc (km)	Correct	Mean	0.03	0.021	0.028	0.031
		S.D.	0.022	0.022	0.022	0.031
	Incorrect	Mean	0.035	0.037	0.051	0.166
		S.D.	0.028	0.027	0.058	0.153
Next closest arc (km)	Correct	Mean	0.00	0.00	0.00	0.00
		S.D.	0.00	0.00	0.00	0.00
	Incorrect	Mean	0.00	0.00	0.00	0.00
		S.D.	0.00	0.00	0.00	0.00
Serial correlation	Correct	Mean	9.67	15.6	10.8	21.5
		S.D.	2.14	3.17	2.58	4.48
	Incorrect	Mean	5.54	5.9	3.19	9.77
		S.D.	2.78	2.25	1.10	4.00
Closest intersection (km)	Correct	Mean	0.067	0.076	0.065	0.289
		S.D.	0.045	0.048	0.046	0.379
	Incorrect	Mean	0.027	0.027	0.036	0.125
		S.D.	0.037	0.03	0.063	0.153

of small errors in the map matched location. In other cases, the directions change dramatically. Hence, work needs to be done both on more robust path-finding algorithms and on varying the map matching algorithm in different situations.

Acknowledgements

This research was sponsored by the NJ Center for Transportation Information and Decision Engineering (www.njtide.org) and the NJ Commission on Science and Technology. The authors would like to thank the anonymous referees for helpful comments on an earlier draft of this paper.

References

Abousalem, M.A., Krakiwsky, E.J., 1993. A quality control approach for GPS-based automatic vehicle location and navigation systems. In: Proceedings of the Vehicle Navigation and Information Systems Conference, pp. 466–470.

Bentley, J.L., Maurer, H.A., 1980. Efficient worst-case data structures for range searching. Acta Inf. 13, 155–168.

Blackwell, E.G., 1986. Overview of differential GPS methods. Global Positioning Sys. 3, 89–100.

Collier, W.C., 1990. In-vehicle route guidance systems using map matched dead reckoning. In: Proceedings of IEEE Position Location and Navigation Symposium, pp. 359–363.

Degawa, H., 1992. A new navigation system with multiple information sources. In: Proceedings of the Vehicle Navigation and Information Systems Conference, pp. 143–149.

Deretsky, Z., U. Rodny, 1993. Automatic conflation of digital maps: how to handle unmatched data. In: Proceedings of the Vehicle Navigation and Information Systems Conference, pp. A27–A29.

Fuchs, H., Kedem, Z.M., Naylor, B.F., 1980. On visible surface generation by a priori tree structures. Comput. Graphics 14, 124–133.

Hofmann-Wellenhoff, B., Lichtenegger, H., Collins, J., 1994. GPS: Theory and Practice. Springer, Berlin.

Iwaki, F., Kakihari, M., Sasaki, M., 1989. Recognition of Vehicle's Location for Navigation. In: Proceedings of the Vehicle Navigation and Information Systems Conference, pp. 131–138.

Jo, T., Haseyamai, M., Kitajima, H., 1996. A map matching method with the innovation of the Kalman filtering. IEICE Trans. Fund. Electron. Comm. Comput. Sci. E79-A, 1853–1855.

Kim, J.-S., 1996. Node based map matching algorithm for car navigation system. In: Proceedings of the International Symposium on Automotive Technology and Automation, pp. 121–126.

Krakiwsky, E.J., Harris, C.B., Wong, R.V.C., 1988. A Kalman filter for integrating dead reckoning, map matching and GPS positioning. In: Proceedings of IEEE Position Location and Navigation Symposium, pp. 39–46.

Mattos, P.G., 1994. Integrated GPS and dead reckoning for low-cost vehicle navigation and tracking. In: Proceedings of the Vehicle Navigation and Information Systems Conference, pp. 569–574.

Scott, C.A., Drane, C.R., 1994. Increased accuracy of motor vehicle position estimation by utilizing map data, vehicle dynamics and other information sources. In: Proceedings of the Vehicle Navigation and Information Systems Conference, pp. 585–590.

Schiff, T.H., 1993. Data sources and consolidation methods for creating, improving and maintaining navigation databases, In: Proceedings of the Vehicle Navigation and Information Systems Conference, pp. 3–7.

Shibata, M., 1994. Updating of digital road map. In: Proceedings of the Vehicle Navigation and Information Systems Conference, pp. 547–550.

Tanaka, J., 1990. Navigation system with map-matching method. In: Proceedings of the SAE International Congress and Exposition, pp. 45–50.

Watanabe, K., Kobayashi, K., Munekata, F., 1994, Multiple sensor fusion for navigation systems. In: Proceedings of the Vehicle Navigation and Information Systems Conference, pp. 575–578.

PERGAMON

Transportation Research Part C 8 (2000) 109–127

TRANSPORTATION RESEARCH PART C

www.elsevier.com/locate/trc

A query resolution engine to handle a path operator with multiple paths

Michel Mainguenaud *

Laboratoire Perception, Système et Information, Institut National des Sciences Appliquées (INSA), Site du Madrillet, Avenue de l'Université, 76800 Saint Etienne du Rouvray, France

Abstract

An end-user query of a Geographical Information System (GIS) can formally be defined as the application of a set of operators (spatial or not). Geographical Information Systems used for Transportation (GIS-T) must provide a path evaluation operator. For example, an end-user query may involve a selection based on alphanumeric criteria, an evaluation of path, and a spatial intersection. This composition of operators and the fact that the evaluation of a path may not provide a unique result impose the definition of a query resolution model or a database query language able to support this composition. In this paper we present a query resolution model. The use of multi-criteria analysis and the definition of aggregates in a query (nearly mandatory) may involve ambiguities in the final presentation of the results to an end-user. The formal modeling of query results must take into account this risk. The philosophies of a query definition and the presentation of the results may be different (e.g., formular vs visual). The management of query results must take into account the data model associated with the query results, the results themselves (metabase/database), and the interpretation. The interpretation avoids errors due to visual ambiguities of an operator with an aggregate function.

Keywords: Database; Query language; Path management; Query resolution engine; Geographical information system

1. Introduction

Many efforts are being undertaken to elaborate innovative solutions for the representation and exploration of complex database applications (Medeiros and Pires, 1994). In the context of geographical databases, several spatial data models have been identified. Many proposals have been advanced to query these databases (Aufaure-Portier and Trepied, 1996; Di Loreto et al.,

* Fax: +33-2-32-95-97-08.

E-mail address: michel.mainguenaud@insa-rouen.fr (M. Mainguenaud).

0968-090X/00/$ - see front matter
PII: S0968-090X(00)00018-8

1996; Egenhofer, 1994; Kirby and Pazner, 1990; Meyer, 1992; Tsou and Buttenfield, 1996; Woodruff et al., 1995). Graph structures are introduced into geographic database models to handle networks. Network structures are particularly useful to represent physical or influence relationships in space. Physical networks include the distribution of electricity, gas, water or telecommunication resources. Influence networks describe economical or social patterns in space. Network structures can be also used to represent spatial navigation processes in space and time (i.e., a movement, generally a human one, between several locations in space). Such processes are represented throughout cognitive representations of space that integrate complementary levels of abstraction (i.e., large and local scales according to Kuiper, 1978). In this case, network nodes represent symbolic and discrete locations in space and edge displacements between these locations. A classic logical architecture of a Geographical Information System (GIS) is presented in Fig. 1. Three main levels and their interactions are presented in the figure.

The User Interface (UI) level is responsible for the query definition process and the result display. These two tasks may follow different approaches. As an example, the query definition process may be performed through an interface based on click boxes (e.g., a formular). The result display process may be performed through an interface based on a visual language. There is no reason to constrain the query definition process to be performed on the same hardware as the query display (e.g., a query defined with real buttons and a visualization of the result on a screen). The Query Resolution Engine (QRE) is responsible for transforming an end-user query into orders that can be understood by the Data Base Management System (DBMS). The allowed expressive power (i.e., the classes of queries an end-user can define) of the UI must be greater than the DBMS is able to accept. Otherwise, the UI level is not a UI level but a simple interface to the DBMS. Naturally, the two expressive powers (i.e., the query language of the UI and the database query language) must be compatible. A solution based on a UI with operators and a predicate-based query language for the DBMS does not make sense. For example, the UI may accept in a single order the composition of operators. Because the DBMS does not accept the composition, the QRE is in charge of providing the software interface that simulates the composition. Let us consider for the time that the DBMS is only in charge of storing data and providing tools to

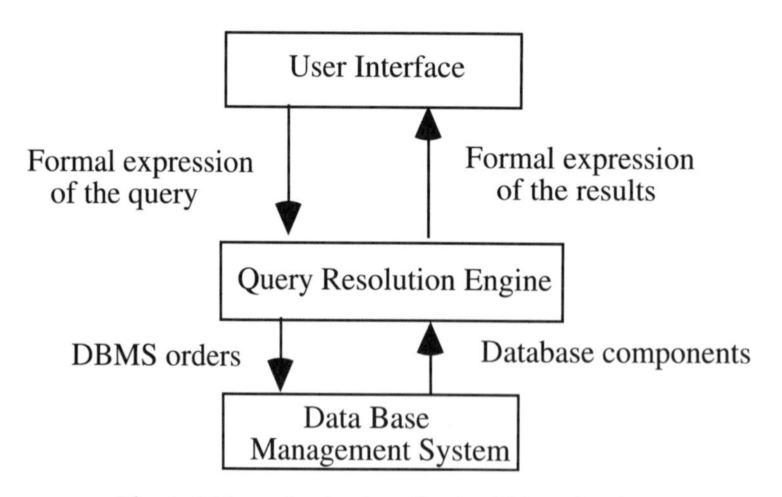

Fig. 1. Different levels of studies for GIS applications.

retrieve this data. We are concerned in this research with the links between the UI level and the QRE.

Geographical Information System for Transportation (GIS-T) represents a class of GIS problems. This kind of application focuses on geographic data that can be organized with or without a spatial representation (i.e., the management of a network). Applications can manipulate a network without any knowledge (or need of knowledge) of the spatial locations (i.e., spatial coordinates) of the different entities. The manipulation is based on the concept of graphing without a spatial component. This property allows for the development of applications on very cheap hardware devices since a unique vision of alphanumeric data is required. For example, planning and management of a public bus system involves a list of buses and connections linked to a timetable and does not require a two-dimensional (2D) representation. Most of the time, a 2D representation of a network is schematic and does not respect scale. In richer applications where scale and positional accuracy are important, the logical representation of a network and its environment are merged. The quantity of data available is more important and provides the opportunity to define more complex queries than "I would like to go from place A to place B". In this case, an end-user query can formally be defined as the application of a set of operators (spatial or not). GIS-T must provide a path evaluation operator. For example, a query may be the application of a selection based on alphanumeric criteria (e.g., a route with the characteristic "highway"), an evaluation of a path (e.g., from a town named "Paris" to a town named "Nice"), and a spatial intersection operator (e.g., the path must cross a forest) to provide a unique end-user query (e.g., a route from Paris to Nice which only uses a highway and crosses a forest).

Several spatial database models and query languages have been proposed to represent the properties of networks – whether in space or not (Angellacio et al., 1990; Car and Frank, 1994; Christophides et al., 1996; Cruz et al., 1987; Erwig and Güting, 1994; Timpf et al., 1992). The definition of a data manipulation language that operates, organizes, and presents a set of network queries within their geographic context remains an important research challenge to date. Database query languages are mainly based on a logic of predicates that restrict a query result to a set of tuples or objects. The projection operator restricts a query result to some of the attributes. (Note that a projection operator is also defined within object-oriented database query languages as an operator which restricts resulting attributes, hides and eventually renames or redefines object properties.) To extend the semantics of the projection operator, aggregate functions have been proposed (e.g., sum, average, maximum, minimum, count). These functions can be integrated within the projection operator in order to extend the semantics delivered by the query results. In a GIS context, a query result combines network, spatial, and alphanumeric properties. For example, routes of air flights present a network with a map as a background. Nodes have relevant geographic locations but the edges are symbolic in the sense that they do not represent the real traveling routes of the planes.

A first extension proposed to handle spatial data in query languages is the introduction of spatial predicates (e.g., in the "where" clause of an extended SQL-like query language). However, the semantics of these languages is not adapted to the complexity of geographic applications. In this domain, operations are often oriented toward the spatial and logical manipulations of entities. The introduction of spatial-oriented (or network-oriented) operators improves the benefit of spatial queries. New alphanumeric and spatial semantics may be derived from their application

(Barrera and Buchmann, 1981; Mainguenaud, 1994; Claramunt and Mainguenaud, 1999). In this paper, we are concerned with the composition of operators as a query.

The consequences of the composition of operators influence two levels: the query modeling process and the result modeling process. The scope of the work reported here is limited to the conceptual modeling of such interactions, at the exclusion of the physical organization with various technologies such as Corba.

The composition of operators may lead to a situation in which the same operator with the same argument(s) is used several times in the same query. For example, a route from Paris to Nice crosses a forest and borders a polluted area (i.e., the same operator – the path operator), or a polluted area must border the non-forest part of a town as well as the forest part of this town (i.e., the same arguments – the town and the forest). Such queries can no longer be modeled with an algebraic tree (as it is in conventional DBMS) so that one must resort to a Directed Acyclic Graph (DAG). Even with the current extensions defined in SQL3, we cannot express the fact that the argument may be used by several operators during the evaluation: for instance, the interweaving of an SQL statement in a "from" clause of another SQL statement, while the inner statement does not have a "from" clause. Two solutions may be advanced to take this constraint into account. In the first solution, the composition is introduced in the query language of the database. The composition is provided by the "from" clause (Haas and Cody, 1991) or in the "select" clause (Larue et al., 1993). This approach enables optimization at the database level. The second solution is the definition of a query resolution mechanism to handle the composition outside the DBMS (i.e., a QRE). The latter proposition is more portable than the former in view of the fact that the DBMS may not support the composition. This approach is followed in this research.

The QRE manages a set of operators. The expressive power of the underlying graph model used to represent a network is not considered. This graph model is very general in that it can manage planar or non-planar graphs, acyclic graphs or non-acyclic graphs, loops or non-loops, and as few or as many levels of abstraction as needed. Whereas the expressive power of the graph model is DBMS-dependent, the QRE is graph model independent. Therefore, several different formalisms can be used to define the graph model (e.g., object-oriented, extended relational with abstract data types).

The expressive power of the underlying path operator used to evaluate a path needs not be considered. The path operator can manage simple paths or complex paths, as well as variable levels of abstraction. The expressive power of the path operator is DBMS-dependent but the QRE is path operator-independent. One imposes the single restriction that the path operator must provide zero, one or several paths as an answer.

The conceptual approach of a QRE is therefore more powerful than an approach based on a single answer path operator (e.g., a shortest path). The difference stems from the fact that multiple paths exist between a given path origin and a given path destination. To provide only a unique path as an answer does not change the conventional querying of a GIS. From a conceptual point of view, the path operator is indeed similar to a spatial intersection: from a pair of arguments, a single answer is provided.

Queries commonly use aggregate measures (e.g., the total duration of a trip must be less than 10 units). These aggregates and the use of multi-criteria analysis may involve ambiguities in the final presentation of the results to an end-user. The true result is a sub-set of the Cartesian product of data involved in the global result due to the aggregate functions. The formal modeling of query results must take into account this risk (e.g., inter-dependent applications). The philosophies of a query definition and the presentation of the results may be different (e.g., formular vs visual). The management of query results involves with three components. The first is the data model associated with the query results. The second entails the results themselves (metabase/database). The last is the interpretation to avoid errors due to visual ambiguities of an operator with an aggregate function.

The remainder of this paper is organized as follows. The second section presents the query modeling. This is the first step to enable GIS-T database manipulations. The modeling of the query results and their basic manipulations is discussed in the following section. This is the first step to achieve data visualization to end-users. Conclusions of this work are drawn in Section 4.

2. Query modeling

An end-user query involves several (spatial or network-oriented) operators. An operator of the user's level may require the application of several database operators since the UI and the DBMS do not manipulate at the same level of abstraction. The initial query may require the process of re-writing rules to transform the abstraction level of the initial query into the abstraction level of the DBMS (e.g., the application of an overlay operator).

A very convenient way of representing a query is the use of a functional language. For instance, in SQL, a Select From Where statement can be used in the "Where" clause of another "Select From Where" statement. The formal representation of a query is an algebraic tree: in the case of a relational algebra, leaves are the relations, intermediate nodes are the operators and the root is the result.

In our case, this tree structure must be extended to handle a composition of operators. Therefore, the structure to model a query is no longer a tree but a DAG since the same operator with the same arguments may be used several times in the same query.

Let us consider a GIS-T query applied to a database. A database schema models a transportation network with a graph data structure. To handle huge graphs, several levels of abstraction can be defined (Car and Frank, 1994; Güting, 1991; Mainguenaud, 1995). We do not consider here the expressive power of such a graph data model nor the formal model used to represent a graph (e.g., object-oriented, extended relational model with abstract data types). Application data and their manipulations (e.g., the results of a path operator) are defined on such a structure. An end-user query is transformed into a (set of) database query(ies) that involves one (or mainly several) database manipulation primitive(s). The query modeling is the formal representation merging the three previous concepts (graph, application data and database primitives).

2.1. Graph

Without any loss of generality, we consider here a graph with a single level of abstraction. We require a graph structure to illustrate the examples on transportation data. The way this graph is represented (the data model) and acquired (constructor operators) are not important at all since we are only concerned with the composition of operators in a query: this graph structure is used by a path operator. Therefore, let us use a simple definition of a graph: a graph is formally defined by a set of labeled vertices called nodes and labeled edges. Data can be associated with a graph as

G (N, E, ν, ε, Ψ, γ)
N is a set of nodes (i.e., labeled vertices), $N = \{n_1, \ldots, n_p\}$
E is a set of edges, $E = \{ (n_i, n_j) / n_i \in N, n_j \in N \}$
ν is a labeling function for vertices
(n_i is said to be the head and n_j the tail of an edge (n_i, n_j))
ε is a labeling function for edges
Ψ is an incident function E -> N x N
γ is a labeling function for a graph

Fig. 2. Formal definition of a graph.

a whole. Fig. 2 presents the formal definition of a graph *G* (Cruz et al., 1987). This definition also serves to model a query, in which case nodes are the operators and edges model the arguments (i.e., a conventional representation of a functional expression).

2.2. Application data and their manipulations

The notion of a graph (an abstract data type) will guide us to model the application data, whatever the data model and the semantics of the application are. The formal modeling of application data (database schema) can be defined by one of several approaches (Peckham and Maryanski, 1988; Smith and Smith, 1987; Stonebraker and Moore, 1996; Zdonik and Maier, 1990).

Let us use a traditional complex object notation to denote a sample database: [ ] for aggregations, { } for sets, and ( ) for lists and conventional types integer, string, and float. Let us define an abstract data type (Seshadri et al., 1997; Stemple et al., 1986) to model the cartographic representation of a node (resp. an edge) named SpatialRepresentation_type. Fig. 3 presents a

```
Node_type       =   [   name : string ]
Edge_type       =   [   name : string,
                        head : Node_type, tail : Node_type ]
Graph_type  =   [   name : string,
                        nodes : { Node_type}, edges : {Edge_type} ]
Town_type   =   [   name : string,
                            spatialRepresentation : SpatialRepresentation_type]
Forest_type  =  [   name : string,
                            spatialRepresentation : SpatialRepresentation_type ]
PollutedArea_type  =  [  name : string,
                            spatialRepresentation : SpatialRepresentation_type ]
Road_type       =   [   name : string,
                            origin : Town_type, destination : Town_type,
                            duration : float,
                            spatialRepresentation : SpatialRepresentation_type ]
TransportNetwork_type = [  name : string,
                            nodes : {Town_type}, edges : {Road_type},
                            context : [    forests : {Forest_type},
                                           pollutedAreas : {PollutedArea_type} ] ]
```

Fig. 3. Sample database schema.

Town: { Paris, Dijon, Lyon, Avignon, Marseille, Toulon, Nice },
Forest: {Fontainebleau, Melun}, PollutedArea: {PA1, PA2},
Road

Name	Origin	Destination	Duration
1.1	Paris	Dijon	4
1.2	Paris	Dijon	4
2	Paris	Lyon	2
3	Dijon	Lyon	2
4	Lyon	Avignon	1
5	Avignon	Toulon	2.5
6	Avignon	Marseille	1
7	Toulon	Nice	2.5
8	Marseille	Nice	2

Fig. 4. Database instances (alphanumeric part).

sample database with towns, forests, roads, and polluted areas, while Fig. 4 presents the instances of this database (alphanumeric part). Without any loss of generality, we assume that all links are unidirectional. Fig. 5 presents the logical view of the transportation graph, that is, the relevant relative spatial location between nodes is missing. In this example, one mandatory link exists between Lyon and Avignon in a path from Paris to Nice, whereas several alternative paths can be followed to join Paris to Nice. Hence this is a fairly typical situation in transportation.

In the formal definition of a graph, ν is a function that provides spatialRepresentation to a node; ε is a function that provides duration and spatialRepresentation to an edge. Town_type is a subtype of Node_type, and Road_type is a subtype of Edge_type. TransportNetwork_type is a subtype of Graph_type. γ is a function that provides a visualization context for a transportation network (e.g., with some forests and some polluted areas).

A double spatial representation of transportation data is handled with the ε function. This allows for duration or travel time to be different. The logical representation of a given (origin, destination) pair and of the double spatial representation associated to it is depicted in Fig. 6.

Data manipulations in a GIS-T database can be reduced to four classes. Class (a) conventional database manipulations (i.e., querying ν or ε function) are based on a selection with alphanumeric

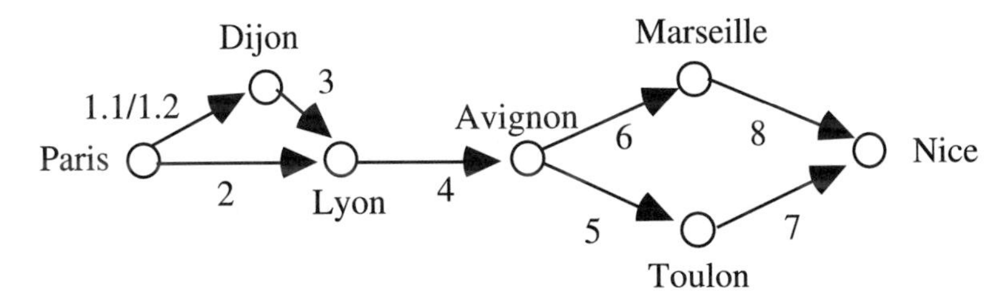

Fig. 5. Logical graph.

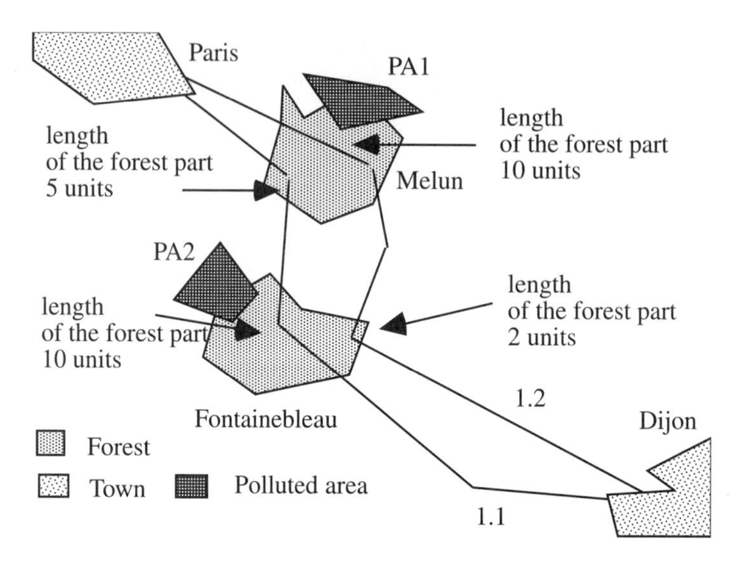

Fig. 6. Spatial representation of the edge (Paris, Dijon).

criteria: for instance, a town with a given name, "what portion of road network *N* is maintained by political entity *P*?" Class (b) conventional GIS data manipulations are based on spatial operators: for instance, an intersection operator for "a road must cross a forest", "Do the bounding boxes of two networks overlap?" This class involves the spatial representation. Class (c.1) is the evaluation of paths with a finite state automata-based query: the road classification must be "secondary", then "highway", and finally "secondary" (Cruz et al., 1987). For this query type, the constraint is evaluated at each edge of a path. Class (c.2) is the evaluation of paths with one or more aggregates: for instance, the total duration is less than 10 units. The constraint is here evaluated for the path as a whole.

Within this classification, class (b) is the only one that requires a spatial representation. Classes (c.1) and (c.2) involve a path evaluation. A path may have two representations, namely a logical one – a graph, and a spatial one. The spatial representation of a path may be used in a class (b) query.

In this paper, we are not concerned with queries of class (a) since they correspond to conventional database manipulation operators. Furthermore, classes (c.1) and (c.2) can be considered as one same class (c) since we are not concerned with the query definition process (e.g., what is the kind of UI? how many paths are required? what is the value of the aggregate?) or the query evaluation process (e.g., what algorithm should be used to evaluate a path? what is the "best" optimization?). To illustrate transportation manipulations on our sample database, let us define the following set of queries to tackles the problems of GIS-T:

Query Q_1: I would like to join Paris to Nice.

Query Q_2: I would like to join Paris to Nice with a total duration less than 10 units.

Query Q_3: Show me a map with a road(s) that crosses, with at least a 10-units length, a forest (with a polluted area) in its non-polluted part?

Query Q_4: Show me a map with a route(s) from Paris to Nice that crosses a forest and borders a polluted area?

Query Q_1 requires the evaluation of a transitive closure on a graph G defined in the database (i.e., class (c)). Query Q_2 has the same requirement extended with an aggregate function (i.e., path evaluation under constraints). Query Q_3 requires the composition of two spatial operators (an intersection applied on the result of a spatial difference), i.e., class (b). Query Q_3 can be generalized as a mix with query Q_1 to involve the path operator (road vs path from Paris to Nice). Query Q_3 represents a vertical composition (i.e., the algebraic representation is a tree). Query Q_4 is also a query with a composition of operators. It represents a horizontal composition (i.e., the algebraic representation is a DAG).

In the following, we consider that some data types are available (e.g., Town_type), and some queries are defined to illustrate the different configurations (e.g., query Q_3).

2.3. Database primitives

The output of the QRE to the DBMS depends on the expressive power of the database query language. The introduction of predicates or operators as primitives of a spatial database query language leads to different expressive powers (and therefore some queries may or may not be expressed).

A spatial predicate (e.g., spatial difference) applied to data provides a Boolean result (e.g., a basic manipulation in a "Where" clause in SQL). As an example, query Q_3 cannot be formulated with a predicate-based language since it requires the results of the spatial difference operator to be able to evaluate whether an intersection exists or not (i.e., using the result of an operator). The application of a GIS to transportation data introduces several consequences for the query language definition:

Operator-based language: The extension to transportation management requires the definition of an operator-based language. The important point in the evaluation of a path between Paris and Nice (e.g., query Q_1) is the components (i.e., edges and nodes) of the path. Information, such as a path that exists (i.e., evaluation with a predicate) is not relevant for GIS-T application (i.e., the result of the path operator instead of a Boolean result). To manage transportation data, GIS must handle other graph manipulation operators such as node/edge/graph manipulations.

Closedness of operators: The underlying notion of a labeled graph requires that a result of a graph manipulation should be compatible with the notion of a labeled graph. This requirement is an extension of spatial data manipulations. Geographic data is composed of at least one pair of alphanumeric and spatial data. A GIS operator must provide as a result a datum with the same structure (a pair of alphanumeric and spatial data). As an extension, a GIS datum for transportation is a geographical datum with an associated graph. The result of an operator must be a geographical datum (i.e., alphanumeric and spatial data) and an associated graph. In the context of a network, the spatial part may be a symbolic representation (i.e., the real world coordinates are not mandatory).

Expressive power: Once the query language provides the concept of operator, the next problem is to define data on which operators are applied. Traditional database query language (e.g., relational algebra) is based on the first order logic without function. To provide a realistic expressive power to develop applications, query languages (e.g., SQL) introduce some basic functions/operators (e.g., count). GIS-T must provide spatial and graph manipulation operators. In parallel with GIS databases, two levels can be defined: with and without a combination of operators. To guarantee the consistency of query results, the combination of operators must be provided with a single database order. Otherwise, an application routine must be provided to simulate a single database order (i.e., the QRE).

In the following, we consider that the expressive power of the database query language is at least an operator-based language providing the closedness of operators and not allowing the composition of operators in the equivalent of the "Select/From" clause.

2.4. Query modeling

In parallel with database languages, two levels of interaction can be defined: logical and operational. A logical order is independent of implementation. An operational order takes into account the different strategies of the database query resolution engine (e.g., indexes, statistics). We are here only concerned with a logical order since operational orders are hardware and software dependent – even if providing an optimized query is a very challenging task.

Several formalisms can be used to model a query, namely functional, procedural, and declarative. Let us use an intuitive functional approach and simplified grammar. A label (i.e., an integer – to be able to use the same operator with the same arguments several times in the same query) identifies an operator. This operator has a name and a list of arguments. For example, query Q_3 can be formalized by:

$$\text{Query } (1 \cap (2 \text{ DB (Road)}, 3\Delta \ (4 \text{ DB (Forest)}, \ 5 \text{ DB (PollutedArea)})))$$

where $\cap$ denotes the spatial intersection operator, Δ is the spatial difference operator, DB is the querying operator of a basic database component, and "*i*" is the label of an operator. Such a formalism can easily be represented with a DAG, without isolated node. With this approach, data and query are both represented with the concept of graph. Fig. 7 presents the definition of a query.

Figs. 8 and 9 portray the algebraic expressions associated with query Q_3 and query Q_4, respectively. The border operator is represented by "$\langle \ \rangle$" and the path operator is represented by "$\rightarrow$".

The definitions of a transportation database and a query are similar. They correspond to a Graph_type type: Database_type = Graph_type and Query_type = Graph_type. The query

$$G\,(N, E, \nu_q, \varepsilon_q, \Psi, \gamma_q)$$

N, E and Ψ have the same definitions than previous

ν_q denotes the labeled function for nodes (e.g., operator, database component constraint)

ε_q denotes the labeled function for edges (e.g., an order in the list of arguments to handle non-symmetric operator such as the path operator)

γ_q denotes the labeled function for a query (e.g., the required database model as a result).

Fig. 7. Definition of a query.

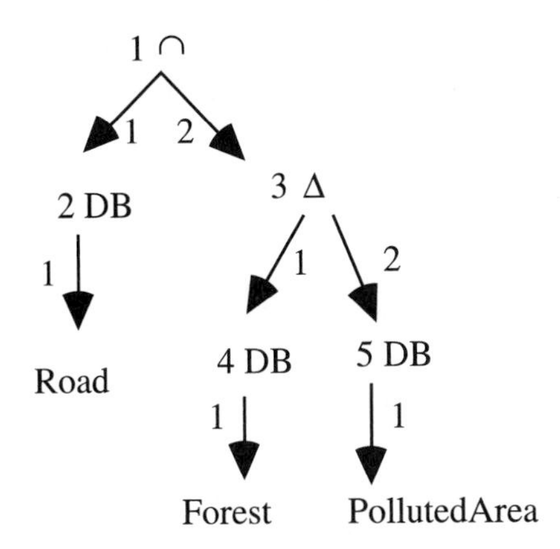

Fig. 8. Algebraic expression of query Q_3.

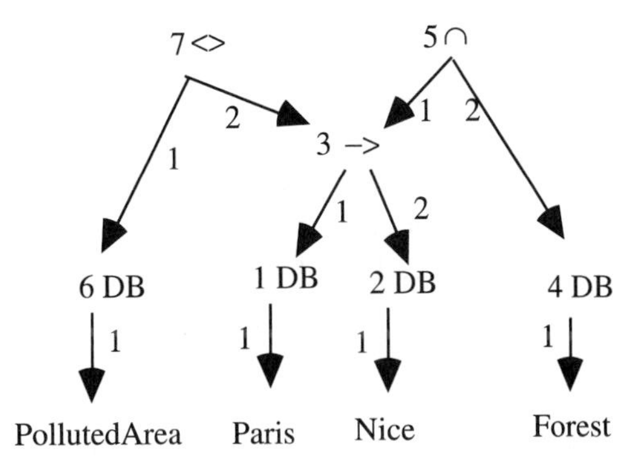

Fig. 9. Algebraic expression of query Q_4.

modeling represents the input of the QRE. The result modeling represents the output. The added value of the QRE is introduced in the output.

3. Result modeling

The evaluation of a path operator in a GIS-T database should provide a set of paths instead of a single path as a result. The QRE is not responsible for the definition of the relevant number of paths to be evaluated. This task belongs to the interface level or the database operator. For example, in a GIS application for tourism management and information, a shortest path may be far from desirable since visitors are more motivated by landmarks and other sightseeing attractions that by efficiency considerations. Furthermore, the less expensive path may also be the longest. In the context of a multi-criteria analysis, it does not seem realistic to define a priori query. Human interaction may be necessary to make a choice from a small set of propositions. To reduce the volume of possible answers, a wide range of aggregate functions ought to be provided during the query definition process (e.g., total travel time under 10 units).

Aggregate functions may involve some ambiguities while the end-user interprets the results of a query. The query result modeling must provide a structure to avoid such ambiguities. To manipulate this structure, some operators must be defined.

3.1. Consequences of aggregate functions

If the result of a path operator in a GIS-T query is considered as a unique graph, modeling the set of paths (i.e., a set-oriented approach) may lead to confusion when an aggregation function is involved in the query. The result must therefore be a set of graphs.

A path is itself a graph, that is a set of nodes and a set of edges, along with some characteristic data (γ function). Let us consider the result presented to the end-user in a single graph (G_r). This graph is built as the union ($\cup$) of the sets of nodes (resp. edges) of each path. Let n be the number of paths $G_i(N_i, E_i, \nu, \varepsilon, \Psi, \gamma_i)$, where $i = 1, \ldots, n$. The result of the path operator is

$$G_r = (\cup N_i, \cup E_i, \nu, \varepsilon, \Psi, \gamma_i) \quad \forall\, i = 1, \ldots, n.$$

This formalism can be provided when no aggregate function is involved in the query (e.g., query Q_1). It cannot be used anymore once an aggregate function is involved (e.g., query Q_2). As an example, the following paths are provided while query Q_2 is evaluated (Fig. 4):

(1) Paris–Dijon–Lyon–Avignon–Marseille–Nice: Duration 10,

(2) Paris–Lyon–Avignon–Marseille–Nice: Duration 6,

(3) Paris–Lyon–Avignon–Toulon–Nice: Duration 8.

The following path:

(4) Paris–Dijon–Lyon–Avignon–Toulon–Nice

does not fulfill the requirements since its total duration is 12, which is greater than 10.

The generation of a graph built with the union of the different paths that fulfill the requirement leads to two difficulties:

The answer is wrong. The union provides the same graph as the one presented in Fig. 5 and therefore, one can infer that path (4) is relevant, This impossibility has been lost by the application of the union operator.

The result for the aggregate function is lost. It must be re-computed as soon as a path is required since the γ function cannot individualize each path. This disappearance is due to the union operator.

A set of paths results from the path operator. Each path in this set is defined independently of the others. The result of a path operator with multiple paths as an answer cannot be managed as a shortest path would be. A path operator is applied between Paris and Nice in our example. Three paths are provided as an answer. This problem can be generalized. The only restriction is to provide a realistic computational time (i.e., several starting places for a single arriving place or the opposite).

3.2. Query results

Query results depend on the expressive power of the language used to define a query. A predicate-based query language leads to a result built as a set of database components (e.g., relations for relational-based DBMS, classes for object oriented DBMS). An operator-based query

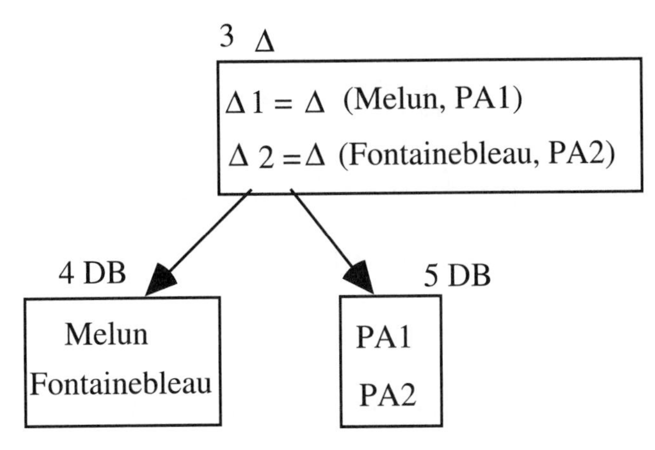

Fig. 10. Query Q_3: a sub-graph for the results.

language leads to a result built as a set of database components and the results of operators defined in the query. The query and the result are both graphs that are isomorphic. The leaf nodes (that is, a node such as it does not exist an edge with this node as a head) of the graph modeling the results are the database components (Fig. 8). The non-leaf nodes are the results of the operators. Each node represents a set of instances. Fig. 10 presents a sub-graph of query Q_3 in Fig. 8 with data from Fig. 6.

A conventional interpretation of the principle is provided with the following rule for a binary operator (Fig. 11): "An instance number k of the result of an operator (e.g., spatial difference, Δ) labeled K is built with an instance number i of an operator labeled I (e.g., database query, DB) and an instance number j of an operator labeled J (e.g., database query, DB)". The sub-query is

$$K\Delta(I\ \mathrm{DB}(\cdots), J\ \mathrm{DB}(\cdots)).$$

The same sub-query may be used several times in the same query (i.e., the graph modeling a query is not a tree). For example, Fig. 9 presents a horizontal composition with an operator. Fig. 12 presents a horizontal composition with a leaf, that is a simplified graph associated with the following query: "a road crosses the non-polluted part of a forest and borders "⟨ ⟩" this polluted area" – i.e., an extension of query Q_3. The functional representation is

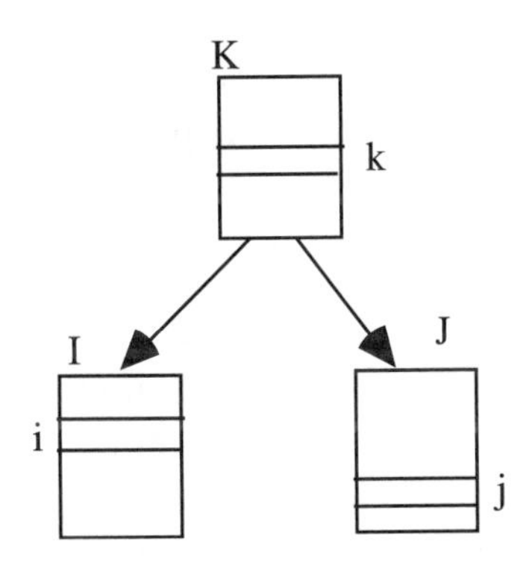

Fig. 11. Example of conventional interpretation (sub-graph).

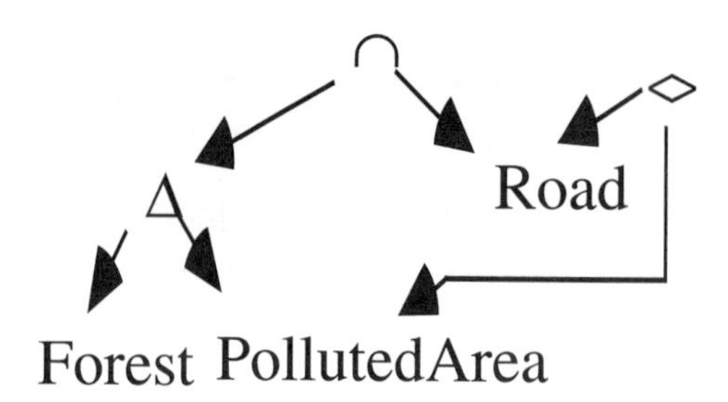

Fig. 12. Query graph.

Query (1 ∩ (2 DB (Road), 3Δ(4 DB (Forest)), 5 DB (PollutedArea)),
6⟨ ⟩(2 DB (Road), 5 DB (PollutedArea))).

The evaluation of such a query is similar to a conjunction of queries. Therefore, the evaluation mechanism will end since only deletions are involved. The deletion of non-relevant data is a recursive application of the evaluation since the query is modeled with a DAG instead of being modeled with a tree. Some arguments may have a reduced set of relevant data when they share the same operator: for instance, a forest may be deleted from the relevant data set if it has a polluted part, but the polluted area is not bordered by a road that crosses this forest.

The interpretation is different for a path operator whose answer is multiple paths because the application of the path operator on an instance number *i* (the origin) and on an instance number *j* (the destination) does not provide a single result (i.e., an instance number *k*) but a set of instances (a set of paths). Since aggregate functions are almost unavoidable in a path evaluation, the result must be a set of graphs (and not a unique graph). This restriction leads to some troubles while defining in the query language the signatures of operators manipulating the result of a path operator since the result is a set of graphs.

In conventional GIS interpretation (Fig. 10), a given pair of entities (say, Melun and PA1) provides a unique answer (resp., $\Delta 1$). When aggregate functions are involved in a query, the result of a path operator can no longer be defined as the union of the different paths. Therefore, from a given pair, several results are provided.

Two modes can be defined depending on the expressive power of the query language: the union mode (a graph as a result) and the elementary-at-a-time mode (a set of graphs as a result). The union mode is appropriate to predicate-based query language or to a query in which no aggregate function is involved. It is also appropriate with a path operator providing a unique result by definition (e.g., a shortest path operator). As discussed above, when an aggregate function is involved in a query (i.e., in a path operator or on the result of a spatial operator), the Union mode can no longer be used: the elementary-at-a-time mode becomes mandatory.

A mode is independent of a path operator. A query is defined as a graph with labeling function γ. The γ function is in charge of defining the relevant mode since its determination can be made before the evaluation or after the evaluation if the result is a unique path. Visual ambiguities may arise from the path operator or from a different operator in the query. Query Q_2 and the generalization of query Q_3 (i.e., with the query Q_1 that does not require an aggregate function) are examples of such queries. Query Q_2 involves an aggregate function in the path operator. The mode must be Elementary-at-a-time to ensure a set of graphs as a result. The generalization of query Q_3 involves a constraint on a spatial intersection operator (i.e., at least a 10-unit length). Fig. 6 illustrates the visual ambiguity that the definition as a union mode arises. The two forests are relevant and so are the two roads cross these forests. But for each road, one of the intersections is not relevant because too short: for the pair (1.1, Melun) the intersection is only 5 units, while for the pair (1.2, Fontainebleau)), it is 2 units.

Since ambiguities may arise, an interpretation structure must be provided to define the relevant instances used as arguments. An interpretation structure defines the relevant instances of the query result (i.e., instances of the database components) and the associated results for each operator. Its organization is therefore similar to a query and can be modeled by a set of graphs (isomorphic to a result graph). Each graph of this set defines an unambiguous combination of database instances that fulfills the query requirements. The manipulation of an instance in the interpretation structure is similar to the manipulation of a result graph in a union mode (they can be therefore substituted). A node contains a pair (label of the operator, identifier of an instance). The semantics of an edge is similar to a result graph one (i.e., an edge links an instance of the result to the instances used as an argument of an operator – Fig. 11). Fig. 13 presents an interpretation structure for query Q_3 (Figs. 8 and 9) with data from Fig. 6. Interestingly, in these graphs, a link does not exist between the road labeled 1.1 (resp. 1.2) and the forest named Melun (resp. Fontainebleau) since the length of the intersection is not long enough. The result is a subset of the Cartesian product of data involved in the global result (see Fig. 10). The definitions of a transportation database, a query, and associated results can be extended as follows:

Result_type = [Data : Database_type,
Interpretation : (Graph_type)].

A query and the query results can be modeled with a graph. The links between the UI and the QRE and vice versa are based on the graph concept. From the UI an exchange structure is defined as follows:

FromUIto QRE_type = [Query : Query_type].

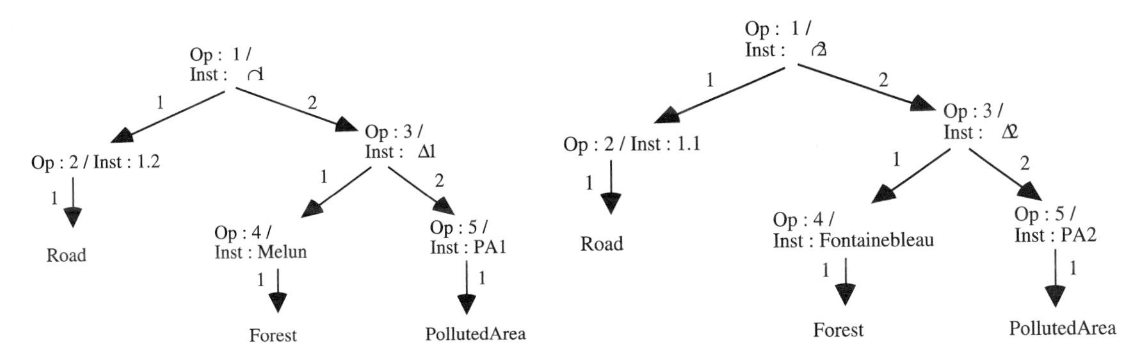

Fig. 13. Interpretation graphs for query Q_3.

From the QRE, an exchange structure is defined as follows:

FromQREtoUI_type = [Query : Query_type,
Result : Result_type].

From this structure of communication, logical manipulations can be defined.

3.3. Logical manipulations

It is useful to provide the formal representation of a query in the FromQREtoUI_type type since the philosophy of the query definition process and the query result display may be different (and the restitution devices, too). This allows a UI in its part of display result to establish a link with the query. For example, the user may query the map obtained as a result but can also do the opposite and have get information from the query (e.g., show me the intersection of a forest and a polluted area in query Q_3).

Logical manipulations depend on the mode of a query. A union mode is similar to a GIS that provides a path operator with a unique result (e.g., a shortest path). The result is a single map. An elementary-at-a-time mode requires the definition of new primitives to manipulate query results. The manipulation is based on the definition of the FromQREtoUI_type type. Within this structure, the UI must be able to manipulate the interpretation field.

A recursive decomposition of the UI level is presented in Fig. 14. The human interface is in charge of communicating with an end-user. The UI Resolution Engine (UIRE) interprets and manages a query or a result manipulation. A result manipulation can be assimilated to a query on a reduced set of data. As the query belongs to the FromQREtoUI_type type, it can be presented with a result. The user can define a "new query" as "I would like to see the results of this sub-part of the query": for example, show me the intersection between the forest and the polluted area in query Q_3. The data management layer handles the FromQREtoUI_type type.

The interaction between the human interface level and the UIRE can be formalized with a conventional set of GIS manipulations, such as zoom or change of the legend. The interpretation graph (Fig. 12) of a query involving manipulations with aggregate function provides several possibilities as a result. From an ordered set of graphs (i.e., a list), conventional manipulations consist of being able to initialize the process (Init), being able to present the next solution (Next),

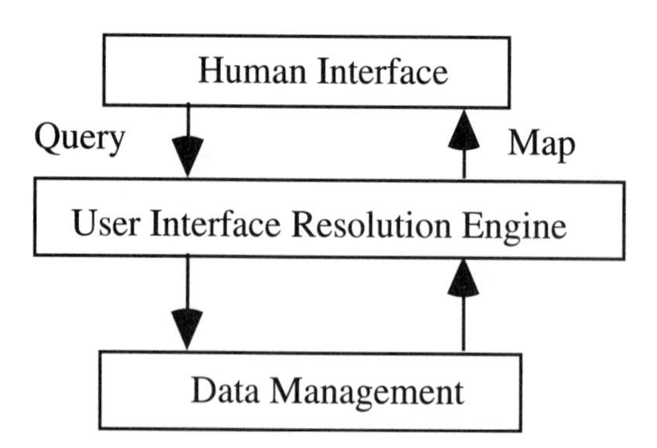

Fig. 14. Decomposition of the UI level.

being able to return to the previous solution (Previous), being able to retain a solution (Handled), being able to select a retained solution (Select), and being able to cancel a "handled" solution (Cancel). A manipulation type, Manipulation_type, is defined to model the relevant operations

$$\text{Manipulation\_type} = (\text{Init, Next, Previous, Handled, Select, Cancel}).$$

Whenever an ambiguous query is presented, several solutions have to be displayed. The basic manipulation has the following signature:

$$\text{DisplayAmbiguousResult} : \text{FromQREtoUI\_type} \times \text{Manipulation\_type} \rightarrow \text{Graph\_type}.$$

The graph obtained by the DisplayAmbiguousResult function can be processed as if the query was in mode Union since no visual ambiguity may occur. The context is similar to a manipulation with a GIS that provides a shortest path operator (i.e., a single path as a result).

4. Conclusion

The large diffusion of spatial database applications in scientific, planning, and business domains leads to the emergence of new requirements in terms of data representation and derivation. Particularly, current spatial data models and query language operations have to be extended in order to integrate a more complete semantic representation of complex domains. The integration and representation of graph structures within spatial data models is of particular interest for many application areas involved in the management of transportation networks.

Transportation data are modeled with the concept of a labeled graph. In this paper, the starting point is an end-user query defined as a set of operators. This query is modeled with the notion of graph. The result of a query is modeled with a graph, which is isomorphic to the query graph. This choice allows an homogeneity between the modeling of the different components of a GIS database manipulation: data, query, and results are modeled with the same concept. To avoid visual ambiguities that may arise once an aggregate function is involved in a query, an interpretation structure is provided to define the relevant instances of query results. This structure is also modeled with a graph (isomorphic to a result graph). The approach is of great significance in transportation applications because aggregate functions are nearly mandatory to reduce the amount of data and to provide a realistic computation time.

The decomposition of a GIS logical architecture into three independent layers (a UI, a QRE and a DBMS) allows for greater flexibility in the management of the expressive power. A formal communication between these components (a graph structure between the interface and the query resolution engine, an extended SQL-query language from the QRE to the DBMS) allows for easily changing one of them. An end-user query is a composition of operators (spatial or not). The aim of a QRE is to fill the gap between the level of abstraction of the UI and the level of abstraction of the DBMS (without composition of operators, with composition that does not allow the same operator with the same arguments to be used several times, and with a total composition). The introduction of a path operator that provides several results requires changing the interpretation of the final result once an aggregate function is involved in the query. Two modes are defined: the union mode (a set oriented approach) and the elementary-at-a-time mode (path by path). New manipulation functions are associated to the elementary-at-a-time mode to handle the risk of visual ambiguities.

The introduction of a GIS as a decision tool requires the opportunity to propose several solutions. The added value of a human being is the decision. The UI must be able to present different solutions without any visual ambiguity. Using the framework of the QRE, several concrete interfaces may be provided depending on the end-user (obviously it requires that the path operator of the database provide several paths as a result instead of a single path).

References

Angellacio, M., Catarci, T., Santucci, G., 1990. QBD*: a graphical query language with recursion. IEEE Transaction on Software Engineering 16 (10), 1150–1163.

Aufaure-Portier, M.A., Trepied, C., 1996. A survey of query languages for geographic information systems. In: Third International Workshop on Interfaces to Databases, Edinbugh, UK, 8–10 July.

Barrera, Buchmann, 1991. Schema definition and query language for a geographical database system. Transactions on Computer Architecture: Pattern Analysis and Image Database Management 11, 250–256.

Car, A., Frank, A., 1994. General principles of hierarchical spatial reasoning – the case of wayfinding. In: Sixth International Symposium on Spatial Data Handling, Edinburgh, Scotland, 5–9 September.

Christophides, V., Cluet, S., Moerkotte, G. 1996. Evaluating queries with generalized path expressions. In: Proceedings of ACM SIGMOD Conference, Montreal, Canada.

Claramunt, C., Mainguenaud, M., 1999. A revisited database projection operator for network facilities in a GIS. Informatica 23, 187–201.

Cruz, I.F., Mendelzon, A.O., Wood, P.T., 1987. A graphical query language supporting recursion. In: Proceedings of ACM SIGMOD Conference, San Francisco, USA, 27–29 May.

Di Loreto, F., Ferri, F., Massari, F., Rafanelli, M., 1996. A pictorial query language for geographical databases. In: International Workshop on Advanced Visual Interfaces, Gubbio, Italy, 27–29 May.

Egenhofer, M., 1994. Spatial SQL: a query and presentation language. IEEE Transactions on Knowledge and Data Engineering 6 (1), 86–95.

Erwig, M., Güting, R.H., 1994. Explicit graphs in a functional model for spatial databases. IEEE Transaction on Knowledge and Data Engineering 6 (5), 787–804.

Güting, R.H., 1991. Extending a spatial database system by graphs and object class hierarchies. In: International Workshop on Database Management System for Geographical Applications, Capri, Italy, May.

Haas, L.M., Cody, W.F., 1991. Exploiting extensible DBMS in integrated geographical information systems. In: Second Symposium on Large Spatial Databases, Zurich, Switzerland, August 1991, Lecture Notes in Computer Science 525, Springer, Heidelberg, pp. 423–450.

Kuiper, B., 1978. Modelling spatial knowledge. Cognitive Science 2, 129–153.

Kirby, K.C., Pazner, M., 1990. Graphic map algebra. 4th International Symposium on Spatial Data Handling, Zurich, Switzerland, 22–28 July.

Larue, T., Pastre, D., Viémont, Y., 1993. Strong integration of spatial domains and operators in a relational database system. In: Abel, D.J., Ooi, B.C. (Eds), Advances in Spatial Databases, Lecture Notes in Computer Science 692. Springer, Singapore, pp. 53–71.

Mainguenaud, M., 1994. Consistency of geographical information system query result. Computers, Environment and Urban Systems 18, 333–342.

Mainguenaud, M., 1995. The modeling of the geographic information system network component. International Journal of Geographical Information Systems 9 (6), 575–593.

Medeiros, C.B., Pires, F., 1994. Databases for GIS SIGMOD record 23 (1), 107–115.

Meyer, B., 1992. Beyond icons: towards new metaphors for visual query languages for spatial information systems. In: Cooper, R. (Ed.), First International Workshop on Interfaces to Database Systems. Springer, Heidelberg, pp. 113–135.

Peckham, J., Maryanski, F., 1988. Semantic data models. ACM Computing Surveys 20 (3), 153–189.

Seshadri, P., Livny, M., Ramakrishnan, R., 1997. The case for enhanced abstract data types. In: 23rd International Conference on Very Large Databases, Athens, Greece.

Smith, J.M., Smith, D.C., 1987. Database abstraction: aggregation and generalization. ACM Transaction On Database System 2 (2), 105–133.

Stemple, D., Sheard, T., Bunker, R., 1986. Abstract Data Types in databases: specification, manipulation and access. In: Proceedings of the Second International Conference on Data Engineering, Los Angeles, CA, USA, 6–8 February.

Stonebraker, M., Moore, D., 1996. Object-Relational DBMSs: The Next Great Wave. Morgan Kaufmann, San Mateo, CA.

Timpf, S., Volta, G.S., Pollock, D.W., Egenhofer, M., 1992. A conceptual model of wayfinding using multiple levels of abstraction. In: Frank, A.U., Campari, I., Formentini, U. (Eds.), Theories and Methods of Spatio-Temporal Reasoning in Geographic Space. Springer, Berlin, pp. 349–367.

Tsou, M.H., Buttenfield, B., 1996. A direct manipulation interface for geographical information processing. In: Seventh International Symposium on Spatial Data Handling, Delft, The Netherlands, 12–16 August.

Woodruff, A., Su, A., Stonebraker, M., Paxson, C., Chen, J., Aiken, A., Wisnovsky, P., Taylor, C., 1995. Zooming and tunneling in Tioga: supporting navigation in multidimensional space. In: Spaccapietra, S., Jain, R. (Eds.), Visual Database Systems 3: Visual Information Management. Chapman & Hall, London, pp. 360–371.

Zdonik, S.B., Maier, D., 1990. Readings in Object-oriented Database Systems. Morgan Kaufmann, San Mateo, CA.

PERGAMON

Transportation Research Part C 8 (2000) 129–146

TRANSPORTATION RESEARCH PART C

www.elsevier.com/locate/trc

Development of a transit network from a street map database with spatial analysis and dynamic segmentation

Keechoo Choi [a,*], Wonjae Jang [b]

[a] *Department of Transportation Engineering, Ajou University, 5 Woncheon-Dong, Paldal-Ku, Suwon 442-749, South Korea*

[b] *Department of Civil and Environmental Engineering, University of Wisconsin-Madison, USA*

Abstract

This paper presents an integrated transit-oriented travel demand modeling procedure within the framework of geographic information systems (GIS). Focusing on transit network development, this paper presents both the procedure and algorithm for automatically generating both link and line data for transit demand modeling from the conventional street network data using spatial analysis and dynamic segmentation. For this purpose, transit stop digitizing, topology and route system building, and the conversion of route and stop data into link and line data sets are performed. Using spatial analysis, such as the functionality to search arcs nearest from a given node, the nearest stops are identified along the associated links of the transit line, while the topological relation between links and line data sets can also be computed using dynamic segmentation. The advantage of this approach is that street map databases represented by a centerline can be directly used along with the existing legacy urban transportation planning systems (UTPS) type travel modeling packages and existing GIS without incurring the additional cost of purchasing a full-blown transportation GIS package. A small test network is adopted to demonstrate the process and the results. The authors anticipate that the procedure set forth in this paper will be useful to many cities and regional transit agencies in their transit demand modeling process within the integrated GIS-based computing environment. © 2000 Elsevier Science Ltd. All rights reserved.

Keywords: Transit network development; GIS; Digital map; Spatial analysis; Dynamic segmentation

1. Introduction

As evidenced by the series of proceedings of the Geographic Information Systems for Transportation (GIS-T) Conference published since 1988 by the American Association of State

* Corresponding author. Tel.: +82-331-219-2538; fax: +82-331-215-7604.
E-mail address: keechoo@madang.ajou.ac.kr (K. Choi).

PII: S0968-090X(00)00013-9

Highway and Transportation Officials (AASHTO), and as indicated by Spear and Lakshmanan (1998) and McCormack and Nyerges (1997), the vast majority of GIS-T applications have used GIS as a platform for integrating and displaying data on fixed transportation infrastructure, such as roads and bridges. While facility management is clearly an important area for applying GIS for transportation (see TRB, 1990 for more information), it represents only a small fraction of those transportation problems to which GIS could be successfully applied. In urban transportation, especially, planners, policy makers, and transit managers are just beginning to comprehend how important a tool GIS can be in their work. Spear and Lakshmanan (1998) included such promising application areas as travel demand forecasting by Golledge (1998) and activity-based travel analysis by Goodchild (1998), Greaves and Stopher (1998), Miller (1998), and McNally (1998); transit route planning and market analysis (Racca, 1998); vehicle monitoring and real-time operational control; customer information systems; paratransit and emergency vehicle dispatching; and congestion management.

Some other GIS-T applications include the following: transportation/land-use planning (You and Kim, 1998), traffic analysis zone design (You et al., 1997a,b), safety benefit evaluation (Affum and Taylor, 1997), transit service area analysis (O'Neill et al., 1992), data attribution and network representation in GIS-T (Sutton, 1997), the integration of GIS concepts into transportation network data structure (Horowitz, 1997), linking urban models and GIS (Johnston and de la Barra, 2000), and impact assessment.

Some enhancements, however, need to be made to the functional capabilities of current GIS software to accommodate such areas as travel demand forecasting and transit planning. This paper discusses such enhancements, which are needed to address emerging urban transit applications of GIS. More specifically, the objective of this paper is to generate link and line data of a transit network, which are used with the conventional urban transportation planning systems (UTPS) type transit demand modeling packages like TRANPLAN, MINUTP, and EMME/2, with the help of GIS databases.

This paper is a sequel to previous work that deals with highway and transit network developments using GIS topological databases (Choi, 1993; Choi and Kim, 1994). Choi and Jang (1997) presented the issue of transit network development (for demand modeling) from the GIS database. There has been similar research to create bus routes in a CAD-based environment (BusMAP project, see Wakefield, 1997) for bus stop mapping, analysis, and planning for the purpose of bus route management of BC-British Columbia-Transit and a better communications between its many departments and employees. However, the approach undertaken was a bit different in the sense that it used CAD's polyline join command (instead of a GIS application) to join the actual road segments making up the bus routes so that the routes fell exactly over the road network. Albeit Wakefield's concept and approach can be used in transit demand modeling as well, it seems to be most useful toward the facilitation of stop management, such as new bus stop installations and spatial changes of existing stops and bus route map production for the public who may ride the bus.

Choi and Jang (1997) included the derivation of an algorithm for automatically generating link and line data sets for an urban street from a GIS database. The assumption made in their 1997 work, however, was that transit stops are located at the intersections represented by nodes in highway networks, as we may find in the case of subway networks. The reality, however, is sometimes different, especially in the case of bus networks. That is, stops are sometimes located in the middle of highway links, represented by arcs in GIS. One motivation of this paper is to overcome these types of discrepancies and difficulties.

The purpose of this research is to derive a more realistic transit network data to be used in transit demand modeling from the street GIS map databases. The proposed approach makes use of spatial analysis and dynamic segmentation. Using spatial analysis, the nearest (bus) stops are to be identified as nodes, while the topological relation between nodes/links and a series of line data sets are computed using the dynamic segmentation. In this paper, both the procedure and algorithm to generate transit network data are presented along with a concrete implementation and a discussion of the advantages of the algorithm.

2. Transit network for demand modeling: problems and development

As stated before, the basic data used in travel demand forecasting are origin–destination matrices and network data. The origin–destination data represent the demand side of quantity. The network data describe the supply side that accommodates the demand. Among the criticisms identified by Choi (1993) and Choi and Kim (1994) in making those data sets for transportation planning and demand forecasting, labor-intensiveness is considered as one of the most serious problems, especially in preparing transit network data.

During the preparation of the transit network and related data set needed, transportation planners have to prepare maps to describe study areas and actual transportation networks. As O'Neil (1991) states, network generation requires extensive data collection and integration efforts. Furthermore, the generated networks are frequently modified to reflect changes, such as (1) changes in study area boundaries, (2) changes in zone delineation due to the land use change, (3) modified networks (link shape and node location change) for testing alternative network scenarios, and (4) link attribute changes like speed limit or capacity. In such cases, data requirements are extensive. Two typical problems in transit network preparation are described first, along with some possible problem solving activities.

2.1. Typical problem I

A directed graph, $G = \{N, A\}$, consists of a set N of nodes and a set A of arcs whose elements are ordered pairs of distinct nodes. A network is a directed graph whose nodes and/or arcs have associated numerical values, typically costs, capacities, and/or supplies and demands (Ahuja et al., 1993). In everyday life, transportation networks are perhaps the most common and the most readily identifiable classes of networks among physical networks. Focusing on the transit network can identify typical problems encountered in the preparation of the transit network data.

Unlike the highway network, stops in a transit network are not spatially coincident with intersections (represented by nodes in GIS) of highway network and are sometimes located in the middle of links (Fig. 1). Therefore, it is impossible to use the vector topology database provided by the highway GIS network database, and some steps are required to transform the highway database to the data type needed for transit network data generation and modeling.

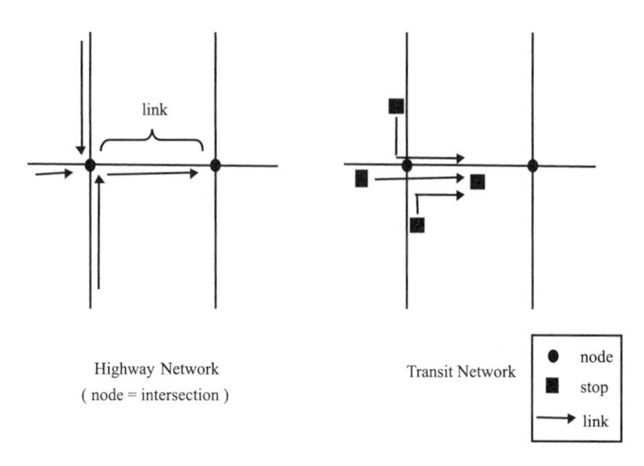

Fig. 1. Highway network vs. transit network. An arrow indicates the link direction.

2.2. Typical problem II

Besides node and link data, a complete transit network requires additional input for transit demand modeling. This is called line data. It is composed of an ordered set of stops on top of link data. The most prevalent problem encountered during the network preparation is the lack of consistency between link and line data. Fig. 2 illustrates such a situation where link 102–111 exists in the line data set but is absent from the link data set.

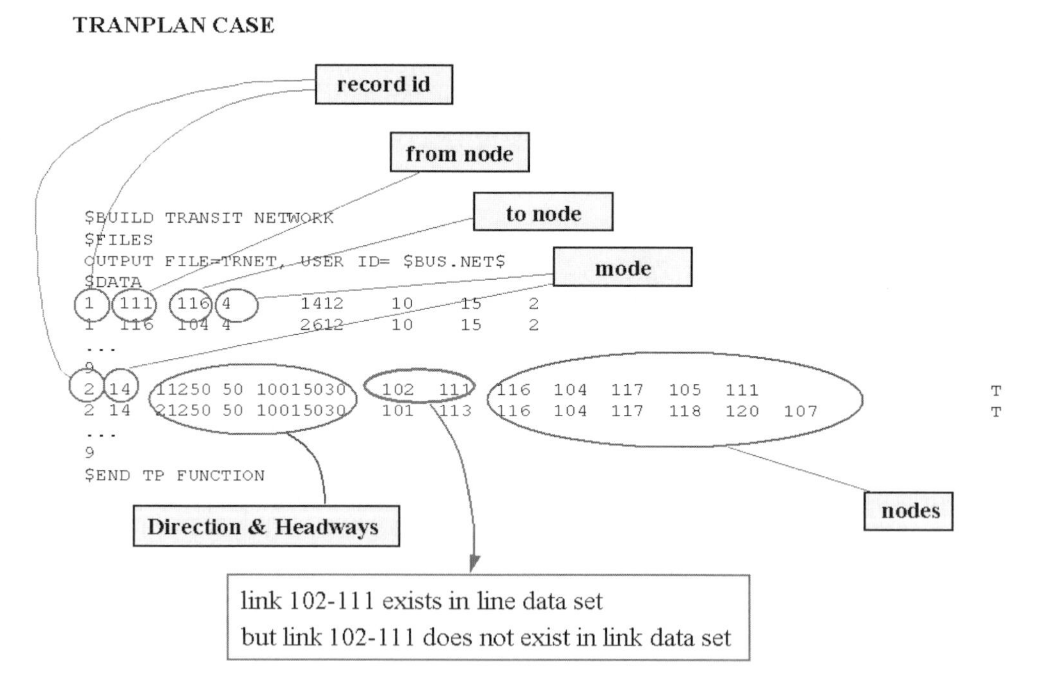

Fig. 2. Lack of consistency.

2.3. Two types of stop data

Transit stops can be represented either by a point or by a node. Both are primitives of geographic elements representing reality. Nodes are assigned to intersections where two or more lines meet, whereas points can be positioned along the line. If a stop position is fixed and connected to arcs, as in the case of a subway, it is preferable to represent it with a node connecting to incoming and outgoing arcs (Fig. 3). But bus stops, which might be moved frequently, are better represented by points, and dynamic segmentation and other GIS capabilities should be used to derive link and line data sets.

2.4. Node-based vs. point-based transit networks

Choi and Jang (1997) proposed node-based transit network development procedures in which all stops are assumed to be located at physical highway intersections and are represented as nodes. But in actual transit networks, stops are normally represented as points along arcs rather than nodes. The node-based transit network development procedure is rather straightforward to implement. The data preparation relationship in node-based transit network development procedures is shown in Fig. 4. See Choi and Jang (1997) for more details. In Fig. 4, we use a terminology consistent with the use of ESRI's ARC/INFO GIS application. The RAT, AAT, NAT, SEC, and AML acronyms stand for route attribute table, arc attribute table, node attribute table, section table, and arc macro language, respectively (see Environmental Systems Research Institute, 1993a,b, and 1995 for more detail). On the other hand, the point-based procedure requires more complex steps to derive link and line data. The process will be discussed in more detail in the rest of the paper. Table 1 summarizes the major differences between the two procedures.

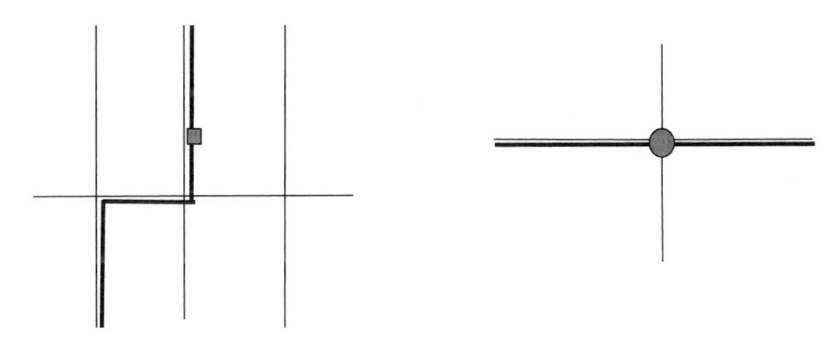

Fig. 3. Transit stop data types.

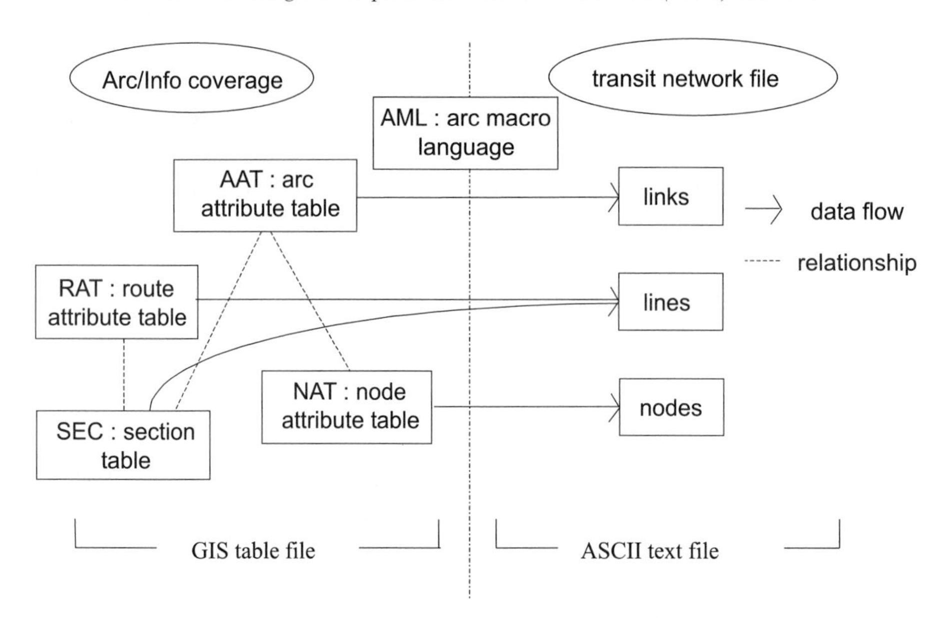

Fig. 4. Node-based transit network development.

Table 1
Node-based vs. point-based transit network development

Aspects	Node-based	Point-based
Stop is represented by	Node	Stop
Line data are	Series of nodes	Ordered set of stops
Link length	Arc length (embedded in AAT)	Need to be calculated
Application	Not practicable	Practicable

3. Data preparation

Before proceeding with the description of the main algorithm for generating link and line data, processes involved in stop positioning, stop–arc relationship building, and link distance calculation are introduced in this section.

3.1. Positioning stops

When the stop is positioned during stop data preparation, care must be taken to maintain the relationship between arcs and stops. As is the case in Fig. 5, the stop position should be inserted more closely to the link to which it is eventually related. For example, in Fig. 5, stop 2 must be positioned closer to link A than to link B to represent that stop 2 is located along the link A. This kind of relationship is needed to build the route system and to generate transit network building during the steps described later.

The stop's direction is categorized as one-way or two-way. Two-way means that the stop can be used for both directions of traffic, as is the case with a normal subway station. Stop 1 in Fig. 5 is

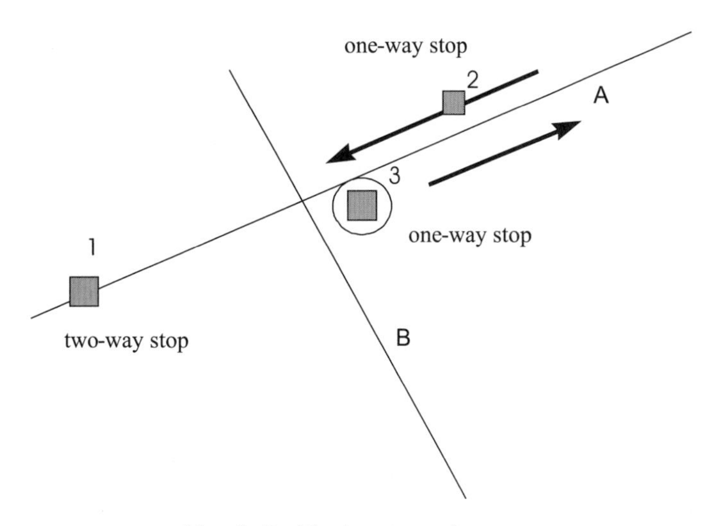

Fig. 5. Positioning stops along arcs.

the typical example of a two-way stop. In this situation, the stop is positioned on the centerline. But a one-way stop should be positioned at an offset on one side or the other (stop 2). One more important attribute of the stop data is the mode availability at that stop. Every available mode should be listed under the stop's attribute data.

3.2. Building the relationship between stops and arcs

The relationship between stops and arcs does not exist at first (after the stop data preparation). However, the relationship can be constructed easily by GIS functions and Fig. 6 shows this

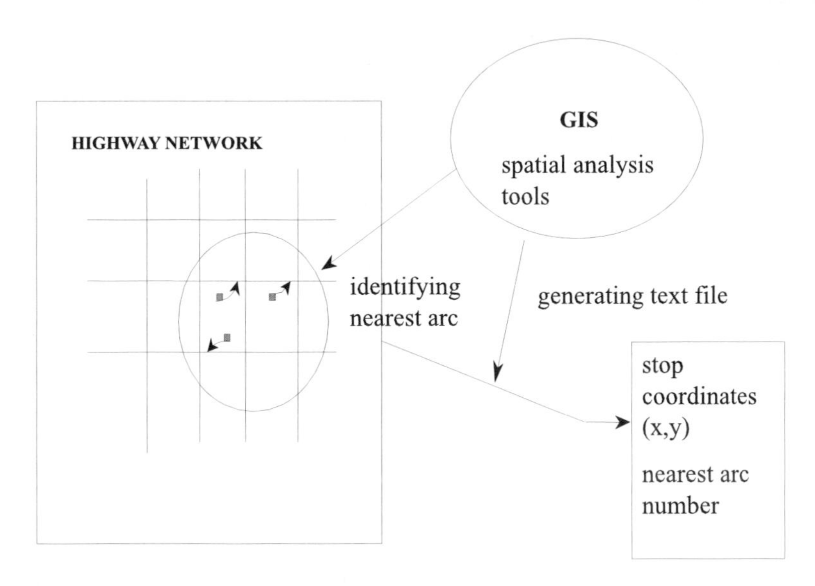

Fig. 6. Relationship between stops and arcs.

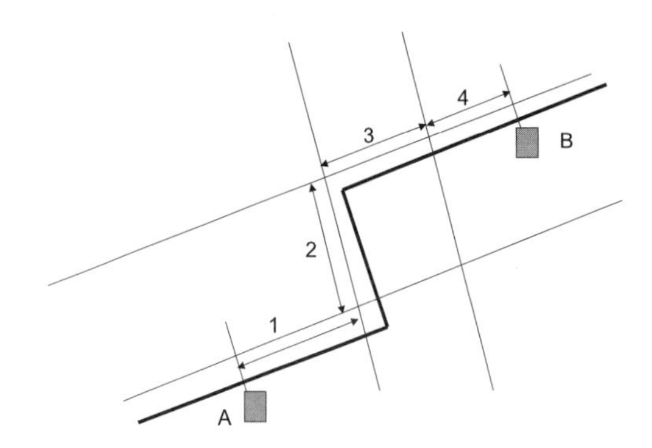

Fig. 7. Stop-to-stop distance: 1, 2, 3, and 4 denote sections.

procedure. In Fig. 6, stops represented as points have no relational information about arcs nearby. However, by computing the shortest distance between a node and an arc, we can identify the relationship, that is, which stops are lying on which arc.

3.3. Calculation of stop-to-stop distance

Link data information in a transit network consists of a pair of stops, link distance, and other related attribute data of the link. In a node-based transit network development procedure, the distances between stops are embedded in the topological information of the arc-node relational database. But in a point-based procedure, there is no direct link distance (stop-to-stop distance along the route) information available. Therefore, a separate procedure/algorithm must be developed to calculate transit link distances.

As shown in Fig. 7, if either stop is located in the middle of an arc, the algorithm has to calculate the partial length of the arc where the stop is located. Four steps are required to achieve this. First, find the perpendicular points from the stops to the nearest arc segments using GIS' nearest arc finding capability from a node representing stop using spatial analysis. Second, calculate the distance from the beginning stop of the link to the first node encountered along the route (using the dynamic segmentation). Third, calculate the distance from the ending stop of the link to the first node encountered back through the route using the arc–node and section tables, as in the second step. Fourth, sum the results from second and third steps and all intervening arc lengths as illustrated in Fig. 8. If the beginning stop is located exactly at the node position, the second step is not needed. The same is true of the third step.

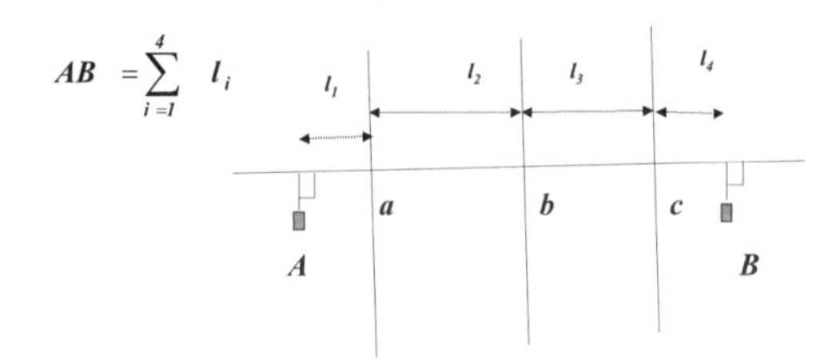

Fig. 8. Summing up intervening arc lengths: l_1, l_2, l_3, and l_4 are corresponding section lengths.

4. The algorithm

4.1. Algorithm outline

Fig. 9 shows the schematic flow of the procedure algorithm, whereas Fig. 10 depicts the flow chart of the proposed algorithm. The upper part of the algorithm in Fig. 9 depicts the stop positioning on top of the existing GIS layer, building a route system with stops positioned using

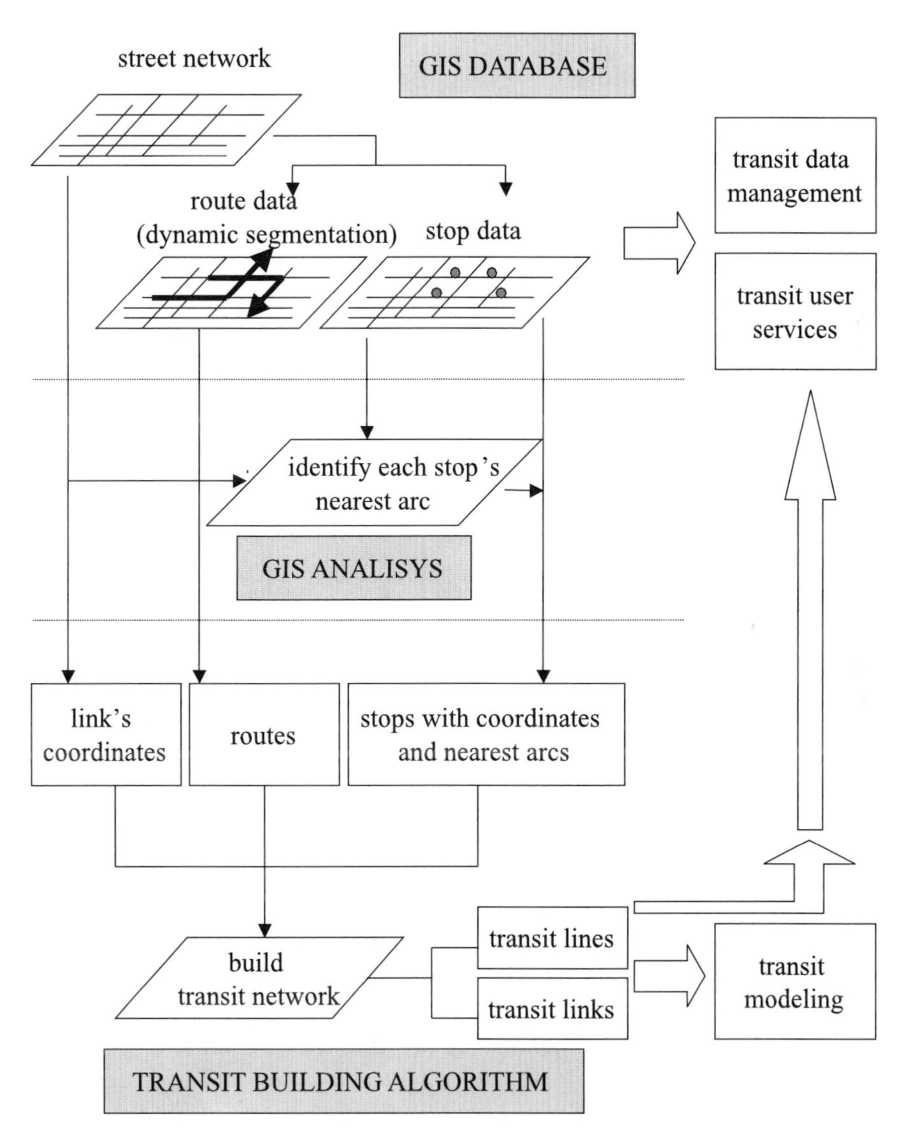

Fig. 9. Overall flow of algorithm.

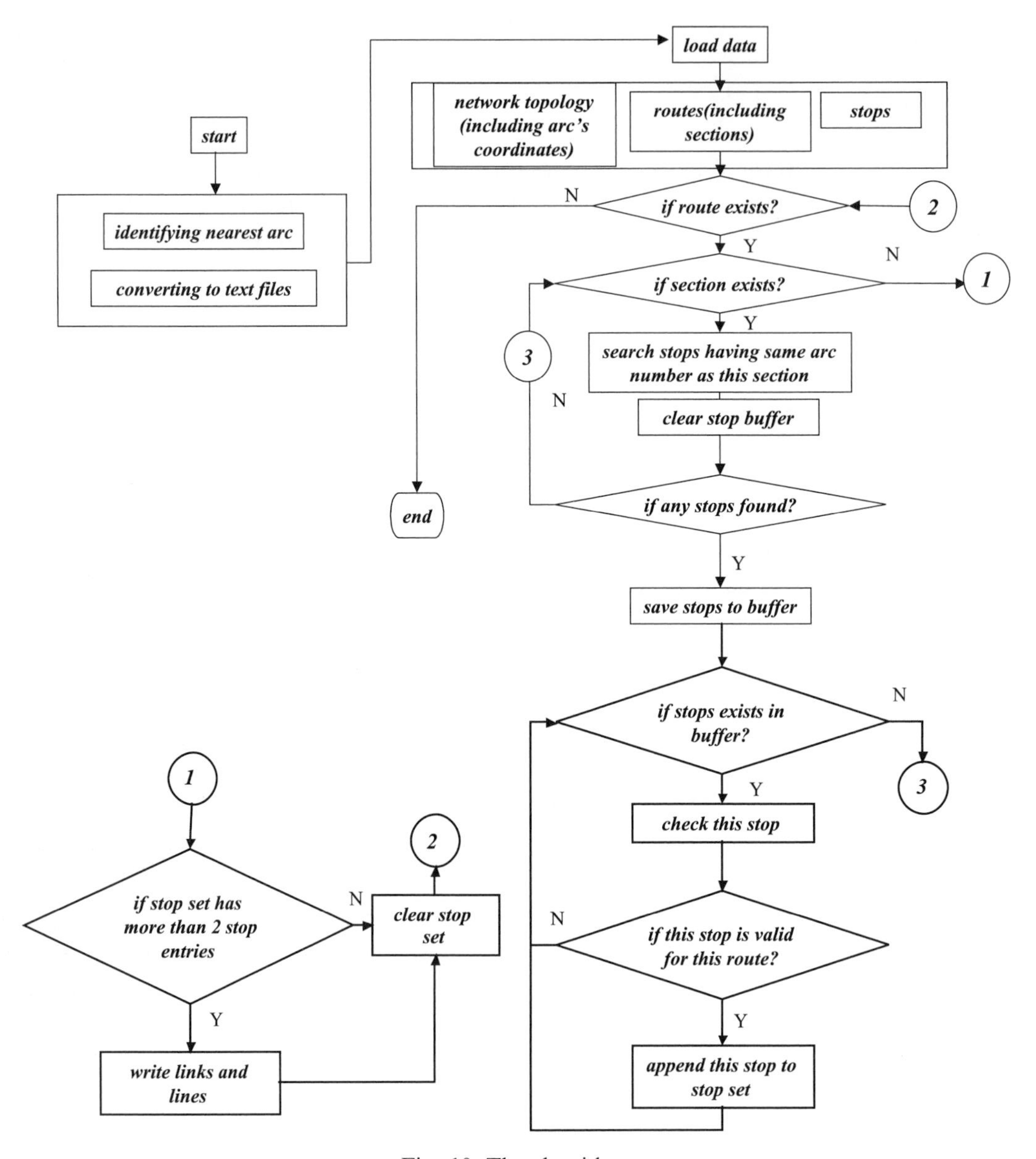

Fig. 10. The algorithm.

dynamic segmentation and nearest arc finding process using GIS spatial analysis. The lower part of Fig. 9 describes the link and line data derivation using the prescribed data set in the upper part of the algorithm.

The algorithm converts route data (generated by the GIS dynamic segmentation model) to point-based links and lines using the relationship between stops and arcs described earlier. The algorithm converts one route at a time and examines all sections in that route. At the section examining stage, it searches for all stops that have the same arc number. If any stops are found, they are checked to see if the route passes through them. Once the stop is found to be on the route, it is added to the temporary stop buffer with its cumulative length from the beginning point of the route. After examining all sections, stops stored in the temporary buffer are sorted by cumulative length, and link and line data are extracted. After a route system is processed in this fashion, the algorithm goes to the next route and so on until all the routes are processed.

To describe the algorithm algebraically, the following notation is employed.

A: set of all arcs
A^i: arc corresponding section i
R: set of all routes
T^k: set of ordered sections of route k
S: set of all stops
S^i: set of stops having arc A^i as nearest arc
B: set of stops
L: set of transit lines
P: set of transit links
D: total distance from the beginning of route

The most fundamental three steps of the algorithm are described hereunder.
$d(k,i)$: length of section i of route k
$D(k,j)$: total distance from the beginning of route k to the stop j along the route k
$d(k,i,j)$: distance from the beginning of section i of route k to the stop j along the route k
$Q_j =$ jth element in set Q
$|Q| =$ number of elements in set Q

step (0)
Read R, S, A from files
$L = \Phi$, $Lk = \Phi$
step (1)
for each T^k, $k = \{1, \dots, |R|\}$
for each T_i^k, $i = \{1, \dots, |T^k|\}$
$S^i = \{$l: l's nearest arc $= A^i\}$
$B = \Phi$, $D = 0$
(1) for each S_j^i in $j = \{1, \dots |S^i|\}$
if S_j^i is within section i and (S_j^i is on the right of route or S_j^i is a bi-directional stop) then

$$D(k,j) = D + d(k,i,j), \; B = B \cup \{S_j^i\}$$
$$D = D + d(k,i)$$

(2) sort B by $D(k,j)$
(3) $L = L \cup B$
if $|B| \geqslant 2$ then $P = P \cup \{(B_j, B_{j+1})\}$, $j = \{1, \dots, |B| - 1\}$
step (2) stop

5. Demonstration with test network

To demonstrate the working of the algorithm described above, a small test network has been arranged. Fig. 11 shows the test network and Fig. 12 depicts the associated stop and section

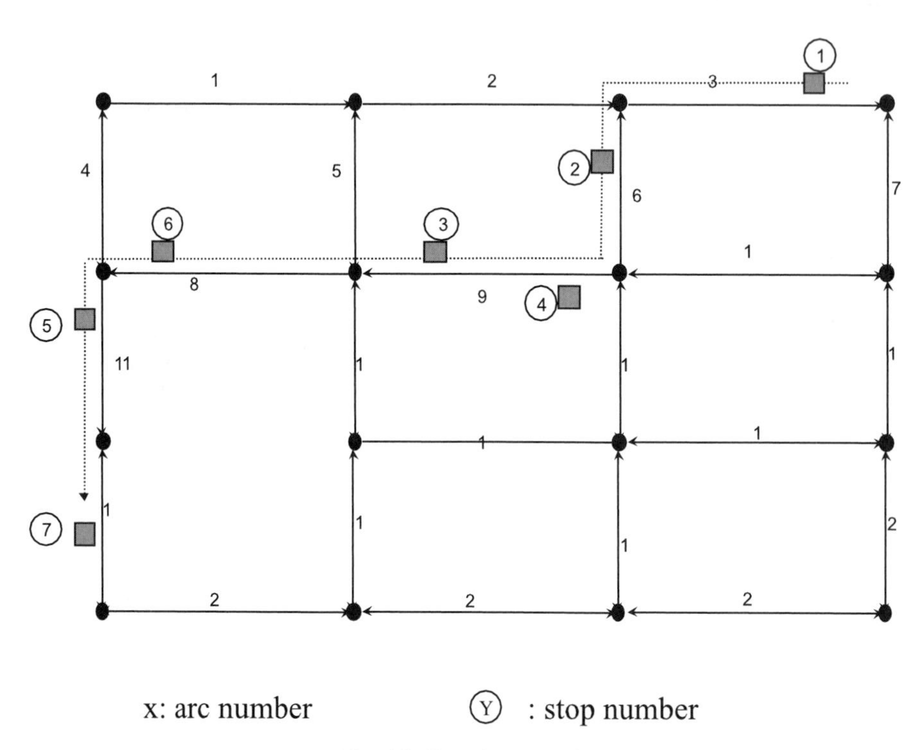

Fig. 11. Sample network.

sections

SECNO	ARC#	FROMPOS	TOPOS
1	3	90	0
2	6	100	0
3	9	0	100
4	8	0	100
5	11	0	100
6	17	100	65

stops

STOP-ID	X	Y
1	38	42
2	29	37
3	23	31
4	27	29
5	9	18
6	13	32
7	8	15

Fig. 12. Section and stop tables: "SECNO" means section number and "ARC#" means arc number. "FROMPOS" specifies the from-position where the section starts on the arc (%), whereas "TOPOS" denotes the to-position where the section ends on the arc (%).

tables. The network is composed of 7 stops and 1 route based upon a street network having 23 arcs and 16 nodes. The dotted line in Fig. 11 shows a transit route under consideration, governing 5 full arcs and 2 partial arcs.

5.1. Data preparation

Above all, each stop's nearest arc should be identified as shown in Fig. 13. The far right column in the table shows the result. After identifying the (ARC#) nearest arcs, the following data sets are converted to text files: highway network topology, arc coordinates, route and section tables, stop table including nearest arc number, and related attribute data.

5.2. Transit network (link-line data) generation

The converted text files prepared in the data preparation are input data to the algorithm. These data sets are loaded into memory as arrays. For each element in a route array, each section in that route is examined in order. First, the arc number of the section is identified via the section–arc relationship, which is ascertained with the help of dynamic segmentation. Then, all stops having the same arc number are searched. If any stops are found, each one is examined to see whether the route passes through, following the two sub-steps:

1. Check whether the stop is to the right of an arc along the route or a two-way stop; if not, process the next found stop.
2. Check if the stop is within the boundaries of a section; if not, process the next stop.

Once every stop passes the above examination process, the next step is to calculate the total length from the beginning of the route to arrange the stops in sequential order, and save the associated

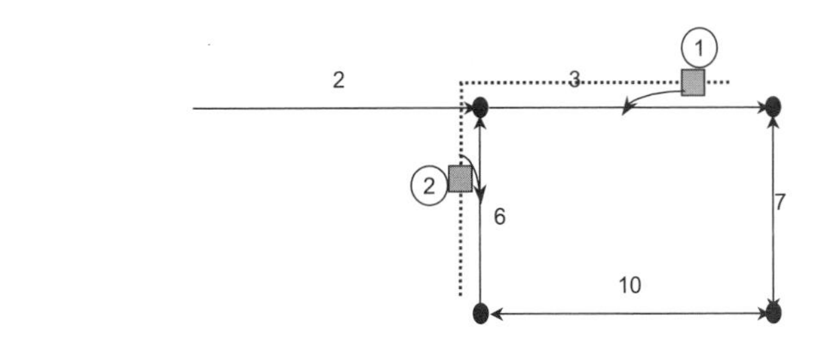

stops

STOP-ID	X	Y	ARC#
1	38	42	3
2	29	37	6
3	23	31	9
4	27	29	9
5	9	18	11
6	13	32	8
7	8	15	17

Fig. 13. Identifying nearest arc.

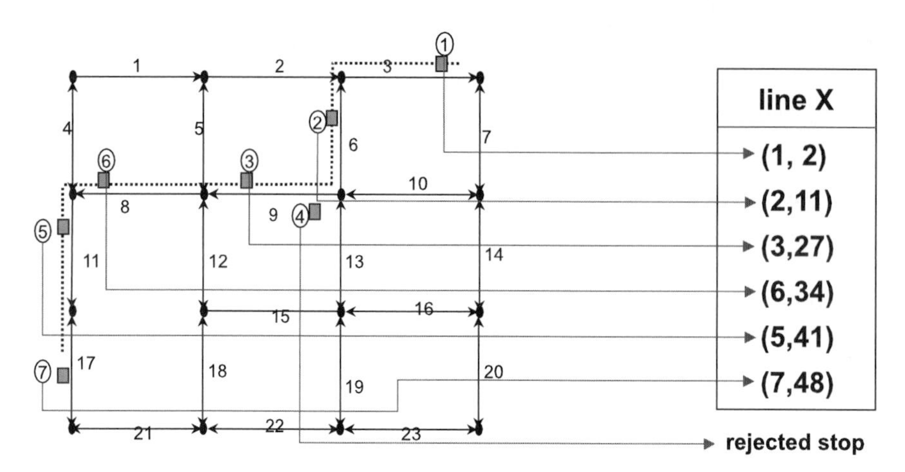

Fig. 14. Identified stops with lengths (from the start of line) via section examining.

data (stop number, total length) to the temporary stop buffer. If any sections remain, continue to examine the next section.

The final step is to fetch the link and line data sets from the buffer. Again, a link is a pair of stops and a line is an ordered set of stops. Before writing down the link and line data, stops saved in the buffer must be sorted by length (in ascending order) to secure the order of stops from the start of the line. The line data for line X, in this example in Fig. 14, is composed of stops 1, 2, 3, 6, 5, and 7 in this order and the associated link length can also be generated.

While processing a route, the route is transformed into a series of stops, which are stored in the temporary buffer. A single line and its associated multiple links are created at the same time, maintaining the consistency in the relationship between link and line data pointed out earlier. The algorithm should process all the lines in this manner before termination as shown in Fig. 15.

6. Application to real street network

We applied the procedure discussed above to the actual transit network in the city of Suwon, Korea (Fig. 16). Although the city has more than 70 bus lines, only four bus lines, each having 20–30 stops in both directions, are considered, for the sake of simplicity.

As stated in the introduction, the goal is to extract the link and line data for each bus line and feed it into transit demand modeling packages, such as TRANPLAN and EMME/2, without incurring the additional costs of purchasing new GIS packages for transit demand modeling. The extracted data are based on a network composed of 55 links and 4 lines. Although the results of the transit demand modeling have been omitted in this paper, the extracted link and line data provided seamless and error-free transit network (supply) data. The extracted link and line data are shown in Table 2.

STOP-ID	cumulative Length
1	2
2	11
3	27
6	34
5	41
7	48

(a) building line data : line X=1,2,3,6,5,7

STOP-ID	cumulative Length	link length
1	2	
2	11	9
3	27	16
6	34	9
5	41	7
7	48	7

from	to	length
1	2	9
2	3	16
3	6	9
6	5	7
5	7	7

(b) building link data :

Fig. 15. Building lines and links.

Fig. 16. Actual bus route network over street network.

7. Summary and conclusion

With the advent of the information age, GIS is playing a pivotal role in the integration of transit agencies' legacy information systems, such as payroll and management information systems, in order to increase both efficiency and effectiveness. Wakefield (1997) also pointed out that

Table 2
Results of link and line data for real bus network

Link no.	Stop from	Stop to	Length
Number of links: 55			
1	1	2	95.423
2	2	3	85.280
3	3	4	79.837
4	4	5	78.021
5	5	6	121.233
6	6	7	119.670
7	7	8	102.990
8	8	59	84.632
9	8	71	145.028
10	18	6	146.430
.	.	.	.
.	.	.	.
Number of lines: 4			
Route 1	27, 25, 24, 22, 20, 18, 6, 7, 8, 59, 56, 95, 91, 97, 99, 102, 104, 106, 108		
Route 2	82, 80, 78, 70, 68, 66, 64, 44, 41, 40, 38, 36, 33, 32, 30, 62		
Route 3	1, 2, 3, 4, 5, 6, 7, 8, 71, 73, 75, 77		
Route 4	84, 86, 88, 90, 92, 94, 59, 55, 53, 51, 49, 60, 46, 62		

public transportation agencies are rapidly adopting GIS as an efficient, cost-effective tool for tracking bus-route planning and scheduling information. In view of this situation, the aim of this paper is to utilize the capability of GIS for transit demand modeling using conventional UTPS type transportation modeling systems like TRANPLAN (or the pure FORTRAN-oriented modeling programs). Although there are already transportation modeling packages with GIS function embedded, the method described in this paper might reduce overall costs in the long run and improve the quality of output as well, if used and integrated appropriately.

We have described a procedure to generate a transit network data file from a street centerline based on GIS databases. The procedure presented in the paper is concerned with the generation of a more realistic and consistent transit network file to be used for transit demand modeling. More specifically, the core purpose is to devise a mechanism to automate the process of preparing the link and line data sets using the GIS database, spatial analysis, and dynamic segmentation. The advantage of this approach lies in the fact that the street and/or highway base map represented by a centerline, normally available with little cost, can be directly used. A small test network has been used to illustrate the example of an application to a real network.

As shown in Table 3, the GIS-based transit network development is expected to be better in terms of consistency (between link and line data), accuracy (of data), and labor-intensiveness.

Table 3
Comparison of two methods

Criteria method	Consistency	Accuracy	Labor-intensiveness
Conventional method	Low	Low	High
Using GIS database	High	High	Low

Compared with the conventional method of using an analog map with a text editor to prepare the input transit network data, the GIS-oriented approach seems to be superior. The procedure set forth in this paper should be helpful to transit agencies and city governments facing transit problems. For example, the procedure can be applied to a case of the city of Seoul's transit system composed of 90 transit companies, more than 430 lines, and more than 9000 buses. In addition, the algorithm can be expected to be applicable to the construction of mobility databases, such as activity travel pattern and point-to-point trip pattern for each purpose and mode, to the issue of geocoding of travel data, and to the preparation of inputs of the microsimulation modeling process, such as TRANSIMS or an equivalent that is currently under construction.

Acknowledgements

Partial support of the Ajou University's research fund (Gongdongkiki Center) is greatly appreciated. The authors also would like to thank to Prof. Jean-Claude Thill for his constructive comments and Ms. Louise M. Malloy for her painstaking revision of the manuscript from the early draft.

References

Affum, J.K., Taylor, M.A.P., 1997. SELATM-A GIS based program for evaluating the safety benefits of local area traffic management schemes. Transportation Planning and Technology 21, 93–120.

Ahuja, R.K., Magnanti, T.L., Orlin, J.B., 1993. Network Flows. Prentice-Hall, New York.

Choi, K., 1993. The implementation of an integrated transportation planning with GIS and expert systems for interactive transportation planning. Unpublished Ph.D. Dissertation, University of Illinois at Urbana-Champaign, Urbana, IL.

Choi, K., Jang, W., 1997. Transit network development using arc–node topological database. In: Proceedings of the 1997 Geographic Information Systems for Transportation Symposium, pp. 357–364.

Choi, K., Kim, T.J., 1994. Transportation planning with GIS: issues and prospects. Journal of Planning Education and Research 13 (3), 199–207.

Environmental Systems Research Institute, 1993a. User's Guide of Dynamic Segmentation, Redland, CA.

Environmental Systems Research Institute, 1993b. User's Guide of Network Modeling, Redland, CA.

Environmental Systems Research Institute, 1995. Workshop Proceedings of the 1995 ESRI User Conference, Redland, CA.

Golledge, R.G., 1998. The relationship between GIS and disaggregate behavior travel modeling. Geographical Systems 3 (5), 9–18.

Goodchild, M.F., 1998. Geographic information systems and disaggregate transportation modeling. Geographical Systems 3 (5), 19–44.

Greaves, S., Stopher, P., 1998. A synthesis of GIS and activity-based travel-forecasting. Geographical Systems 3 (5), 59–90.

Horowitz, A.J., 1997. Integrating GIS concepts into transportation network data structures. Transportation Planning and Technology 21, 139–153.

Johnston, R.A., Barra, T., 2000. Comprehensive regional modeling for long-range planning: linking integrated urban models and geographic information systems. Transportation Research 34 (2), 125–136.

McCormack, E., Nyerges, T., 1997. What transportation modeling needs from a GIS: a conceptual framework. Transportation Planning and Technology 21, 5–23.

McNally, M.G., 1998. Activity-based forecasting models integrating GIS. Geographical Systems 3 (5), 163–188.

Miller, H.J., 1998. Emerging themes and research frontiers in GIS and activity-based travel demand forecasting. Geographical Systems 3 (5), 189–198.

O'Neill, W.A., 1991. Developing optimal transportation analysis zones using GIS. ITE Journal 61 (11), 33–36.

O'Neill, W.A., Ramsey, R.D., Chou, J., 1992. Analysis of transit service areas using geographic information systems. Transportation Research Record 1364, 131–138.

Racca, D., 1998. Using GIS to identify markets for transit in delaware. In: Proceedings of the 1998 Geographic Information Systems for Transportation Symposium. Salt Lake City, UT, pp. 270–291.

Spear, B.D., Lakshmanan, T.R., 1998. The role of GIS in transportation planning and analysis. Geographical Systems 3 (5), 45–58.

Sutton, J., 1997. Data attribution and network representation: issues in GIS and transportation. Transportation Planning and Technology 21, 25–44.

Transportation Research Board (TRB), 1990. Transportation Research Record 1261: GIS 1990, Washington, DC.

You, J., Nedovic-Budic, Z., Kim, T.J., 1997a. A GIS-based traffic analysis zone design: technique. Transportation Planning and Technology 21, 45–68.

You, J., Nedovic-Budic, Z., Kim, T.J., 1997b. A GIS-based traffic analysis zone design: implementation and evaluation. Transportation Planning and Technology 21, 69–91.

You, J., Kim, T.J., 1998. An integrated land use-transportation modeling system with GIS. In: Proceedings of the 1998 Geographic Information Systems for Transportation Symposium. Salt Lake City, UT, pp. 448–461.

Wakefield, S., 1997. BusMAP brings BC transit up to speed. GeoInfo Systems. May, pp. 29–32.

PERGAMON

Transportation Research Part C 8 (2000) 147–166

TRANSPORTATION
RESEARCH
PART C

www.elsevier.com/locate/trc

Intermodal and international freight network modeling ☆

Frank Southworth *, Bruce E. Peterson

Center for Transportation Analysis, Oak Ridge National Laboratory, P.O. Box 2008, Oak Ridge, TN 37831-6206, USA

Abstract

The authors describe the development and application of a single, integrated digital representation of a multimodal and transcontinental freight transportation network. The network was constructed to support the simulation of some five million origin to destination freight shipments reported as part of the 1997 United States Commodity Flow Survey. The paper focuses on the routing of the tens of thousands of *intermodal* freight movements reported in this survey. Routings involve different combinations of truck, rail and water transportation. Geographic information systems (GIS) technology was invaluable in the cost-effective construction and maintenance of this network and in the subsequent validation of mode sequences and route selections. However, computationally efficient routing of intermodal freight shipments was found to be most efficiently accomplished outside the GIS. Selection of appropriate intermodal routes required procedures for linking freight origins and destinations to the transportation network, procedures for modeling intermodal terminal transfers and inter-carrier interlining practices, and a procedure for generating multimodal impedance functions to reflect the relative costs of alternative, survey reported mode sequences.

Keywords: Intermodal freight; Networks; Geographic information systems; Traffic routing

1. Introduction

The demand for goods has grown steadily over the past half century so that today an essential ingredient of a thriving national economy is a cost-effective freight transportation system. This involves the use of multimodal, including intermodal, transportation options. Intermodal movements are those in which two or more different transportation modes are linked end-to-end in order to move freight and/or people from point of origin to point of destination. Within the

☆ The submitted manuscript has been authored by a contractor of the US Government under contract No. DE-AC05-96OR2464. Accordingly, the US Government retains a nonexclusive, royalty-free license to publish or reproduce the published form of this contribution, or allow others to do so, for US Government purposes.

* Corresponding author.

0968-090X/00/$ - see front matter
PII: S0968-090X(00)00004-8

United States intermodal transportation is associated with the movement of goods through seaports, and increasingly, airports. It also involves the use of an extensive network of truck-rail, truck-barge and rail-barge transfer facilities.

New vehicle, container and cargo handling technologies, including real time inventory tracking and management technologies, continue to increase the intermodal as well as mode specific options for moving commodities from place to place (GAO, 1996). To understand where these new freight technologies and new transportation infrastructures are leading us requires sound methods for measuring, portraying and costing freight movements. Geographic information systems (GIS) technology has begun to play a central role in this understanding. By supporting the mapping of freight movements onto specific transportation routes and through specific transfer terminals, GIS offers cost effective tools for manipulating a wide range of data on both the spatial and non-spatial aspects of commodity transportation logistics (see Crainic et al., 1990; Jourquin and Beuthe, 1996; Southworth et al., 1997). The key to this mapping is a properly constructed and multimodal digital representation of the freight supporting transportation infrastructure.

This paper describes the development and application of a large and detailed example of such a digital transportation network. This network was used to simulate some five million origin to destination freight shipments, including tens of thousands of intermodal shipments both within and across the borders of the United States. Each of these shipments was reported to the United States Bureau of the Census as part of its 1997 Commodity Flow Survey (CFS). The CFS is a congressionally mandated survey of approximately 100,000 freight shippers within the United States, carried out by the Bureau of the Census with sponsorship from the Bureau of Transportation Statistics (BTS). The purpose of the survey is to ascertain how, and how much, shippers are using the nation's multimodal transportation system to move freight, both domestically and for export (the survey does not cover imports). Like the 1993 CFS before it (Bureau of the Census, 1995) the 1997 CFS covers commodity shipments associated with mining, manufacturing, wholesale and selected retail and service activities.

The 1997 survey required shippers to report the origin, destination, commodity type, tonnage, value and mode sequence used to move a sample of their shipments, drawn at four different times during the course of the calendar year. What the survey does not ascertain is the routing and therefore the mileages involved in getting freight from one place to another. To avoid putting such a burden on survey respondents these mileages were simulated on the basis of the mode sequences reported. With these data the Census Bureau produces a national picture of annual freight movements by geographic region, mode combination, dollar value, tonnage and shipment size, for a set of some 500 different commodity types.[1]

The principal task faced by the authors was to develop a representation of the nation's transportation network that would allow routes to be simulated that reproduce the mode sequence reported on each CFS shipment record. Where mode(s) were reported but not sequence, or where a reported sequence was considered incomplete (for example, where "truck, rail" really required a truck–rail–truck sequence to ensure access to the reported destination) the digital intermodal network described in this paper was also used, with appropriate routing algorithms, to infer missing intermodal connections and mileages. The resulting route mileages, suitably

[1] Survey details and results can be found at the BTS website *www.bts.gov*.

expanded to reflect the shipment volumes routed over them, are then used to estimate the national, statewide and regional ton-miles and dollar-miles of freight activity associated with these calendar year 1997 shipments, by modes and commodity types. These mileage statistics support a range of policy analysis, such as an assessment of the ton-miles of truck freight activity passing into, within, out of, and through each of the 50 states (BTS, 1997a,b).

A significant value of the CFS is its treatment of "door-to-door" freight movements in which the true origin of a shipment is tied to the true destination. This is important because most sources of freight movement data collected by the federal government deal with terminal to terminal movements within specific modes. Hence both the true geography of movements as well as the intermodal nature of many shipments is not captured. The CFS fills this data gap: while the use of network-based models to route CFS shipments allows a better understanding of how the different modal infrastructures are being used in getting freight from source to customer – an area identified as needing attention in the recent freight literature (GAO, 1996).

The CFS uses zip code locations to capture both shipment origin and destination locations, while city and country of destination as well as US port of exit are also reported for export shipments. This meant developing a digital network capable of routing freight between some forty-eight thousand different traffic generators and attractors. Given the voluminous number and variety of possible shipments generated by the survey, a number of reasonably generic computer programs had to be developed to accommodate an automated set of shipment distance calculations. These methods are described below under the following headings:

- intermodal network construction: placing components of intermodal freight shipments within a network structure, by merging different modal networks into a single, intermodal network;
- intermodal network access: ways to connect shipment origins and destinations to networks, and the need to evaluate multiple origin as well as multiple destination network access and egress options in order to select most likely shipment routes;
- modeling intermodal terminal transfers and inter-carrier interlining practices within a network structure: including the modeling of trans-oceanic and trans-border export shipments; and
- the use of generalized impedance functions for modeling the trans-global as well as trans-continental routing of domestic and export shipments.

These topics are discussed in turn in Sections 2–5 of the paper. Much of this discussion has an applicability to national and international network analysis that goes well beyond the CFS problem per se. Indeed, considerable value resides in a transportation network data model that can now act as a starting point for a wide range of freight movement studies. Some additional uses for such digital networks are summarized in Section 6 of the paper.

2. Intermodal network construction

2.1. Building blocks: the underlying network and terminal databases

CFS network construction involved merging mode-specific transportation network databases into a single, integrated multimodal network that allows both single and intermodal traffic routing between any pair of zip codes within the United States. This was accomplished by constructing of a single, logical network that can support the identification of any combination or sequence of intermodal paths. The network is "logical" in the sense that a computer program can find a chain of links between all possible origins and destinations. All of these network links represent some reality, whether physical trafficways, or processes that the shipment passes through in sequence, and they all have a geographic location. The resulting "links" in the CFS composite network therefore range from sections of real highway pavement to broad ocean sea lanes, to transfer processes involving cranes, drayage, storage, and repackaging at locations within a large seaport area. Two separate digital intermodal networks were constructed for traffic routing in the 1997 CFS: a truck–rail–waterways (TRW) network, and a truck–air (TA) network. Only the construction and application of the former network is the subject of this present paper. It was built by combining, and where necessary modifying, early 1997 calendar year versions of the following digital databases (see Southworth, 1997 for attribute details; also Southworth et al., 1998 for data sources):

- the Oak Ridge National Laboratory (ORNL) National Highway Network and its extensions into the main highways of Canada and Mexico;
- the Federal Railroad Administration's (FRA) National Rail Network and its extension into the main rail lines of Canada and Mexico;
- the US Army Corps of Engineers National Waterways Network;
- the ORNL constructed Trans-Oceanic Network;
- the ORNL constructed National Intermodal Terminals Database; and
- a database of 5-Digit Zip-Code area locations.

For shipment routing purposes each of these databases may be thought of as a brick in the CFS multimodal network building exercise, while the modeling and data handling techniques described in this paper provide the mortar that was used to integrate them into a coherent network data structure. The intermodal terminals database similarly represents a major database development effort in its own right. Middendorf (1998) describes this database and its construction, including a list of the many different data sources that were used to build it.

Table 1 lists the number of separate link records, and implied network mileages contained in a final version of this merged, multimodal CFS network. Three points are worth noting about this table. First, all mileages are what are commonly referred to as "center-line" mileages. Highway miles for example are not lane-miles. Second, differences exist in both the level of geographic detail and also in the geographic scale of the various modal networks listed in the table. Hence waterway links were significantly longer than other links, while highways were significantly

Table 1
1997 CFS North-American and trans-oceanic network statistics

Link type	# of links	Network mileage
Highway	84,537	489,679
Rail	22,126	236,931
Water (US):		
Inland & Inter-coastal:	4,269	44,116
Great Lakes:	912	13,156
Water (deep sea)	7,278	2,543,672
Intermodal transfers	14,176	
Access/egress	Generated as needed	

shorter, on average, than the links for other modes. While the highway and inland waterway networks were constructed at a scale of 1:100,000, the only rail network containing data on trackage rights was built at a scale of 1:2,000,000. The ORNL deep sea component of the waterways network was much cruder: scaled at approximately 1:6,000,000.

For traffic routing purposes these differences in scale are not important. What is important is the ability to connect networks together at appropriate (terminal) transfer locations. Once the usefulness of geographic detail has been established for a particular multimodal network it may be computationally efficient to simplify one or more of the constituent networks. For this purpose and prior to computing the number of links shown in Table 1 a certain amount of highway end-on link chaining across county and other administrative borders was carried out. Note also that all network access and egress links, which are used to put freight on and off the network, were built "on the fly" by a set of CFS routing algorithms, as needed for specific shipments. This procedure is discussed in Section 3.

2.2. Network modifications for routing purposes

To support traffic routing each of the major mode-specific networks needed some modification and enhancement prior to being merged into the multimodal CFS network. Besides the addition of a few specific links, notably rail spur lines and terminal access connectors not already in these databases at the time, the following structural modifications were needed:

Highways. The ORNL National Highway Network was built specifically for, and has been used continuously in traffic routing studies for over a decade (Southworth et al., 1986; Chin et al., 1989). However, the CFS distinguishes between private and for-hire trucking sub-modes and reports separate statistics for each. Therefore where both private and for-hire trucking was listed in a shipment's mode sequence, efforts were made to identify likely truck (typically, cargo consolidation) terminals as intermediate stops on a route. These within-the highway-mode terminals were simulated as additional network links in a manner similar to intermodal transfer terminals (see Section 4).

Waterways. For traffic routing purposes the National Waterway Network was divided into three different but connected sub-networks, following US Army Corps of Engineers definitions. These are the inland and inter-coastal (largely barge traffic) sub-network, the Great Lakes sub-network, and a trans-oceanic or "deep sea" sub-network. These distinctions are important for waterborne commerce routing because each of these sub-networks uses different vessel types to handle freight, since large, more robust vessels needed to cross large bodies of open water are uneconomical as a method of inland, riverine transport. Hence it was important to capture both the locations and relative costs involved in transferring goods from one vessel type to another. This was done by adding inter-vessel transfer links to the logical CFS network, at locations where such cargo unloading and loading takes place.

Second, for assistance in imputing within the United States export shipment mileages, where the US port of exit was unreported in the CFS, the ORNL trans-global deep sea sub-network was merged with this US waterways network. This deep sea sub-network takes the form of a latticework of open water links supplemented by much longer, more direct links between selected high volume seaport corridors (Southworth et al., 1998). It was linked by manual GIS-based editing to the National Waterways Network principally by adding connector links outside US seaports.

Railroads. Two important aspects of rail traffic routing involve railroad company specific "trackage rights" and between company "interlining" practices. To accommodate these the 1997 version of the FRA Rail Network was also subjected to modification prior to its inclusion within the CFS network. First, the representation of the railroad system as a connected set of individual companies' sub-networks was needed for traffic routing purposes. This required that each railroad link have a list of the companies that can operate over it, railroads that are said to have trackage rights. The FRA network included all trackage rights over a decade-long period, but we needed lists for 1997 only. This was accomplished by adding transition dates to ownership and trackage rights attribute information on each link. These dates were calculated from a model of corporate ancestry and the results were used to populate the geographic database.

Second, we needed to know where traffic was being exchanged between railroad companies. These locations are known as interlines. While the FRA network contained some data on these they were assigned for the most part only in an approximate fashion to the nearest metropolitan area. To obtain the desired level of geographic specificity it was necessary to assign these interlines to specific network locations, by defining them as a set of inter-railroad connector links joining the pair of railroads involved. This link-by-link attribute editing of the rail network was carried out within a commercial GIS.

Finally, both the CFS rail and highway networks were extended to include major Canadian and Mexican rail lines and highways, each tied to the US domestic transportation networks by the addition of transfer links at border crossings. The cost of delays at customs stations can be attached to these transfer links to simulate the relative costs of alternative routes when considering truck and rail export shipments.

2.3. Construction and application of the logical multimodal network

The multimodal CFS network was created by merging the above, now traffic routable single mode networks. This was done by linking them through a series of intermodal truck-rail (TR), truck-water (TW) and water-rail transfer terminals. Fig. 1 illustrates this concept, for the case for a truck-rail-truck (TRT) shipment. Using a suitable "shortest path" route-finding algorithm, a route is generated by first of all accessing the highway sub-network, linking this via a TR terminal to the rail sub-network and returning to the highway network via a second intermodal terminal transfer. In practice, two separate versions of the (same) highway sub-network are invoked in this routing procedure. Each of the three sub-networks shown may be activated or suppressed by suitable, user driven program commands to handle specific shipments, using a common sub-network selection software. This is done by invoking only those parts of the multimodal network database that are necessary for a specific shipment's routing exercise.

The correct mode sequence for intermodal trips is ensured as follows. We can begin by thinking of all of the CFS network's links as being "switched off", and by processing each reported shipment in turn. A copy of the highway portion of the CFS intermodal network is switched on. For the shipment shown in Fig. 1 a set of highway sub-network access links are generated from the traffic's origin (zip code) by the method describe in Section 4 of this paper. This is done for the first copy of the highway sub-network only. Similarly a set of destination egress links are created and indexed only to a second copy of the (structurally identical) highway sub-network. To bring the intermediate rail portion of the route into the picture the rail sub-network is also switched on as is a

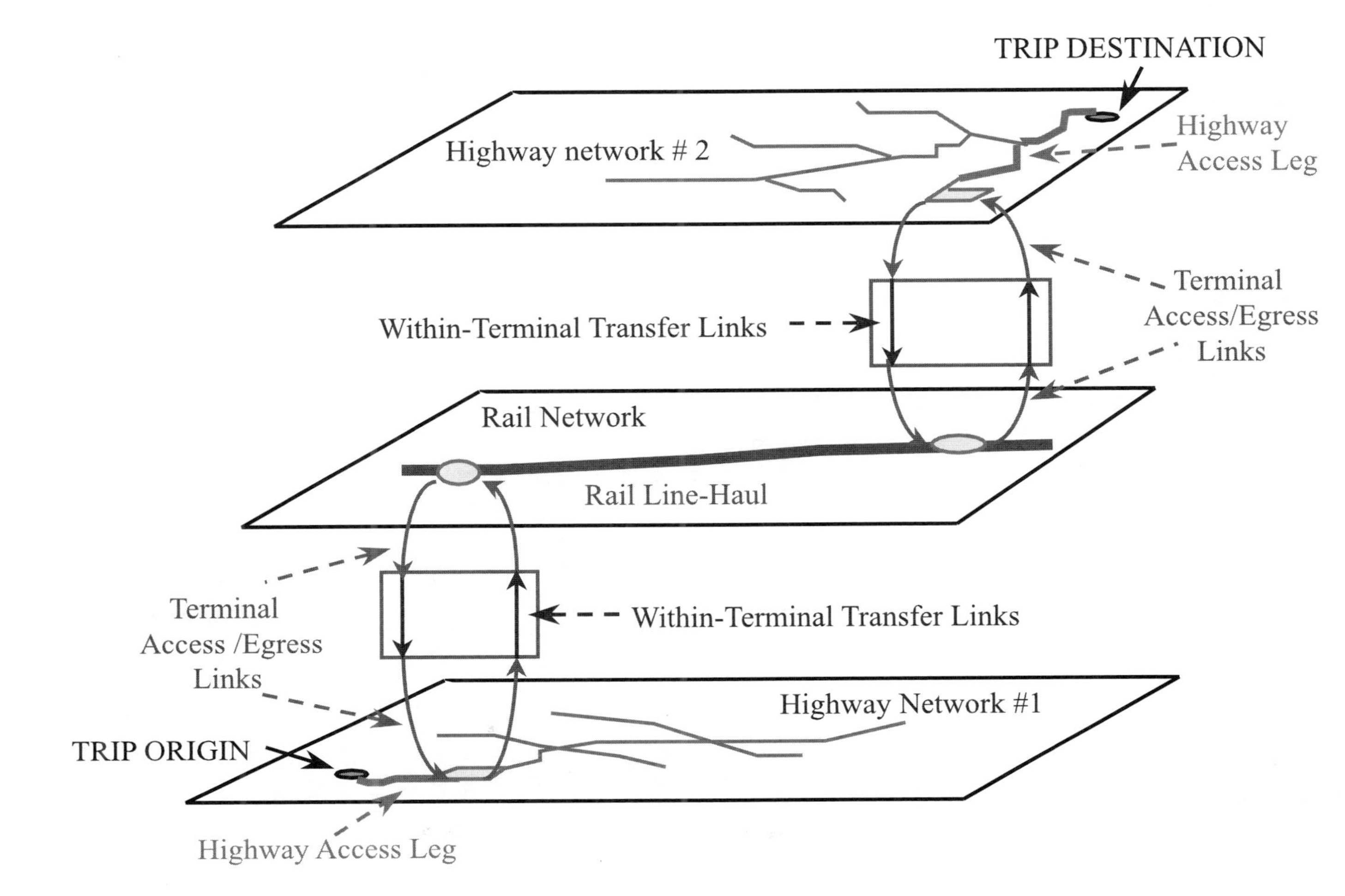

Fig. 1. Construction of a multi-layer intermodal shipment routing.

suitable subset of the CFS network's intermodal truck-to-rail terminal transfer links. All other terminal transfer links, including all rail-to-truck transfers are at this time turned off prior to shipment routing (by assigning them an infinite impedance as a starting default). The remaining, direction-specific terminal transfers then ensure the correct TRT routing sequence reported in the shipper survey. Finally, in making such terminal transfers it was often necessary to also generate, at execution time, a set of local terminal access and egress links not present in any of the modal sub-networks , and notably where the use of trucks was involved. These are also illustrated in Fig. 1.

Efficient organization of the multimodal shipments to be routed made for rapid computer processing on a shipment by shipment basis. Time saving computational procedures were also developed to recognize and store previously computed shipment routes.[2] Once the impedance for a network link or complete origin to destination route has been computed it can be stored for re-use in subsequent routing exercises. How these modal and intermodal impedances and their resulting routes were selected is discussed in Section 4.

2.4. Using the intermodal network within GIS packages

After the intermodal terminals and their local access/egress links to the modal networks are included in the CFS network we have a unified, logical, geographic network with a reduced but common set of attributes necessary for the CFS routing programs. While this network is formally a set of geographic objects, there are a number of features that must be overcome to import and use it within a commercial GIS.

While the individual modal and terminals databases can be readily stored and maintained within commercial GIS software, it was found less efficient to store the composite multimodal network is this fashion. Rather, it is constructed by a custom program that reads and processes the individual databases and produces the data structures required by the CFS routing programs. The impediments in using this or similarly structured networks in a GIS are:

- A large number of logical links are of zero length because they represent a process that is idealized to occur at a point, such as terminal activities and inter-company cargo interlines. Other terminals are co-located with modal network nodes, so their access links appear, geometrically, as points. Most commercial GIS packages do not allow points, even when they are degenerate lines, in a dataset of polyline objects.
- Some other objects exactly overlay each other, such as railroad logical links that represent the trackage rights of different companies on the same physical link, or different logical terminals that handle multiple commodities in the same physical facility.
- Most commercial GIS packages construct network topology geometrically, i.e., if two links share endpoints at the same geographic location, the links will be connected. This will clearly render the CFS network described here analytically useless. (The atypical alternative is to establish connectivity through endpoint nodes identified on each link's attribute list.)

In order to use this network in a standard GIS, two solutions may be implemented during its construction. The first is to slightly perturb logical node locations that would otherwise fall on the

[2] Maintenance of two separate highway sub-networks was also useful for capturing the separate route mileage estimates for private versus for-hire trucks discussed earlier in Section 2.

same point, along with the ending vertices of polylines incident to them. This will simultaneously insure positive line lengths and prevent spurious link connections. A second method (in GIS packages that support it) is to specify infinite turning penalties between links that do not share a logical endpoint node. Since our network routing analyses occurred outside a commercial GIS we did not need to use these solutions. It was necessary, however, for us to load a version of the CFS network into a commercial GIS to display particular routes. The solution we chose was to pick out zero length links and call them points, so that the network would have three classes of objects as separate data layers with corresponding attribute sets: links, nodes, and zero length links. For identification, objects within layers are distinguished further by attributes such as mode or commodity.

3. Getting traffic onto the network

To be made "traffic routable" a network database needs to be connected to a set of traffic origins and destinations, i.e., connected to a set of traffic generators and attractors. The principal technical issue associated with a shipment being placed on the CFS network involved selecting one or more truck, rail or waterway links or nodes that are close to the zip code centroids that represent the shipment's origin and destination. For connections to the highway network this was usually straight forward, since this is a comparatively dense network in both the database and the real world. However, dealing with the less geographically extensive rail or waterways networks required the ability to identify and test a number of different network access *and* egress pairs before eventual route selection(s) can be made. It frequently turned out that the nearest railroad connection to a shipment origin was not selected as part of the least impedance route. This was because an alternative railroad offered either more direct line haul routing to the destination, without recourse to expensive railroad interlining costs along the way, or because it offered a less costly network egress connection at the destination end of the move. Final selection for any shipment routing in all cases depended on the total sum of its network access, plus line haul, plus interlining, plus terminal transfer and network egress costs. The modeler must therefore identify *all* likely alternative routing options, involving all reasonable network access and egress pairs for any shipment, prior to final route selections. While generating and storing these access and egress links is now a feasible option this leads to tens of thousands of additional links being added to the intermodal network. (Plotting these options within a GIS also poses data visualization problems if too many access links are displayed at one time.) The approach adopted for CFS processing was instead to generate these access and egress links "on the fly" for each shipment routing, using the link selection methods described below.

In practice each set of access links generated by the CFS routing algorithms was treated as mode-specific variations on the same general *maximum distance threshold and sectoral search process*, as now described. First, the following formula is used to define a geographic area within which to search for appropriate network access or egress connections (see also Middendorf et al., 1995; Southworth et al., 1997):

$$R_z = \mathrm{RMAX}_z + (2e) + p, \tag{1}$$

where R_z is the local access threshold distance from the centroid of zip code area z, RMAX_z the straight-line distance from the centroid to the farthest point on the zip code area boundary, e the

Table 2
Mode-specific default values of access model parameters[a]

Parameter	Highway	Rail	Inland & Great Lakes	Deep sea
p	0.00 (0.00)	4.97 (8.00)	4.97 (8.00)	6.22 (10.00)
$(2e)$ (Domestic)	0.22 (0.35)	1.86 (3.00)	1.86 (3.00)	1.86 (3.00)
$(2e)$ (Foreign)[b]	6.22 (10.00)	18.65 (30.00)	31.08 (50.00)	31.08 (50.00)

[a] Distances are in miles (kilometers given in brackets).
[b] Includes Canada and Mexico.

root mean square geographic error in the network's representation of link locations, and p is the maximum length of a local access connector (e.g., a local road or rail spur line) not included in the CFS mode specific network databases.

Table 2 lists the initial, mode specific values of e and p used in the above formula.

Using Eq. (1) a mode specific maximum access distance from a zip code centroid is defined and all network links and nodes within that distance are searched as possible network access or egress points. How this search occurs and how specific access links are subsequently selected and linked to the CFS network varied by mode of transportation, as follows:

Highway access: In the case of the comparatively dense highway network the above local access formula p is initially set to zero. In this case such access represents local street mileage, or possibly local minor arterials not included in the national network database. Highway access link distances were nearly always quite short. These access distances are approximated by finding the straight-line distance from the zip centroid to the nearest point on the network (link or node) in each of three mutually exclusive, 120° sectors which meet at the zip centroid. As shown in Fig. 2, these

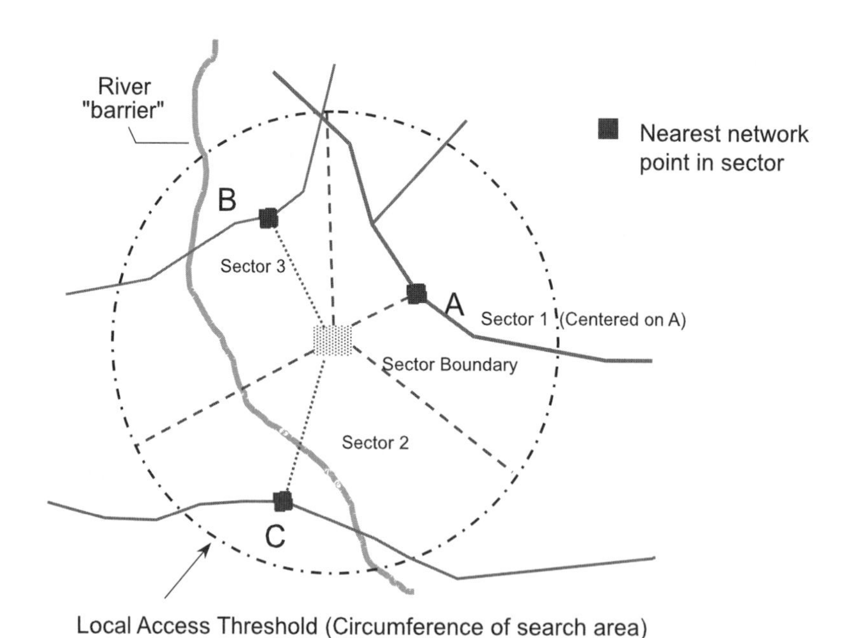

Fig. 2. Highway access modeling using distance threshold and sector search.

sectors are defined initially by finding the nearest point on the highway network and using it as the center of the first of these sectors. Also shown in Fig. 2 is the presence of a river barrier. Where such barriers were known to exist an access link had its length increased to recognize the need to cross the barrier at a point other than that bisected by its straight line approximation, with additional length penalties imposed if a road was of the limited access type. Also, access was not allowed to occur within a ferry link. The resulting network access distances were then multiplied by 1.2 to represent local highway network circuity (i.e., the ratio of network distance to Great Circle distance). In the rare case that no network links existed within R_z the single closest highway link was selected as the access point so that no zip code area was without highway access. For route selection purposes the impedances assigned to access links were those of low level urban or rural collector roads.

Rail access: The 1997 CFS used a modified version of the Federal Railroad Administration's 1:2,000,000 scale network database. This meant there was about a 95% chance that the geographic error in line placement was within $(2 \times e) = 1.86$ miles (3 km): while the maximum length of an industrial rail spur, p, was set at almost 5 miles (8 km). The left half of Fig. 3 shows how Eq. (1) is used. Access links were typically attached to the interior of a rail network link by segmenting it into two or more subdivisions called "shadow links" (because they logically lie beneath the original link). The right half of this figure shows how two shadow links are created. These links are built from the point at which the zip code centroid connector meets the network (point X) to each end-point (nodes A and B) of the real rail link. Each of these shadow links is given a distance equal to the length of that portion of the rail network link it represents, in effect creating a notional node at point X for access modeling purposes. This approach was used in order to retain as much of the rail network's link distance detail as possible.

An alternative to this shadow link approach would be to connect two notional links directly from the traffic generating or attracting centroid to the end-points of the nearest real rail link. A problem with this more data efficient method (i.e., two new links are added instead of three) is the chance that access connectors from more than one centroid may connect to a real network node, causing routes to be built from one set of notional access links to another without traversing any real network links. Fig. 4 shows how the two approaches would work in this case. Of the two

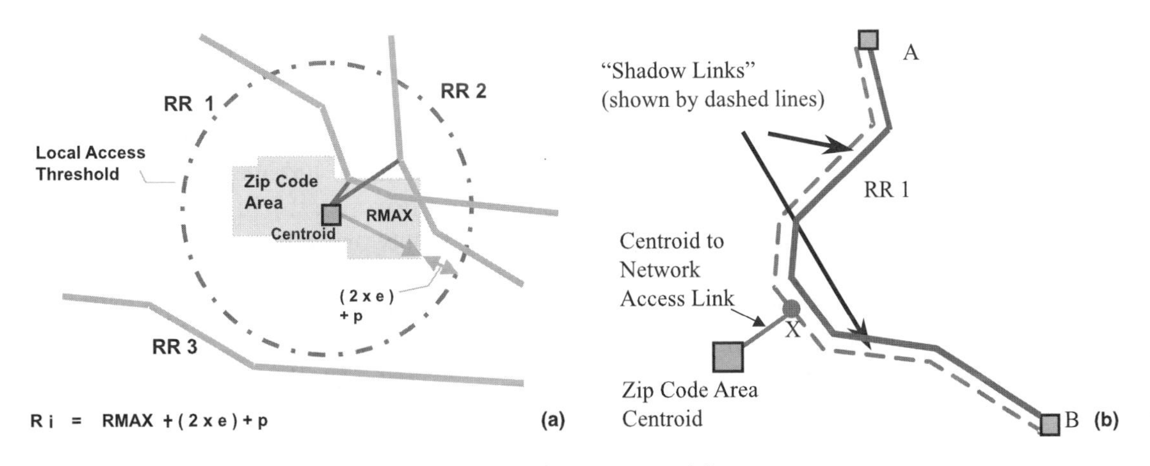

Fig. 3. Rail line access modeling.

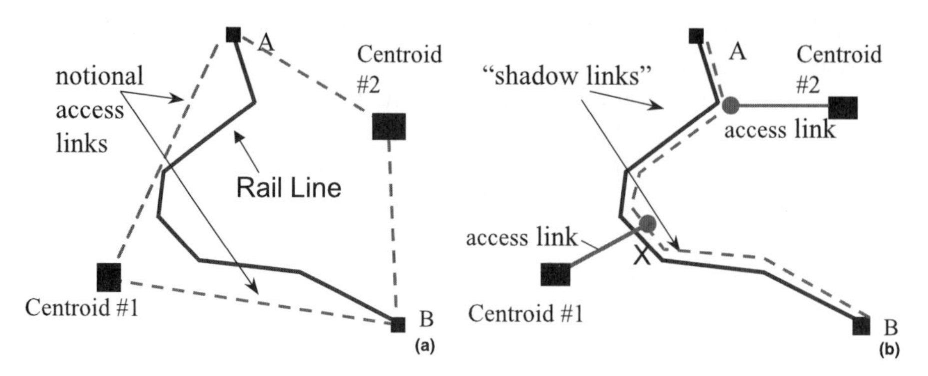

Fig. 4. Alternative rail network connection methods.

approaches only the shadow link approach allows short movements between traffic generators adjacent to the same (rail) network link to receive the correct network distances.

Water access: Navigable waterways access distance is determined in a manner similar to that used for rail access, but with two notable differences. First, given that many of the longer waterway links meander across the landscape, an additional procedure was added to ensure selection of the most likely access location. Once a set of centroid-to-waterway notional access links has been constructed such connectors have their impedance multiplied by 5.0. This value was derived by experiment and generally ensures that the access link with the lowest value of ($5.0 \times$ notional connector straight line distance + network distance along a connected waterway link to its nearest real network node) is selected for routing purposes, i.e., selection strongly favors that point on the river system closest to the zip code centroid. Second, the value for spur length, p, is re-set to zero after its use in the threshold search process, i.e., before the mileage computation is recorded. That is, the distance from the zip code centroid to the nearest waterway link is not included in the CFS distance estimate. This is because shipments reported by the survey to either start or end by waterway are assumed to originate or terminate at a dock located *alongside* the waterway network.

4. Representing intermodal terminal transfers

To derive suitable intermodal routes from different combinations, *and sequences*, of single mode networks requires a network merging to occur. That is, functional linkages are required at locations in the real world where intermodal freight transfers take place. This is accomplished by modeling the operation of intermodal terminals within a network context. Fig. 5 shows two approaches. The approach shown in the left half of Fig. 5 is termed the bi-modal connections model, since each intermodal transfer is represented as a single network link between two different modes of transport. This is the method that ORNL used to model intermodal transfers in the 1993 CFS (Middendorf et al., 1995; Southworth et al., 1997). This is a straightforward solution that allows the modeler to assign a direction-specific transfer cost to each terminal link in the database for the purpose of traffic routing. It can be implemented by adding a series of bi-modal "notional" links to a network's database of single mode links. However, this solution has the limitation that

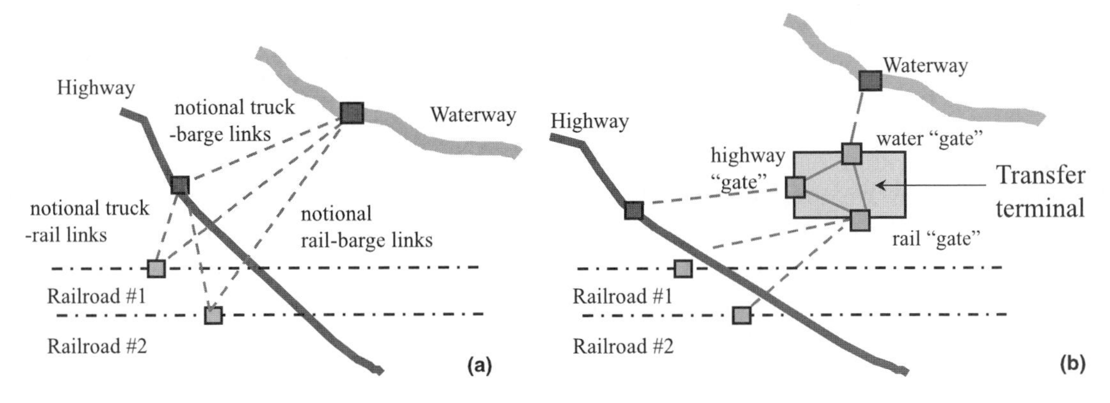

Fig. 5. Two ways to model intermodal terminal transfers.

only a single, catch-all transfer cost or impedance value is assigned to each bi-modal, direction-specific linkage: and changing any single component of transfer cost would involve a re-calculation of each composite bi-modal link cost.

For the 1997 CFS, making use of a detailed intermodal terminals database developed since 1993 (Middendorf, 1998), this bi-modal link transfer model was replaced with the one represented in the right half of Fig. 5. In this approach an explicit transfer facility at a specific geographic location is identified, along with explicit representation of local network-to-terminal access (and egress) links. The local terminal access links shown as single dashed lines in Fig. 5 are actually represented in the database as two, uni-directional linkages for each of a series of terminal access "gates". There may also be more than one such gate per primary mode of transportation. For example, in Fig. 5(b) two separate railroad company connections are shown. The same procedure is used to for every other mode, varying only those parameters specific to a modal sub-network's characteristics.

The more elaborate access model shown in Fig. 5(b) is especially useful where the main-line modal networks are comparatively sparse and therefore need quite long access links to bring terminals into the network: especially where more than one geographical direction of mode specific entry/exit exists to a large terminal complex. Indeed, it is often at these local access and transfer points that major delays, and hence costs, occur in today's freight movement system. This approach also allows a more realistic representation of within-terminal versus outside terminal operating and maintenance costs should such a logistical analysis prove to be of interest. Finally, it allows terminal and network representation issues to be de-coupled, so that multiple terminal sets or models may be used with the same networks, and vice versa. It must be noted, however, that data at the level shown in Fig. 5(b) are far from being universally available at the present time (Middendorf, 1998).

5. Intermodal route selection methods

Putting CFS shipments onto the CFS network for the purpose of estimating mode and commodity specific ton-miles and dollar-miles of freight activity required a method or methods for first of all generating sensible single and multi-modal routes, and where more than one route was likely to be used, a method for assigning percentages of shipment volumes to each of these candidate routes. For consistency with the 1993 CFS single route truck freight modeling was used to compute 1997 CFS shipment distances. Single path waterway routing was also the norm. However, where rail dominated a route's mileage (both rail only and rail-inclusive intermodal routing) the situation was more complicated. More than one rail carrier-specific route was often both plausible and likely, and therefore each route needed to be both generated and assigned a portion of the origin-to-destination volume. Rail shipment volumes were then spread across a limited number of highly likely rail routes using a logit assignment model calibrated roughly to the tonnages carried on the high volume traffic corridors reported in the Surface Transportation Board's annual railcar waybill sample (AAR, 1998).

A "good" route, for CFS purposes, is a route that reproduces the shipper reported mode sequence and can either be validated using other data sources, or in the absence of such sources can stand up to some common sense rules associated with the economics of freight movement. Recourse to the literature on multimodal freight routing practices, including the work of Friesz et al. (1986), Harker (1997) and Guelat et al. (1990), indicates a complex set of factors influencing actual routes taken, involving carrier as well as shipper decisions, and one that also varies by commodity type. However, empirical validation of a large number of such mode and commodity specific route selection models was beyond the resources of the study. Nor would such a thing be easy to accomplish given the current state of freight movement data across the nation as a whole, and notably so for movements involving trucks (see Southworth, 1997). To ensure the selection of sensible routes, therefore, link specific impedances were developed to represent the generalized cost of different en route activities, including the costs of:

- local access to major traffic ways and terminals;
- within terminal transfer activities including loading and unloading between modes, vehicles, and railroad companies;
- negotiation of border crossings; and
- the line haul costs in different corridors.

In all cases one or more routes through the CFS network are identified by a shortest path routine. Path length is determined here on the basis of a set of modal impedances This process starts with a set of what we term "native link impedance functions", i.e., native to the mode in question. In the case of the highway network these native impedances are assigned based on a number of link attributes, notably distance and urban and rural functional class, with default link traversal speeds modified on the basis of traffic conditions, access controls, the presence of a toll or a truck route designation and whether the highway is divided or not. The native impedance for highways is therefore a surrogate travel time impedance. Route selection over the railroad network, in contrast, is determined by an evaluation of line importance called "main line class". Though primarily based on traffic volumes (e.g., "branch" lines carry less than five million gross tons/year and "A-main" lines more than 30 million), we subjectively modified these classes on the basis of operating conditions and the principal commodities carried. Our routing procedure also required the identification and assignment of an impedance to those "interline" points where railcars may be transferred between separate railroad companies, data that is now a part of the rail network database. Waterway routings in contrast were comparatively straight-forward for the most part, as only a single waterway route was typically available and competitive. However, where Great

Lakes transport was an option the differences in impedances, as well as the costs of transferring cargo from or to shallow draft barges needed to be incorporated into the network, requiring additional, within-mode cargo transfer links.

Given these native link impedances the next step is to determine the relative costs of transport between the different modes. This depends in reality on a number of shipment characteristics – value and weight, the importance of service reliability, ease of facility access and cargo handling, among others. Fortunately the CFS problem is made much simpler because the modes used are already known to the analyst. The routing problem is then one of locating these likely transfer points between modes. By and large we presumed that if a less expensive mode was used at all it was used preferentially for as large a proportion of the trip as practicable, relegating more expensive modes to an access role.[3] This lead us to factor native modal impedances to ensure that the lowest cost mode would be used predominately, other things being equal. First, native impedances on each mode were scaled so that one mile of travel on the best type of facilities of that mode would incur one impedance penalty unit.[4] These native impedances were then multiplied by the following *relative modal impedance factors* to produce a unified network with consistent intermodal impedances:

Highway	1/1.0
Railroad	1/3.5
Inland water	1/5.8
Great Lakes	1/6.6
Ocean	1/7.0

In this approach the truck mode acts as the "base" or highest impedance mode. For example, one would accept a path that increased rail mileage 3.5 miles in order to reduce the highway portion of the trip by one mile.

It should be noted that these relative modal impedance factors are not intended to estimate relative freight transport costs per se, or even generalized costs. They are used simply to force realistic route selections from which sensible mileage estimates can then be drawn. The general effect of using these impedance factors was to place the vast majority of the mileages on the least expensive mode. If water was used it dominated the route miles. Otherwise rail dominated, with highway usually acting as the mode of terminal access and/or egress where a great deal of intermodal routing was concerned. Once a set of intermodal routes had been generated a number of additional checks were carried out. In particular, a specific intermodal, terminal inclusive route was considered to be unreasonable if one or more of the following criteria was met:

- the route circuity factor was too high,
- there was an unlikely split between the different modal mileages, or
- there existed contradictory expert knowledge.

[3] There are a few exceptions, such as truck or container-sized shipments to Hawaii, which were presumed to use a Pacific coast port in preference to passage through the Panama Canal.

[4] For highways such a "best" facility was a rural interstate road. For rail it was an "A-Main" rail line. For waterborne commerce, with few places where any route choice existed, all links were treated as equal within each of the three vessel types.

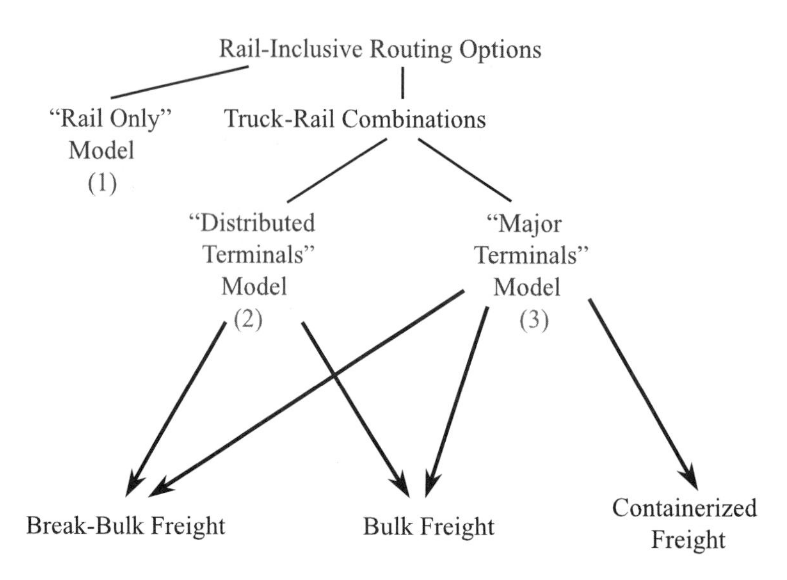

Fig. 6. Family of rail-inclusive freight shipment routing models.

Unlikely splits between modal mileages occur when the routing algorithm selects paths with long mileages on a more expensive mode relative to a less expensive one. This latter can also occur when the algorithm selects mode-specific mileages by going through a transfer terminal that produces mileages that are much longer than a direct trip by a single mode would be. With a little computer programming it was possible to pick out these questionable routings from among very long data lists and investigate these cases in more detail, subsequently using a GIS package to display questionable routes.

Alternative TR routing models. Many of the dubious cases identified by the above route validation criteria involved TR intermodal moves. A majority of the intermodal shipments reported in the 1997 CFS involve these two modes. To address these issues two different TR routing models were developed. These models are termed respectively the "major terminals" model and "distributed terminals" model. In particular, a distinction was made between containerized and non-containerized (bulk and break-bulk) freight. As shown in Fig. 6, freight designated as containerized by 1997 CFS shippers was handled by the major terminals model, and specifically by allowing TR transfers to occur only at those terminals where containerized traffic was known to be handled.[5] Where non-containerized shipments were concerned the ORNL terminals database provided the first set of candidate intermodal transfer locations tried by the rail-inclusive routing algorithms. If the resulting circuity was found to be unacceptably high for a specific origin–destination shipment, or if the resulting allocation of highway-to-rail mileage was deemed too high to warrant expensive rail-based intermodal transfers, then the alternative "distributed

[5] For the continental United States, and from among the more than 2900 terminals contained within the ORNL intermodal terminals database at that time (mid-1997), some 256 TR containerized cargo terminals were identified, involving either Trailer-on-Flatcar or Container-on-Flatcar operations.

terminals" model was applied. In such cases a "major terminals" routing alternative was considered suspect when either of the following conditions was violated:

- when the route circuity is more than 2.5 times the Great Circle Distance,
- when the highway proportion of the entire origin-to-destination route length is greater than 25%.

A GIS is a valuable tool here for examining suspect terminal-inclusive routes. Rejection of a route led to the use of the "distributed terminals" model. This model assumes that for certain types of TR intermodal movement there will be a team track or other rail transfer facility within a reasonable distance of the shipment origin or destination (depending on the TR, RT, or TRT mode sequence involved). Without knowing where all of these terminals are located the model posits a TR transfer facility at the single closest node on each rail company's sub-network, for all rail nodes within a 90 mile search radius of the truck end of the trip. Once located, a highway route between a zip code traffic generator and these "ad hoc" terminals is then constructed. If the result obtained from this distributed terminals model was deemed significantly better, in the sense of the route being noticeably less circuitous than that supplied by the major terminals model, it was accepted. As a practical matter, access links to all major terminals are constructed by the ORNL routing procedures at network generation time. Access links to ad hoc terminals under the distributed terminals model are constructed at model execution time.

Handling export shipments. Routing export shipments within the 1997 CFS required data on the US seaport of exit as well as the domestic origin and foreign destination of the movement. Where US port of exit data was missing from otherwise useful shipment records a method for imputing the most likely port of exit was devised. This was done by adding deep sea impedances to within-US truck, rail and/or waterway impedances associated with each export shipment. The resulting US port-inclusive, relative origin-to-destination impedances were then used to estimate a set of travel impedance-discounted comparative port attraction factors, with the most attractive port(s) being assigned the export shipments (Southworth et al., 1998). In terms of ton-mileage and other distance calculations the non-US portions of these routes are not reported by the CFS, so that the principal value of the routes to the survey is to identify the US origin to US port of exit mileages involved. Fig. 7 shows an example trans-oceanic shipment routing, using a commercial GIS to pan and zoom in along three specific sections of the route. This particular shipment[6] begins by rail in East St. Louis with cargo transfer in the New York–New Jersey port area followed by trans-Atlantic shipping to the port of Amsterdam in the Netherlands. Expert knowledge of the network options and port terminals involved can be verified quickly with the aid of the GIS.

6. Summary

This paper has described the development of a large and detailed multimodal network, created and stored in digital form for use in a specific freight traffic routing study: the 1997 United States Commodity Flow Survey (CFS). While commercial GIS software was found to be invaluable for

[6] Note, this is an example route, not an actual shipment route drawn from the 1997 CFS database, for which detailed information may not be divulged.

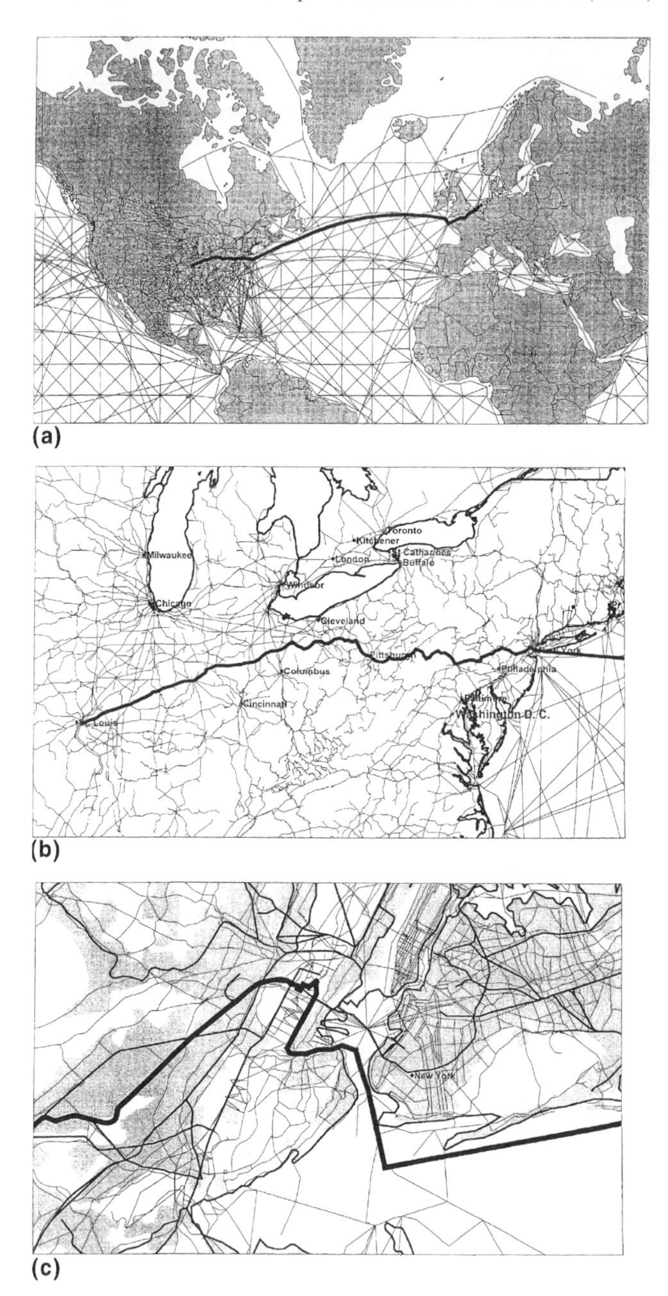

Fig. 7. Three views of a rail-water trans-oceanic route: (a) pan view of route, (b) US portion of rail route, and (c) zoom in on rail-water port transfer.

displaying, checking and editing the network, it was also found to be most efficient to construct and process shipment routes outside this environment. Among other benefits this approach allows different mode specific line haul networks to be linked together via more than one data representation for transportation terminals, and using more than one approach to defining local network access and egress. The procedures described in the paper are for the most part the generic or default methods used to automate this process. The reader should note that adjustment and manipulation of the various network parameter settings reported in the paper were often needed to accommodate specific route selections, where available empirical evidence suggested that they were warranted. Most of this post-model development adjustment activity is currently necessary because there exists no nationally representative sample of how either trucks or intermodal shipments move around the United States against which to calibrate such routing models. For rail and water movements this situation is somewhat better, if still far from ideal. Given these cautions, geographically referenced digital databases such as the CFS network now offer a starting point from which to model freight activity in more detail, on a route or corridor specific basis.

As long as such activities as cargo handling and intermodal transfer can be translated into link-specific measures of relative modal impedance, they can be used within the network modeling framework described in this paper to investigate the implications for future freight movements from introducing new *and location-specific* freight movement technologies. This includes the simulation of often far reaching congestion impacts caused by in-transit delays at major seaports or other traffic bottlenecks (see, for example, Port of Long Beach, 1994). While validation of many CFS-generated routes is currently problematic due to the limitations of alternative data sources, the generation of sensible traffic routing options, based on shipper reported mode sequences, provides a good deal of insight into the infrastructure-constrained options available for intermodal transportation within the continental United States. This is a topic of growing importance as the globalization of trade puts competitive pressures on national economies to increase their freight transportation carrying capacity. A detailed geography of freight transportation networks will make it easier to anticipate as well as understand the need for these new capital investments, both at home and abroad, and within increasingly inter-dependent national and international freight movement systems.

Acknowledgements

This work was made possible by funding from the Bureau of Transportation Statistics, US Department of Transportation. Methods described benefited from the insight of a number of ORNL staff within the Center for Transportation Analysis, including Drs. Shih-Miao Chin, Cheng Liu, David Middendorf, Mike Bronzini and Ms. Jane Rollow.

References

AAR, 1998. User Guide to the 1997 Surface Transportation Board Carload Waybill Sample. Association of American Railroads, Washington, DC 20001.

Bureau of the Census, 1995. 1993 Commodity Flow Survey. United States. TC92-CF-52. US Department of Commerce, Washington, DC 20233.

Bureau of Transportation Statistics, 1997a. Transportation Statistics Annual Report 1997. US Department of Transportation, Washington, DC 20590.

Bureau of Transportation Statistics, 1997b. Truck movements in America: shipments from, to, within, and through states. Transtats 1. US Department of Transportation, Washington, DC 20590.

Chin, S-M., Peterson, B.E., Southworth, F., Davis, R.M., Scott, R.G., 1989. Graphics display of convoy movements. In: Proceedings, ASCE Conference on Microcomputers in Transportation Planning. Oak Ridge National Laboratory, San Francisco, July.

Crainic, T.G., Florian, M., Guelat, J., Spiess, H., 1990. Strategic planning of freight transportation: STAN, an interactive graphic system. Transportation Research Record 1283, 97–124.

Friesz, T.L., Gottfried, J.A., Morlok, E.K., 1986. A sequential shipper-carrier network model for predicting freight flows. Transportation Science 20, 80–91.

GAO, 1996. Intermodal Freight Transportation. Projects and Planning Issues, United States General Accounting Office, GAO/NSIAD-96-159.

Guelat, J., Florian, M., Crainic, T.G., 1990. A multimode multiproduct network assignment model for strategic planning of freight flows. Transportation Science 24, 25–39.

Harker, P.T., 1997. Predicting Intercity Freight Flows. VNU Science Press, Utrecht.

Jourquin, B., Beuthe, M., 1996. Transportation policy analysis with a geographic information system: the virtual network of freight transportation in Europe. Transportation Research C 4 (6), 359–371.

Middendorf, D., 1998. Intermodal terminals database: concepts, design, implementation, and maintenance. Report Prepared for the Bureau of Transportation Statistics by Oak Ridge National Laboratory, Oak Ridge, TN 37831.

Middendorf, D., Bronzini, M.S., Peterson, B.E., Liu, C., Chin, S-M., 1995. Estimation and validation of mode distances for the 1993 Commodity Flow Survey, In: Proceedings of the 37th Annual Transportation Research Forum. Reston, VA 22090, pp. 456–473.

Port of Long Beach, 1994. The national economic significance of the Alameda corridor. Report Prepared for the Alameda Corridor Transportation Authority, Long Beach, CA 90801.

Southworth, F., 1997. Development of data and analysis tools in support of a national intermodal network analysis capability. Bureau of Transportation Statistics, US Department of Transportation, Washington, DC (http://www.bts.gov/gis/reference/develop/develop.html).

Southworth, F., Peterson, B.E., Chin, S-M., 1998. Methodology for estimating freight shipment distances for the 1997 Commodity Flow Survey. Report Prepared for the Bureau of Transportation Statistics, US Department of Transportation, Washington, DC 20590.

Southworth, F., Peterson, B.E., Davis, R.M., Chin, S-M., Scott, R.G., 1986. Application of the ORNL highway network data base to military and civilian transportation operations planning. Papers and Proceeding of Applied Geography Conferences, vol. 9. pp. 217–227.

Southworth, F., Xiong, D., Middendorf, D., 1997. Development of analytic intermodal freight networks for use within a GIS. In: Proceedings of the GIS-T 97 Symposium. American Association of State Highway and Transportation Officials Conference, pp. 201–218.

PERGAMON

Transportation Research Part C 8 (2000) 167–184

TRANSPORTATION RESEARCH PART C

www.elsevier.com/locate/trc

A new framework for the integration, analysis and visualisation of urban traffic data within geographic information systems

C. Claramunt [a,*], B. Jiang [b], A. Bargiela [a]

[a] *Department of Computing, The Nottingham Trent University, Nottingham NG1 4BU, UK*
[b] *Division of Geomatics, Institutionen för Teknik, University of Gävle, SE-801 76 Gävle, Sweden*

Abstract

Current geographical information systems (GIS) are not well adapted to the management of very dynamic geographical phenomena. This is due to the lack of conceptual and physical interoperability with real-time computing facilities. The research described in this paper is oriented towards the identification and experimentation of a new methodological and applied framework for the real-time integration, manipulation and visualisation of urban traffic data. It is based on proactive interaction between the spatio-temporal database and visualisation levels, and between the visualisation and end-user levels. The proposed framework integrates different spatial and temporal levels of granularity during the analysis of urban traffic data. Urban traffic behaviours are analysed either by observation of the movements of several vehicles in space, or by changes in urban network properties (i.e., micro- versus macro-modelling). Visualisation and interaction tools together constitute a flexible interface environment for the visualisation of urban traffic data within GIS. These concepts provide a relevant support for the visual analysis of urban traffic patterns in the thematic, spatial and temporal dimensions. This integrated framework is illustrated by an experimental prototype developed in a large town in the UK. © 2000 Elsevier Science Ltd. All rights reserved.

Keywords: GIS; Urban traffic; Real-time; Visualisation

1. Introduction

Recent developments in information technology are having a major effect on the way in which systems are designed and used in many application fields. Geographical information systems

* Corresponding author.
E-mail addresses: clac@doc.ntu.ac.uk (C. Claramunt), bin.jiang@hig.se (B. Jiang), andre@doc.ntu.ac.uk (A. Bargiela).

PII: S0968-090X(00)00009-7

(GIS) have been adopted as a successful solution by a wide range of disciplines such as environmental planning, business demographics, property management and urban studies to mention some examples. Currently, one of the most important challenges for GIS is to generate a corporate resource whose full potential will be achieved by making it accessible to a large set of end-users. In the urban domain, an important issue is the development of a co-operative traffic GIS that integrates static urban data with dynamic traffic flows (Pursula, 1998). Such a system will be of great interest for many applications related to the monitoring and analysis of urban traffic in which represented vehicles or network properties are changing in a fast and almost continuous mode. Recent advances in traffic systems include the development of graphical interfaces as a new functional level of monitoring tasks (Peytchev et al., 1996; Kosonen et al., 1998; Barcielo et al., 1999) and real-time traffic interfaces in the World Wide Web that display traffic conditions on a regular basis (Dayley and Mayers, 1999; Feng et al., 1999). Nevertheless, the functions provided by these solutions are quite limited in terms of the analysis and visualisation of urban traffic data. Furthermore, we believe that the full potential and benefit of traffic databases still need a closer integration with GIS that will facilitate the integration of traffic data as a component of urban and transport planning and environmental and health studies. However, current GIS software and interfaces do not provide the set of functions to make this technology compatible with traffic systems used for monitoring and simulation purposes. First, the integration of GIS and traffic systems is likely to be a challenging and worthwhile objective for user communities whose needs are not satisfied by a loosely connected set of existing systems. This poor level of integration is often the result of the different paradigm used within GIS and modelling systems and the fact that many integrated solutions often imply the re-design of existing solutions (Abel et al., 1992). Secondly, despite recent progress in the development of temporal GISs (e.g., Langran, 1992; Peuquet, 1994; Claramunt and Thériault, 1995), current GISs are still not adapted to the management of very dynamic geographical phenomena due to the lack of interoperability with real-time computing facilities. Moreover, the development of GIS applications, characterised by a high frequency of changes, implies a reconsideration of the modelling, manipulation, analysis and visualisation functions as GIS models and architectures have not been preliminarily designed to handle the properties of very dynamic phenomena.

The research described in this paper is oriented to the identification and experimentation of a new methodological framework for the real-time integration, manipulation, visualisation and animation of urban traffic data within GIS. We characterise a very dynamic GIS (VDGIS) as a GIS application which has a high frequency of change (e.g., real-time traffic databases, simulated traffic databases). Changes include modifications to located properties (e.g., traffic flows within an urban network) and moving properties of one to several geographical objects (e.g., vehicle positions within an urban network). A high frequency of change corresponds to a small temporal unit of change, which can be generally quantified in seconds or minutes. The scope of VDGIS is relatively large. It also includes, for example, real-time applications that monitor a large volume of urban or environmental data, and simulation systems that attempt to predict the future states of a real-world system. VDGIS objectives are multiple, from the control of the geographical locations of one to several moving objects, to the support of analysis tasks oriented to the identification of complex spatial behaviours. For example, traffic monitoring and simulation applications integrate real-time locations of vehicles on a second, or less, temporal granularity basis. The integration of GIS capabilities within these engineering systems is a promising challenge to explore as it could expand the current data management and visualisation functions of these systems, and further increase the potential benefits of the large information flows generated. Our research explores some experimental methods for the real-time integration (i.e., pre-processing), manipulation, visualisation and animation of dynamic phenomena within VDGIS. We analyse the representation of urban traffic data by either an observation of the movements of several vehicles in an urban network (microscopic level), or changes in the traffic properties of an urban network (macroscopic level). Moreover, due to the nature of VDGIS, urban traffic behaviours are represented at different levels of spatial and temporal granularities. We propose an integrated framework based on the interaction between spatio-temporal data and visualisation tasks on the one hand, and visualisation and end-users on the other. This two-step approach supports a flexible and interactive visualisation of urban network properties either for presentation, analysis or for exploration purposes. As such, this framework constitutes a proactive environment as defined in Buttenfield (1993). It supplies a dynamic component to GIS that allows the derivation of dynamic data from the successive states of a real or simulated urban traffic system. It also provides a multi-layered approach that combines thematic, spatial and temporal queries and visualisations that together constitute a flexible and powerful communication resource for the understanding and analysis of traffic events and patterns within an urban network.

The remainder of this paper is organised as follows. Section 2 introduces a brief review of the integration of the temporal dimension within GIS and cartography. Section 3 presents the methodological principles of a multi-layered approach that supports the visualisation of urban traffic properties. Section 4 introduces the roles of the pre-processing, visualisation and interaction tools in the development of our framework. Section 5 presents an application of these concepts in the context of a traffic GIS used in a large town in the UK. Finally, Section 6 draws some conclusions and outlines for further work.

2. GIS, cartography and time

A definition of temporal map has been suggested as a representation or abstraction of changes that support the understanding and analysis of dynamic phenomena (Kraak and MacEachren, 1994). Within GIS, a map can be considered as a visualisation and interactive tool generally oriented to the representation of a spatial configuration at a specific instant in time, or a spatial configuration valid for an interval of time. Cartography has a long tradition of representing spatial information in time. An early example is the Minard map realised in 1869 to illustrate Napoleon's march to Moscow (Tufte, 1983, p. 176). Nowadays, with the availability of a large volume of spatio-temporal data, both GIS and modern cartography face the challenge of exploring new methods and techniques to analyse patterns and structures in space and time (Buttenfield, 1993).

In conventional temporal maps, two approaches are often used in visualising dynamic phenomena (Szego, 1987; Campbell and Egbert, 1990). The first one, the map series approach, is composed of successive maps valid for an instant or a period of time (Fig. 1(a)). A map series along the time line provides a global view of change that is useful for the understanding of overall patterns. However, this approach is limited to studies oriented to the analysis of local changes and interactions in space and time. The second approach is more oriented to the representation and

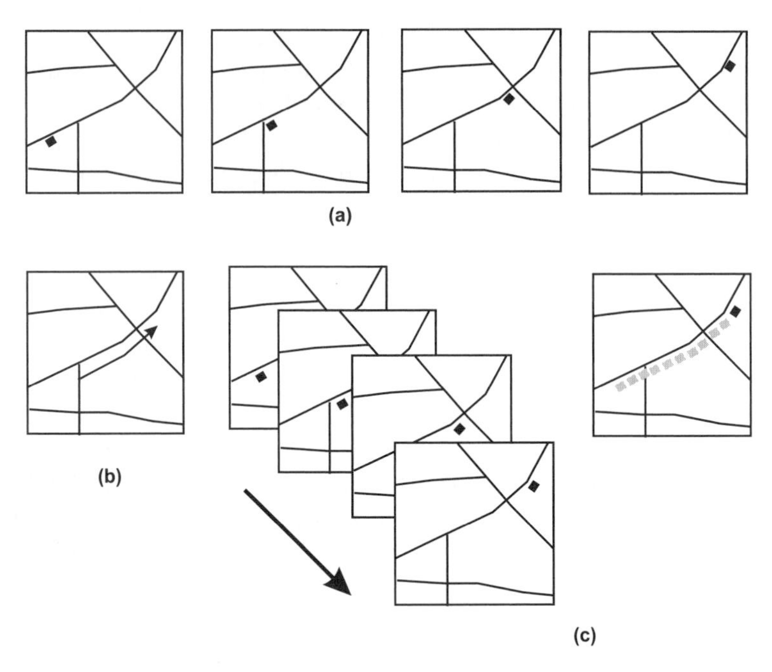

Fig. 1. Visualisation of changes with: (a) map series; (b) map symbol; (c) computer animation.

visualisation of local changes at the individual object level; cartographic symbols such as an arrow line, for example, are commonly used to indicate the trace of a dynamic object in space (Fig. 1(b)).

In order to facilitate the representation and analysis of dynamic phenomena and change patterns, cartographers have developed temporal aggregation mechanisms to reduce the number of snapshots of a map series, and to provide a level of temporal analysis adapted for the particular needs of an application domain. These concepts can be illustrated with the study of population migration in urban areas using temporal maps (Szego, 1987), or at the local level, by the map representation of daily individual activities (Parkes and Thrift, 1980). Animation techniques (Fig. 1(c)) also play an important role in the analysis of dynamic geographical data, as they provide a form of temporal continuity, thus facilitating an understanding of processes and changes. An animated map is a cartographic statement that occurs in time; its interpretation is based on the human sensitivity to detect movement or changes in a graphic display (Peterson, 1995). One of the first cartographic animations was in fact the urban growth simulation in the Detroit region developed by Tobler (1970). The combination of GIS and animations provide powerful platforms to simulate very dynamic phenomena. These can be used for the analysis of real-time systems (Valsecchi et al., 1999) or the simulation of human behaviours (Jiang, 1999).

The availability of a large volume of spatio-temporal data has stimulated research interest in visualisation and exploration of dynamic phenomena (Robertson, 1988; Campbell and Egbert, 1990; Kraak and MacEachren, 1994; MacEachren, 1994; Jiang, 1996). A map, or an interactive map, supports the visualisation of spatio-temporal objects, or properties in space, individually, but also more interestingly in a logic way, from which map users can perceive spatial relationships, density, arrangements, trends, connectivity relationships, hierarchies and spatial associations (Muehrcke, 1981). Efforts have been made on implementing strategies for the cartographical exploration of time-series data (Monmonier, 1990), proactive graphics (Buttenfield, 1993), and the identification of dynamic variables for the visualisation of changes (DiBiase et al., 1992; MacEachren, 1994). Recent advances have explored time-series animation of urban growth (Buziek, 1997), geological changes (Bishop et al., 1999), and socio-economical changes (Andrienko and Andrienko, 1998). MacEachren et al. (1999) have investigated the integration of geographic visualisation and data mining for knowledge discovery in the context of spatio-temporal environmental data. Various new terms have been used to reflect this evolution of cartography such as 'animated cartography' (Peterson, 1995), and 'exploratory cartography' (Kraak, 1998). Techniques for visualising time and change in cartography and GIS also benefit from related research areas such as information visualisation, human–computer interaction and data mining (McCormick et al., 1987; Schneiderman, 1994; Card et al., 1999; Chen, 1999). In particular, visualisation techniques used for the analysis of communication traffic in large computing networks are of interest as they are based on comparable network models. Let us mention among others the development of filtering, interactive manipulation of visualisation parameters, and the interactive use of temporal (e.g., selection of appropriate temporal periods) and spatial operations (e.g., zooming) for displaying telecommunication network traffic (Eick, 1996).

3. Visualisation of very dynamic phenomena: a multi-layered approach

Due to the limitations of current solutions, temporal GISs are often oriented to a cartographical visualisation and analysis of real-world phenomena that have a relatively low frequency of changes that cannot be considered as very dynamic according to our definition of VDGIS. Moreover, traditional maps show various limitations for the representation of very dynamic phenomena, due to the fact that traditional maps serve as both a visual representation and information repository of a real-world system. Nowadays, with the development of GIS and visualisation and animation techniques, maps are often considered as a proactive derivation of GIS data from the database level.

In the context of VDGIS, a visualisation can be considered as a result of several functional tasks, which depend on the database and cognitive levels. Firstly, from a technical point of view, visualisation functions are dependent on the database model and query processing characteristics, i.e., visualisation as an interface to a database. The expressive power and the flexibility of a visualisation are partly dependent on the database properties such as the expressiveness of the underlying database model and the data manipulation language. Secondly, a visualisation is subject to the end-users' perception, i.e., whether or not certain visualisation schemes are efficient in conveying relevant information. Therefore, an analysis of the principles that support the visualisation of geographical phenomena leads to our proposal of a multi-layered architecture that makes a distinction between: (1) the database and processing levels, (2) the visualisation interaction level, and (3) the end-user interaction level. These three components are characterised as follows.

Database level: The expressiveness of the database level is qualified by the database model used (e.g., relational, geo-relational, object-oriented), the query language implemented (e.g., SQL, geographical query languages), and additional analysis and statistical functions available. Very dynamic geographical data are characterised by an important volume of data generated (in the spatial, thematic and temporal dimensions) which constrains manipulation and visualisation operations.

Visualisation level: In the context of this research, the visualisation level represents derived and displayed geographical data presented through an interface that supports visual presentations of very dynamic phenomena. The visualisation level includes interaction tools, that is, the set of logical and physical computer facilities that allow users to act on the visualisation level (e.g., using menus, messages, keyboard and mouse actions).

End-user level: End-user backgrounds vary from novice, intermediate and expert users. They all intend to use a visualisation as a facility to enhance the understanding and perception of very dynamic phenomena.

These three component levels of a very dynamic system, oriented to the visualisation of real-world phenomena, provide a highly interactive environment. These three layers interact in successive order: the first interaction level represents the manipulation functions in which spatial, thematic and temporal operations apply on the database level; the second interaction level represents the control functions on the visualisation tasks operated by final users (Fig. 2). Due to huge information flows between these different levels, this framework also requires a flexible interface between the database and visualisation tasks in order to select appropriate database operations (e.g., spatio-temporal queries), and the visualisation tasks and the user level in order to achieve flexible and frequent user-interactions (e.g., visualisation and animation functions).

Current spatial database models do not consider the visualisation level as a proper part of the system but rather as an interaction level given to the final users. The main components considered in geographical database modelling and design are the integration, representation and querying of spatio-temporal data. However, recent research suggests that visualisation and user-interaction functions can be integrated as a proper component of the database modelling level. This includes the integration of graphic parameters within the query language and the modelling of query operations (Claramunt and Mainguenaud, 1996), description and manipulation of a visualisation as a set of derived objects (Voisard, 1991), and connections of different visualisations defined at complementary levels of granularity using semantic relationships (Stonebraker et al., 1993). We believe that these concepts are particularly relevant in the context of very dynamic visualisations

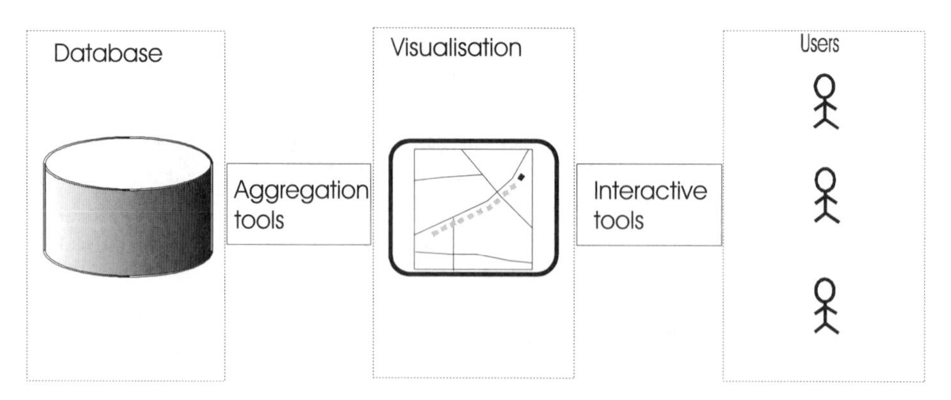

Fig. 2. A framework for visualising very dynamic phenomena.

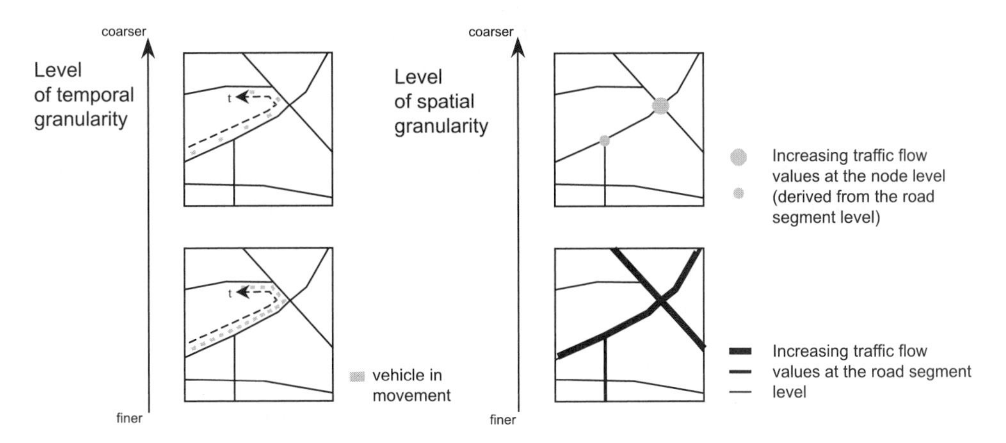

Fig. 3. Temporal versus spatial levels of granularity.

in which the information flows between the database, visualisation and user levels are particularly intensive. Such concepts provide a solution to the derivation of spatio-temporal data, as a query language that supports the expression of user-defined spatio-temporal queries on the one hand and as an interaction language between users and the visualisation level on the other.

We promote a flexible view of this multi-layered approach to the visualisation of very dynamic phenomena, which is independent of any data model or query language. The set of principles explored defines a user-defined level between a spatio-temporal database and the range of visualisation mechanisms required for the manipulation of very dynamic databases. Within the proposed framework, proactive aggregation tools are illustrated by the composition of derived views based on different levels of granularity in the temporal and spatial dimensions (Fig. 3). This interactive visualisation level allows for the manipulation of this user-defined level using different visualisation mechanisms (e.g., animated map, animated chart) depending on the objective of the end-user(s).

4. Pre-processing, visualisation and interaction

Within VDGIS, visualisations require a reconsideration of manipulation, animation and analysis functions. This is due to the very dynamic nature of the geographical phenomena represented. In the context of urban traffic data, a visualisation integrates the properties and behaviours of dynamic vehicles within their environment. It is widely recognised that the aim of visualisation is not only for visual representation but also for information exploration and discovery. Therefore, visualisation is supported by interactive and proactive functions such as basic display functions, navigation and browsing manipulations, query operations, integration of different granularities, re-classification of data, and combination of multiple views and animations (Kraak, 1998). Such a proactive environment facilitates the user's understanding and perception of dynamic phenomena.

A visualisation is different from the concept of traditional maps in various ways. Firstly, a visualisation can integrate multiple views; e.g., a geographical representation associated to a set of interconnected maps defined, for example, using different scales, and additional data media (e.g., photographs). Secondly, a visualisation goes further by integrating animated scenes that represent the temporal evolution of a region of interest and/or thematic property changes. A visualisation is then more than a static map as it also integrates a dynamic component, which can be user-controlled depending on the properties of the phenomenon represented. The structure of a visualisation is not a linear one but rather a complex one composed of different levels of granularity in both the spatial and temporal dimensions. Thirdly, a visualisation is completed by interactive tasks for the manipulation of its visual components (e.g., pan and zooming functions). As such, a visualisation is supported by a set of interactive facilities offered to end-users, which can then manipulate different visual components in order to develop a user-oriented perception of real-world phenomena.

In the context of VDGIS, visualisations combine geographical and thematic data along the temporal line from different media and sources using visual communication techniques. A visualisation integrates multiple components such as maps, charts and tables. Interactive functions, such as spatial operations or temporal brushes allow for the manipulation of visualisations, and in fact of the underlying GIS database. Several complementary aspects need to be analysed for the development of VDGIS visualisations:

- underlying properties of very dynamic geographical data;
- pre-processing functions for filtering large volumes of data;
- query and visualisation operations;
- interactive functions for animation and interface manipulation.

These considerations lead to the analysis of the nature of the dynamic data and changes to be displayed. Changes have been categorised by DiBiase et al. (1992) as changes in either: (a) spatial location; (b) spatial location and/or attributes; and (c) classification of objects within the attribute space (e.g., re-classification). Dynamic objects or spatial properties are difficult to evaluate on an individual basis. Therefore, the analysis and presentation of a set of dynamic objects often require the use of parsing, aggregation and/or statistical techniques (Kraak and MacEachren, 1994). This is illustrated, for example, by the pre-processing of the spatial, thematic and temporal properties of very dynamic objects. Often, passing from one spatial or temporal level of granularity to a coarser one provides a complementary insight for the analysis and understanding of spatial phenomena. Within the time dimension, temporal operators could be used for changing the temporal granularity of the phenomena represented. As such, these operations allow phenomena visualisation from a hierarchical point of view. Propagation of temporal constraints between spatio-temporal processes represented at complementary levels of granularity can be realised using constraint propagation algorithms (Claramunt and Bai, 1999). Reasoning and manipulation in temporal systems have been widely studied in temporal logic (Allen, 1984; Bestougeff and Ligozat, 1992; Badaloni and Benati, 1994). Formal temporal languages and operators are of particular interest for the temporal aggregation of very dynamic data. Generally, composition operations, based on the manipulation of temporal intervals, allow the representation of phenomena at a coarser level of granularity using temporal operators that aggregate temporal periods (e.g., from an hour to a day frequency of change). Within the spatial and thematic dimensions, aggregational, statistical and relational operations can also be applied. These operations constitute a set of pre-processing functions that cover the different dimensions of geographical phenomena.

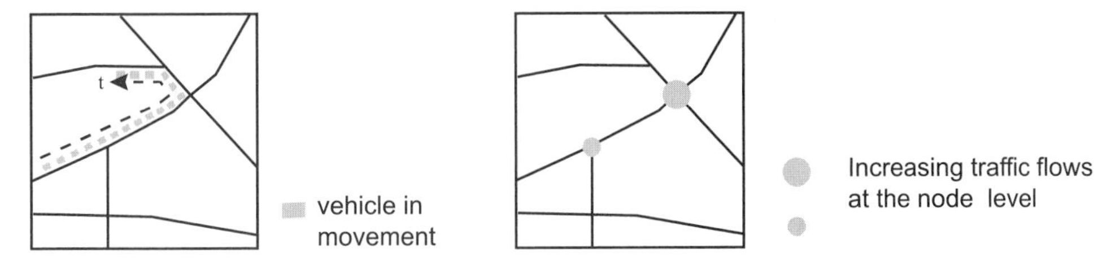

Fig. 4. Individual vehicle versus network property changes.

The information contained within a VDGIS visualisation includes dynamic objects and their changing properties on the one hand and the relatively static environment on the other. The latter acts as a visual background for presentation and animation purposes. For static data, a single visual representation is generally used, it is bounded in time by the temporal validity of the geographical data visualised which is either a time instant or interval. The static component of a visualisation provides support for the interactive exploration of changes as it gives a geographical reference to the dynamic phenomena analysed. The duration of a temporal visualisation scene (i.e., animation) can be proportional to the magnitude of the phenomenon represented. For example, the temporal progression of animation slows down as changes visualised are increasing in intensity. This type of animation technique is referred to as pacing (DiBiase et al., 1992). This also requires an analysis of the visual properties that support the presentation of dynamic objects

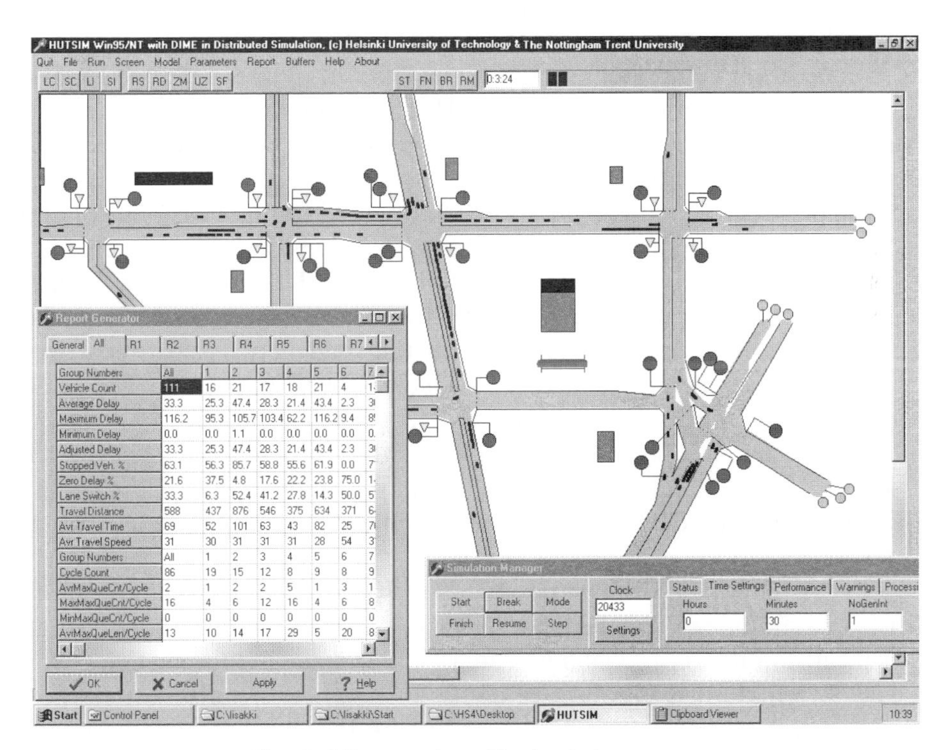

Fig. 5. Microscopic traffic simulation system.

within a visualisation. In particular, the design of a visualisation needs an integration of the visual constraints that affect the graphical layout. For example, the perception of the successive locations of individual objects is a difficult task in very dynamic visualisations (e.g., many vehicles moving several times per second as illustrated in Fig. 5). In very dynamic visualisations, moving vehicles are generally stable in the thematic dimension. In this case, the analysis of the thematic dimension is not of particular interest. This provides an interface environment for the development of actions at the interaction level (e.g., animating a visualisation using a temporal browser).

We make a distinction between the visualisation of moving vehicles and network properties considered at an individual level and the aggregation of moving objects and network properties used to represent the dynamism of an observed system at a coarser level of granularity. In other words, we may analyse changes in movements from place to place or differences in local phenomena from time to time (Vasiliev, 1996) (Fig. 4). Very dynamic individual objects are not active in terms of interface interaction. On the contrary, aggregated objects, temporarily and/or spatially, can be manipulated using interaction tools.

5. Visualisation of very dynamic GIS: application to urban traffic data

In the context of traffic applications, microscopic models represent a first modelling approach oriented to the monitoring and simulation of individual vehicle behaviours (i.e., vehicle displacements) (Kosonen et al., 1998) (Fig. 5). Approaches to microscopic models involve the representation and/or optimisation of complementary data such as driver behaviour, vehicle characteristics and performance, road components, lane structure and geometry, and their relationships. A simulation attempts to minimise the amount of data required to give reasonably accurate simulation results. For example, the HUTSIM micro-simulation system provides a flexible and detailed model potentially adapted to different traffic system configurations (Kosonen 1999).

Macroscopic models represent a second approach oriented to the measurement and estimation of traffic conditions at fixed points such as junctions, stop-lines or lanes (i.e., change of thematic properties at fixed network locations) (Peytchev et al., 1996). The main network entities used for the management of incoming traffic data are nodes that represent an intersection of roads within the network, road segments that describe a part of the road network between two nodes, and incoming lanes that represent an oriented lane which arrives at a node. This model makes a distinction between static entities that describe the geographical properties of the network (e.g., network, node, road segment) and dynamic properties that represent the behaviour of the traffic system (e.g., length of a traffic queue, number of cars per traffic light cycle). These entities and properties allow the representation of traffic flows at different levels of granularity (i.e., lane, road segment and node).

The framework of the VDGIS encompasses a full spectrum of current transport and traffic telematics (TTT) tasks ranging from the basic real-time information provision through optimising control and 'what-if' simulation modelling. The TTT (an amalgamation of two words telecommunications and informatics) is a collective name for a range of information and automation services that have been developed specifically on the basis of modern telecommunication technologies. Some of the most actively developed TTT services include traffic and travel information (urban, rural, motorway). The research prototype reported here relates to our work with the SCOOT traffic monitoring and control system which provides a basis for a whole range of telematics applications including microscopic and macroscopic traffic simulation and portable traffic information systems (Bargiela and Berry, 1999). The SCOOT traffic management system retained for the development of this project model a part of the city of Mansfield, a mid-sized city in the UK. The temporal granularity of incoming traffic data provided by the memory management system is given on a second basis. Such a frequency of communication flow leads to a huge volume of traffic data (about one million traffic data messages per day). In order to reduce such a huge volume of data, we decided to aggregate incoming traffic data to half an hour time interval samples. This resolution largely reduces the amount of traffic data generated, and is still relevant for the objectives of an analysis of traffic conditions. The applications have been interfaced using a generic inter-process communication facility developed at Nottingham Trent University, the distributed memory environment DIME (Argile et al., 1996). This communication environment is based on a TCP/IP protocol and a client-server architecture. This system has been developed and tested in conjunction with the SCOOT system. With the aid of DIME, the VDGIS can appear to SCOOT as another application that performs complex data aggregation and visualisation tasks while essentially maintaining its autonomy.

We illustrate these concepts in the context of the OSIRIS prototype oriented to the development of an inter-*O*perable *S*ystem for the *I*ntegration of *R*eal-time traff*I*c data within a GI*S* (Etches et al., 1999; Valsecchi et al., 1999; Grzywacz and Claramunt, 2000). The database method used to support the description of the traffic system is based on an object-relationship model (Etches et al., 1999). For the purposes of our prototype, the database design has been mapped to a geo-relational model. The resulting model supports both object and attribute versioning, thus allowing a flexible representation of temporal properties. The OSIRIS prototype extends the current capabilities of traffic monitoring systems in terms of database functions and develops a user-oriented interface based on the integration, aggregation, manipulation, visualisation and animation of traffic conditions within an urban network. Such a system complements the monitoring functions provided by real-time traffic systems. It is oriented towards urban studies that integrate traffic conditions as a parameter. Within OSIRIS, traffic data are imported from an urban traffic control system that optimises the split, cycle, and offset times of traffic signals. This traffic system is a macroscopic traffic system which is therefore not oriented towards the modelling of individual cars but rather traffic conditions within a road network (e.g., queue lengths). The OSIRIS implementation is realised on top of MapInfo GIS using C++, Delphi (a windows GUI editor and Pascal compiler), and MapBasic programming languages. The urban network component of the database is based on ordnance survey centre alignment of roads (OSCAR) data.

Changing the level of granularity in the representation of any real-world phenomenon has an impact on both the spatial and temporal dimensions. In the temporal dimension, the granularity of very dynamic data needs to be pre-defined according to user needs. Within a traffic system, incoming data based on a very dynamic frequency of change can be pre-processed according to the minimal time interval of interest (Valsecchi et al., 1999). In order to analyse traffic conditions at complementary levels of granularity, several spatial and temporal aggregation mechanisms have been developed. At the spatial level, the aggregation of traffic data is based on three complementary levels that provide different representations of traffic data flows, from the finest spatial level of granularity to the coarser spatial level of granularity, i.e., incoming lane, road segment and node, respectively (Fig. 6).

Additionally, a user-defined level allows the aggregation of traffic data on pre-selected routes (e.g., set of road segment ends). At the temporal level, the source temporal granularity provided by DIME (i.e., 1 s) is aggregated on a half an hour basis by the pre-processor (i.e., averages and maximum of traffic data values). A user-oriented temporal granularity is also selected during aggregation analysis according to application needs. The pre-processing functions of incoming traffic data have been implemented through a visual user interface. The pre-processing of a visualisation requires the definition of the temporal parameters (i.e., time interval, period of aggregation), the definition of incoming traffic attributes (either based on a maximum or average basis), and the levels of spatial and temporal granularity (Fig. 7). For example, pre-processing functions calculate averages and maximums (Fig. 7(a)) of queue lengths, traffic light periods, and node saturation. These functions are applied on either maximum or average incoming traffic data attributes (Fig. 7(b)). For the analysis of very dynamic phenomena such as traffic flows, spatial and/or temporal aggregations provide different levels of analysis. At a coarser level of granularity, aggregated behaviours are identified. Coarser temporal and/or granularity levels allow the identification of global changes. On the other hand, the analysis of local changes requires finer temporal and spatial granularity levels. Browsing throughout different spatial and temporal levels of granularity is an important functional requirement for the development of successful VDGIS in order to support a large range of user functions that cover both the study of local properties and the analysis of general trends within the urban traffic network.

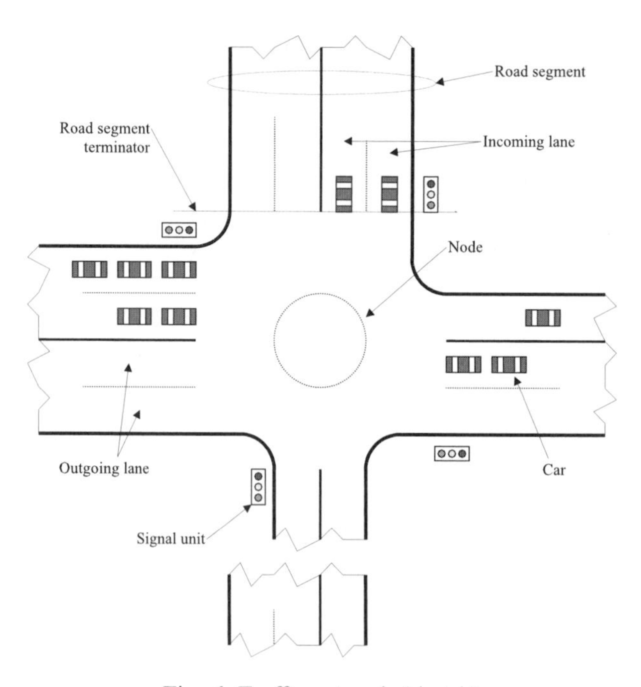

Fig. 6. Traffic network (sketch).

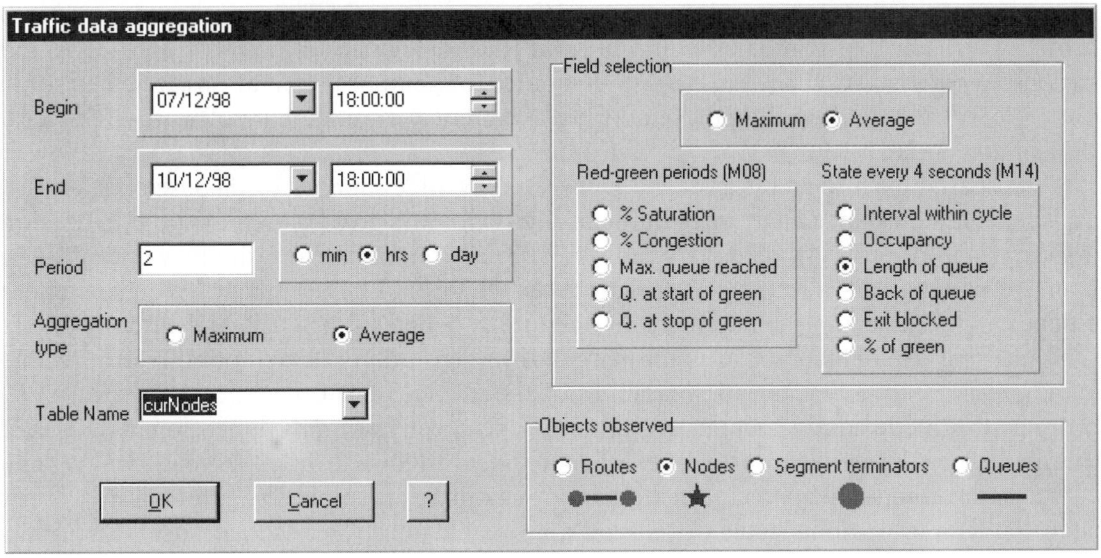

Fig. 7. Pre-processing: configuration parameters.

The query components of the OSIRIS prototype have been completed by the implementation of temporal operations that extend current relational and spatial operations provided by a GIS system (Grzywacz and Claramunt, 2000). Temporal operations are embedded within a query interface that integrates thematic, spatial and temporal operations. The temporal functions represent the extension developed. The graphic user interface (GUI) extends the current MapInfo query interface by integrating temporal predicates within the WHERE clause and temporal functions within the SELECT clause (e.g., Valid( ), Cast(Valid( ) as interval). A temporal operation wizard allows the user to create a temporal predicate (Fig. 8). Each temporal predicate consists of two operands and an operation in-between. The system controls the choices of

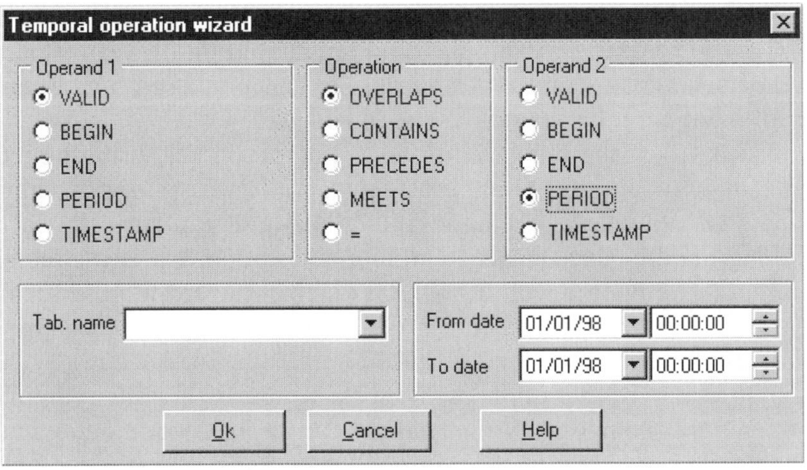

Fig. 8. Pre-processing: temporal operations.

operands according to the temporal operation selected by the user. This implementation is based on a dual approach that combines a first normal form (1NF) approach with TSQL2 temporal operations, which is the current database standard for temporal operations. Such a solution presents the advantage of being compatible with current geo-relational software architectures, which is a constraint of our prototype environment. Typical query examples are as follows: (1) display the spatial extents and deliver the identifiers, average number of passing cars, and valid times of the lanes that have an average number of passing cars greater than or equal to 25, during periods that end after 11:45 on 12 December 1998; (2) return the maximum value of average numbers of passing cars, for periods of time after 10:30 on 12 December 1998. This temporal manipulation interface implements the main operations defined in TSQL2, and extends the range of GIS querying capabilities towards the temporal dimension. These temporal operations complete the pre-processing and query capabilities of the OSIRIS prototype.

In order to provide complementary visualisation perspectives, multi-dimensional visualisation techniques reflect the dynamic properties of incoming traffic data (e.g., thematic chart, spatial chart, thematic animation, spatial animation). For example, an animation allows users to browse through the temporal traffic states of selected and aggregated traffic values within a considered period of time. Such functions enrich the user perception of traffic data through time and act as an exploratory tool that can be used to identify traffic patterns in space and time. These visualisations can be used to detect incidents in order to identify critical nodes, or for the analysis of traffic patterns within the traffic network. In the context of our project, the visualisation of very dynamic geographical data implies a high level of interaction that supports complementary user-defined tasks:

- definition of complementary temporal and spatial levels of granularity (Fig. 7);
- derivation of traffic data using query language capabilities (Fig. 8);
- combination of different dimensions in order to analyse patterns in the spatial, temporal and thematic dimensions (Fig. 9).

Within the scope of the OSIRIS prototype, different visualisation and animation techniques have been used:

- Map animations that present the variation of traffic properties located in the network, using different spatial (lane, road segment or node) and temporal aggregations (i.e., different temporal granules). Fig. 9(a) presents an example of spatial animation that can either simulate traffic behaviours at the queue, road segment or node levels. The animation can be controlled through the GUI with an interaction box that is user-controlled.
- Animated graphs that describe the variation of traffic properties, using different spatial (i.e., lane, road segment or node) and temporal aggregation levels (different temporal granules). Fig. 9(b) presents an example of thematic animation that simulates the variation of traffic queue values along the time line thanks to an interaction box that is user-controlled.
- Charts that present the temporal evolution of a traffic parameter for a user-defined route or set of road elements (i.e., lane, road segment or node). Fig. 9(c) presents an example of variation of traffic queue values along the time line for a set of traffic network nodes.
- Animations that present the evolution of the distribution of a traffic parameter for a user-defined set of temporal components (i.e., lane, road segment or node). Fig. 9(d) presents an animation that illustrates the distribution of traffic values for a set of traffic network nodes.

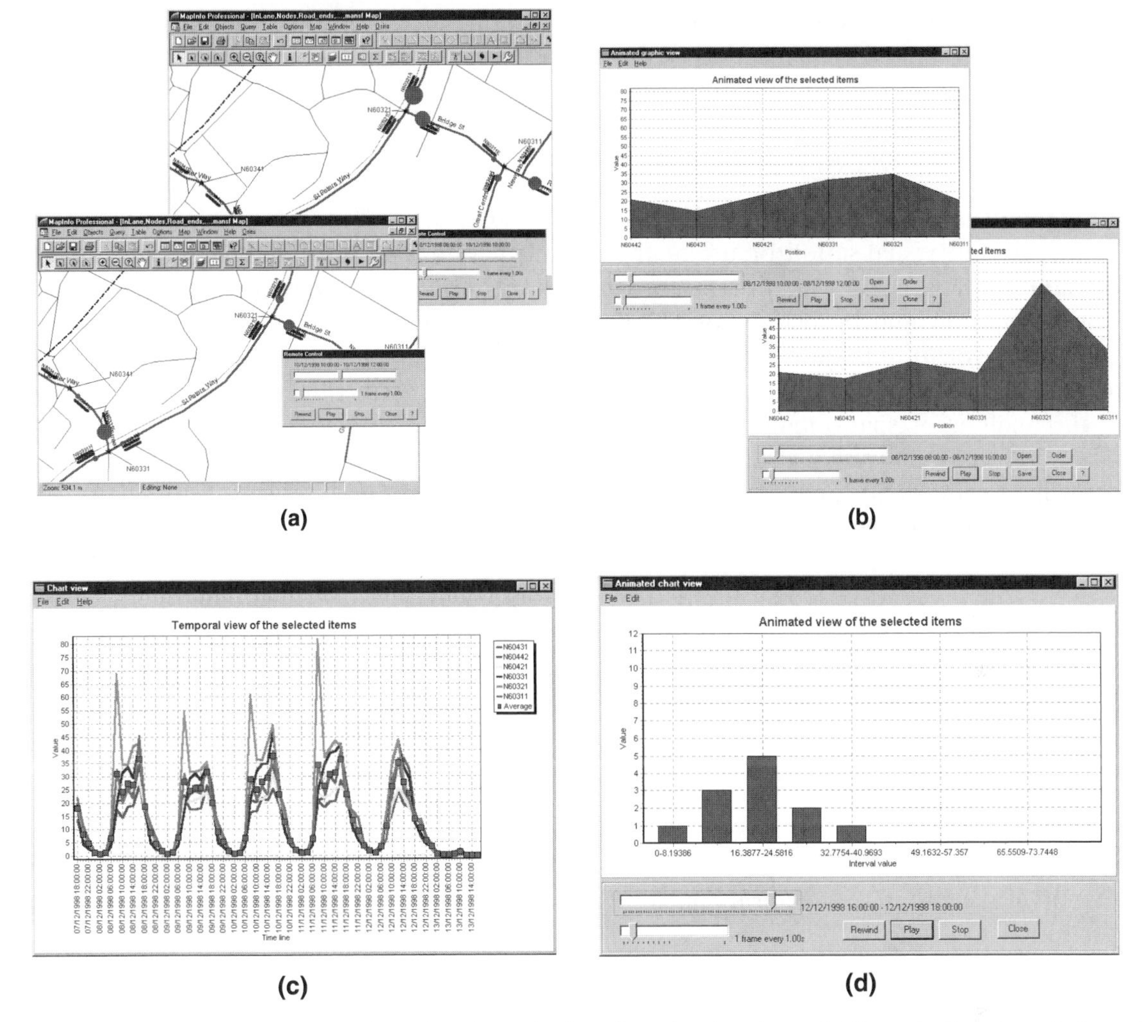

Fig. 9. (a) Spatially oriented animation. (b) Thematically oriented animation. (c) Thematically oriented chart. (d) Distribution-oriented animation.

The following screen snapshots – examples taken from the OSIRIS prototype – illustrate the concepts of visualisation and animation tools that integrate complementary graphical and cartographical techniques. The interaction level is given by a set of actions that support temporal browsing functions within the different visualisations. Different levels of granularity are user-defined during the aggregation of data selected for the visualisation process. All together, these visualisation and interaction tools provide a suitable platform that allows users to explore urban traffic data from various perspectives and to generate a set of dynamic visual representations that give an overview of traffic flows. Such functions enrich the user perception of traffic data through time and act as an exploration tool that can be used to identify traffic incidents and patterns in space and time. For example, OSIRIS visualisations can be used to detect the impact of an accident on the network, to identify critical nodes, or for the analysis of traffic patterns within a road network.

6. Conclusion

The experimental research presented in this paper develops a new framework for the integration, analysis and visualisation of urban traffic data within VDGIS. The integration of urban traffic data within VDGIS requires a sequence of manipulations that include pre-processing functions, selection and derivation of traffic data, and visualisation and animation tasks. In particular, the constraints of a VDGIS imply the development of pre-processing functions that aggregate incoming traffic data in both the spatial and temporal dimensions. These functions allow the analysis of spatio-temporal phenomena at complementary levels of granularity. The manipulation and analysis of urban traffic is based on several complementary levels: pre-processing, visualisation and interaction tools that allow users to analyse urban traffic data within GIS. The presented framework has been illustrated and validated in the context of a VDGIS for a real-time traffic system. The method proposed and the implementation realised with the prototype OSIRIS are original as the proposed architecture combines: (1) a dynamic integration of traffic data, (2) pre-processing of traffic data at complementary levels of granularity, (3) the integration of temporal operations within a GIS query language, and (4) an interface that supports visualisations and animations in the thematic, spatial and temporal dimensions. Further work includes the development and prototyping of a real-time traffic GIS for simulation purposes.

Acknowledgements

We would like to thank the anonymous reviewers and Prof. Jean-Claude Thill for their most helpful comments and suggestions.

References

Abel, D.J., Yap, S.K., Ackland, R., Cameron, M.A., Smith, D.F., Walker, G., 1992. Environmental decision support system project: an exploration of alternative architectures for GIS. International Journal of Geographic Information Systems 6 (3), 193–204.

Allen, J.F., 1984. Towards a general theory of actions and time. Artificial Intelligence 23, 123–154.

Andrienko, G.L., Andrienko, N.V., 1998. Visual data exploration by dynamic manipulation of maps. In: Poiker, T., Chrisman, N. (Eds.), Proceedings of the Eighth International Symposium on Spatial Data Handling, Vancouver, pp. 533–542.

Argile, A., Peytchev, E., Bargiela, A., Kosonen, I., 1996. DIME: a shared memory environment for distributed simulation, monitoring and control of urban traffic. In: Proceedings of European Simulation Symposium ESS'96, SCS, vol. 1, Genoa, pp. 152–156.

Badaloni, S., Benati, M., 1994. Dealing with time granularity in a temporal planning system. In: Proceedings of the First International Conference on Temporal Logic. Springer, Berlin, pp. 101–116.

Barcielo, J., Ferrer, J.L., Martin, R., 1999. Simulation assisted design and assessment of vehicle guidance systems. International Transactions in Operational Research 6, 123–143.

Bargiela, A., Berry, R., 1999. Enhancing the benefits of UTC through distributed applications. Traffic Technology International, February, pp. 63–66.

Bestougeff, H., Ligozat, G., 1992. Logical Tools for Temporal Knowledge Representation. Ellis Horwood, UK.

Bishop, I.D., Ramasamy, S.M., Stephens, P., Joyce, E.B., 1999. Visualisation of 8000 years of geological history in Southern India. International Journal of Geographic Information Science 13 (4), 417–427.

Buttenfield, B.P., 1993. Proactive graphics and GIS: prototype tools for query, modelling and display. In: Proceedings of Auto Carto 11. ACSM/ASPRS, Minneapolis, pp. 377–385.

Buziek, G., 1997. The design of a cartographic animation – experiences and results. In: Proceedings of the 18th International Cartographic Conference. ICA, Stockholm, pp. 1344–1351.

Campbell, C.S., Egbert, S.L., 1990. Animated cartography/Thirty years of scratching the surface. Cartographica 27 (2), 24–46.

Card, S.K, Mackinlay, J.D., Schneiderman, B., 1999. Readings in Information Visualisation: Using Vision to Think. Morgan Kaufmann, San Francisco.

Chen, C., 1999. Information Visualisation and Virtual Environments. Springer, Berlin.

Claramunt, C., Bai, L., 1999. A multi-scale approach to the propagation of temporal constraints in GIS. Journal of Geographic Information and Decision Analysis 3 (1), 9–20.

Claramunt, C., Mainguenaud, M., 1996. A Spatial representation and navigation model. In: Kraak, M.J., Molenaar, M. (Eds.), Advances in GIS II. Taylor & Francis, Delft, Netherlands, pp. 767–784.

Claramunt, C., Thériault, M., 1995. Managing time in GIS: an event-oriented approach. In: Clifford, J., Tuzhilin, A. (Eds.), Recent Advances in Temporal Databases. Springer, Berlin, pp. 23–42.

Dayley, D.J., Mayers, D., 1999. A statistical model for dynamic ride-matching in the World Wide Web. In: Proceedings of the ITSC'99 Conference. The IEEE Computer Society, Tokyo, pp. 154–165.

DiBiase, D.A., MacEachren, M., Krygier, J.B., Reeves, C., 1992. Animation and the role of map design in scientific visualisation. Cartography and Geographic Information Systems 19 (4), 201–214.

Eick, S.G., 1996. Aspects of network visualization. Computer Graphics and Applications 16 (2), 69–72.

Etches, A., Claramunt, C., Bargiela, A., Kosonen, I., 1999. An interoperable TGIS model for traffic systems. In: Gittings, B. (Ed.), Innovations in GIS 6, Integrating Information Infrastructures with GI Technology. Taylor & Francis, London, pp. 217–228.

Feng, C., Wei, H., Lee, J., 1999. WWW-GIS strategies for transportation applications. In: Proceedings of the 78th Transportation Research Board, Washington, DC, pp. 234–249.

Grzywacz, M., Claramunt, C., 2000. An implementation of temporal operations within a co-operative traffic system. International Journal of Applied Systems Studies, Special Issue on Applied Co-operative Systems 1 (1) forthcoming.

Jiang, B., 1996. Cartographic visualisation: analytical and communication tools. Cartography, 1–11.

Jiang, B., 1999. SimPed: simulating pedestrian crowds in a virtual urban environment. Journal of Geographic Information and Decision Analysis 3 (1), 21–30.

Kosonen, I., 1999. HUTSIM – Urban traffic simulation and control model: principles and applications, unpublished Ph.D. dissertation. Helsinki University of Technology.

Kosonen, I., Bargiela, A., Claramunt, C., 1998. A distributed information system for traffic control. In: Bargiela, A., Kerckhoffs, E. (Eds.), Proceedings of the Tenth European Symposium in Simulation Systems. Nottingham, pp. 355–361.

Kraak, M.J., 1998. The cartographic visualisation process: from presentation to exploration. The Cartographic Journal 35 (1), 11–15.

Kraak, M.J., MacEachren, A.M., 1994. Visualisation of temporal component of spatial data. In: Waugh, T.C., Healey, R.G. (Eds.), Proceedings of the International Spatial Data Handling Conference SDH'94, Edinburgh, pp. 391–409.

Langran, G., 1992. Time in Geographic Information Systems. Taylor & Francis, London.

MacEachren, A., 1994. Time as a cartographic variable. In: Hearnshaw, H.M., Unwin, D.J. (Eds.), Visualisation in Geographical Information Systems. Wiley, New York, pp. 115–130.

MacEachren, A.M., Wachowicz, M., Edsall, R., Haug, D., 1999. Constructing knowledge from multivariate spatio-temporal data: integrating geographical visualisation with knowledge discovering in database methods. International Journal of Geographic Information Science 13 (4), 311–334.

McCormick, B.H., DeFanti, T.A., Brown, M.D., 1987. Visualisation in scientific computing (special issue). ACM SIGGRAPH Computer Graphics 21 (6).
Monmonier, M., 1990. Strategies for the visualisation of geographic time-series data. Cartographica 27 (1), 30–45.
Muehrcke, P.C., 1981. Maps in geography. Cartographica 18 (2), 1–41.
Parkes, D., Thrift, N., 1980. Times, Spaces, and Places. Wiley, New York.
Peterson, M.P., 1995. Interactive and Animated Cartography. Prentice-Hall, Englewood Cliffs, NJ.
Peuquet, D.J., 1994. It's about time: a conceptual framework for the representation of temporal dynamics in geographic information systems. Annals of the Association of the American Geographers 84 (3), 441–461.
Peytchev, E., Bargiela, A., Gessing, R., 1996. A predictive macroscopic city traffic flows simulation model. In: Proceedings of European Simulation Symposium ESS'96, SCS, vol. 2, Genoa, pp. 38–42.
Pursula, M., 1998. Simulation of traffic systems: an overview. In: Bargiela, A., Kerckhoffs, E. (Eds.), Proceedings of the 10th European Simulation Symposium, pp. 20–24.
Robertson, P.K., 1988. Choosing data representations for the effective visualisation of spatial data. In: Proceedings of the Third International Symposium of Spatial Data Handling, ICA, Sydney, pp. 243–252.
Schneiderman, B., 1994. Dynamic queries for visual information seeking. IEEE Software 11 (6), 70–77.
Stonebraker, M., Chen, J., Natha, N., Paxson, C., Wu, J., 1993. Tioga: providing data management support for scientific visualisation applications. In: Agrawal, R., Baker, S., Bell, D.B. (Eds.), Proceedings of the Very Large Database Conference, Dublin, Ireland, pp. 123–134.
Szego, J., 1987. Human Cartography: Mapping the World of Man. The Swedish Council for Building Research, Stockholm, Sweden.
Tobler, W.R., 1970. A computer movie simulating urban growth in the Detroit region. Economic Geography 46 (2), 234–240.
Tufte, E.R., 1983. The Visual Display of Quantitative Information. Graphics Press, Cheshire, CT.
Valsecchi, P., Claramunt, C., Peytchev, E., 1999. OSIRIS: an inter-operable system for the integration of real time traffic data within GIS. Computers, Environment and Urban Systems 23 (2), 245–257.
Vasiliev, I., 1996. Design issues to be considered when mapping time. In: Wood, C., Keller, C.P. (Eds.), Cartographic Design: Theoretical and Practical Perspectives. Wiley, New York, pp. 137–146.
Voisard, A. (1991). Towards a toolbox for geographical user interfaces. In: Günther, O., Schek, H.-J. (Eds.), Advances in Spatial Databases. Springer, Zurich, pp. 75–98.

PERGAMON

Transportation Research Part C 8 (2000) 185–203

TRANSPORTATION
RESEARCH
PART C

www.elsevier.com/locate/trc

Interactive geovisualization of activity-travel patterns using three-dimensional geographical information systems: a methodological exploration with a large data set

Mei-Po Kwan *

Department of Geography, The Ohio State University, 1036 Derby Hall, 154 North Oval Mall, Columbus, OH 43210-1361, USA

Abstract

A major difficulty in the analysis of disaggregate activity-travel behavior in the past arises from the many interacting dimensions involved (e.g. location, timing, duration and sequencing of trips and activities). Often, the researcher is forced to decompose activity-travel patterns into their component dimensions and focus only on one or two dimensions at a time, or to treat them as a multidimensional whole using multivariate methods to derive generalized activity-travel patterns. This paper describes several GIS-based three-dimensional (3D) geovisualization methods for dealing with the spatial and temporal dimensions of human activity-travel patterns at the same time while avoiding the interpretative complexity of multivariate pattern generalization or recognition methods. These methods are operationalized using interactive 3D GIS techniques and a travel diary data set collected in the Portland (Oregon) metropolitan region. The study demonstrates several advantages in using these methods. First, significance of the temporal dimension and its interaction with the spatial dimension in structuring the daily space-time trajectories of individuals can be clearly revealed. Second, they are effective tools for the exploratory analysis of activity diary data that can lead to more focused analysis in later stages of a study. They can also help the formulation of more realistic computational or behavioral travel models.

1. Introduction

As evident in early time-use and activity-travel studies (e.g. Chapin, 1974; Cullen et al., 1972; Szalai, 1972), a major difficulty in the analysis of human activity-travel patterns is that individual

* Tel.: +1-614-292-9465; fax: +1-614-292-6213.
E-mail address: kwan.8@osu.edu (M.-P. Kwan).

0968-090X/00/$ - see front matter
PII: S0968-090X(00)00017-6

movement in space-time is a complex trajectory with many interacting dimensions. These include the location, timing, duration, sequencing and type of activities and/or trips. This characteristic of activity-travel behavior has made the simultaneous analysis of its many dimensions difficult (Burnett and Hanson, 1982). Often, one has either to focus on a few component dimensions at a time (e.g. Bhat, 1997, 1998; Chapin, 1974; Golob and McNally, 1997; Goulias, 1999; Lu and Pas, 1999; Michelson, 1985; Pendyala, 1997), or to treat the pattern as a multidimensional whole and use multivariate methods to derive generalized activity-travel patterns from a large number of variables (e.g. Bhat and Singh, 2000; Golob, 1985; Hanson and Hanson, 1980, 1981; Janelle and Goodchild, 1988; Koppelman and Pas, 1985; Ma and Goulias, 1997a,b; Pas, 1982, 1983; Recker et al., 1983, 1987).

The development and application of these quantitative methods in transportation research have enhanced our understanding of activity-travel behavior. Through the use of multivariate group identification methods, such as clustering or pattern recognition algorithms, complex patterns in the original data set can be represented by some general characteristics and organized into relatively small number of homogenous classes. Further, once activity-travel patterns are represented, they can be related to a large number of attributes of the individuals or households which generate them and used as a response variable in models of activity-travel behavior (Koppelman and Pas, 1985). While these quantitative methods are useful for modeling purposes and for discovering the complex interrelations among variables, they also have their limitations.

First, few of these methods were designed to handle real geographical locations of human activities and trips in the context of a study area (Kwan, 1997). Often, the spatial dimension is represented by some measures derived from real geographical locations (e.g. distance or direction from a reference point such as home or workplace of an individual). Further, locational information of activities and trips was often aggregated with respect to a zonal division of the study area (e.g. traffic analysis zones). Using such zone-based data, measurement of location and/or distance involves using zone centroids where information about activity locations in geographic space and their spatial relations with other urban opportunities is lost (Kwan and Hong, 1998). As point-based activity-travel data geocoded to street addresses have gradually become available in recent years, new analytical methods that can handle the location of activities and trips in real geographic space are needed.

Second, since many analytical methods (e.g. log-linear models) are designed to deal with categorical data, organizing the original data in terms of discrete units of space and time has been a necessary step in most analyses of activity-travel patterns in the past. Discretization of temporal variables, such as the start time or duration of activities, involves dividing the relevant span of time into several units and assigning each activity or trip into the appropriate class (e.g. dividing a day into 8 or 12 temporal divisions into which activities or trips are grouped). Discretization of spatial variables, such as distance from home, involves dividing the relevant distance range into several "rings". Since both the spatial and temporal dimensions are continuous, results of any analysis that are based upon these discretized variables may be affected by the particular schema of spatial and/or temporal divisions used. The problem may be serious when dealing with the interaction between spatial and temporal variables since two discretized variables are involved. Visualization may have an important role to play in alleviating this difficulty since the spatio-temporal patterns of the original data can be explored before they are discretized for further analysis or modeling.

Third, as the amount and complexity of activity-travel data increase considerably in recent years (Cambridge Systematics, 1996a), effective methods for exploring these data are also urgently needed (McCormack, 1999). Without them, the researcher may need to model activity-travel patterns without a preliminary understanding of the behavioral characteristics or uniqueness of the individuals in the sample at hand. This can be costly in later stages of a study if the model's specifications fail to take into account of the behavioral anomalies involved. Since exploratory data analysis (EDA) can often lead to more focused and fruitful methods or models in later stages of a study, the recent development and use of scientific visualization for EDA suggest a possible direction for overcoming the problem (Dykes, 1996; Gahegan, 2000). Recent developments in the integration of scientific visualization and exploratory spatial data analysis (ESDA) also indicate the potential of geovisualization for the analysis of activity-travel patterns (Anselin, 1998, 1999; Wise et al., 1999).

This study explores the application of interactive geographical visualization (or geovisualization) in the analysis of georeferenced activity-travel data. It describes several GIS-based three-dimensional (3D) visualization methods for handling the spatial and temporal dimensions of activity-travel behavior that avoid the interpretative complexity of multivariate pattern generalization or recognition methods. These include space-time activity density surfaces, space-time aquariums and standardized space-time paths. These methods are operationalized using interactive 3D GIS techniques and an activity-travel diary data set collected in the Portland (Oregon) metropolitan area in 1994/95. While some of these methods were developed by the author for analyzing a smaller data set collected in Columbus, Ohio (Kwan, 1999a), new methods are developed and explored in this paper. These include the use of GIS-based surface modeling and virtual reality techniques. Further, as these visualization methods are computationally intensive, implementing them for handling a large data set would shed light on their feasibility, value and limitations for the analysis of activity patterns in space-time for transportation researchers.

2. The case for the interactive 3D geovisualization of activity-travel patterns

Visualization is the process of creating and viewing graphical images of data with the aim of increasing human understanding (Hearnshaw and Unwin, 1994). It is based on the premise that humans are able to reason and learn more effectively in a visual setting than when using textual and numerical data (Tufte, 1990, 1997). Visualization is particularly suitable for dealing with large and complex data sets because conventional inferential statistics and pattern recognition algorithms may fail when a large number of attributes are involved (Gahegan, 2000). In view of the large number of attributes that can be used to characterize activity-travel patterns, and given the capability of scientific visualization in handling a large number of attributes, visualization is a promising direction for exploring and analyzing large and complex activity-travel data. Geovisualization, on the other hand, is the use of concrete visual representations and human visual abilities to make spatial contexts and problems visible (MacEachren et al., 1999). Through involving the geographical dimension in the visualization process, it greatly facilitates the identification and interpretation of spatial patterns and relationships in complex data in the geographical context of a particular study area.

For the visualization of geographic data, conventional GIS has focused largely on the representation and analysis of geographic phenomena in two dimensions. Although 3D visualization programs with advanced 3D modeling and rendering capabilities have been available for many years, they have been developed and applied largely in areas outside the GIS domain (Sheppard, 1999). Only recently has GIS incorporated the ability to visualize geographic data in 3D (although specialized surface modeling programs have existed long before). This is so not only in the digital representation of physical landscape and terrain of land surfaces, but also in the 3D representation of geographic objects using various data structures. There are many methods for representing complex geographic objects in 3D (Li, 1994). One is to assign the Z value using attributes available in the two-dimensional (2D) database to produce a "3Dable" geographic database. For example, a 3D representation of a building can be created by extruding the 2D building outline along the Z-axis by the height of the building. This practice is often referred to as $2\frac{1}{2}$-D as there can be only one Z value for any single location (X, Y) on the 2D surface, thus limiting its ability to represent complex geographic objects in 3D. To represent geographic entities as true 3D objects, one has to use other methods. These include solid modeling used in computer-aided design (CAD) software, the voxel data structure that covers 3D space with 3D pixels (voxels), or object-oriented 3D data models (Lee and Kwan, 2000).

Although GIS-based 3D geovisualization has been applied in many areas of research in recent years, its use in the analysis of human activity patterns is rather limited to date. In many early studies, 2D maps and graphical methods were used to portray the patterns of human activity-travel behavior (e.g. Chapin, 1974; Tivers, 1985). Individual daily space-time paths were represented as lines connecting various destinations. Using such kind of 2D graphical methods, information about the timing, duration and sequence of activities and trips was lost. Even long after the adoption of the theoretical constructs of the time-geographic perspective by many researchers in the 1970s and 1980s, the 3D representation of space-time aquariums and space-time paths seldom went beyond the schematic representations used either to explain the logic of a particular behavioral model or to put forward a theoretical argument about human activity-travel behavior. They were not intended to portray the real experience of individuals in relation to the concrete geographical context in any empirical sense.

However, as more georeferenced activity-travel diary data become available, and as more GIS software has incorporated 3D capabilities, it is apparent that GIS-based 3D geovisualization is a fruitful approach for examining human activity-travel behavior in space-time. For instance, Forer (1998) and Huisman and Forer (1998) implemented space-time paths and prisms based on a 3D raster data structure for visualizing and computing space-time accessibility surfaces. Their methods are especially useful for aggregating individuals with similar socioeconomic characteristics and for identifying behavioral patterns. However, since the raster data structure is not suitable for representing the complex topology of a transportation network, the implementation of network-based computational algorithms is difficult when using their methods. On the other hand, Kwan (1999a, 2000a) implemented 3D visualization of space-time paths and aquariums using vector GIS methods and activity-travel diary data. These recent studies indicate that GIS-based geovisualization has considerable potential for advancing the research on human activity-travel behavior. Further, implementing 3D visualization of human activity-travel patterns can be an important first step in the development of GIS-based geocomputational procedures that are applicable in many areas of transportation research. For example, Kwan (1998, 1999b), Kwan and Hong (1998) and Miller (1991, 1999) developed different network-based algorithms for computing individual accessibility using vector GIS procedures.

The use of GIS-based 3D geovisualization in the analysis of human activity patterns has several advantages. First, it provides a dynamic and interactive environment that is much more flexible than the conventional mode of data analysis in transportation research. The researcher not only can directly manipulate the attributes of a scene and its features, but also can change the views, alter parameters, query data and see the results of any of these actions easily. Second, since GIS has the capability to integrate a large amount of geographic data in various formats and from different sources into a comprehensive geographic database, it is able to generate far more complex and realistic representations of the urban environment than conventional methods (Weber and Kwan, 2000). The concrete spatial context it provides can greatly facilitate exploratory spatial data analysis and the identification of spatial relations in the data. Results can also be exported easily to spatial analysis packages for performing formal spatial analysis (Anselin and Bao, 1997). Third, with many useful navigational capabilities such as fly-through, zooming, panning and dynamic rotation, as well as the multimedia capabilities to generate map animation series such as 3D "walk-throughs" and "fly-bys", the researcher can create a "virtual world" that represents the urban environment with very high level of realism (Batty et al., 1998). Lastly, unlike quantitative methods that tend to reduce the dimensionality of data in the process of analysis, 3D geovisualization may retain the complexity of the original data to the extent that human visual processing is still capable of handling.

3. Data

The data used in this research are from the Activity and Travel Survey conducted in the Portland (Oregon) metropolitan area in the spring and autumn of 1994 and the winter of 1995 (see Cambridge Systmatics (1996b) for details of the survey). The survey used a two-day activity diary to record all activities involving travel and all in-home activities with a duration of at least 30 min for all individuals in the sampled households. Of the 7090 households recruited for the survey, 4451 households with a total of 10,084 individuals returned completed and usable surveys. The data set logged a total of 128,188 activities and 71,808 trips.

Besides the information commonly obtained in travel diary survey, this data set comes with the geocodes (*xy* coordinates) of all activity locations, including the home and workplace of all individuals in the sample. This greatly facilitates its incorporation into a geographic database of the study area. Besides the activity diary data, geographic information about the Portland metropolitan region are also used in this study. This includes data on various aspects of the urban environment and transportation system. This contextual information allows the activity-travel data to be related to the geographical environment of the region during visualization.

The next two sections explore several methods for the 3D geovisualization of human activity-travel patterns in space-time. For the implementation of these methods, various segments of a subsample of the original respondents are used. The subsample consists of individuals who were identified as head of household or spouse or partner and are employed full-time or part-time. It provides information for 4,744 individuals who are working adults in households with at least one adult. All geoprocessing is performed using ARC/INFO and ArcView GIS, while the 3D interactive geovisualization is conducted using ArcView 3D Analyst.

4. Geovisualization of activity density patterns in space-time

The conceptual basis for the GIS-based 3D geovisualization methods discussed below is the time-geographic perspective formulated by Hägerstrand (1970) and his associates. In time-geographic conception, an individual's activities and trips in a day can be represented as a daily space-time path within a 'prism' defined by a set of constraints (Hägerstrand, 1970; Lenntorp, 1976; Parkes and Thrift, 1975). This time-geographic conception is valuable for understanding activity-travel behavior because it integrates the temporal and spatial dimensions of human activity patterns into a single analytical framework. Although time, in addition to space, is a significant element in structuring individual activity patterns, past approaches mainly focus on either their spatial or temporal dimension. The significance of the interaction between the spatial and temporal dimensions in structuring individual daily space-time trajectories are often ignored. Yet, using the concepts and methods of time-geography that focus on the 3D structure of space-time patterns of activities, not only this kind of interaction can be examined, but also many important behavioral characteristics of different population subgroups can be revealed. As a result, many transportation researchers have found the time-geographic perspective useful for understanding human activity-travel behavior (Kitamura et al., 1990; Kondo and Kitamura, 1987; Kostyniuk and Kitamura, 1984; Kwan, 2000b; Recker et al., 1983). This section focuses on interactive 3D geovisualization methods for representing activity intensity in space-time. Geovisualization of space-time paths will be discussed in the next section. Color versions of the figures are provided on Web site http://geog-www.sbs.ohio-state.edu/faculty/mkwan/gis-t/Links.htm.

A note on the quality of the figures is warranted at this point. Since the 3D patterns involved are highly complex and the figures are non-scalable raster images produced from screen captures, there are many limitations on producing clear illustrations using 2D graphics. The reader may find it hard to follow the discussion simply by looking at these figures since the text is based on observations enabled by the computer-aided interactive 3D visualization environment (which is not available to the reader). The difficulty the reader may have in "seeing" these 2D images clearly is the inevitable outcome of the need to present results of the color display of complex 3D patterns in the form of 2D graphics. The visual quality of the 2D figures in the paper, therefore, should not undermine the argument that interactive 3D geovisualization is useful. Further, as these figures cannot convey the same amount and quality of information enabled only by interactive 3D visualizations, it is best to treat them as illustrations of what one might see when performing the interactive 3D visualizations. Their purpose is to give the reader a feel for what the interactive visualization does. To fully appreciate the value of the methods discussed in the paper, one needs to go through the real computer-aided interactive geovisualization sessions instead of looking at their black-and-white 2D representations.

4.1. Simple activity patterns in space-time

A simple method for visualizing human activity patterns in 3D is to use the 2D activity-travel data provided in the original data set and convert them to a format displayable in three dimensions. An important element in this conversion process is to identify the variable in the original data file that will be used as the Z value, which represents the value of a particular activity in the vertical dimension (besides the geographical coordinates X and Y). For a meaningful representation of activity patterns in space-time, the Z variable in this study represents the time dimension of activities and trips. In this particular example, activity start time is used as the Z variable in the conversion process. Using this Z value, each activity is first located in 3D space as a point entity using its geographic location (X, Y) and activity start time (Z). To represent the duration of each activity, the activity points in 3D are extruded from their start times by a value equal to the duration of the activity.

Fig. 1 shows the result of using this method for the 14,783 out-of-home, non-employment activities performed by the 2157 European–American (white) women in the subsample. In the figure, activity duration is indicated by the length of the vertical line that represents the temporal span of an activity. Activity start time can be color-coded so that the temporal distribution of activities can be viewed during interactive visualization (these color codes are used in the color version of Fig. 1 posted on the Web). A helpful background for relating the activity patterns to various locations in the study area is created through adding several layers of geographic information into the 3D scene. These layers include the boundary of the Portland metropolitan region, freeways and major arterials. For better visual anchoring and locational referencing during visualization, a 3D representation of downtown Portland, which appears as a partially transparent 3D pillar derived from extruding its 2D boundary along the Z dimension, is also added into the scene.

Through interactive geovisualization, it is apparent that the highest concentrations of non-employment activities of the selected women are found largely in areas close to downtown Portland inside the "loop" and areas west of downtown along and south of Freeway 84. Important clusters of non-employment activities are also found in Beaverton in the west and Gresham in the east. Most of the non-employment activities are of very short duration (94% of them with duration under 5 min, and less than 1% have duration over 10 min). Further, most of the non-employment activities that were undertaken during lunch hour are found largely in the high density areas near downtown, while non-employment activities in the more suburban areas tend to take place in the morning, late afternoon or evening (perhaps associated with the commute trip). There is strong spatial association between the location of non-employment activities and the locations of workplace for the selected individuals.

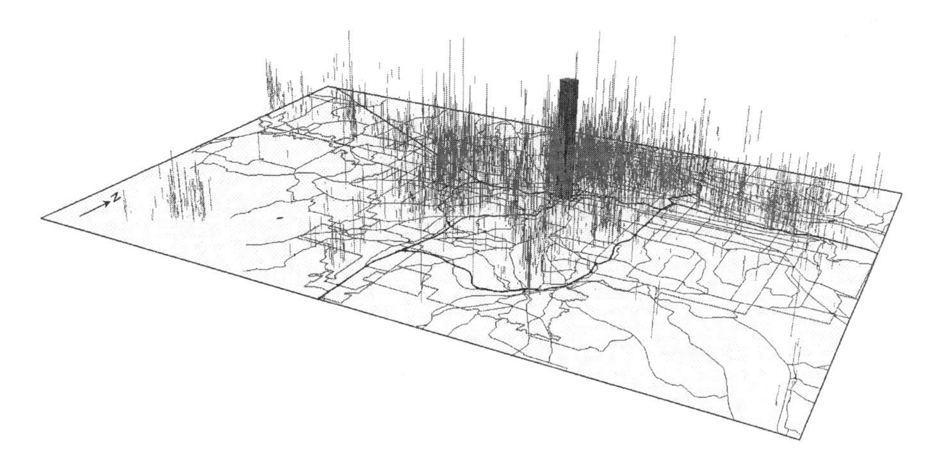

Fig. 1. Simple activity patterns in space-time.

4.2. Activity density patterns in geographic space

Comparison of the patterns of different activities for the same population subgroup or the patterns of the same activity between different population subgroups using this simple representation, however, is difficult. As the number of activities involved will increase considerably when more population subgroups are included and the patterns may be difficult to compare visually, other methods that facilitate inter-group or inter-activity comparisons are needed. This section explores the use of 3D activity density surfaces for representing and comparing the density patterns of different activities in real geographic space. The same group of respondents discussed above is also used here. The purpose is first to represent the spatial intensity of the locations of workplace, home and non-employment activities of these individuals, and then to examine the spatial relationships between these density patterns.

To generate a density surface from a point distribution of n activity locations, a non-parametric density estimation method called kernel estimation is used (Gatrell, 1994; Silverman, 1986). Following Bailey and Gatrell's (1995) formulation, if $\mathscr{R}$ represents the study area, x represents a general location in $\mathscr{R}$ and $\boldsymbol{x}_1, \boldsymbol{x}_2, \ldots, \boldsymbol{x}_n$ are the locations of the n activities, then the intensity or density, $\lambda(\boldsymbol{x})$, at $\boldsymbol{x}$ is estimated by

$$\lambda_h(\boldsymbol{x}) = \frac{1}{\delta_h(x)} \sum_{i=1}^{n} \frac{w_i}{h^2} k\left(\frac{(\boldsymbol{x} - \boldsymbol{x}_i)}{h}\right), \quad \boldsymbol{x} \in \mathscr{R},$$

where $k(\cdot)$ is the kernel function, the parameter $h > 0$ is the bandwidth determining the amount of smoothing, w_i is a weighing factor, and $\delta_h(\boldsymbol{x})$ is an edge correction factor (Cressie, 1993). In this study, the quartic kernel function

$$k(\boldsymbol{x}) = \begin{cases} 3\pi^{-1}(1 - \boldsymbol{x}^{\mathrm{T}}\boldsymbol{x})^2 & \text{if } \boldsymbol{x}^{\mathrm{T}}\boldsymbol{x} \leqslant 1, \\ 0 & \text{otherwise} \end{cases}$$

described in Silverman (1986) is used for generating space-time activity density surfaces. The method is implemented through covering the study area by a 1110×723 grid structure (with 802, 530 cells) and using bandwidths appropriate for the point pattern at hand (2.0–3.6). The density surfaces, originally created in the grid data structure, are then converted to 3D format and added into a 3D scene (Fig. 2).

In Fig. 2, the density surface of non-employment activities is displayed transparently on top of the density surface of home locations of the selected individuals. To help identify the location of density peaks and troughs, three geographic data layers, namely freeways, major arterials and rivers, are draped over the density surface of home locations. The figure shows that the major peak of non-employment activities is centered at downtown Portland within the "loop", whereas the highest density of home locations is found at two peaks in the east and west of downtown Portland.

Fig. 3 provides a close-up view of another 3D scene in which the transparent workplace density surface is displayed on top of the home location density surface for the selected individuals. These two surfaces are vertically closer in this figure than those in Fig. 2 so that the peaks of the home location surface (bottom) pass through the transparent workplace density surface (top) from below, highlighting the spatial relationships between the peaks of the surfaces. Three distinctive

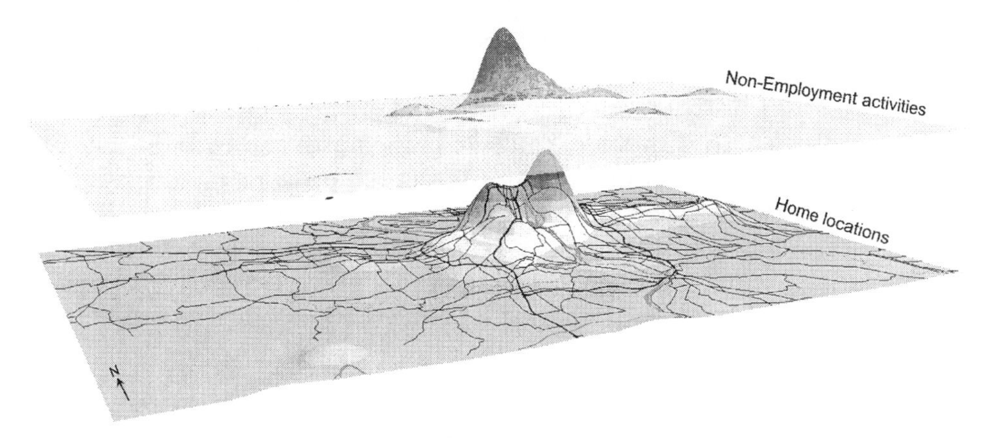

Fig. 2. Activity density patterns in geographic space, with the surface for non-employment activities over the surface of home locations.

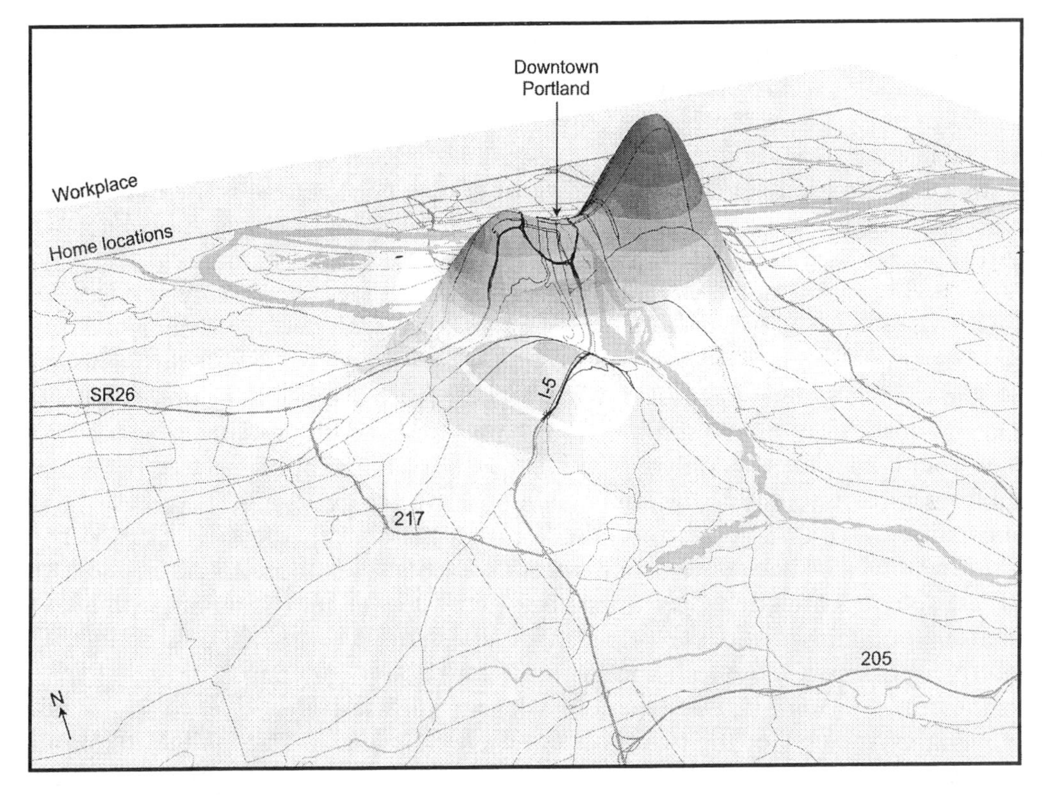

Fig. 3. A close-up view of activity density patterns in geographic space, with the surface for workplace over the surface of home locations.

peaks of home locations can be clearly seen. The most intensive one is in the east of downtown Portland while the other two are in the west of downtown and southwest of downtown along Freeway 5. Not as obvious in the figure is the peak of the workplace surface which is centered at downtown Portland inside the "loop" (it cannot be seen easily because the surface is interrupted from below at the "saddle" between the peaks of the home surface due to its transparency). Another area with high density of workplace is found in areas along Freeway 217 on the southwest between the junctions with State Route 26 in the north and Freeway 205 in the south. During the interactive 3D geovisualization of this scene, the proximity and spatial relationship between the peaks of the two surfaces are striking.

The major advantage of this method is its capability for examining the spatial relationships between different surfaces in their concrete geographical context. However, to explore the temporal dimension and its interaction with the spatial dimension, another visualization method is needed.

4.3. Space-time activity density surfaces

For representing the intensity of activities in space-time, kernel estimation is used again to generate 'space-time activity density surfaces' (Gatrell, 1994; Silverman, 1986). In this implementation, a space-time region $\mathscr{R}$ is established with a locational system similar to an $x-y$ geographic coordinate system. The time-axis of this coordinate system covers a 24-h day from 3 a.m. and the space-axis represents the distance of an activity from home. A fine grid structure of 960×960 space-time grids (with 921, 600 cells) is then created by dividing a day into 960 1.5-min time slices and distance from home into 960 40.2-m (132-foot) blocks. The quartic kernel function described above is also used here with a bandwidth of 0.6. This method is used to generate three space-time activity density surfaces for individuals in the subsample. One is for women employed part-time (Fig. 4); the second one is for men employed part-time (this figure, Fig. 4(b), is provided on the Web); and the third portrays the difference between these two density surfaces (Fig. 5).

The density surface for part-time employed women (Fig. 4) shows that there is a considerable amount of non-employment activities close to home (largely within 8 km) throughout the day from 7:30 a.m. to 10:00 p.m. There are two especially intensive peaks of non-employment activities. One is found at noon about 4 km from home, and the other happens around 2:45 p.m. about 2 km from home. There is only moderate amount of non-employment activities between 8 and 16 km from home, and during the evening hours between 6 and 9 p.m. The density surface for part-time employed men (see Fig. 4(b) on the Web) reveals that the density of non-employment activities throughout the space-time region is not as intensive as those found in the surface for women. These activities are largely performed between 9:00 a.m. and 8:30 p.m. within 6 km from home. There are two fairly distinctive peaks: (a) a very sharp peak at 8:00 p.m. about 3 km from home, and (b) a peak very close to home (within 1 km) about 3:00 p.m.

Fig. 5 shows the difference between these two density surfaces. It is obtained by using the map algebraic operator "minus" for the two surfaces, where the value in a cell in the 20 output grid is obtained by deducting the value of the corresponding cell in the surface for men from the value of the corresponding cell in the surface for women. Peaks in this "difference surface" indicate areas where the intensity of women's non-employment activities is much higher than that of men, and vice versa. The overall pattern suggests that the activity density for the part-time employed women is in general higher than that of the part-time employed men. Women performed more activities throughout the day from 6:00 a.m. to 9:00 p.m. from 4 to 8 km from home. There is a very sharp trough centered at 2:00 p.m. very close to home (within 1 km) indicating that men

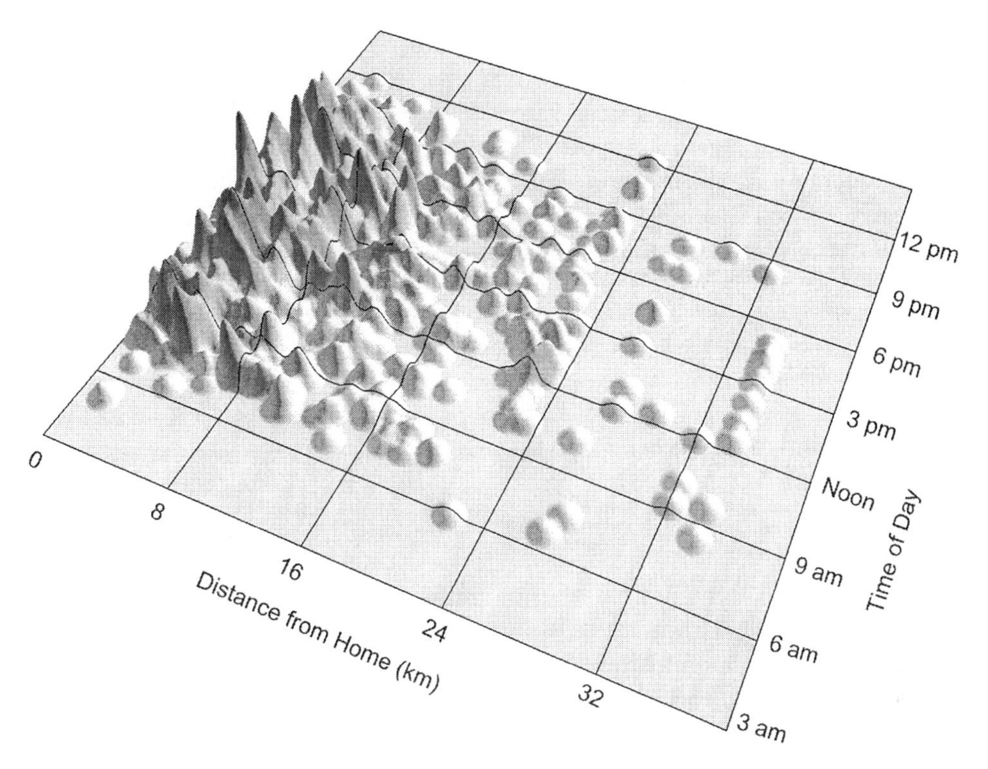

Fig. 4. Space-time activity density of non-employment activities for women employed part-time.

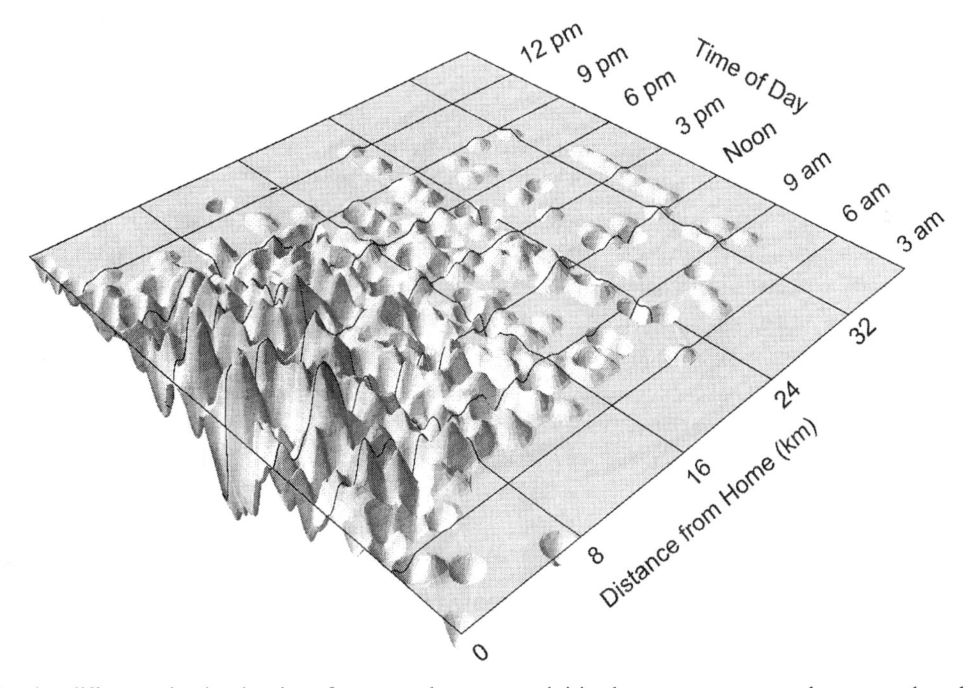

Fig. 5. Gender difference in the density of non-employment activities between women and men employed part-time.

performed many more non-employment activities than women in this space-time area. Another sharp trough is found in the evening hours between 7 p.m. to 8 p.m. about 3 km from home.

There are two major advantages in using these 3D space-time activity density surfaces. First, they reveal the intensity of activities in space and time simultaneously, thus facilitating the analysis of their interaction. Second, the grid-based method is amenable to many map-algebraic operations that can be used to adjust the computed raw density for highlighting the distinctiveness in the activity patterns of a particular population subgroup. It also makes the derivation of a "difference surface" for two population subgroups relatively easy, thus facilitating the examination of inter-group difference. The following section turns to explore the 3D geovisualization of space-time paths.

5. Geovisualization of individual space-time paths

5.1. The space-time aquarium

For the visualization of individual space-time paths, the earliest 3D method is the 'space-time aquarium' conceived by Hägerstrand (1970). In a schematic representation of the 'aquarium', the vertical axis is the time of day and the boundary of the horizontal plane represents the spatial scope of the study area. Individual space-time paths are portrayed as trajectories in this 3D aquarium. Although the schematic representation of the 'space-time aquarium' was developed long ago, it has never been implemented using real activity-travel diary data. The main difficulties include the need to convert the activity data into "3Dable" formats that can be used by existing visualization software, and the lack of comprehensive geographic data for representing complex geographic objects of the urban environment. The recent incorporation of 3D capabilities into GIS packages and the availability of contextual geographic data of many metropolitan regions have greatly reduced these two difficulties.

To implement 3D geovisualization of the space-time aquarium, four contextual geographic data layers are first converted from 2D map layers to 3D shape files and added to a 3D scene. These include the metropolitan boundary, freeways, major arterials, and rivers. For better close-up visualization and for improving the realism of the scene, outlines of commercial and industrial parcels in the study area are converted to 3D shapes and vertically extruded in the scene. Finally, the 3D space-time paths of individuals who are African Americans, Hispanics and Asian Americans from the subsample are generated and added to the 3D scene. These procedures finally created the scene shown in Fig. 6.

The overall pattern of the space-time paths for these three groups shown in Fig. 6 indicates heavy concentration of day-time activities in areas in and around downtown Portland. Using the interactive visualization capabilities of the 3D GIS, it can be seen that many individuals in these ethnic minority groups work in that area and a considerable amount of their non-employment activities are undertaken in areas within and east of downtown Portland. Space-time paths for individuals who undertook several non-employment activities in a sequence within a single day tend to be more fragmented than those who have long work hours during the day. Further, ethnic differences in the spatial distribution of workplace are observed using the interactive capabilities provided by the geovisualization environment. The space-time paths of Hispanics and Asian

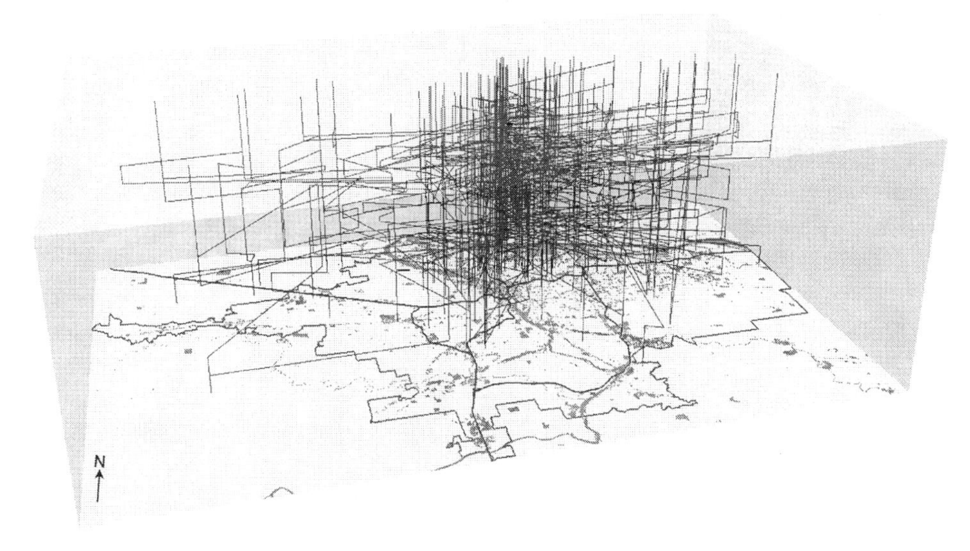

Fig. 6. Space-time aquarium with the space-time paths of African Americans, Hispanics and Asian American in the subsample.

Americans are more spatially scattered throughout the area, while those of the African Americans are spatially restricted, concentrating largely in the east side of the metropolitan region.

A close-up view from the southwest of this interactive geovisualization session is given in Fig. 7, which shows some of the details of downtown Portland in areas around the "loop" and along the Willamette River in the foreground. Portions of some space-time paths can also be seen in this scene. With the 3D parcels and other contextual layers in view, the figure gives the researcher a strong sense about the geographical context through a virtual reality-like view of the downtown area. This interactive virtual environment not only contextualizes the visualization in its actual geographical surrounding but also enables the analysis of local variations at fine spatial scales. For instance, in the color version of the figure provided on the Web, commercial buildings are color-coded orange–brown, while industrial buildings are in green. The use of color codes for distinguishing different types of buildings would give the analyst a sense of the potential interaction space and its context, which can then be compare to activities and paths of the individuals (where activities and stops can also be color coded by activity type). This approach will therefore have considerable potential for the development of person-specific, activity-based methods at fine spatial scales.

5.2. Standardized space-time paths

Although the 'aquarium' is a valuable representational device, interpretation of patterns becomes difficult as the number of paths increases with the number of individuals examined. Further, since the home-work axis for different individuals have different locations and are oriented in different directions, it may be difficult to detect patterns through visualization. One way to overcome these limitations is to plot space-time paths using a standardized or transformed

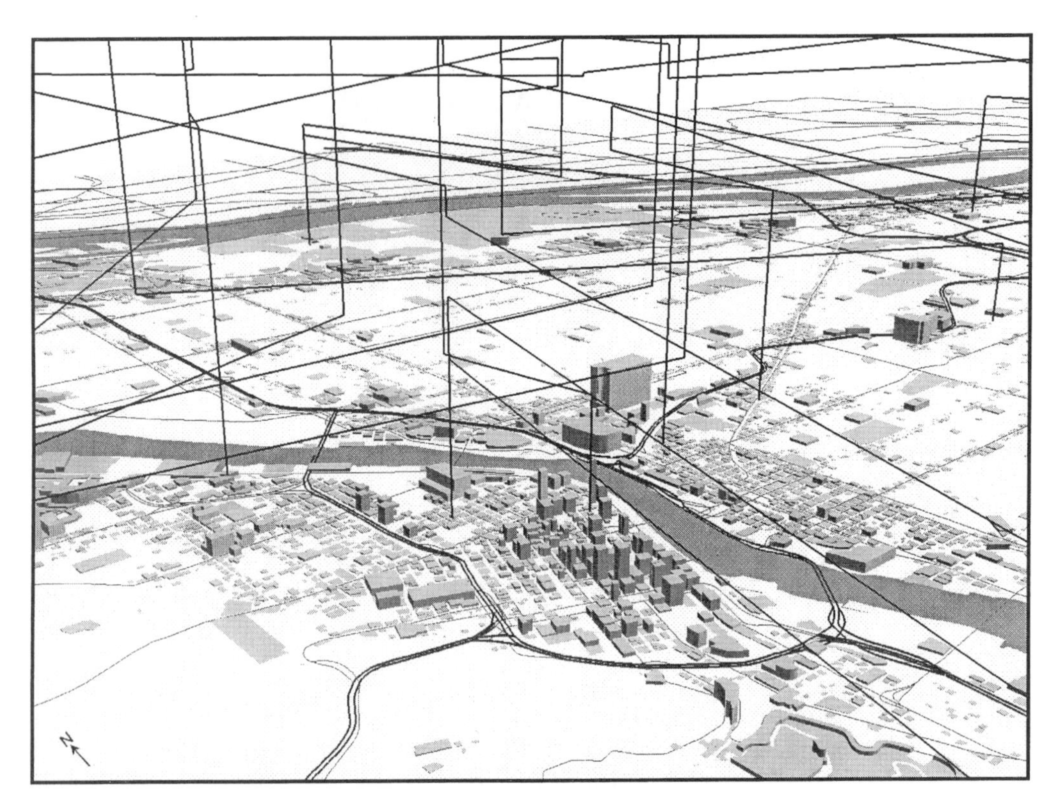

Fig. 7. A close-up view of downtown Portland from the 3D scene shown in this figure.

coordinate system. This can be done by shifting the locational coordinates of all activity sites for an individual so that the home location becomes the origin (0,0) and the home-work axis is rotated until it becomes the positive x-axis. Using these transformed or 'standardized' space-time paths, many distinctive features of the trajectories of a particular population subgroup may still be identifiable even when numerous space-time paths of many individuals are plotted. Fig. 8 shows the standardized space-time paths for the individuals of the three minority groups in the subsample. The vertical plane is the home-work plane where the home location is indicated by the origin (0,0). In the interactive visualization session, it can be seen that there are considerable amount of non-employment activities during the day, and that there are distinctive bundles of work activities at particular distances from home. Further, the spatial distribution of these non-employment activities reflects a bias toward the home-work axis, supporting similar observations in previous studies (e.g., Kitamura et al., 1990; Saxena and Mokhtarian, 1997).

In view of the complex space-time patterns the researcher has to deal with when using these methods, pattern extraction algorithms and other geocomputational procedures can be developed to complement the geovisualization methods discussed in this section (e.g. the GIS-based geocomputation of individual accessibility by Kwan, 1998; Miller, 1999). Thus, these 3D methods enabled by the 3D geovisualization environment can be the basis for developing and formulating quantitative methods for the characterization and extraction of patterns from the large number of space-time trajectories as valuable analytical tools.

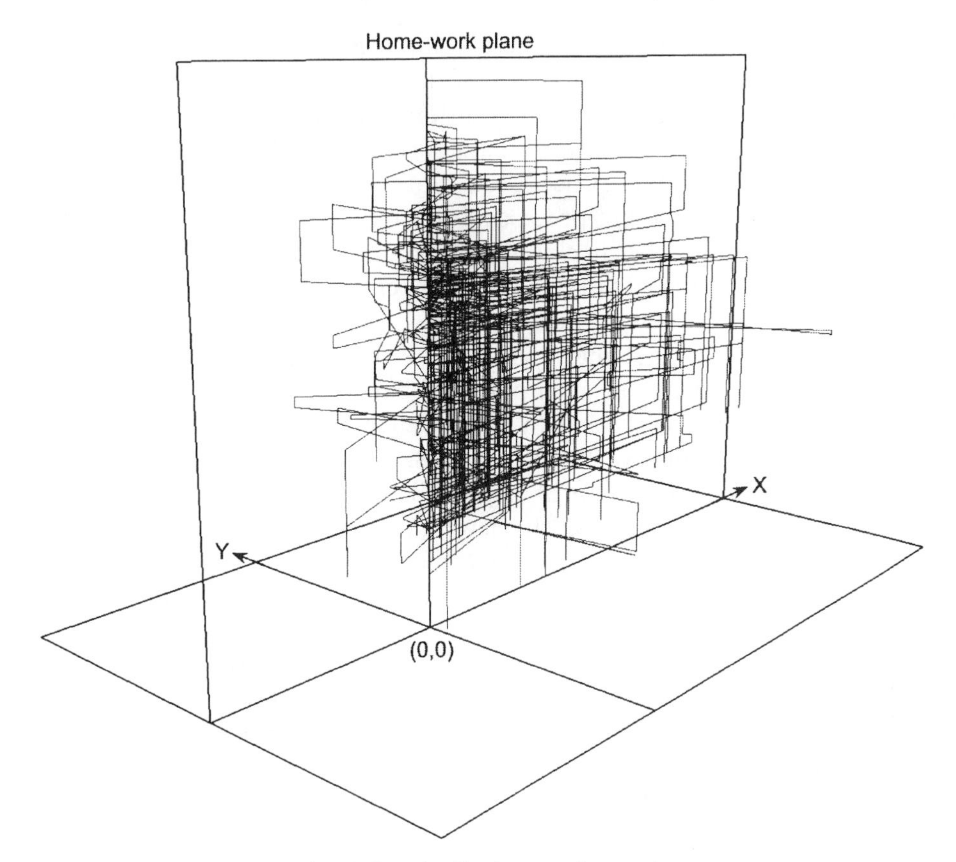

Fig. 8. Standardized space-time paths.

6. Conclusions

The dynamic and interactive GIS-based 3D geovisualization methods discussed in this paper are useful for the exploratory analysis of activity-travel patterns. They allow the researcher to interact, explore and manipulate the 3D scene. Not only the visual properties of objects can be altered to reflect their various attributes, the highly flexible viewing and navigational environment is also a great help to the researcher. As shown by the examples, these methods are capable of revealing many important characteristics of the space-time activity patterns of different population subgroups in relation to the concrete urban environment. They also facilitate the identification of complex spatial relations and the comparison of patterns generated by individuals of different gender/ethnic subgroups. Further, these interactive 3D geovisualization methods may provide the foundation for developing geocomputational algorithms or formulating operational measures of various aspects of activity-travel behavior. As individual-level, geo-referenced data become increasingly available (Kwan, 2000c), the development and implementation of these kind of geovisualization methods is a promising direction for transportation research in the future.

There are, however, several difficulties in the development and use of these 3D methods. First, there is the challenge of converting many types of data into "3Dable" formats for a particular geovisualization environment. Since every visualization software may have its unique data format requirements, and the activity and geographic data currently available are largely in 2D formats, the data preparation and conversion process can be time consuming and costly. For example, considerable data preparation and pre-processing are required for converting the Portland activity-travel data before they can be displayed as 3D space-time paths. Future research should investigate how the effort and time spent on data conversion could be reduced when data from various sources are used.

Second, the researcher may encounter barriers to the effective visualization of large and complex activity-travel data sets. Four such potential barriers identified by Gahegan (1999) are: (1) rendering speed: the ability of the hardware to deliver satisfactory performance for the interactive display and manipulation of large data sets; (2) visual combination effects: problems associated with the limitation in human ability to identify patterns and relations when many layers, themes or variables are simultaneously viewed; (3) large number of visual possibilities: the complexity associated with the vast range of possibilities that a visualization environment provides (i.e., the vast number of permutations and combinations of visual properties the researcher can assign to particular data attributes); and (4) the orientation of the user in a visualized scene or virtual world. Implementation of the interactive 3D methods in this study shows that a geovisualization environment which provides a geographical context for the researcher may considerably alleviate the fourth problem. However, the other three barriers may still remain a significant challenge to researchers who want to use this kind of methods. For instance, rendering the density surface in Fig. 4, which involves 227,041 triangles, can be taxing on the hardware. Further, identifying patterns from the space-time paths covering 129,188 activities undertaken by the survey respondents may push our visual ability beyond its limit. Future research should examine how human cognitive barriers involved in the interpretation of complex 3D patterns may be overcome.

Third, the use of individual-level activity-travel data geocoded to street addresses, given their reasonable degree of positional accuracy, may lead to considerable risk of privacy violation. As Armstrong and Ruggles (1999) demonstrated, although "raw" maps that comprised of abstract map symbols do not directly disclose confidential information, a determined data spy can use GIS technology and other knowledge to "hack" the maps and make an estimate of the actual address (and hence, a good guess of the identify of an individual) associated with each point symbol. This practice, called "inverse address-matching", has the potential for serious confidentiality or privacy violation. As "map hackers" may be able to accurately recover a large proportion of original addresses from dot maps, any use of such kind of individual-level geocoded data should be conducted with great concern in protecting the privacy of survey respondents and maintaining the confidentiality of information. As apparent in the 3D geovisualization examples in this paper (e.g., the details in Fig. 7), releasing a 3D scene created from several accurate data themes in virtual reality markup language (VRML) format may lead to significant risk of privacy violation because map hackers may be able to recover the identity of a particular survey respondent. This may further lead to the disclosure of other confidential information. As a result, researchers using the 3D geovisualization methods discussed in the paper should pay particular attention to this potential risk.

Acknowledgements

Support for this research from the College of Social and Behavioral Sciences, the Ohio State University, is gratefully acknowledged. The author would like to thank three anonymous reviewers for their comments on an earlier draft of this paper.

References

Anselin, L., 1998. Exploratory spatial data analysis in a geocomputational environment. In: Longley, P.A., Brooks, S.M., McDonnell, R., MacMillan, B. (Eds.), Geocomputation: A Primer. Wiley, New York, pp. 77–94.

Anselin, L., 1999. Interative techniques and exploratory spatial data analysis. In: Longley, P.A., Goodchild, M.F., Maguire, D.J., Rhind, D.W. (Eds.), Geographic Information Systems, vol. 1: Principles and Technical Issues, second ed. Wiley, New York, pp. 253–266.

Anselin, L., Bao, S., 1997. Exploratory spatial data analysis linking SpaceStat and ArcView. In: Fischer, M., Getis, A. (Eds.), Recent Developments in Spatial Analysis. Springer, Berlin, pp. 35–59.

Armstrong, M.P., Ruggles, A.J., 1999. Map hacking: on the use of inverse address-matching to discover individual identities from point-mapped information sources. Paper presented at the Geographic Information and Society Conference, University of Minnesota, 20–22 June.

Bailey, T.C., Gatrell, A.C., 1995. Interactive Spatial Data Analysis. Longman, New York.

Batty, M., Dodge, M., Doyle, S., Smith, A., 1998. Modelling virtual environments. In: Longley, P.A., Brooks, S.M., McDonnell, R., MacMillan, B. (Eds.), Geocomputation: A Primer. Wiley, New York, pp. 139–161.

Bhat, C.R., 1997. Work travel mode choice and number of non-work commute stops. Transportation Research B 31 (1), 41–54.

Bhat, C.R., 1998. A model of home-arrival activity participation behavior. Transportation Research B 32 (6), 387–400.

Bhat, C.R., Singh, S.K., 2000. A comprehensive daily activity-travel generation model system for workers. Transportation Research A 34 (1), 1–22.

Burnett, P., Hanson, S., 1982. The analysis of travel as an example of complex human behavior in spatially constriained situations: definition and measurement issues. Transportation Research A 16 (2), 87–102.

Cambridge Systematics, 1996a. Scan of Recent Travel Surveys. Cambridge Systematics, Oakland, CA.

Cambridge Systematics, 1996b. Data Collection in the Portland, Oregon Metropolitan Area. Cambridge Systematics, Oakland, CA.

Chapin, F.S. Jr., 1974. Human Activity Patterns in the City. Wiley, New York.

Cressie, N.A.C., 1993. Statistics for Spatial Data. Wiley, New York.

Cullen, I., Godson, V., Major, S., 1972. The structure of activity patterns. In: Wilson, A.G. (Ed.), Patterns and Processes in Urban and Regional Systems. Pion, London, pp. 281–296.

Dykes, J., 1996. Dynamic maps for spatial science: a unified approach to cartographic visualization. In: Parker, D. (Ed.), Innovations in GIS 3. Taylor & Franics, London, pp. 177–187.

Forer, P., 1998. Geometric approaches to the nexus of time, space, and microprocess: implementing a practical model for mundane socio-spatial systems. In: Egenhofer, M.J., Golledge, R.G. (Eds.), Spatial and Temporal Reasoning in Geographic Information Systems. Oxford University Press, Oxford, England, pp. 171–190.

Gahegan, M., 1999. Four barriers to the development of effective exploratory visualization tools for the geosciences. International Journal of Geographic Information Science 13 (4), 289–309.

Gahegan, M., 2000. The case for inductive and visual techniques in the analysis of spatial data. Journal of Geographical Systems 2 (1), 77–83.

Gatrell, A., 1994. Density estimation and the visualization of point patterns. In: Hearnshaw, H.M., Unwin, D.J. (Eds.), Visualization in Geographical Information Systems. Wiley, New York, pp. 65–75.

Golob, T.F., 1985. Analyzing activity pattern data using qualitative multivariate statistical methods. In: Nijkamp, P., Leitner, H., Wrigley, N. (Eds.), Measuring the Unmeasurable. Martinus Nijhoff, Boston, MA, pp. 339–356.

Golob, T.F., McNally, M.G., 1997. A model of activity participation and travel interactions between household heads. Transportation Research B 31 (3), 177–194.

Goulias, K.G., 1999. Longitudinal analysis of activity and travel pattern dynamics using generalized mixed Markov latent class models. Transportation Research B 33 (8), 535–558.

Hägerstrand, T., 1970. What about people in regional science? Papers of Regional Science Association 24, 7–21.

Hanson, S., Hanson, P., 1980. Gender and urban activity patterns in Uppsala. Sweden Geographical Review 70 (3), 291–299.

Hanson, S., Hanson, P., 1981. The travel-activity patterns of urban residents: dimensions and relationships to sociodemographic characteristics. Economic Geography 57, 332–347.

Hearnshaw, H.M., Unwin, D. (Eds.) 1994. Visualization in Geographical Information Systems. Wiley, Chichester, England.

Huisman, O., Forer, P., 1998. Towards a geometric framework for modelling space-time opportunities and interaction potential. Paper presented at the International Geographical Union, Commission on Modelling Geographical Systems Meeting (IGU-CMGS), Lisbon, Portugal, 28–29 August.

Janelle, D.G., Goodchild, M.F., 1988. Space-time diaries and travel characteristics for different levels of respondent aggregation. Environment and Planning A 20 (7), 891–906.

Kitamura, R., Nishii, K., Goulias, K., 1990. Trip chaining behavior by central city commuters: a causal analysis of time-space constraints. In: Jones, P. (Ed.), Developments in Dynamic and Activity-Based Approaches to Travel Analysis. Avebury, Aldershot, pp. 145–170.

Kondo, K., Kitamura, R., 1987. Time-space constraints and the formation of trip chains. Regional Science and Economics 17 (1), 49–65.

Koppelman, F.S., Pas, E.I., 1985. Travel-activity behavior in time and space: methods for representation and analysis. In: Nijkamp, P., Leitner, H., Wrigley, N. (Eds.), Measuring the Unmeasurable, Martinus Nijhoff, Boston, MA, pp. 587–627.

Kostyniuk, L.P., Kitamura, R., 1984. Trip chains and activity sequences: test of temporal stability. Transportation Research Record 987, 29–39.

Kwan, M.-P., 1997. GISICAS: an activity-based travel decision support system using a GIS-interfaced computational-process model. In: Ettema, D.F., Timmermans, H.J.P. (Eds.), Activity-Based Approaches to Travel Analysis. Elsevier, New York, pp. 263–282.

Kwan, M.-P., 1998. Space-time and integral measures of individual accessibility: a comparative analysis using a point-based framework. Geographical Analysis 30 (3), 191–216.

Kwan, M.-P., 1999a. Gender, the home-work link, and space-time patterns of non-employment activities. Economic Geography 75 (4), 370–394.

Kwan, M.-P., 1999b. Gender and individual access to urban opportunities: a study using space-time measures. The Professional Geographer 51 (2), 210–227.

Kwan, M.-P., 2000a. Human extensibility and individual access to information: a multi-scale representation using GIS. In: Janelle, D., Hodge, D. (Eds.), Information, Places, and Cyberspace: Issues in Accessibility. Elsevier, Amsterdam (in press).

Kwan, M.-P., 2000b. Gender differences in space-time constraints. Area (in press).

Kwan, M.-P., 2000c. Analysis of human spatial behavior in a GIS environment: recent developments and future prospect. Journal of Geographical Systems 2 (1), 85–90.

Kwan, M.-P., Hong, X.-D., 1998. Network-based constraints-oriented choice set formation using GIS. Geographical Systems 5, 139–162.

Lee, J., Kwan, M.-P., 2000. A 3D data model for representing spatial entities in built environments. Paper presented at the 96th Annual Meeting of the Association of American Geographers, 4–8 April, Pittsburgh, Pennsylvania.

Lenntorp, B., 1976. Paths in Time-Space Environments: A Time Geographic Study of Movement Possibilities of Individuals. Gleerup, Lund.

Li, R., 1994. Data structures and application issues in 3-D geographic information systems. Geomatica 48 (3), 209–224.

Lu, X., Pas, E.I., 1999. Socio-demographic, activity participation and travel behavior. Transportation Research A 33 (1), 1–18.

Ma, J., Goulias, K.G., 1997a. An analysis of activity and travel patterns in the Puget Sound transportation panel. In: Ettema, D., Timmermans, H. (Eds.), Activity-based Approaches to Travel Analysis. Elsevier, Tarrytown, NY, pp. 189–207.

Ma, J., Goulias, K.G., 1997b. A dynamic analysis of person and household activity and travel patterns using data from the first two waves in the Puget Sound Transportation Panel. Transportation 24 (3), 309–331.

McCormack, E., 1999. Using a GIS to enhance the value of travel diaries. ITE Journal 69 (1), 38–43.

MacEachren, A.M., Wachowicz, M., Edsall, R., Haug, D., 1999. Constructing knowledge from multivariate spatiotemporal data: integrating geographical visualization and knowledge discovery in database methods. International Journal of Geographical Information Science 13 (4), 311–334.

Michelson, W., 1985. From Sun to Sun: Daily Obligations and Community Structure in the Lives of Employed Women and their Families. Rowman and Allanheld, Totowa, NJ.

Miller, H.J., 1991. Modelling accessibility using space-time prism concepts within geographic information systems. International Journal of Geographical Information Systems 5 (3), 287–301.

Miller, H.J., 1999. Measuring space-time accessibility benefits within transportation networks: basic theory and computational procedures. Geographical Analysis 31 (2), 187–212.

Parkes, D.N., Thrift, N., 1975. Timing space and spacing time. Environment and Plannning A 7, 651–670.

Pas, E.I., 1982. Analytically derived classifications of daily travel-activity behavior: description, evaluation and interpretation. Transportation Research Record 879.

Pas, E.I., 1983. A flexible and integrated methodology for analytical classification of daily travel-activity behavior. Transportation Science 17 (3), 405–429.

Pendyala, R.M., 1997. An activity-based microsimulation analysis of transportation control measures. Transportation Policy 4 (3), 183–192.

Recker, W.W., McNally, M.G., Root, G.S., 1983. Application of pattern recognition theory to activity pattern analysis. In: Carpenter, S., Jones, P. (Eds.), Recent Advances in Travel Demand Analysis. Gower, Aldershot, England, pp. 434–449.

Recker, W.W., McNally, M.G., Root, G.S., 1987. An empirical analysis of urban activity patterns. Geographical Analysis 19 (2), 166–181.

Saxena, S., Mokhtarian, P.L., 1997. The impact of telecommuting on the activity spaces of participants. Geographical Analysis 29 (2), 124–144.

Sheppard, S.R.J., 1999. Visualization software bring GIS applications to life. GeoWorld 12 (3), 36–37.

Silverman, B.W., 1986. Density Estimation for Statistics and Data Analysis. Chapman & Hall, London.

Szalai, A. (Ed.) 1972. The Use of Time. Mouton, The Hague.

Tivers, J., 1985. Women Attached: The Daily Lives of Women with Young Children. Croom Helm, London.

Tufte, E.R., 1990. Envisioning Information. Graphics Press, Cheshire, Connecticut.

Tufte, E.R., 1997. Visual Explanations: Images and Quantities, Evidence and Narrative. Graphics Press, Cheshire, Connecticut.

Weber, J., Kwan, M.-P., 2000. The influence of time-of-day travel time variations on individual accessibility. Paper presented at the 96th Annual Meeting of the Association of American Geographers, 4–8 April, Pittsburgh, Pennsylvania.

Wise, S., Haining, R., Signoretta, P., 1999. Scientific visualisation and the exploratory analysis of area data. Environment and Planning A 31, 1825–1838.

VOLUME II

PERGAMON

Transportation Research Part C 8 (2000) 205–229

TRANSPORTATION RESEARCH PART C

www.elsevier.com/locate/trc

Modeling regional mobile source emissions in a geographic information system framework

William Bachman [a], Wayne Sarasua [b,*], Shauna Hallmark [c], Randall Guensler [d]

[a] *Center for Geographic Information Systems, Georgia Institute of Technology, Atlanta, GA 30332, USA*
[b] *Department of Civil Engineering, Clemson University, Clemson, SC 29634, USA*
[c] *School of Civil and Construction Engineering, Iowa State University, Ames, IA 50011, USA*
[d] *School of Civil and Environmental Engineering, Georgia Institute of Technology, Atlanta, GA 30332, USA*

Abstract

Suburban sprawl, population growth, and automobile dependency contribute directly to air pollution problems in US metropolitan areas. As metropolitan regions attempt to mitigate these problems, they are faced with the difficult task of balancing the mobility needs of a growing population and economy, while simultaneously lowering or maintaining levels of ambient pollutants. Although ambient air quality can be directly monitored, predicting the amount and fraction of the mobile source components presents special challenges. A modeling framework that can correlate spatial and temporal emission-specific vehicle activities is required for the complex photochemical models used to predict pollutant concentrations. This paper discusses the GIS-based modeling approach called the Mobile Emission Assessment System for Urban and Regional Evaluation (MEASURE). MEASURE provides researchers and planners with a means of assessing motor vehicle emission reduction strategies. Estimates of spatially resolved fleet composition and activity are combined with activity-specific emission rates to predict engine start and running exhaust emissions. Engine start emissions are estimated using aggregate zonal information. Running exhaust emissions are predicted using road segment specific information and aggregate zonal information. The paper discusses the benefits and challenges related to mobile source emissions modeling in a GIS framework and identifies future GIS mobile emissions modeling research needs.

Keywords: Air quality; Geographic information systems; Mobile source emissions; Travel demand modeling

* Corresponding author. Tel.: +1-864-656-3318; fax: +1-864-656-2670.
E-mail address: sarasua@clemson.edu (W. Sarasua).

0968-090X/00/$ - see front matter
PII: S0968-090X(00)00005-X

1. Introduction

The Clean Air Act, as amended in 1990, and other federal legislation and regulations require metropolitan areas with unacceptable air quality to develop strategies for reducing air pollution. In planning for attainment of these standards, metropolitan areas establish emissions 'budgets' that provide benchmarks for gauging attainment progress. Meeting emissions budget limits in target years often becomes difficult. Metropolitan areas must accommodate the needs of a growing population and economy, while simultaneously lowering or maintaining levels of ambient pollutants. Therefore, growing urban areas must continually develop creative strategies to curb increased pollutant production. Transportation systems contribute significantly to carbon monoxide (CO), nitrogen oxides (NO_x), and hydrocarbon (HC) emissions in urban areas. Estimates for the amount of pollutants produced by motor vehicles vary from 33% to 50% of NO_x, 33% to 97% of CO, 40% to 50% of HC, 50% of ozone precursors, and at least one-fourth of volatile organic compounds (VOC) (Chatterjee et al., 1997; SCAQMD, 1996; USEPA, 1995; CARB, 1994; USDOT, 1993). Developing measures of effectiveness and subsequent predictions of the overall impact of control strategies requires an ability to model the relationships between observable transportation system characteristics and their resulting emissions. In addition, models that incorporate these relationships must balance input data availability and quality with predictive power.

Motor vehicle emission rates correlate with a variety of vehicle and engine characteristics (weight, engine size, transmission type, emission control equipment, etc.), operating modes (idle, cruise, acceleration and deceleration), and transportation system conditions (road grade, pavement condition, etc.) (Barth et al., 1996; Guensler, 1994). Emission rates used in models are estimates of the rate at which different pollutants are emitted in grams per activity unit, such as grams of CO/s or CO/mile. Different vehicle activities (starting an engine, accelerating, cruising, etc.) result in different emission rates. Exhaust pollutants produced from starting a vehicle are correlated to the vehicle's engine characteristics and duration of the engine cool down time between starts. Running exhaust emissions require additional estimates of dynamic engine conditions that vary with the way the vehicle is driven. Estimating motor vehicle emissions requires the ability to predict or measure these activity parameters for an entire region at a level of spatial and temporal aggregation fitting the scope of anticipated control strategies.

Traditional motor vehicle emissions modeling involves four separate modeling regimes: travel demand forecasting models, mobile source emissions rate models, photochemical models (for emission inventories and resulting regional air quality), and microscale models. Travel demand forecasting models use characteristics of the transportation system and socioeconomic data to estimate road-specific traffic volumes. Emission rate models employ fleet characteristic data, operating environment characteristics, and assumptions related to emission control programs to predict emission rates for the on-road fleet. The travel demand estimates are linked with the outputs of the emission rate models to predict mobile source mass emissions. Analysts spatially allocate the mobile source emissions estimates along with stationary source estimates, to a regional grid as input to photochemical models. The photochemical models employ emissions estimates (from all sources) and meteorological data to predict ambient pollutant levels in space and time. Microscale models such as CALINE or FLINT employ mobile source emissions estimates and ambient estimates to predict pollutant levels near specific transportation facilities such as signalized intersections. CO is usually analyzed on a microscale because of the immediate localized health effects, whereas HC and NO_x are most often analyzed on a regional scale since they are precursors to ozone formation.

Several problems inherent in the four-model system limit effective evaluation of motor vehicle emission control strategies. First, the estimates of vehicle activity (vehicle-miles traveled and average speed) lack the accuracy and spatial resolution needed to evaluate control measures (Chatterjee et al., 1997; Stopher, 1993). Second, the mobile source emission rate modeling process uses highly aggregate fleet estimates and biased emission rates, which more recent research has shown to be inaccurate (Barth et al., 1996; Guensler, 1994; LeBlanc et al., 1995). Third, the modeling process is not oriented to the needs of the transportation planners and engineers who design and implement emissions control strategies These users require more feedback from typical transportation system improvement strategies (e.g., lane additions, optimized signal timing, signal-coordination, and peak-hour smoothing) than is provided in the four-model system.

A modeling approach that provides significant improvements to these three issues would be desirable. This paper focuses on using GIS as a platform for modeling mobile source emissions and the potential improvements it offers over the current regulatory approach. While GIS does not implicity improve the ability to forecast travel, nor improve the accuracy of existing spatial data, it does provide a mechanism for storing and maintaining complex socio-economic and infrastructure data. Using GIS, these data, as well as a variety of other types and resolutions of spatial data required for emissions modeling, can be brought together into an integrated modeling environment. Furthermore, many transportation agencies already use GIS to maintain the inputs into landuse and travel demand forecasting models (Vonderohe et al., 1993). Intrinsically, GIS fits the character of emission science as well as the technical environment of the expected end users.

One drawback to the integrated GIS modeling approach is that it requires the development and integration of new data that takes a great deal of time and effort to produce. Thus, the costs associated with implementing a comprehensive GIS-based emissions model are large because of the resources needed for model development, standardization, and integration of new data sources. However, since a metropolitan area's ability to demonstrate conformity depends on accurate emissions modeling, the disadvantages of current emissions modeling methods significantly overshadow the benefits of simplicity and lower modeling cost.

2. Establishing a foundation for modeling emissions in a GIS environment

Mobile emissions are intrinsically spatial. Among other things, emission rates vary by location and engine activity. Fig. 1 shows conceptually how emissions vary as a vehicle operates in space and time. Each 'block' of elevated or reduced emissions has different predictor variables. When the engine is off, emissions continue but at a reduced level (evaporative mode). When a vehicle starts (engine start mode), emissions are high due to the nature of catalysts (they need to reach an elevated temperature before operating efficiently). After a few minutes, emissions are low unless interrupted by a sharp power demand from hard acceleration or grade-induced engine load (running exhaust mode). All of these conditions vary in spatial terms as a vehicle moves along its path. This relationship between emission rates and location is a primary argument for using GIS in mobile emissions model development.

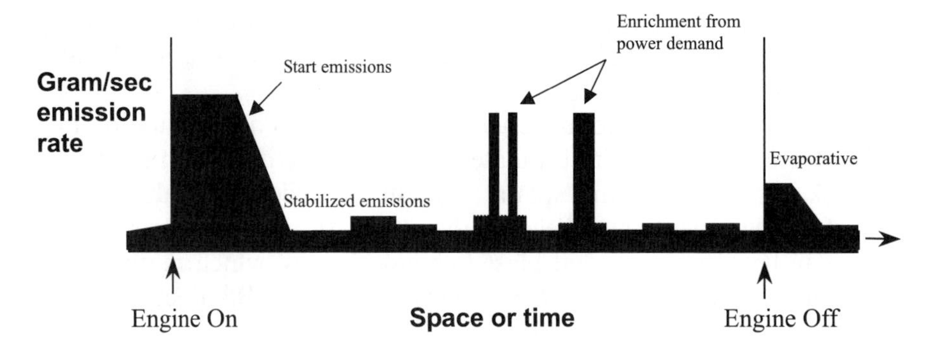

Fig. 1. Conceptual diagram of HC emission rate fluctuation for a typical vehicle.

2.1. Considering the spatial variability of mobile emissions related data

Mobile emissions model inputs can be divided into three general categories: fleet activity, fleet characteristics, and operating conditions. Most regions model emissions using regional aggregations of all these inputs, and then use estimates of vehicle miles traveled (VMT) calculated for each grid cell to disaggregate each of the totals. Unfortunately this practice does not account for the spatial variability of each of the inputs. For example, current regulatory emissions models assume a uniform distribution of the vehicle fleet across a region in calculating their emission estimates. This is not a reasonable assumption in most areas. Fig. 2 shows the census block group distribution of average vehicle model years in the Atlanta metropolitan area. Clearly, significant spatial variability of automobile ownership exists across this region. Improving the spatial scale of emissions modeling would be a major advance that could benefit from the many geoprocessing capabilities of a robust GIS modeling platform.

2.2. Literature review of GIS activities in air quality modeling

Many research efforts have taken advantage of the spatial environment of a GIS to perform various aspects of transportation-related air quality modeling. Bruckman et al. (1992) identified many benefits of using GIS in the preparation of mobile emissions model input data including fleet activity, fleet characteristics, and operating characteristics. Souleyrette et al. (1992) acknowledged the potential of a GIS to manage the complex spatial data often required to address transportation and air quality analysis problems. They developed a model that investigated the relationship between CO concentrations and traffic characteristics such as vehicle miles traveled, location of refueling stations, and wind patterns. Another research effort created an activity-based model for travel demand forecasting that integrated household activities, landuse patterns, traffic flow, and regional demographics in a GIS. Although not directly developed as an emissions model, the model was able to provide output for analyzing issues related to the Clean Air Act Amendments (Stopher et al., 1996). Hallmark and O'Neill (1996) capitalized on the use of a GIS for localized air quality modeling. They describe development of a model that combines the microscale air quality model (CAL3QHC) with a GIS. Medina et al. (1994) presented the framework for an air quality analysis model that integrates CADD, GIS, transportation, and air

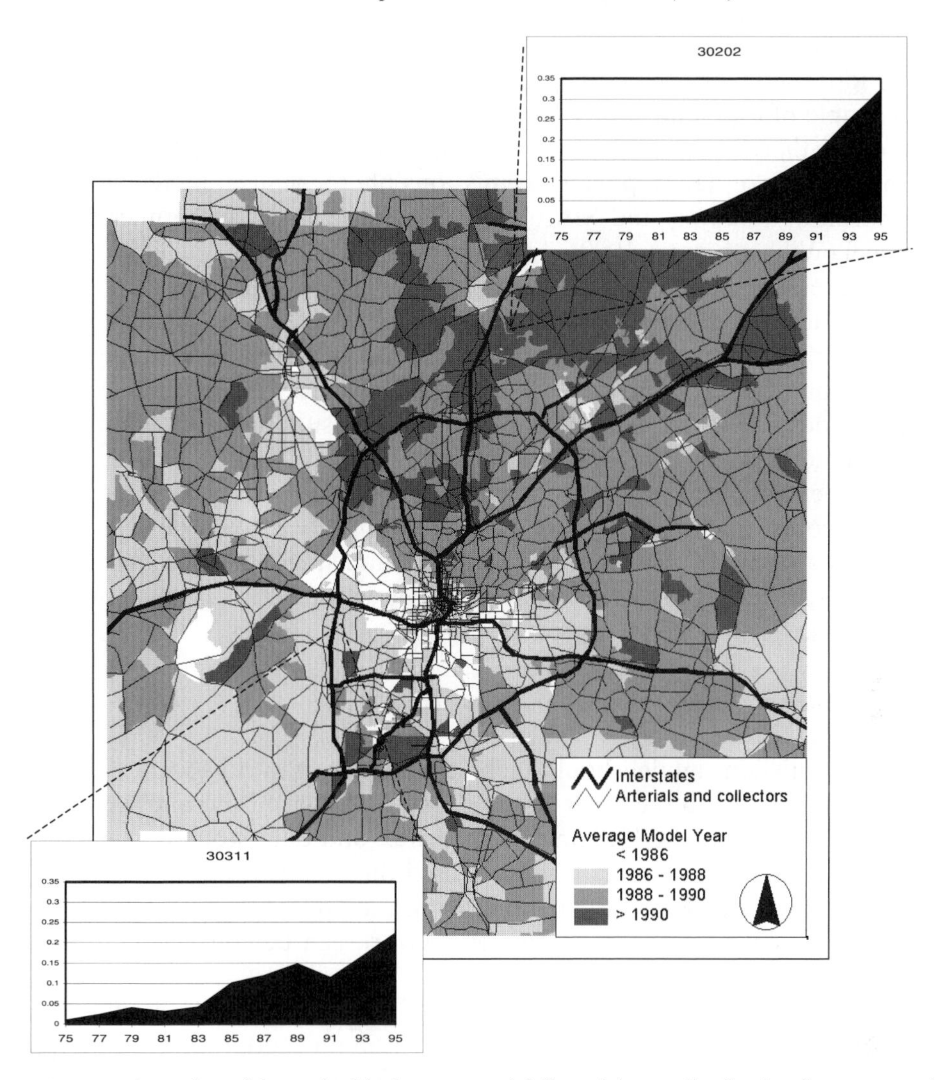

Fig. 2. Atlanta mean registered model year by block group and full model year distribution for two sample ZIP codes.

quality models linking traffic information with a GIS to produce synthesized databases for use in vehicle emission and air dispersion models. Barros et al. (1998) developed a methodology to develop a GIS-based traffic emission inventory for Portugal, useful for estimating both area and line sources. Briggs et al. (1997) described the use of a GIS combined with least squares regression analysis for mapping traffic-related air pollution to generate predictive models of pollution surfaces based on monitored pollution data and exogenous information. Anderson et al. (1996) also described the use of a GIS as a tool to illustrate the spatial patterns of emissions and to visualize the impact congestion has on emissions. The model consisted of an integrated urban landuse model that interfaced with the emissions rate model MOBILE 5C. The integrated model allowed the impact of transportation and landuse policy changes to be simulated in terms of their air quality impact.

One common theme of all of these efforts is the use of GIS as a tool to prepare or process data related to emissions modeling. None of these earlier efforts used GIS as an integrated modeling environment capable of estimating emission rates at a user-defined grid cell level.

A research effort that does not involve a commercial GIS platform but has focused on improving the spatial scale of data inputs into emissions photochemical models is the development of the TRansportation Analysis and SIMulation System (TRANSIMS) microsimulation travel forecasting model (Williams et al., 1999). TRANSIMS processes, stores, and manipulates spatial data through the use of a powerful spatial database engine with explicit network topology. It uses an advanced approach of vehicle microsimulation using synthetic populations. The TRANSIMS approach dramatically improves the spatial scale issue by modeling individual synthetic vehicles on a second-by-second basis. Unfortunately, TRANSIMS model inputs far exceed even the most detailed regional models. Further, the TRANSIMS model is still under development with many of its components undergoing calibration and validation. Until TRANSIMS becomes widely accepted and implemented, there will be continual reliance on existing modeling frameworks. A GIS-based macroscopic modeling approach that does not rely heavily on regional aggregations for inputs provides a robust alternative to the current modeling regime as well as future regional microscopic model such as TRANSIMS. In Section 2.3, we introduce a conceptual framework for a GIS-based macroscopic modeling approach.

2.3. Conceptual framework for a GIS-based emissions model

Elements of any emissions model should include data important to accurately predict emission rates, data that are available to the expected user, and data that fit the mitigation tools available to transportation professionals. GIS, computing power, and data storage capabilities allow this model design to expand from historical ones focused on simplicity, to one focused on comprehensiveness, usefulness, and flexibility. By removing the concern of processing time and disk storage, a robust emissions model conceptual framework can be conceived without fear that it cannot be implemented.

Research suggests that the conceptual framework for a robust model should be modal in nature. A GIS framework is ideally suited for implementing a modal modeling approach, where emission rates are a function of specific modes of vehicle operation (engine starts, running exhaust, enrichment, etc.). Modal models hold the most significant promise for improving model accuracy and eliminating the various shortcomings of the current highly aggregated approach (Washington, 1995; Barth et al., 1996). Conceptually, a GIS can be used to estimate mobile source emissions for different operating modes and store these results on individual layers. The layered information could then be aggregated to grid cell layers that are compatible with photochemical models. Once aggregated, total estimates can be calculated by summing across layers. The summing process in itself would not contribute to the error of the individual layer estimates because the cell polygons whose attributes are summed across corresponding layers would be identical in shape, size, and location. Thus the main contributor to error from a spatial processing standpoint is in the disaggregation (and to some extent the aggregation) of data to grid cells.

In 1995, Bachman et al. (1996a) developed a conceptual framework for a modal GIS-based emissions modeling regime. The conceptual framework benefited from lessons learned in earlier attempts using a GIS to assist in modeling mobile source emissions. Fig. 3 illustrates this original

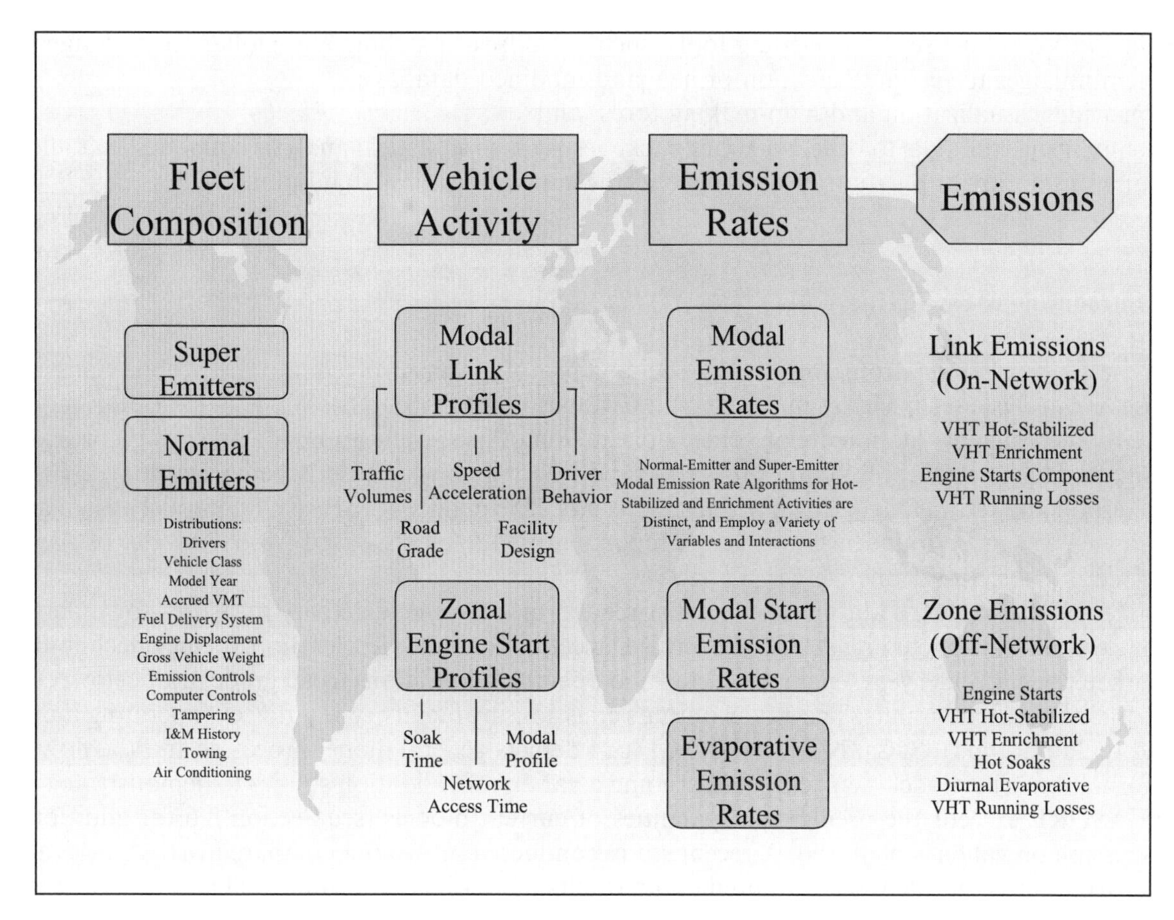

Fig. 3. GIS-based emissions model framework (Bachman et al., 1996a,b).

conceptual model. The framework attempted to improve on the spatial resolution of inputs, while adding additional elements that had been shown to have significant impact on emissions but were not currently considered in regulatory emissions models. For example, several research efforts have shown that roadway grades are directly related to elevated emissions because they have significant impacts on engine loads (Cicero-Fernandez et al., 1997; Pierson et al., 1996). Unfortunately, grades have been largely ignored in current regulatory models.

Portions of the original conceptual framework were implemented into a working prototype model that included experimental emission rates designed to identify the impacts of changes in acceleration and engine load. While the prototype was limited (e.g., it did not consider fleet distribution) it did illustrate several benefits of using a GIS platform for modeling emissions estimates:

- efficiently manages spatially referenced parameters that affect emissions,
- provides manipulation tools to calculate emissions from the modal parameters,
- allows a 'layered' approach to individual vehicle activity estimation,
- can efficiently aggregate emission estimates into grid cells for input to photochemical models using topologic overlay capabilities,

- includes a robust set of geocoding tools, such as address matching and global positioning system linkages to facilitate creation of new and modified databases,
- provides visualization and map-making tools, and
- contains useful links to other software packages such as statistical analysis software, that allow analysis and manipulation of data beyond the capabilities of a stand-alone GIS.

3. Introducing MEASURE

The GIS conceptual framework and associated prototype has evolved into the Mobile Emission Assessment System for Urban and Regional Evaluation (MEASURE). While MEASURE model development and evaluation are ongoing, a discussion of its specific approach and design provides insights into how the production GIS-based model for public release is being developed. Early validation and testing have shown promising results. The details of MEASURE model design and architecture can be found in the USEPA report entitled "A GIS-Based Modal Model of Automobile Exhaust Emissions" (USEPA, 1998).

MEASURE includes automobile exhaust-related modal vehicle activity measures for different vehicle conditions including starts, idle, cruise, acceleration, and deceleration. Vehicle technology characteristics (model year, engine size, etc.) and operating conditions (road grade, traffic flow, etc.) are included as data inputs. The model outputs emissions by facility type, grid cell, operating mode, and pollutant type (VOCs, NO_x, and CO). Fig. 4 depicts general model processes and the flow of information. The model is currently undergoing a variety of validation studies to demonstrate that MEASURE exceeds the predictive capabilities of current models (see Section 3.6 for additional discussion on validation efforts). A recent study conducted in Atlanta compared MEASURE and MOBILE emission rate models to predict the results of 16 different running exhaust emission test cycles (bag tests). MEASURE proved to be substantially better than MOBILE in predicting the cycles (which varied in cycle speeds and accelerations) (Fomunung et al., 2000).

This description focuses on the model components and procedures that demonstrate innovative uses of GIS and spatial analysis. Other crucial elements, such as new modal emissions rates, are discussed briefly; in depth discussions have been published elsewhere (Washington, 1995; Washington et al., 1997; Fomunung et al., 2000). While not all the indirect relationships are modeled (e.g., landuse change), an effort was made to include these parameters in anticipation of future research findings.

Required data necessary to develop a MEASURE model can be divided into five categories: spatial character, temporal character, vehicle technology, modal activity, and trip generation. These data are identified as follows:

Spatial character:
- landuse boundaries,
- US census block boundaries,
- traffic analysis zone boundaries,
- roads,
- travel demand forecasting network,
- output grid cell boundaries (user-defined).

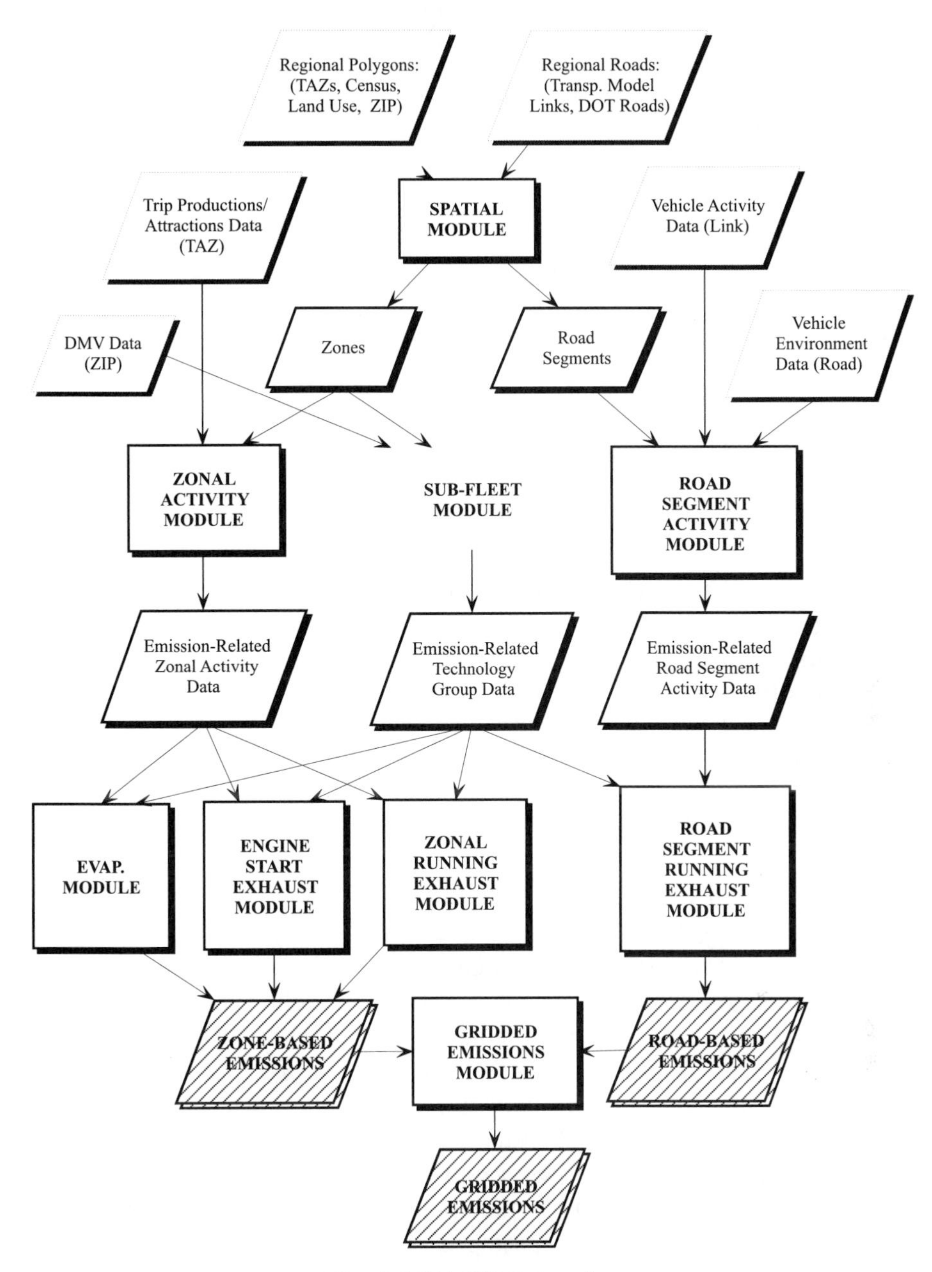

Fig. 4. MEASURE data flow.

Temporal character:
- hour of the day.

Vehicle technology:
- model year,
- engine displacement,

- transmission type,
- fuel delivery technology,
- supplemental air injection system,
- catalyst configuration,
- exhaust gas recirculation.

Modal activity:
- idle,
- cruise,
- acceleration,
- deceleration,
- starts,
- engine off.

Trip generation:
- landuse,
- housing units,
- socioeconomic characteristics (for spatial allocation only),
- home-based work trips,
- home-based shopping trips,
- home-based university trips,
- home-based grade school trips,
- home-based other trips,
- non-home-based trips.

3.1. Organizing the spatial environment

The organization of the MEASURE spatial environment is a function of the format of input data provided from other sources. Historically, emissions modeling regimes have divided exhaust emissions into start and non-start (running exhaust) emissions. Most prognostic travel models provide a traffic analysis zone (TAZ) estimate of the number of trip origins and a line (link) estimate of road volume and average speed. By defining an engine start as being synonymous with a trip origin, TAZs become the base spatial entity used for estimating engine start emissions within MEASURE. Running exhaust emissions occur on the road network, suggesting the network 'link' as the base spatial entity. Improvements in the spatial resolution of the zonal estimates can be made outside the travel model by disaggregating trips to smaller zones. In MEASURE, trip origins are disaggregated by census landuse data and by US census block group household data. Census blocks can be used to disaggregate trips even further. For example, TAZ trips leaving from home are spatially allocated to the portions of the TAZ that contain residential land uses and further weighted by household density determined from census blocks. This process does not affect the total TAZ home-based origins; the origins are simply allocated to smaller zones within the TAZ based on those factors. The final 'zone' used for emissions modeling is created through a series of GIS polygon overlays that intersects TAZs, census block and census block group boundaries, and US postal code (ZIP code) boundaries.

The spatial accuracy of the line estimate is improved by conflating the travel demand forecasting network 'links' to a comprehensive and accurate road database. Travel demand forecasting model networks are frequently abstract stick networks without intermediate shape points. Conflating improves the spatial resolution of roadway links and helps to clearly identify the network access points from each TAZ. Bachman et al. (1996b) describe a GIS conflation procedure for improving the spatial accuracy of travel forecasting networks. This procedure uses the GIS's built-in rubber sheeting tools as well as a series of heuristics (rules of thumb) to guide the automated conflation. A one-to-one correspondence table is also necessary to accurately conflate complicated road configurations, such as interchanges and closely spaced roads. The Bachman paper also discusses the potential impacts which conflating a travel demand forecasting network would have on hot stabilized emissions when the final estimates are aggregated to 1- and 5-km grid cells suitable for photochemical modeling.

3.2. Estimating vehicle fleet characteristics

Although many different emissions modeling approaches are being developed around the country, all indicate that identifying the emission-significant components of the operating fleet is important to emission rate accuracy (Siwek, 1997). Current regulatory emission models use model year distributions to describe the fleet, taking these data from registration or inspection and maintenance databases. However, many other vehicle characteristics hold significant explanatory capability for predicting emission rates, such as type of emission control equipment, engine size, vehicle weight, and transmission type (Fomunung et al., 1999; Barth et al., 1996; Stopher, 1993). Furthermore, to improve on the highly aggregate approach currently used, spatially resolved subfleet characterization is important.

In MEASURE, regional vehicle registration data are used to define emission-related vehicle characteristics. MEASURE's fleet module develops estimates of vehicle technology distributions for each of the zone and line representations. Every ZIP code has a 'technology group' distribution. Technology groups are combinations of vehicle characteristics and operating conditions that have been identified in regression tree analysis and linear regression analysis as being the most predictive. Regression models were developed for each of the three pollutants of interest (CO, HC, and NO_x) by analyzing a data set of more than 13,000 hot-stabilized laboratory treadmill tests on 19 driving cycles (specific speed versus time testing conditions) and 114 variables describing vehicle, engine, and test cycle characteristics. A total of 44 such technology classes were defined out of 2560 technology rules (Fomunung et al., 1999). Engine start technology groups only include vehicle characteristics. For each emission-significant combination of vehicle characteristics, an associated gram per start emission rate is identified. Running exhaust technology groups include vehicle characteristics and/or modal operating parameters (idle, cruise, acceleration, etc.). Unlike engine start groups, running exhaust technology groups can have different emission rates based on modal operating conditions.

The technology group distributions by ZIP code are assigned to every zone directly using a simple relational database procedure. Attaching distributions to roadway links is a little more difficult. Tomeh (1996) conducted a study of on-road vehicle distributions. In the study, vehicle registration data were used to define vehicle distributions by census block. Vehicle license tags at several interstate ramp locations were observed, and when distributions of the regional fleet and distributions of a local fleet (census blocks within a 3-mile radius of the ramp) were combined, the fleet distribution at each ramp location could be predicted. While Tomeh's approach to predicting on-road vehicles proved valid, his research results also indicated that a more accurate prediction would result if the local fleet could be better defined. Instead of a 3-mile radius, the actual spatial pattern of observed vehicle home locations was skewed based on the network configuration and the time of day. If the ramp was an 'off-ramp', there was an upstream concentration of local vehicles in the afternoon and a regional distribution in the morning (the opposite held true for on-ramps). In its current form, MEASURE relies on Tomeh's original strategy (3-mile search radius), but future efforts may include a more advanced approach based on his observations. In addition, MEASURE adjusts the weightings of regional versus local vehicle distributions based on road classification. Interstates are assigned a higher regional fraction (assuming that interstates serve more of a regional set of vehicles), while local roads are assigned a higher local fraction (Bachman et al., 1998; Tomeh, 1996).

3.3. Estimating emission-specific vehicle activity

The core prognostic capability of the model rests on the ability of travel demand forecasting models to accurately predict regional travel. The emission related vehicle activity estimates provided by regional travel models are the number and location of peak hour (or daily) trip origins, the road segment volumes, and the volume-to-capacity-based average speeds (later post-processed in estimating speed/acceleration distributions). Important activities not provided by most current models are temporal travel behavior and modal (idle, cruise, acceleration, and deceleration) operations. In MEASURE, the usable travel model information is translated into the emissions modeling environment and any missing emission-related parameters are estimated based on data available.

3.3.1. Engine start activity

Engine starts are equivalent to trip origins determined at the TAZ level by the trip generation component of travel demand forecasting models. These TAZs represent a spatial unit for aggregating socioeconomic data and the resulting trip generation (trip production and trip attraction) estimates. This is typically done by the developers of the travel demand forecasting models by aggregating data from smaller census blocks and block groups. Census information that falls across two TAZs is typically divided by examining landuse density within the census block or block groups. Estimates of trip generation are made for each TAZ for a variety of trip purposes. Trip purposes usually include home-based work (to and from the workplace), home-based shopping, home-based school, home-based other, and non-home-based trips. While these trips are estimated to begin or end in certain TAZs, the trip type definitions imply that they are home (residential) going to work (non-residential) or a trip originating from work going home. Likewise, home-based shopping trips are to or from a commercial landuse. For the engine start component, external trips that originate outside the study area are not included. However, any trip that has an origin in the study area is included. A table of temporal distributions of trip origins for the region is used to estimate the time of day that the engine start occurred.

The US Census Bureau maintains zonal databases developed for the decennial census. The smallest zonal designation is a block, usually an area bounded by roads or other line features (cadastral, hydrologic, etc.). Census blocks typically include 50–200 dwelling units. The 1990 estimates of the number of households are available at the census block level. Although these estimates are dated, they can provide clues to housing density within the TAZ and landuse designations. This information is used to further spatially disaggregate trips originating from residential areas. With good landuse and socioeconomic data, various trips can be disaggregated to smaller zones. Even if the landuse designations are as broad as "residential" and "non-residential", the spatial resolution of trip generation estimates is improved. This allows for an improved spatial resolution for engine start estimates.

3.3.2. Intra-zonal running exhaust activity

Travel time is a key variable in predicting running exhaust emissions (preferably broken down by operating mode) because emissions are directly related to hours of vehicle operation. Other than evaluating the size of the zone, travel times for intra-zonal trips (and inter-zonal travel off the major roads) are often unaccounted for in the travel demand modeling process. However, disaggregate trip generation estimates allow the development of travel time estimates using the digital road network and spatial analysis tools provided by the GIS.

Many GISs provide tools that allow the determination of the optimal network path between two points based on a shortest path algorithm. The disaggregated trip generation estimates provide a trip origin location. The closest intersection of an aggregately modeled local road with an explicitly modeled major road provides a destination location where the trip becomes on-network. The shortest network path between the two points provides an estimate of the travel distance. Averaging all these travel distances within a TAZ provides an estimate of the typical intra-zonal travel distance that occurs before vehicles reach the modeled network for a particular TAZ. Using this intra-zonal travel distance along with a typical average speed for local road travel provides an estimate of the average intra-zonal travel time. Although the strategy described above is somewhat crude, the method is more palatable than the alternatives of leaving the estimates out or assuming travel times based on TAZ area. The intra-zonal travel times can be improved if the average of actual spot speeds is used rather than relying on speeds based purely on roadway functional classes.

3.3.3. Modal activity

Modal vehicle activity is characterized by cruise, idle, acceleration, and deceleration operations. Research has clearly identified that modal activity is a better indicator of emission rates than average speed. Determining regional modal operation is not possible using current travel demand forecasting models alone. Travel models can forecast traffic volume (±15%) and average speed (±30%) (USDOT and USEPA, 1993). Because average speed estimates from travel models are relatively inaccurate, they should be used with caution. However, the average speed could be accurate enough to determine differences in levels of service (LOS) *E* and *F*, where forecast volume to capacity (v/c) ratios approach or surpass 1.0.

Research by Grant et al. (1996) and Hallmark and Guensler (1999) has identified methodologies to relate speed and acceleration distributions for vehicle activity as a function of road classification, level of service, and other Highway Capacity Manual parameters. Given a roadway's volume and capacity (and signal timing if evaluating an intersection), an estimate of congestion can be developed. Subsequently, speed and acceleration distribution tables that contain accurate fractions of vehicle activity in the high power demand areas are selected to estimate the fraction of emission-specific modal behavior occurring in those instances. Modal profiles have been developed for interstates, ramps (suspected as high power demand areas), arterials, and signalized intersections. Sample profiles are shown in Fig. 5. Because vehicle emissions for most speed and acceleration conditions are relatively constant, it is only important that the assigned

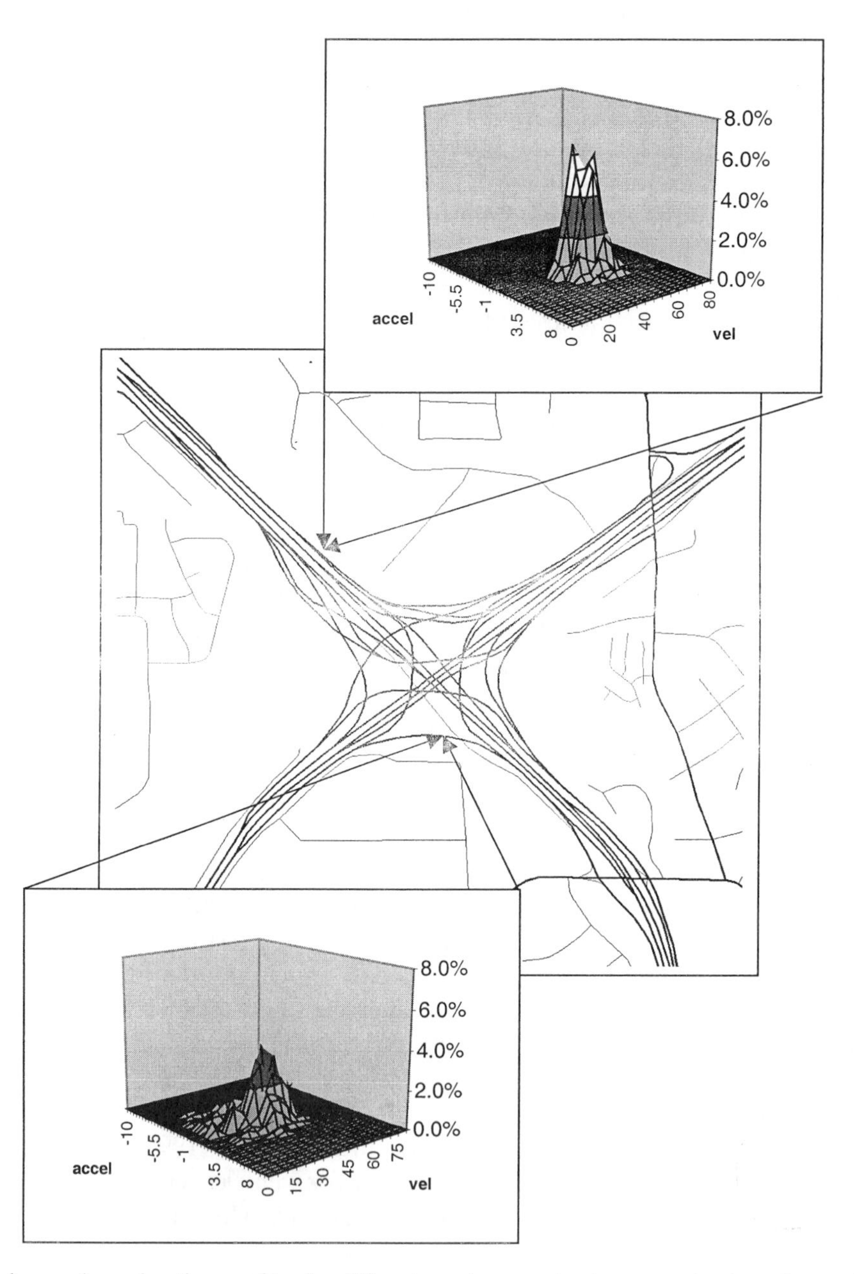

Fig. 5. Sample speed–acceleration profiles for different road segments along a major interchange in Atlanta.

profiles accurately reflect the fraction of vehicle activity under crucial high-emission modes. Hence, although these speed acceleration profiles do not perfectly reflect the entire range of speed and acceleration operations, they do accurately predict the fraction of activity occurring under high acceleration and high power demand conditions for different levels of congestion (Hallmark and Guensler, 1999). When this process is conducted for every road segment section, the distribution of modal behavior can be estimated in space and time.

3.4. Predicting facility-level emissions

Roadway facilities are divided into zones and lines corresponding to the previously mentioned emission modes of engine starts and running exhaust (respectively). Facility activity estimates are used to allocate emission production to those vector spatial data structures currently used by transportation planners. By typing emission production estimates to facilities, tasks associated with research, reporting, validation, or control strategy development are easier. Emission rates for each portion were developed by reanalysis vehicle emission tests from a variety of sources (Wolf et al., 1998; Fomunung, et al., 1999).

3.4.1. Engine start zonal facility estimates

Elevated emissions at engine start occur over a period of one to three minutes while the catalytic converter warms up. These elevated exhaust emissions are modeled as a 'puff' (all engine start emissions allocated to the trip origin zone). While start emissions are actually dispersed through the local network as a vehicle travels, research has not identified a practical strategy for spatial allocation to local roads. Furthermore, it is more useful for planners and/or researchers to have engine start emissions tied to the point of origin, allowing linkages to zonal characteristics that may be crucial in identifying mitigation strategies.

The calculation of emissions for a single zone is

$$E = \sum_{n=1}^{N} (TG_n \times ER_n) \times O,$$

where E is the emissions for single facility in grams (CO, HC, or NO_x), N the number of technology groups for pollutant of interest, TG the fraction of registered vehicles in the zone in the specified technology group, ER the gram per start emission rate for the specified technology group and pollutant, and O is the number of vehicle trip origins.

The resulting emissions of CO, HC, and NO_x are usually reported for a typical weekday (Tuesday–Thursday) on an hourly basis. The typical weekday limitation is a result of the travel demand modeling process, as few models are currently setup to predict weekend or Friday travel. MEASURE allows other time aggregations to occur since vehicle activity is a direct input into the model.

3.4.2. Minor road zonal facility estimates

Minor road zones are used to spatially represent the portion of running exhaust emissions that occur between the trip origin and the major roads modeled by the travel demand forecasting network. For these zones, the total travel time, the typical local road speed and acceleration profile, and the zonal technology characteristics are used as inputs to the emission rate algorithm (discussed in Section 3.4.3).

3.4.3. Line facility estimates

Line facilities are the major roads that are modeled in the travel demand forecasting model. On-road fleet distributions and predicted traffic flow parameters are used to generate road segment specific estimates of CO, HC, and NO_x. For each road segment and each hour, modal variables are determined based on the speed and acceleration characteristics identified in the situation-specific profiles. Road segment technology group fractions and modal variables are combined to develop the fraction of activity occurring with each specific emission rate (grams per second). Total hourly travel time is calculated and segmented by the fraction of the vehicles with each emission rate. Combined running exhaust emissions estimates from minor and major roads provide total on-road running exhaust emissions.

The calculation of emissions for a single road segment is

$$E = \sum_{n=1}^{N} (TG_n \times B_n \times I_n) \times F_{\mathrm{p}} \times T,$$

where E is the emissions for single facility in grams (CO, HC, or NO_x), N the number of technology groups for the pollutant of interest, TG the fraction of registered vehicles on the road in the specified technology group, B the mean FTP Bag2 emission rate in g/s for the specified technology group, I the interaction factor for specific technology combination and estimated modal conditions, F the constant for each of the three pollutant, and T is the total seconds of travel time for that road segment.

More details on the equations and their validation can be found in Fomunung et al. (1999, 2000).

3.5. Generating the mobile emissions inventory

The role of the emissions inventory module is to convert the facility-based emission estimates into gridded estimates. Procedurally, the user selects a grid cell size; the software creates a polygon database of grid cell boundaries, allocates each zone or line (or parts of zones or lines) to its corresponding grid, sums all emissions for each cell, and finally converts the results to raster data structures.

Grid cell size is optional for the user but usually is dictated by the subsequent photochemical model. The gridded results from MEASURE are inputs into other models that predict ambient pollutant concentrations. Most photochemical models use grid cell sizes of 4–5-km; however, new designs plan to use 1-km grid cells. MEASURE creates a grid cell boundary database (polygons, not raster cells) for the study area. Emissions at each facility (zone or line) are converted to a rate based on the area (zones) or length (roads). Resulting values are therefore in g/square km, or g/km. The grid cell boundaries are used as 'cookie cutters' to identify which facilities or parts of facilities fall in each grid cell. The facility emissions are then converted back to grams by multiplying the rates by the new areas or lengths.

The polygon grid cells are then converted to raster data structures with raster cells equivalent to the size and position of the original polygon. This conversion does not contribute to model error because each raster cell has the exact same shape and location of the corresponding polygon grid cell. The conversion is conducted because the raster database is more efficient at storing gridded information. The final raster datasets are individual 'layers' of each pollutant emission 'mode' (totals, engine starts, etc.). MEASURE includes a customized user interface for querying and visualizing two- and three-dimensional images of the various input and output databases.

3.6. Sample model run for Atlanta

MEASURE was written using 'C' code and ARC/INFO AML. Each of the modules described previously is controlled by a 'Makefile'. The model requires ARC/INFO software to be resident on the system, but handles all software access and syntax. A sample model was developed that predicts grams of CO, HC, and NO_x, for all zones, lines, 100-m cells, 250-m cells, 500-m cells, and 1-km cells. The study area was the 13 county, non-attainment area in Atlanta, Georgia. The following input datasets were used:

- 1995 Atlanta Regional Commission (ARC) LandUse Data,
- 1990 US Census Summary Tape File (STF) 3a,
- 1994 US Census Topologically Integrated Geographic Encoding and Referencing (TIGER) File,
- 1995 Updated TIGER Road Database,
- 1996 ARC ARCMAP Road Database,
- 1995 ARC Traffic Analysis Zones,
- 1995 ARC Travel Demand Forecasting Network,
- 1995 ARC Temporal distributions by trip type,
- 1996 Georgia Department of Motor Vehicles Registration Dataset, and
- 1996–97 Georgia Tech Speed and Acceleration Profiles.

The following output files were created by the model:

- Zonal Vehicle Characteristics,
- Road Segment Technology Group Distributions,
- Zonal Vehicle Activity,
- Road Segment Vehicle Activity,
- Zonal Start Emissions (Fig. 6),
- Zonal Running Exhaust Emissions (Fig. 6),
- Road Segment Running Exhaust Emissions, and
- Gridded Emissions (Fig. 7).

Sample model run outputs for Atlanta are demonstrated in Figs. 6 and 7. Spatial input data for the model run were organized under a single datum and projection system. Major data preparation steps that were conducted outside the MEASURE domain include geocoding of the vehicle registration database to ZIP codes, vehicle identification number (VIN) decoding of

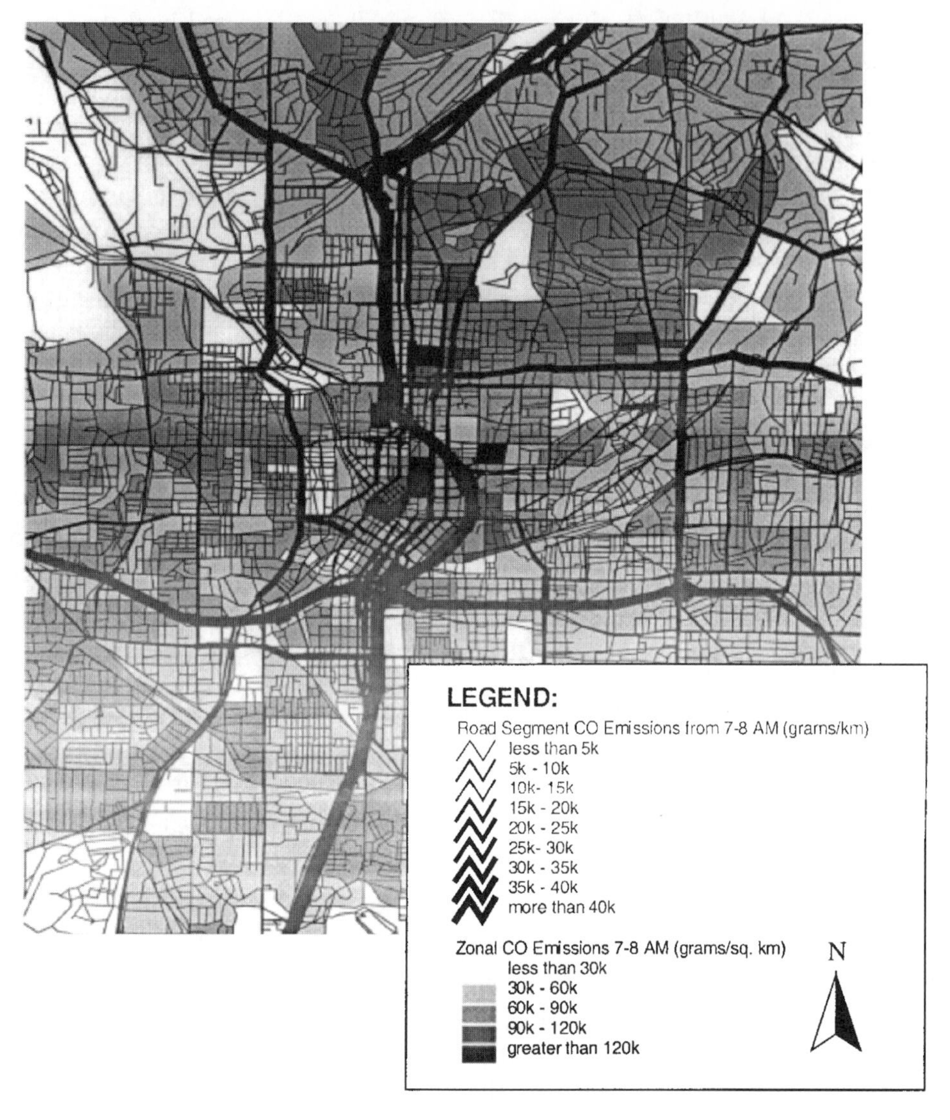

Fig. 6. Road segment and zonal emission estimates for 7–8 a.m. in Atlanta.

the vehicle registration database, generating block group polygons from TIGER, and adding US Census STF3a housing unit data to the block groups. The travel demand forecasting network was converted using a software utility developed specifically for that purpose. Some "clean-up" of this network was necessary using some of the automated techniques described previously.

The model code ('C', and ARC/INFO AML) is organized in a 'make' routine that verifies code updates, cleans temporary files, and initiates the modules in the appropriate sequence. The shortest path routines that allocate each engine start zone to the closest major intersection took the longest to process. The entire run, which took approximately 25 h of continuous processing,

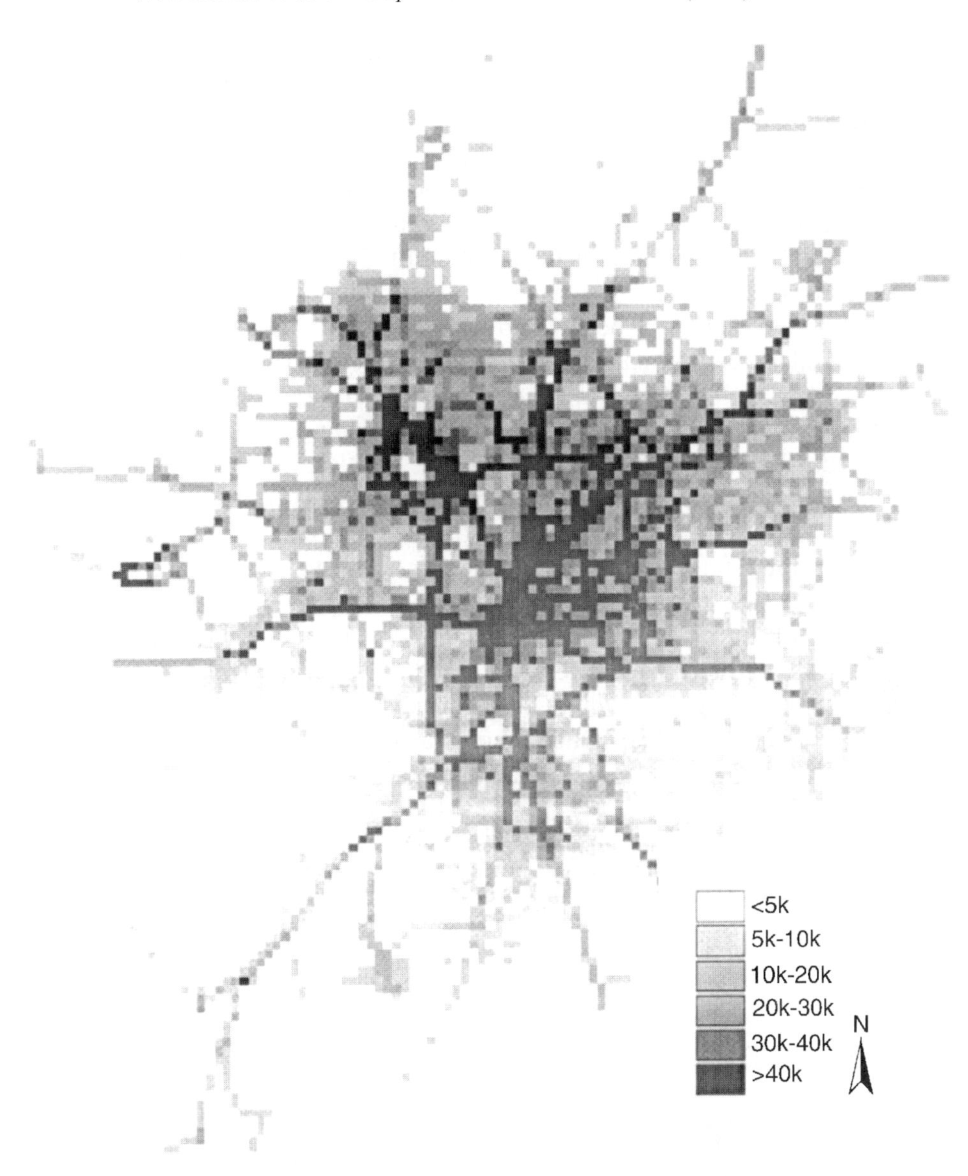

Fig. 7. Total grams of CO mobile emissions in Atlanta from 7–8 a.m. (1-km grid cells).

was conducted on a Dell dual-processor 400 MHz Pentium II operating Windows NT with ARC/INFO resident.

This initial run for Atlanta did not include modal activity for intersections since signalized intersection data were not available. Instead, a speed/acceleration profile for a typical signalized arterial road was used. This 'typical' arterial speed/acceleration profile included observed data aggregated from several signalized and unsignalized arterial roads. All of the sites observed were multi-lane facilities. In the model run, all non-interstate, non-ramp multi-lane roads were assigned this profile as an estimate of modal activity.

The Atlanta model is being used as a basis for validating MEASURE components. One effort currently underway is a vertical pollutant flux study that involves detailed monitoring of a metered ramp system on I-75, north of downtown Atlanta. The research team collected traffic flow data, vehicle classification, fleet technology characteristics (by monitored license plate data and later decoding the registration VINs), and speed acceleration profiles with laser guns on the four ramps and the adjacent freeway links. The research team is currently analyzing the 18 days of vehicle activity data and will use the vertical pollutant flux results to compare predicted and measured emission rates.

Initial comparisons have already been conducted between MEASURE and MOBILE5a to explore and identify differences in their emission rates. Fig. 8 shows the NO_x g/s emission rates for MOBILE5a and MEASURE by speed and acceleration bin (using an Atlanta regional fleet). The charts indicate that MEASURE is much more sensitive to changes in acceleration, particularly at high operating speeds. This is not unexpected because MOBILE emission rates are primarily a function of average speed, while MEASURE rates are directly influenced by the relationship between speed and acceleration. In this comparison, the regional fleet distribution was used to estimate emissions for interstate LOS A–F. Mean emission rates in g/s were within 20% of each method for LOS A. MEASURE emission rates were 50% higher than MOBILE for LOS B and C, and twice as high for LOS D and E. Emission rates were back to within 20% for LOS F. Because MEASURE is sensitive to accelerations at high speed, its emission rates were much higher for moderate congestion levels where high speeds and variable acceleration resulting from increased vehicle interaction is typical. Once there is a breakdown, a traffic flow resulting in low speeds with variable acceleration, MEASURE emission rates drop to levels comparable to MOBILE.

The significant differences in emission rates between MEASURE and MOBILE, especially at moderate LOSs, magnify the need for further validation studies.

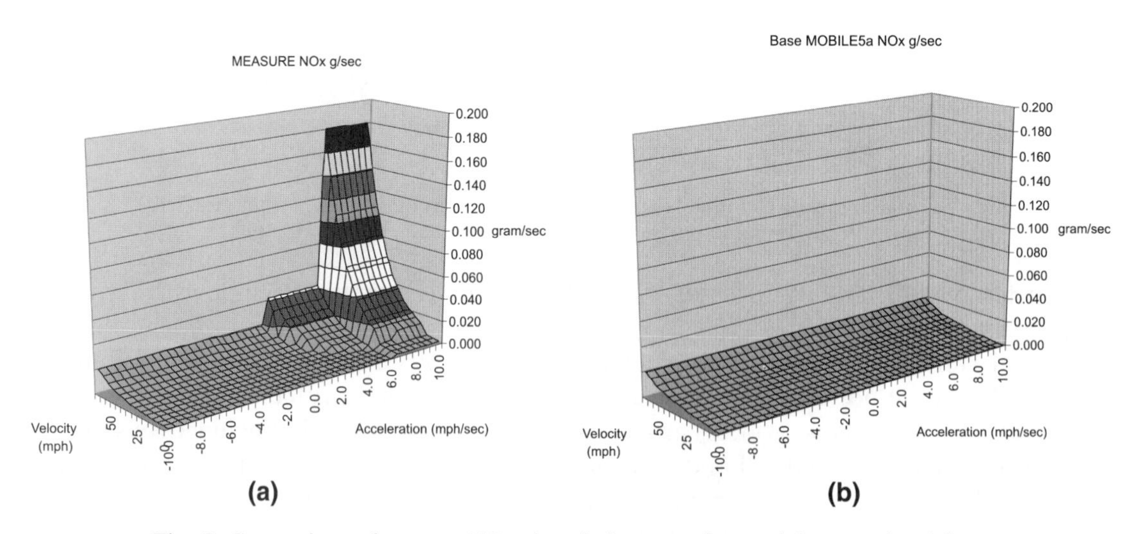

Fig. 8. Comparison of average NO_x g/s emission rates for an Atlanta regional fleet.

3.7. Potential policy impacts

If EPA approves a GIS-based modal emissions model, such as MEASURE, for regulatory use, the types of mitigation strategies available to local and state governments will change dramatically. Under the current modeling system, transportation planners and engineers have only three ways to reduce mobile emissions: depend upon EPA to pass new vehicle certification standards, reduce vehicle miles of travel, or optimize average speeds to ranges where emissions based on average speed estimates are reduced. The choices usually result in reducing the mobility and accessibility desired by the transportation system users.

If spatially resolved modal models are developed, much more diverse and creative strategies become assessable. Any strategy that reduces the number of high-emitting vehicles or reduces the occurrence of hard accelerations and decelerations is expected to reduce mobile emissions. However, current modeling regimes cannot accurately assess the impact of those changes to regional air quality. A model framework, such as MEASURE's, makes two significant improvements in this area: spatial variability becomes an important component of any mitigation strategy and the modal characteristics allow the assessment of variability in traffic flow, not just average speed. Reducing traffic volumes may be less important than improving traffic flow through ITS strategies, signal timing, or even lane additions. The new modal approaches may show that (at least in the short term) mobility and accessibility can increase as mobile emissions decrease (Hallmark et al., 2000).

Spatially resolved emissions estimates allow planners to prioritize certain locations for mitigation strategies because of their disproportional contribution to regional ozone formation. This is the real value of improved spatial variability. The disproportional contribution may be the result of topography, landuse, or climatic factors. Regardless, a dollar spent mitigating mobile emissions in one part of the region may not result in the same reduction if spent in another part. The spatially resolved estimates at proper resolutions also allow local transportation planners and traffic engineers to develop sub-regional strategies that help to improve regional air quality.

4. Future research

Some specific GIS-oriented model design and implementation research projects could improve the accuracy of MEASURE estimates. While model validation is important in confirming current capabilities, addressing these theoretical issues could provide significant modeling benefits over the short term.

- *Fraction of total vehicle operation by vehicle type*: The registration dataset represents all light-duty vehicles that are licensed to operate on the road. The actual operating fleet may look quite different. Older third and fourth vehicles in a household are not expected to be used to the same extent.
- *On-road vehicle distribution search pattern*: Additional studies have indicated that the radial search pattern used in the model could be significantly improved for determining a local operating fleet. Research into the size and shape of the search pattern will significantly improve the capability of predicting the on-road fleet distribution.

While the conceptual design of MEASURE is comprehensive, the actual working model is not. The current model scope is limited to automobile exhaust emissions. Moving to a complete mobile emissions model involves adding much more information and data. Some of the major items are listed below.

- *On- and off-network grade distributions and impacts*: Because road grade has spatial variability and has significant impact on the load on an engine, it should be included in the research design. This may mean moving to more detailed modal emission rates that model emissions as a direct function of engine load rather than as a function of load surrogates.
- *More speed/acceleration matrices*: Currently, the model is limited to approximately 20 different profiles. Further refining the subroutines that define speed and acceleration profiles for all road types and configurations (accounting for influences of weaving sections and other physical parameters) will provide a more comprehensive view of modal activity.
- *Other motor vehicle types*: A comprehensive mobile source model must include all vehicle classifications. Currently, all light-duty vehicles are modeled as automobiles because there are few vehicle emission tests for sports-utility vehicles and light-duty trucks under a wide variety of operating conditions. Heavy-duty truck modeling components (load-based) are currently being added.
- *Load-based approach*: A new engine load-based modeling approach to predicting emissions will allow enrichment emissions to be separately identified, an original model design objective.
- *Non-exhaust mobile emissions*: Exhaust emissions only make-up a portion of the overall mobile emission modes. Evaporative emissions are currently being adapted directly from the MOBILE5a model and will be upgraded in future models.
- *External/internal trips*: Currently, external/internal trips are excluded from the models' predictions of start activity and evaporative emissions.

At the time MEASURE development first began, a number of GIS platforms were evaluated. ARC/INFO was chosen because of its robust set of spatial analysis tools as well as its widespread use at Metropolitan Planning Organizations (MPOs) throughout the US. The emergence and evolution of Arc View and related components, Geomedia, and MapObjects offers attractive and more affordable GIS platforms for a future version of MEASURE once the ongoing validation phase is complete. A thorough evaluation of these and other GIS software will need to be conducted to determine if they are suitably equipped to accommodate the complex spatial modeling requirements of MEASURE.

5. Conclusion

Traditional transportation emission models have been shown to suffer from problems, such as highly aggregated datasets, non-representative emission factors, and lack of spatial resolution. Consequently, there is a desire to move towards a modal approach which relates activity-specific emission production with corresponding vehicle activity. The special capabilities of a GIS can greatly simplify the procedures associated with creating, combining, and manipulating the spatial databases necessary for implementing a modal approach. Further, the GIS can be used to establish emissions estimates as a function of a number of variables associated with points and areas (off-network activities) or roadway links (on-network activities). Once established, these improved spatial estimates can be aggregated to grid cells that are compatible for input into regional photochemical models. The major modeling deficiencies that exist in current models are addressed with the implementation of the MEASURE model. The limitations of MEASURE revolve mostly around the intensity of data required. MEASURE was designed and developed as a research model that considers all relevant data at the best resolution possible. As research and validation continue, relationships between variables may be identified which would reduce the quantity and variety of data. This is critical if MEASURE is to be widely accepted and implemented on a large scale.

While tailpipe emission reductions for the fleet-at-large will continue to be an important control strategy for years to come, there is an increasing interest in other methods of controlling mobile sources. Future mobile-source emission models must be equipped to predict the impact of such methods as improved traffic flow, improved inspection and maintenance programs, targeted enforcement of super-emitters, traffic restrictions, and use of alternative fuels. MEASURE has already been shown to be capable of analyzing these types of policies. For example, MEASURE is currently being used in Atlanta by the Georgia Department of Transportation to evaluate the emission impacts of proposed ITS projects. Even with great strides in mobile emission reductions, there will always be a need to gather comprehensive spatial and temporal distributions of emissions for urban areas. A GIS-based emissions modeling framework makes this more practical than ever before.

References

Anderson, W.P., Kanaroglou, P.S., Miller, E.J., Buliung, R.N., 1996. Simulating automobile emissions in an integrated urban model. Transportation Research Record 1520, Transportation Research Board, Washington, pp. 71–80.

Bachman, W., Sarasua, W., Guensler, R., 1996a. Geographic information framework for modeling mobile-source emissions. Transportation Research Record 1551, Transportation Research Board, Washington, pp. 123–133.

Bachman, W., Sarasua, W., Washington, S., Guensler, R., Hallmark, S., Meyer, M., 1996b. Integrating travel demand forecasting models with GIS to estimate hot stabilized mobile source emissions. In: Proceedings of the 1996 AASHTO GIS-T Symposium. American Association of State Highway and Transportation Officials, Kansas City, pp. 257–268.

Bachman, W., Granell, J., Guensler, R., Leonard, J., 1998. Research needs for determining spatially resolved subfleet characteristics. Transportation Research Record 1625, Transportation Research Board, Washington, pp. 139–146.

Barros, N., Borrego, C., Lopes, M., Miranda, A., Tchepel, O., 1998. Development of an emissions data base for air pollutants from mobile sources in Portugal. In: Fourth International Conference on Urban Transport and the Environment for the 21st Century, 285–294.

Barth, M., An, F., Norbeck, J., Ross, M., 1996. Modal emissions modeling: a physical approach. Transportation Research Record 1520, Transportation Research Board, Washington, pp. 81–88.

Briggs, D.J., Collins, S., Elliot, P., Fischer, P., Kingham, S., Lebret, E., Pryl, K., Van Reeuwijk, H., Smallbone, K., Van Der Veen, A., 1997. Mapping urban air pollution using GIS: a regression-based approach. International Journal of Geographic Information Science 11 (7), 699–718.

Bruckman, L., Dickson, R.J., Wilkonson, J.G., 1992. The use of GIS software in the development of emissions inventories and emissions modeling. In: Proceedings of the Air and Waste Management Association. Pittsburgh, PA.

California Air Resources Board (CARB), 1994. The land use-air quality linkage: how land use and transportation affect air quality. CARB, Sacramento, CA.

Chatterjee, A., Wholley, T.F., Guensler, R., Hartgen, D.T., Margiotta, R.A., Miller, T.L., Philpot, J.W., Stopher, P.R., 1997. Improving transportation data for mobile source emissions estimates. NCHRP Project 25-7, National Cooperative Highway Research Program, Report 394, Washington.

Cicero-Fernandez, P., Wong, W., Long, J.R., 1997. Fixed point mobile source emissions due to terrain related effects: a preliminary assessment. In: Proceedings of the Air and Waste Management Association's 90th Annual Meeting. Toronto, Canada.

Fomunung, I., Washington, S., Guensler, R., 1999. A statistical model for estimating oxides of nitrogen emissions from light-duty motor vehicles. Transportation Research D 4 (5), 333–352.

Fomunung, I., Washington, S., Guensler, R., Bachman, W., 2000. Performance evaluation of MEASURE emission factors – comparison with MOBILE. Published in CD-ROM of the Proceedings of the 78th Annual Meeting of the Transportation Research Board, Washington.

Grant, C., Guensler, R., Meyer, M.D., 1996. Variability of heavy-duty vehicle operating mode frequencies for prediction of mobile emissions. In: Proceedings of the 89th Annual Meeting of the Air and Waste Management Association. Pittsburgh, PA.

Guensler, R., 1994. Vehicle emission rates and average vehicle operating speeds. Dissertation, Department of Civil Engineering, University of California, Davis.

Hallmark, S., O'Neill, W., 1996. Integrating geographic information systems for transportation and air quality models for microscale analysis. Transportation Research Record 1551, Transportation Research Board, Washington, pp. 133–140.

Hallmark, S.L., Guensler, R., 1999. Comparison of speed–acceleration profiles from field data with NETSIM output for modal air quality analysis of signalized intersections. Transportation Research Record 1664, Transportation Research Board, Washington, pp. 40–48.

Hallmark, S.L., Bachman, W., Guensler, R., 2000. Assessing the impacts of improved signal timing as a transportation control measure using an activity-specific modeling approach. Transportation Research Record, Transportation Research Board, Washington (in press).

LeBlanc, D.C., Saunders, M., Meyers, M.D., Guensler, R., 1995. Driving pattern variability and impacts on vehicle carbon monoxide emissions. Transportation Research Record 1472, Transportation Research Board, Washington, pp. 45–52.

Medina, I.C., Schattanek, G., Nichols, F. Jr., 1994. A framework for integrating information systems in air quality analysis. In: Proceedings of URISA 1994, 32nd Annual Conference. Milwaukee, WI, p. 339.

Pierson, W.R., Gertler, A.W., Robinson, N.F., Sagebiel, J.C., Zielinska, B., Bishop, G.A., Stedman, D.H., Zweidinger, R.B., Ray, W.D., 1996. Real-world automotive emissions – summary of studies in the Fort McHenry and Tuscarora mountain tunnels. Atmospheric Environment 30 (12), 2233–2256.

Siwek, S.J., 1997. Summary of Proceedings, EPA-FHWA Modeling Workshop. Ann Arbor Science, Ann Arbor, MI.

South Coast Air Quality Management District (SCAQMD), 1996. Current Air Quality. Diamond Bar, CA.

Souleyrette, R.R., Sathisan, S.K., James, D.E., Lim, S., 1992. GIS for transportation and air quality analysis. In: Proceedings of the National Conference on Transportation Planning and Air Quality. ASCE, New York, NY, pp. 182–194.

Stopher, P., 1993. Deficiencies of travel-forecasting methods relative to mobile emissions. Journal of Transportation Engineering, ASCE 119 (5), 723–741.

Stopher, P.R., Hartgen, D.T., Li, Y., 1996. SMART: simulation model for activities, resources and travel. Transportation, vol. 23. Kluwer Academic Publishers, Dordrecht, Netherlands, pp. 293–312.

Tomeh, O., 1996. Spatial and temporal characterization of the vehicle fleet as a function of local and regional registration mix. Dissertation, School of Civil and Environmental Engineering, Georgia Institute of Technology, Atlanta, GA.

United States Department of Transportation and US Environmental Protection Agency, 1993. Clean air through transportation: challenges in meeting national air quality standards.

United States Environmental Protection Agency, 1998. A GIS-based modal model of automobile exhaust emissions. Report number EPA-600/-98-097, Research Triangle Park, NC.

United States Environmental Protection Agency, 1995. National air pollutant emission trends, 1900–1994. Office of Air Quality Planning and Standards, Research Triangle Park, NC.

Vonderohe, A.P., Travis, L., Smith, R., Tsai, V., 1993. National Cooperative Highway Research Program Report 359. Transportation Research Board, Washington.

Williams, M.D., Thayer, G.R., Smith, L., 1999. A comparison of emissions estimated in the TRANSIMS approach with those estimated from continuous speeds and accelerations. Published in CD-ROM of the 78th Annual Meeting of the Transportation Research Board, Washington.

Wolf, J., Washington, S., Guensler, R., Bachman, W., 1998. High emitting vehicle characteristics using regression tree analysis. Transportation Research Record 1641, Transportation Research Board, Washington, pp. 58–65.

Washington, S., 1995. Estimation of a vehicular carbon monoxide modal emission model and assessment of an intelligent transportation technology. Dissertation, Department of Civil Engineering, University of California, Davis.

Washington, S., Leonard, J., Roberts, C., Young, T., Botha, J., Sperling, D., 1997. Forecasting vehicle modes of operation needed as input to modal emissions models. In: Proceedings of the Fourth International Scientific Symposium on Transport and Air Pollution. Lyon, France.

PERGAMON

Transportation Research Part C 8 (2000) 231–256

TRANSPORTATION
RESEARCH
PART C

www.elsevier.com/locate/trc

Development and evaluation of a hybrid travel time forecasting model

Jinsoo You *, Tschangho John Kim

Department of Urban and Regional Planning, University of Illinois at Urbana-Champaign, 111 Temple Buell Hall, 611 East Lorado Taft Drive, Champaign, IL 61820, USA

Abstract

The purpose of this paper is to develop and evaluate a hybrid travel time forecasting model with geographic information systems (GIS) technologies for predicting link travel times in congested road networks. In a separate study by You and Kim (cf. You, J., Kim, T.J., 1999b. In: Proceedings of the Third Bi-Annual Conference of the Eastern Asia Society for Transportation Studies, 14–17 September, Taipei, Taiwan), a non-parametric regression model has been developed as a core forecasting algorithm to reduce computation time and increase forecasting accuracy. Using the core forecasting algorithm, a prototype hybrid forecasting model has been developed and tested by deploying GIS technologies in the following areas: (1) storing, retrieving, and displaying traffic data to assist in the forecasting procedures, (2) building road network data, and (3) integrating historical databases and road network data. This study shows that adopting GIS technologies in link travel time forecasting is efficient for achieving two goals: (1) reducing computational delay and (2) increasing forecasting accuracy.

Keywords: GIS; Travel time forecasting; Non-parametric regression; Historical database; Machine learning; Parameter adjustment

1. Introduction

Many researchers have endeavored to develop reliable travel time forecasting models using various methods including historical profile approaches, time series models, neural networks, non-parametric regression models, traffic simulation models, and dynamic traffic assignment (DTA) models (Sen et al., 1997; Ben-Akiva et al., 1995; Gilmore and Abe, 1995; Peeta and Mahmassani,

* Corresponding author. Tel.: +1-217-244-5369; fax: +1-217-244-1717.
E-mail addresses: j-you1@uiuc.edu (J. You), t-kim7@uiuc.edu (T.J. Kim).

PII: S0968-090X(00)00012-7

1995; Ben-Akiva et al., 1994; Mahmassani et al., 1991; Davis et al., 1990). Lessons learned from these experimental efforts evince that future travel times are difficult to estimate using a single forecasting method.

Therefore, the main purpose of this study is to develop and evaluate a hybrid travel time forecasting model in a comprehensive framework by adopting geographical information systems (GIS) technologies. Upon reviewing various functions of GIS-based applications (Nygard et al., 1995; Choi and Kim, 1994; Azad and Cook, 1993; Gillespie, 1993; Ries, 1993; Loukes, 1992; Abkowitz et al., 1990), the three important roles of GIS, data management, technology management, and information management, have been employed to support the operation of the hybrid forecasting model:

- *Data Management*. The hybrid model requires various types of traffic data including traffic speed, traffic volume, occupancy rate, number of lanes, and so forth. Therefore, GIS should be able to provide mechanisms for data aggregation and manipulation.
- *Technology Management*. The hybrid model requires various functions including display, computation, and analysis. Therefore, GIS should provide flexible tools to integrate and customize the required functions. Moreover, GIS should support travel time forecasting models to deal with various types of raw traffic data from loop detector, mobile communication, and global positioning systems (GPS).
- *Information Management*. After forecasting future travel times, the results are validated and classified so that traffic management and information centers could utilize them to manage network traffic. It is expected that GIS will eventually become an information gateway to the public, through various types of communication tools, such as the internet, telecommunications, broadcasting, etc.

For a more reliable and operational travel time forecasting model, this study focuses on pursuing the following objectives:

- developing a historical database using traffic data collected from loop detectors and probe vehicles;
- devising a possible way of integrating historical databases with road network data, and;
- developing and evaluating a hybrid travel time forecasting model, which adopts non-parametric regression techniques in conjunction with GIS technologies.

2. Integration of transportation models with GIS: a review

In this study, three distinctive types of travel time forecasting models are considered in conjunction with GIS technologies as shown in Fig. 1. They are: (1) Type I – the forecasting model "including" GIS, (2) Type II – the forecasting model "connected to" GIS, and (3) Type III – the forecasting model "within" GIS. In general, these three integration methods fall within the category of close combination of information systems, which technically combines subsystems in an integrated framework. However, these integration methods show different levels of system performance, integration difficulty, and system modification (Kriger and Schlosser, 1992).

Type I represents a forecasting model that includes GIS functions. In general, it is believed to be difficult to build customized computer programs for GIS functions due to the complicated mechanisms of GIS data. To ease the difficulty of developing Type I applications, many vendors

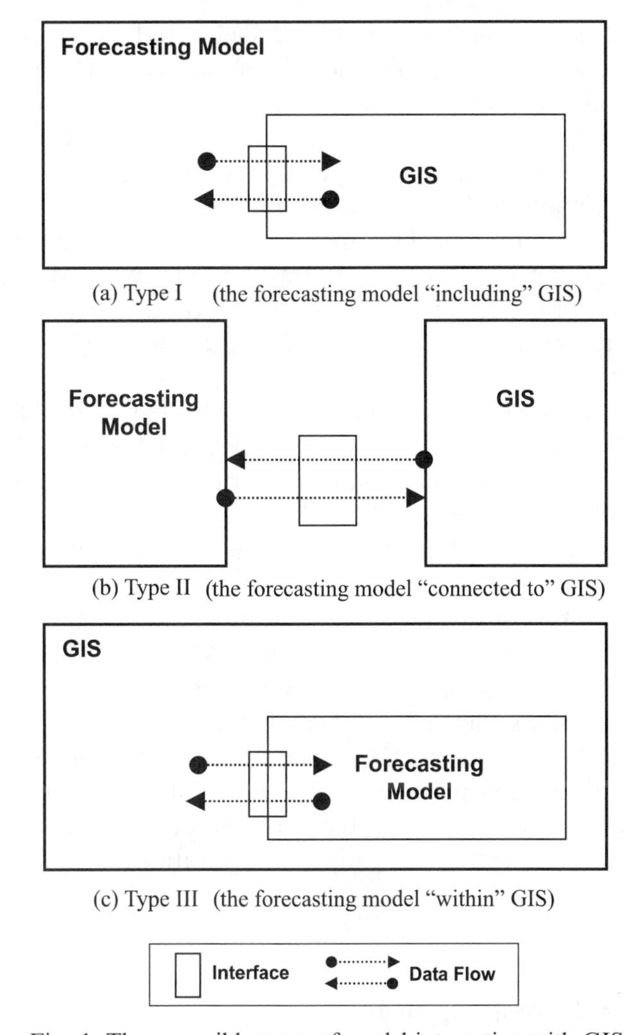

(a) Type I (the forecasting model "including" GIS)

(b) Type II (the forecasting model "connected to" GIS)

(c) Type III (the forecasting model "within" GIS)

Fig. 1. Three possible types of model integration with GIS.

are currently offering object-oriented GIS function libraries and ActiveX controls. MapObjects, NetEngine (ESRI), MGE (Intergraph), GISDK (Caliper Corporation), MapX, and MapBasic (MapInfo) are some examples of this category. These object libraries and controls allow users to perform spatial and attribute-based queries, communicate with external applications, and build custom application interfaces in conjunction with object-oriented development environments such as Visual Basic, Visual C++ (Microsoft), and Delphi (Inprise Corporation). In developing Type I applications, it is possible to adopt necessary GIS components selectively. As a result, applications in this category could be more efficient than other methods, although this would be a difficult task to implement.

By combining GIS and a forecasting model, Type II could be implemented by using customized software interfaces. These customized interfaces can be developed using any software development tools, as seen in Choi and Kim (1994; 1996), and You and Kim (1999a). There are several transportation modeling software packages that have such interfaces to access GIS data formats, such as ESRI's coverage and shape files. TP+ and Tranplan DBC (Urban Analysis Group) can directly access ESRI's shape files. This is done in the same manner as ARC/INFO and ArcView GIS can access EMME/2 (INRO) data (Lussier and Wu, 1997). Moreover, several transportation agencies have developed customized hybrid systems that interface between GIS and transportation models. Santa Clara County, CA, for example, has integrated Tranplan with ARC/INFO to accomplish transportation demand modeling and analysis (Lockfeld and Speed, 1993). Likewise, Mesa County, CO has developed a countywide traffic model with MINUTP (Urban Analysis Group) and ARC/View. To utilize car and public transport time matrices and population statistics to perform an accessibility analysis, both ARC/INFO and TRIPS (MVA) have been combined (Holm and Stavanger, 1997). A research group at Los Alamos National Laboratory (LANL) has developed the TRansportation ANalysis and SIMulation System (TRANSIMS), which is a part of transportation model improvement program (TMIP). TRANSIMS has been developed to provide transportation planners accurate, complete information on traffic impacts, congestion, and pollution (Bush, 1999; Nagel et al., 1998). Although TRANSIMS utilizes ARC/INFO and ArcView, it largely uses the C++ development environments for simulation programs (Berkbigler et al., 1997).

Type III applications can be implemented by developing a forecasting model within GIS using built-in macro languages, such as arc macro language (AML) in ARC/INFO, Avenue in ArcView GIS, and GISDK (Caliper Cooperation). MGE macro language (Intergraph), unlike the built-in macro languages, utilizes the PERL macro language which can be used on most hardware platforms, such as UNIX, NT, and DOS. Nevertheless, it is understood that these macro languages are less effective in comparison to object-oriented development environments. In addition, there are some commercially available software packages that include transportation modeling programs, such as TransCAD (Caliper Corporation) and UfosNET (RST International).

3. System design of a hybrid forecasting model

The hybrid model is designed as illustrated in Fig. 2. It includes five modules: graphic user interface (GUI), real-time traffic data collection, database, forecasting, and machine learning (ML) modules.

3.1. GUI

The GUIs accept user input, such as origin and destination, model parameters, and display model output. Users execute the forecasting tasks and monitor outputs through GUIs.

3.2. Real-time traffic data collection

The real-time traffic data collection module collects real-time traffic data transmitted from such devices as loop detectors and probe vehicles and sends them to the database module.

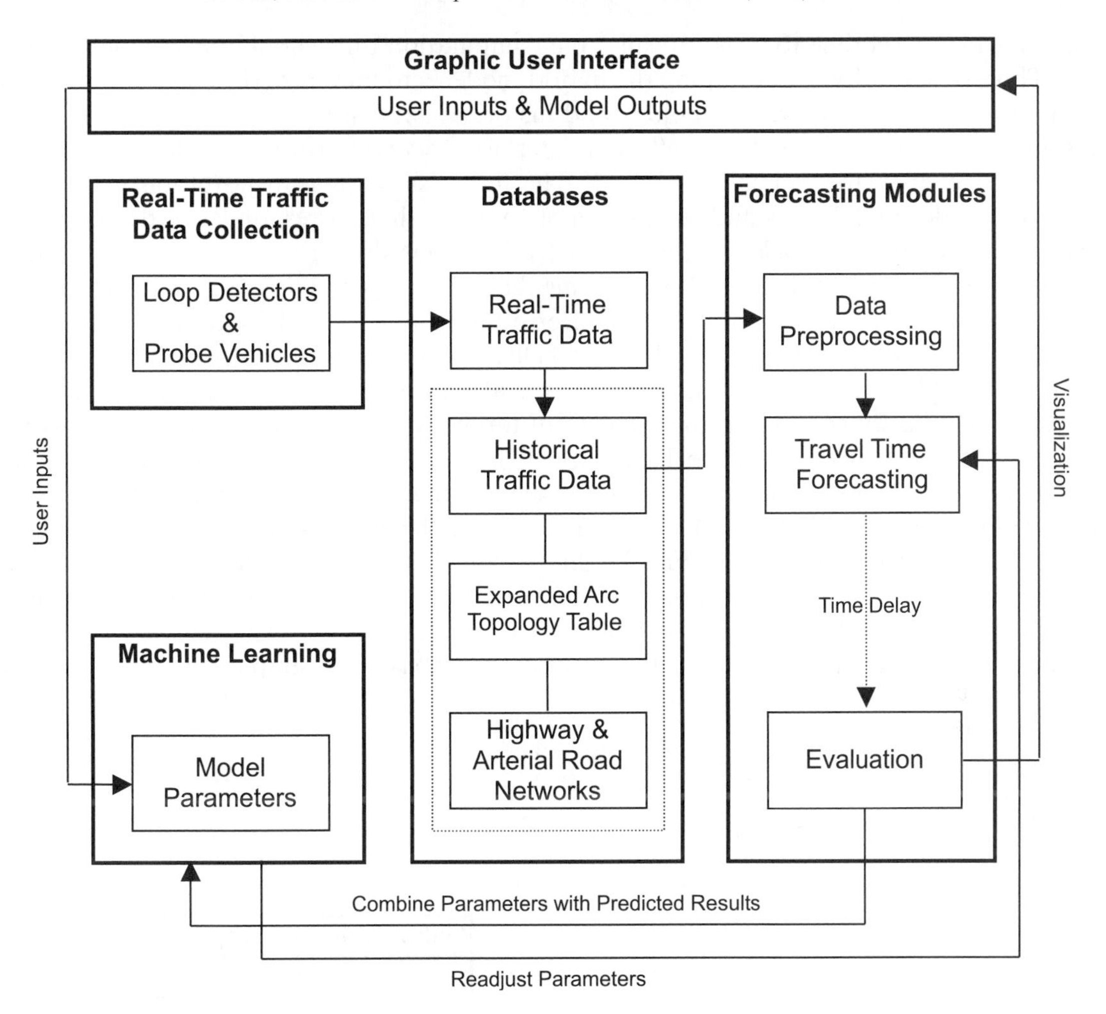

Fig. 2. A schematic structure of hybrid forecasting model.

3.3. Database

3.3.1. Real-time data

The transmitted raw traffic data are initially stored in the real-time traffic data storage, an intermediate data storage before updating the historical database. Data transmission intervals are every 30 s for highway data and 5 min for arterial data. Thus, the intermediate storage becomes necessary because there is no other place to store the data during the forecasting process, and the historical database is not allowed to alter or update when a travel time forecasting process is activated.

3.3.2. Historical data and network

Since the hybrid forecasting model is designed to predict future travel times for a period of 15–60 min, we assume that it is unacceptable if a computational delay is greater than 15 min for a

"15-min ahead" travel time forecast. Ideally, the computation time should be less than a minute or two. Thus, our priority in developing the hybrid model is to minimize the computation time of interaction between GIS network topology and historical databases.

A simpler approach to create an efficient interface between historical databases and road networks is shown in the example in Fig. 3. The figure shows a transportation network that consists of 6 nodes and 7 links and the arc topology table that corresponds to the network. By observing the arc table, it is acknowledged that `Link 1` consists of `Node 1` and `Node 2`, and its direction is from `Node 1` to `Node 2`. If we assume that `Link 1` is a two-way road, there must be a record for the other direction of `Link 1` from `Node 2` to `Node 1`. If all 7 links are bi-directional, the historical traffic database requires 14 record fields to store the traffic information as shown in Table 1, where $V_{T_{-n}}^{L_1}$ is the traffic volume on `Link 1` at time T_{-n}, and $V_{T_0}^{L_1}$ is the traffic volume on `Link 1` at time T_0. By combining the typical arc topology table (as shown in Fig. 3) with the historical traffic database (as shown in Table 1), only half of the traffic data would be connected to the arc topology table. In order to avoid this problem, an interfacing algorithm is devised to make the connection possible among the 14 links in Table 2 and the arc topology table in Fig. 3. Unlike the conventional arc topology table (Fig. 3), this expanded arc topology table (Table 2) includes the bi-directional information of the sample network, and thus it can be connected to the historical database using the identification field, `Historical_DB_ID`. The search algorithm for connecting GIS road network data to historical databases is simple enough to identify

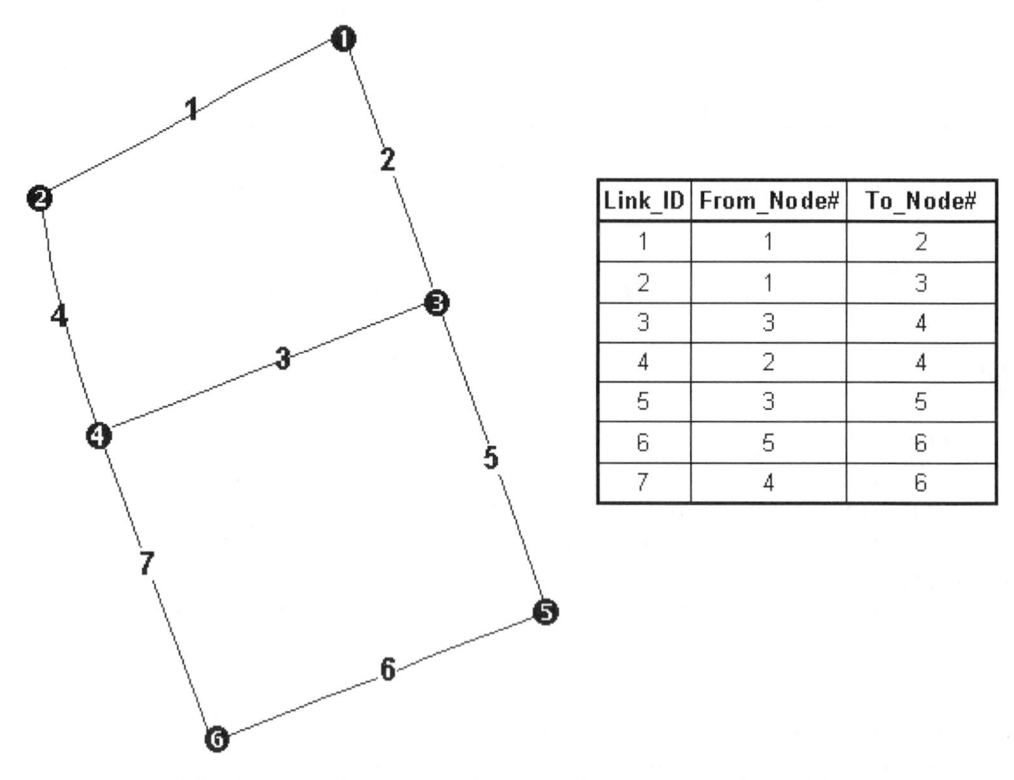

Link_ID	From_Node#	To_Node#
1	1	2
2	1	3
3	3	4
4	2	4
5	3	5
6	5	6
7	4	6

Fig. 3. A sample transportation network with arc topology table.

Table 1
Historical traffic data corresponding to the sample network in Fig. 3

Time (T)	Link (L)						
	L_1	L_2	L_3	...	L_{12}	L_{13}	L_{14}
T_{-n}	$V_{T_{-n}}^{L_1}$	$V_{T_{-n}}^{L_2}$	$V_{T_{-n}}^{L_3}$	...	$V_{T_{-n}}^{L_{12}}$	$V_{T_{-n}}^{L_{13}}$	$V_{T_{-n}}^{L_{14}}$
⋮	⋮	⋮	⋮	⋮	⋮	⋮	⋮
T_{-1}	$V_{T_{-1}}^{L_1}$	$V_{T_{-1}}^{L_2}$	$V_{T_{-1}}^{L_3}$	...	$V_{T_{-1}}^{L_{12}}$	$V_{T_{-1}}^{L_{13}}$	$V_{T_{-1}}^{L_{14}}$
T_0	$V_{T_0}^{L_1}$	$V_{T_0}^{L_2}$	$V_{T_0}^{L_3}$	...	$V_{T_0}^{L_{12}}$	$V_{T_0}^{L_{13}}$	$V_{T_0}^{L_{14}}$

Table 2
Expanded arc topology table

Link_ID	Historical_DB_ID	From_Node#	To_Node#
1	L_1	1	2
2	L_2	1	3
3	L_3	3	4
4	L_4	2	4
5	L_5	3	5
6	L_6	5	6
7	L_7	4	6
1	L_8	2	1
2	L_9	3	1
3	L_{10}	4	3
4	L_{11}	4	2
5	L_{12}	5	3
6	L_{13}	6	5
7	L_{14}	6	4

corresponding links between the two (network data and historical data) as shown in Fig. 4. Since the algorithm is efficient for computation, we did not have to use any other linear referencing systems, such as dynamic segmentation, for this particular application.

3.4. *Forecasting module*

3.4.1. *Data preprocessing*

Before initiating the travel time forecasting process, the data preprocessing module screens and filters noise from raw traffic data to increase forecasting accuracy. Users can choose either the wavelet transformation technique (You and Kim, 1999b) or a robust estimation technique, such as an outlier detection algorithm. In general, it is useful to apply the wavelet transformation technique to continuous time series data that include noise, such as highway loop detector data. The wavelet transformation technique partially utilizes the moving average method, which is often utilized to analyze time series data. On the other hand, a robust estimation technique is particularly useful to deal with extreme outliers. In this research, an outlier detection algorithm has been developed by utilizing the median absolute deviation (MAD), which is useful for screening outliers

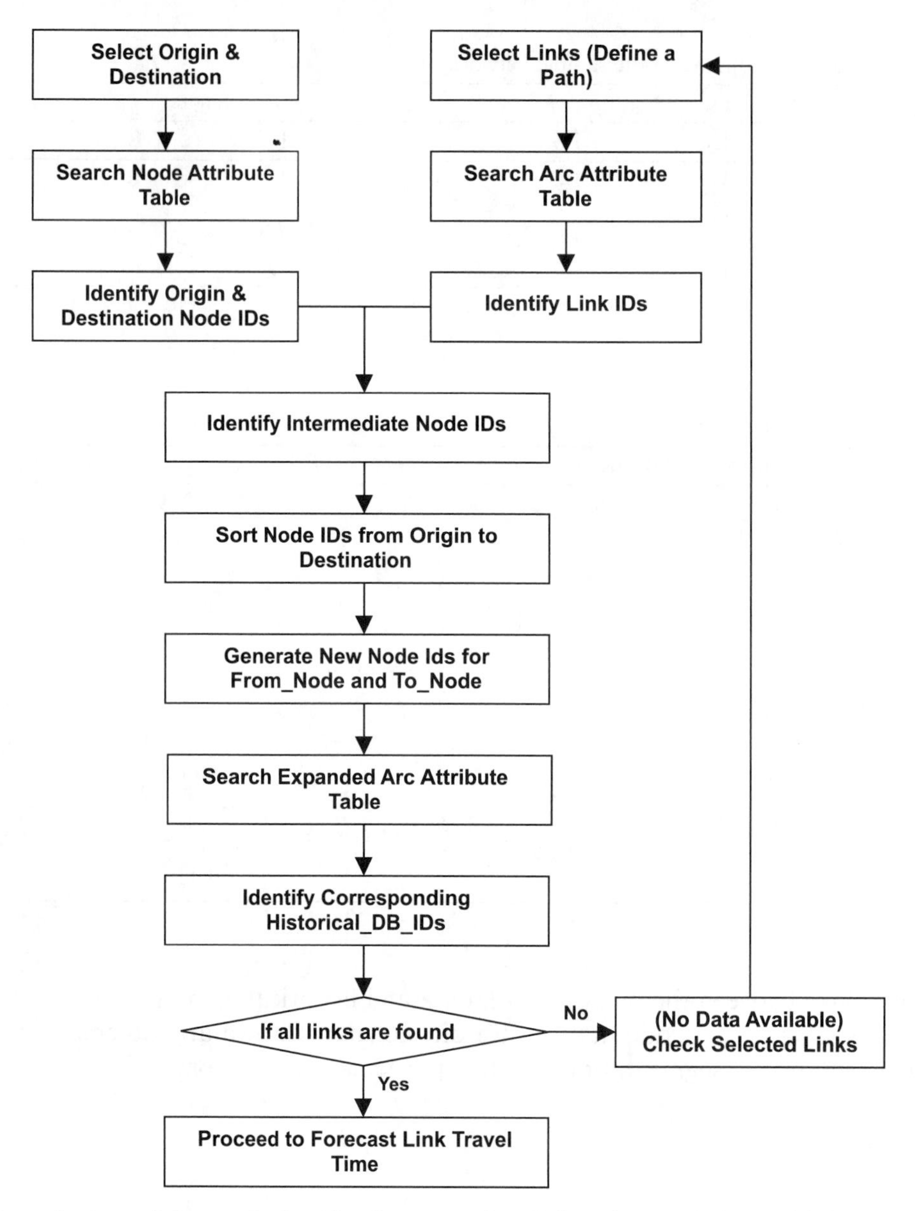

Fig. 4. A schematic chart of the search algorithm for connecting GIS road network data to historical traffic database.

among probe vehicle data samples. One problem associated with data samples is that some probe vehicles are delivery trucks that occasionally report exceptionally large values of travel times (i.e., outliers) between intersections due to delivery and pick-up of packages.

3.4.2. *Non-parametric regression analysis module*

A non-parametric regression model, developed in an earlier study (You and Kim, 1999b), has been adopted for the development of the hybrid travel time forecasting model described in this

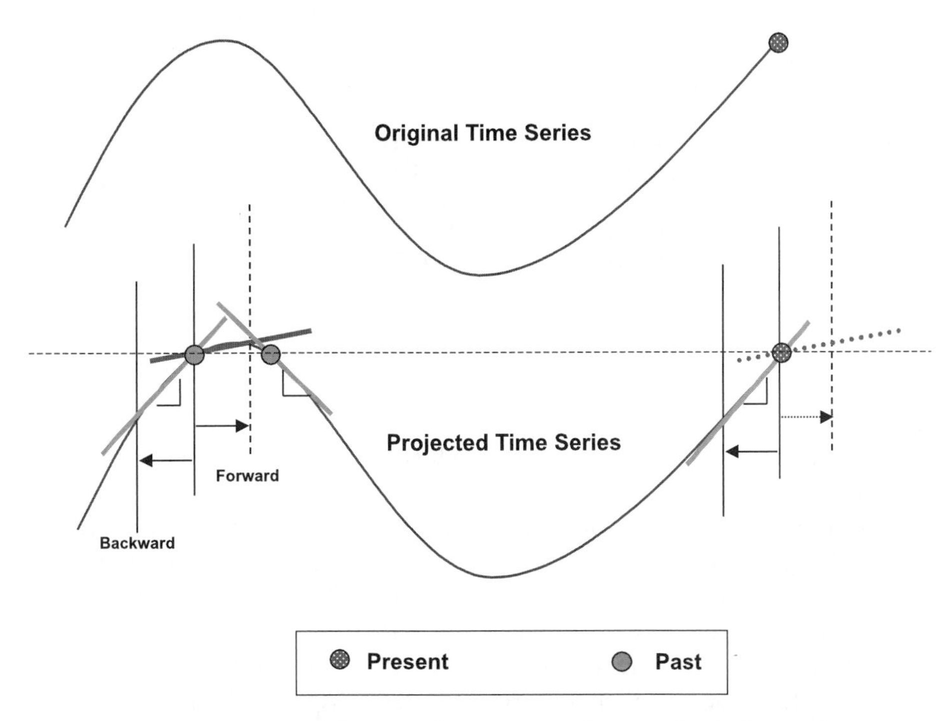

Fig. 5. A conceptual diagram of non-parametric regression in forecasting.

section. Fig. 5 illustrates a simplified concept of non-parametric regression, which adopts the *k-Nearest Neighbor* (*k-NN*) smoothing method. A brief explanation of the model is given below.

Suppose if there is a part of time series data representing a non-linear system, by segmenting the original time series into a number of finite levels, it is possible to obtain local linear subsystems. A group of similar past cases to the present condition is identified (parameter K in required input data), and each slope of the past cases is compared with the slope of the present condition (parameter k in required input data). By having k sets of similar cases, the future conditions of travel time could be estimated.

3.4.3. Required inputs to the hybrid forecasting model

The proposed hybrid forecasting model requires various parameters as input, viz. forecasting ranges, search data segment lengths, day of the week, search ranges, K and k values in the *k-NN* smoothing method of the non-parametric regression module, local estimation method, and data preprocessing. The domain of each parameter is shown in Table 3 and they are defined as follows:

- *Forecasting range* determines the prediction length in minutes, i.e., 15, 30, 45, and 60 min. To effectively monitor the forecasting performance, only pre-specified forecasting ranges are used.
- *Search data segment length* defines the length of segments for searching past traffic patterns. For instance, 15, 30, 45, or 60 min past traffic data segments could be used to forecast a "15-min ahead" travel time.
- *Day of the week* is utilized to separate the weekday and weekend data. It can reduce the search time for similar patterns without decreasing the forecasting accuracy.

Table 3
Domains of model parameters

Parameters	Type	Domain	Unit
Forecasting range	Discrete	{15, 30, 45, 60}	Minute
Search data segment length	Discrete	{15, 30, 45, 60}	Minute
Day of the week	Binary	{Consider, Ignore}	–
Search range	Discrete	{1, 2, 3}	Hour
Large K	Discrete	{1, 2, 3, 4, 5, 6, 7, 8, 9, 10}	–
Small k	Discrete	{1, 2, 3, 4, 5, 6, 7, 8, 9, 10}	–
Local estimation method	Binary	{Local averaging, Local fitting}	–
Data preprocessing	Binary	{Wavelet, Outlier detection}	–

- *Search range* is used to limit the size of the search space in historical databases. The larger the search range, the longer the search time. Domain ranges are ± 1, ± 2, or ± 3 h.
- *Large K* is used to limit the number of similar patterns selected from past data for forecasting in the *k-NN* smoothing method of non-parametric regression module within the search range. 'K' sets of similar cases are selected based on the similarity between the current traffic condition and the historical database. Domain ranges from 1 to 10.
- *Small k* is the subset of the large K. It is used to limit the number of similar patterns to the current condition within 'K' sets of similar cases. The large 'K' sets of similar cases are re-ranked after comparing the similarity between the two: the slope of the current traffic pattern and each of the slopes of 'K' sets of similar cases. Domain ranges from 1 to 10, but k values should not be larger than the K value.
- *Local estimation method* is used to calculate travel times from 'k' sets of similar cases. Basically, each similar case is composed of a number of continuous data points. There are two methods for estimating the travel time representing the 'k' sets. Local averaging method uses only the last set of data points of traffic data for estimating the mean, ignoring all other data points in the set, while local fitting method uses all data points to delineate a local linear trend line using the least-squares estimation.
- *Data preprocessing* assists in filtering noise from traffic data. Users can select either the wavelet transform technique or an outlier detection algorithm of a robust estimation method.

Table 4
Data structure of training samples

Training sample	Parameter					Estimated travel time (F)	Observed actual travel time (O)
	P^1	P^2	...	P^{n-1}	P^n		
1	p_1^1	p_1^2	...	p_1^{n-1}	p_1^n	F_1	O_1
2	p_2^1	p_2^2	...	p_2^{n-1}	p_2^n	F_2	O_2
3	p_3^1	p_3^2	...	p_3^{n-1}	p_3^n	F_3	O_3
...	...	...	...	...	...	...	...
$m-1$	p_{m-1}^1	p_{m-1}^2	...	p_{m-1}^{n-1}	p_{m-1}^n	F_{m-1}	O_{m-1}
m	p_m^1	p_m^2	...	p_m^{n-1}	p_m^n	F_m	O_m

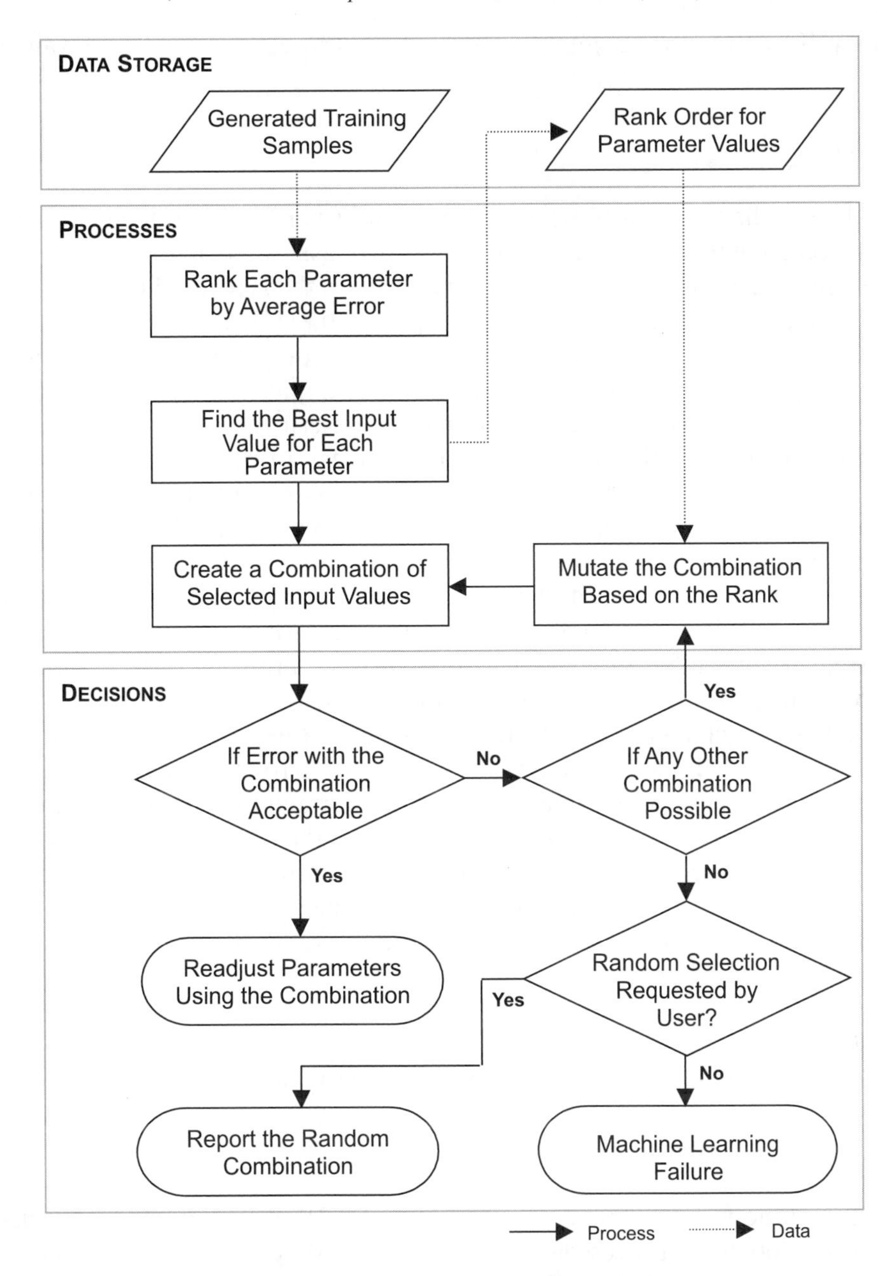

Fig. 6. Flow chart of parameter learning.

3.4.4. *Evaluation*

The evaluation module will be activated when observed actual travel times become available. There is a time delay because the observed actual travel times are not available at the time when a user orders a forecasting task. Based on the evaluation result, if a priori given acceptable error margin has not been met, the hybrid forecasting model activates the ML module to readjust parameter values. If the result is acceptable, it completes a forecasting task and waits for the next task.

3.5. The ML module

The ML module in the hybrid forecasting model readjusts model parameter values to produce more accurate forecasting results. With this objective, two tasks are performed: generating training samples beforehand and learning parameters afterward with the generated training samples. In the first instance, training samples are generated as input to the ML module. The training samples contain estimated forecasting errors with different combinations of parameter values, as shown in Table 4, where p_m^n denotes the nth parameter value of mth sample, F_m the mth estimated travel time, and O_m denotes the mth observed actual travel time. As described before, when the hybrid model detects an unsatisfactory prediction result, training samples can be generated.

Upon generating training samples, the learning process is initiated to identify the lowest forecasting error from each parameter value. In the ML module, as shown in Fig. 6, the input training samples are ranked by the average errors in ascending order. By combining parameters with the least errors, a combination of parameter values (i.e., a linear string of numbers) is created, and the error for this newly combined parameter is estimated. If the forecasting error is within a priori given acceptable margin, then the learned parameter values are applied to the hybrid model, and the learning process is terminated. If the forecasting error is greater than the acceptable error margin, changing one of the parameter values mutates the combination. The order of mutation is based on the size of average forecasting errors within parameter values under consideration. In this manner, the learning processes are iterated until a combination that satisfies the decision rule (i.e., the error level should be smaller than the acceptable error margin) is found. Finally, if this iterative process fails to find any combination that meets the decision rule, the learning process attempts to produce a randomly selected combination among the training samples with relatively small forecasting errors. Once the random generation is completed, the user has the choice to accept or reject them.

4. The hybrid forecasting model: implementation and performance evaluation

4.1. Development tools for system integration

For the implementation of a hybrid forecasting model, a Windows NT-based workstation with an Intel Pentium-II 366-MHz processor and 192 MB RAM has been utilized. To develop customized computer programs, ESRI's ARC/INFO GIS software package with several computer-programming languages, including Microsoft's Visual C++ and Visual BASIC, has been used.

In this study, the Type I integration is implemented by adopting an object-oriented programming method as shown in Table 5. In the table, various functions that are needed for Type I integration are listed along with their corresponding development tools. In order to build an object-oriented forecasting system, the conventional built-in GIS functions need to be separated from their software packages so that they can be included as objects in an object-oriented application. In the scheme, a GIS software package no longer exists independently. Major GIS

Table 5
Development tools for Type I integration

Functions	Development tools
Supporting user inputs and system controls	Visual BASIC
Displaying networks	ODE ARCPLOT ActiveX Control
Managing real-time and historical data	ODE ARCPLOT ActiveX Control
Representing traffic networks	ODE ARCPLOT ActiveX Control
Simulating real-time data inputs	Visual BASIC
Screening and filtering raw traffic data	MATLAB: wavelet transform, Visual BASIC: robust estimation
Performing predictions	Visual BASIC and Visual C++
Managing forecasting model executions	Visual BASIC
Evaluating predicted results	OLE MSGraph object

functions are separated, and they become a part of object-oriented applications. With ARC/INFO, an object-oriented application can be developed using Open Development Environment (ODE), which is a collection of individual GIS function objects.

4.2. GUI

The main GUI for the forecasting model is shown in Fig. 7. There are three other important GUIs in the model. GUIs for evaluation are shown in Figs. 8 and 9, and GUI for the ML module is shown in Fig. 10. Two different methods were carried out to evaluate the performance of the model: forecasting based on continuous time points and forecasting based on randomly selected discrete time points. As shown in Fig. 8, the continuous forecast simulates uninterrupted predictions for all points of time from a prespecified simulation starting time. On the other hand, the random forecast assists in predicting travel time using randomly generated points of time as shown in Fig. 9. Both GUIs utilize OLE objects to draw charts and show output data. Fig. 10 shows the ML GUI that assists in enhancing the hybrid model's forecasting accuracy.

4.3. Data

For the implementation of the hybrid model, raw traffic data that come from loop detectors installed and managed by the Korea Highway Corporation (KHC) are used. KHC is currently collecting various real-time traffic data through loop detectors and other traffic surveillance systems through an information highway network. In this study, a 30-s interval real-time traffic data, transmitted from 8 loop detectors between 11 and 17 March 1996, are used. Collected traffic data include traffic volumes, occupancy rates, and traffic speeds. These 8 detectors cover a total length of 114.3 km between PanGyo I/C and ChungWon I/C, as shown in Fig. 11.

The hybrid model has also been applied for arterial road networks in forecasting travel times. Raw arterial traffic data have been acquired from the LG Traffic Information Systems (LGTIS), Seoul, Korea. LGTIS has developed an advanced traveler information system called Road Traffic

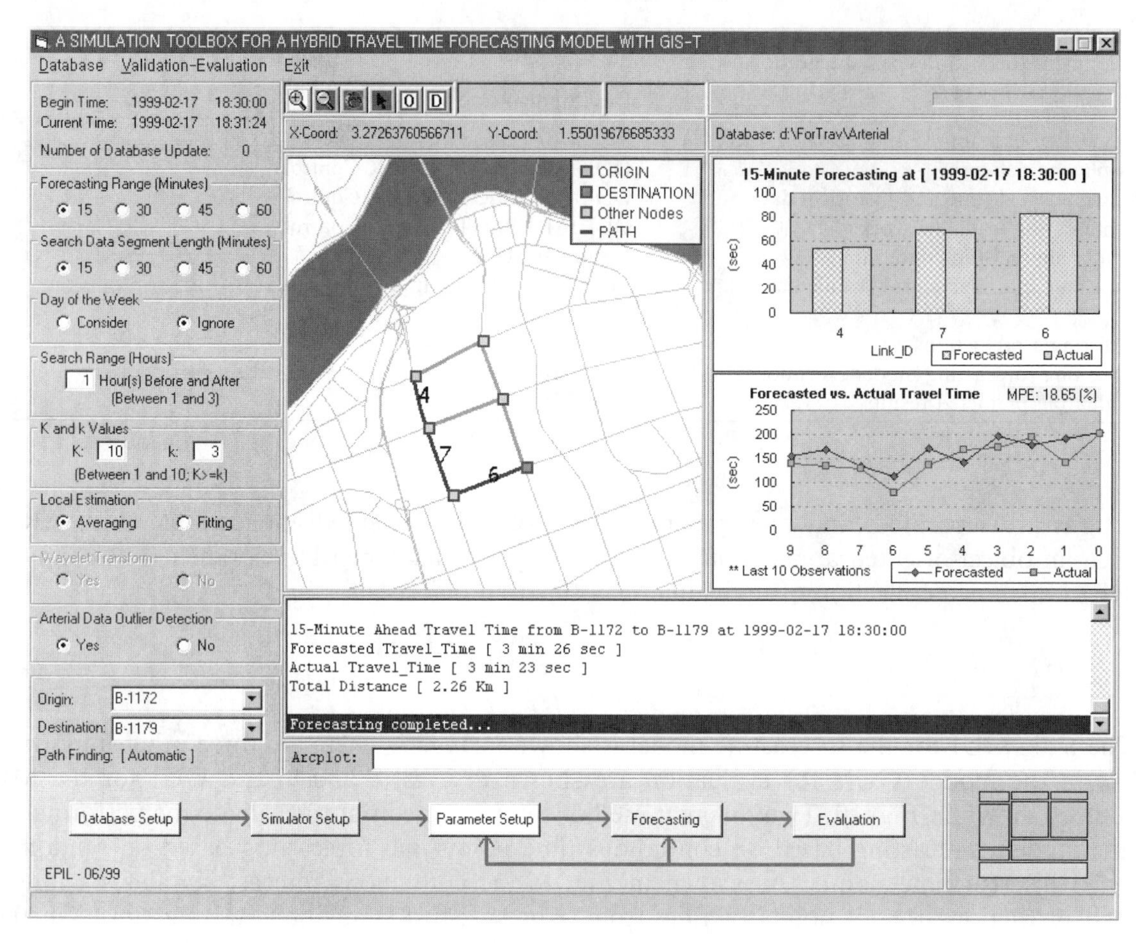

Fig. 7. The main GUI for the forecasting model.

Information Systems (ROTIS) and has collected real-time arterial traffic information using probe vehicles and roadside beacons since 1998. As of 1999, there are approximately 2500 roadside beacons on major intersections in Seoul, Korea. These roadside beacons relay traffic information that is transmitted from probe vehicles to the company's traffic information center. In this study, arterial traffic data come from 6 roadside beacons between 13 and 19 February 1999. These 6 beacons cover the total length of 5.3 km for 7 road links on 5 major arterial roads. Transmitted arterial data include probe vehicles' driven distances, driven times, and stopped times for each link. The locations of roadside beacons are shown in Fig. 12 along with other roadside beacons managed by the company.

4.4. Real-time traffic data simulator

A real-time traffic data simulator simulates traffic data transmissions from the traffic data collection module. For the performance evaluation, a real-time traffic data simulator is devised to simulate as if online traffic data are transmitted from both loop detectors and roadside beacons.

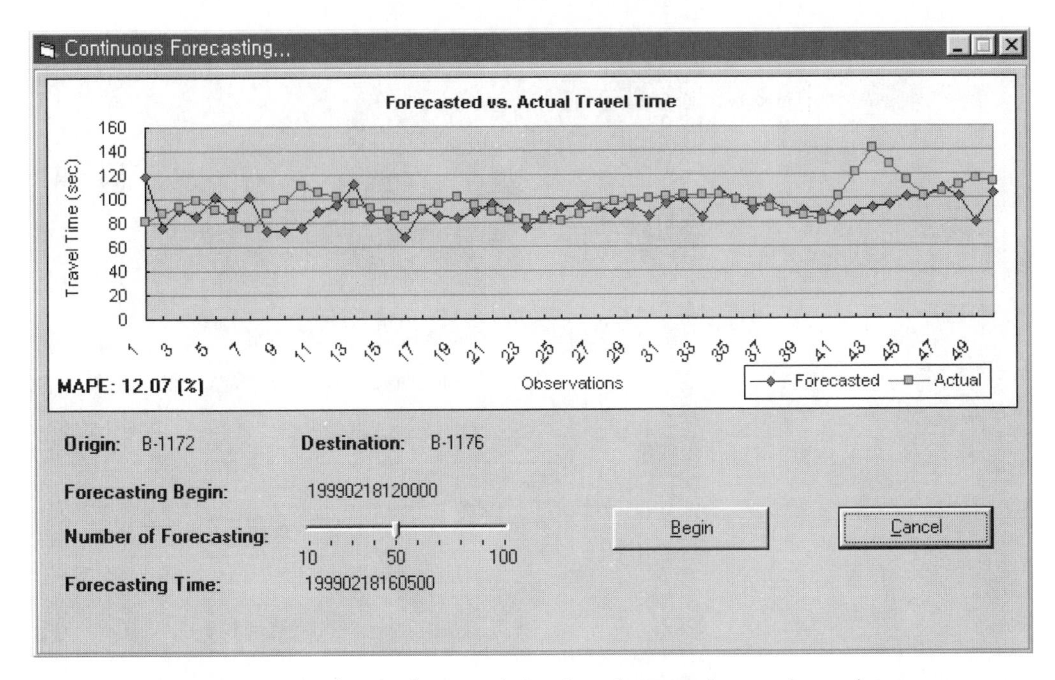

Fig. 8. GUI for forecasting travel time based on continuous time points.

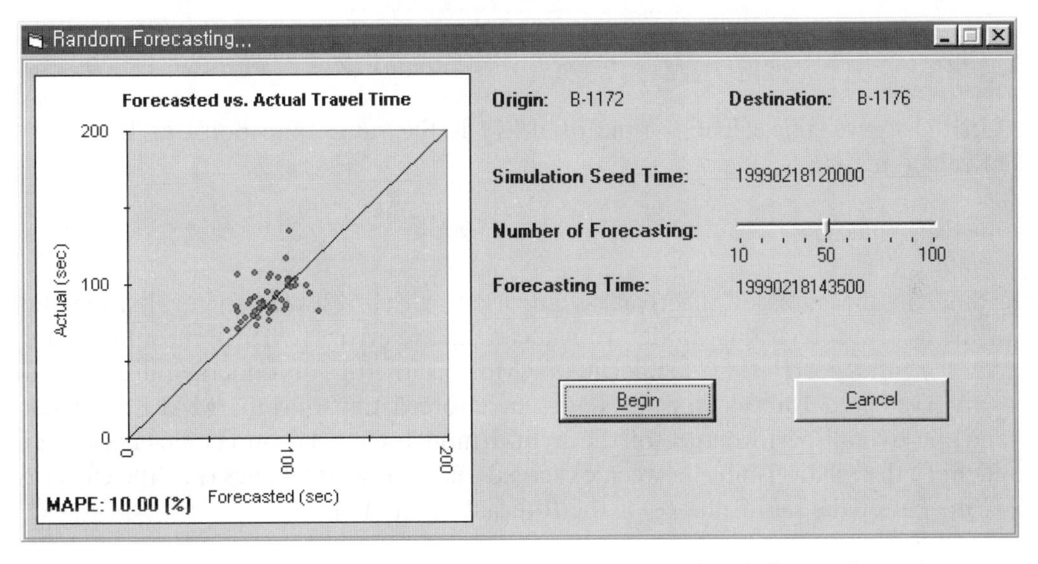

Fig. 9. GUI for forecasting travel time based on randomly selected discrete time points.

Thus, it replaces the traffic data collection module in Fig. 2. For the simulation of real-time traffic data, parts of historical data are taken and assumed as if these are real-time data for both arterial and highways. In other words, the system has been implemented using parts of the historical databases to predict the condition at time t using data from earlier times of $t(-1)$, as shown in

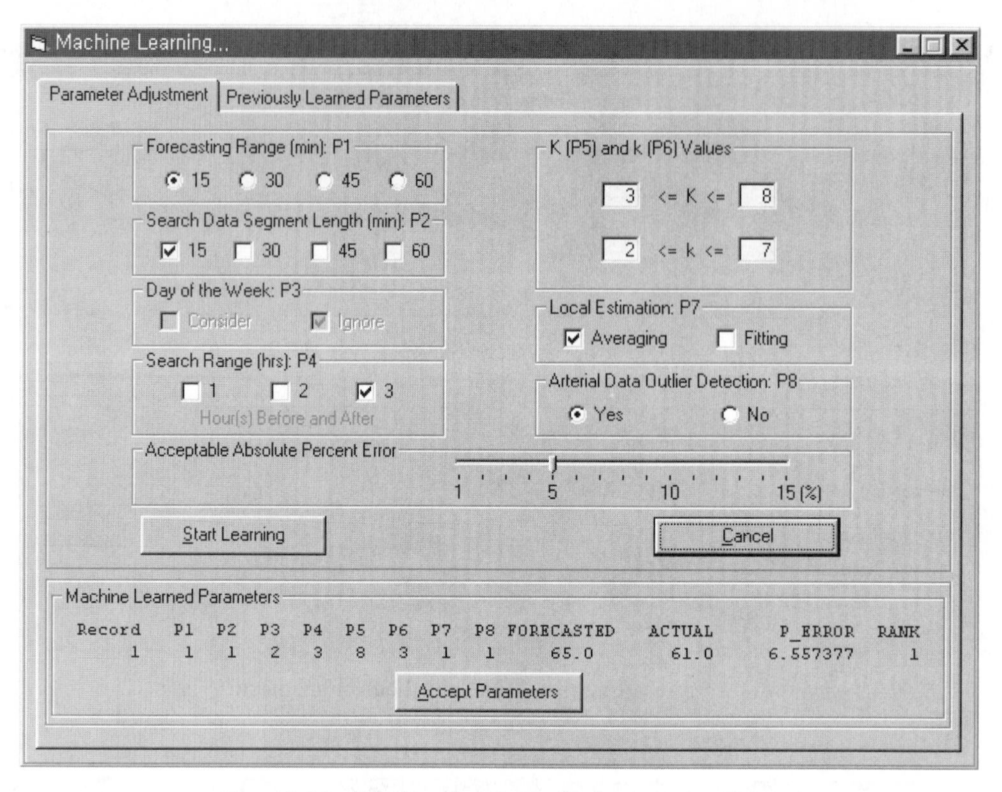

Fig. 10. Machine learning GUI: parameter adjustment.

Fig. 13. In the figure, the calendar time of $t(-1)$ is the same time point as the time t at the simulated calendar time.

4.5. Performance measures

Three different measures of effectiveness are used in this research for evaluating the performance of the hybrid forecasting model: root mean square error (RMSE), mean absolute percent error (MAPE), and correlation coefficients. RMSE is useful for understanding the deviation between observed and forecasted values for each forecasting output. MAPE is calculated to understand the overall performance of the hybrid model. Correlation coefficients (ρ) are calculated to identify the relationships between observed and forecasted values (i.e., the closer the value of ρ to 1, the better the performance of the forecasting model).

$$\text{RMSE} = \sqrt{\frac{1}{n}\sum_{i=1}^{n}\left(\hat{x}_i - x_i\right)^2},$$

$$\text{MAPE} = \frac{1}{n}\left(\sum_{i=1}^{n}\frac{\left|\hat{x}_i - x_i\right|}{x_i} \times 100\right),$$

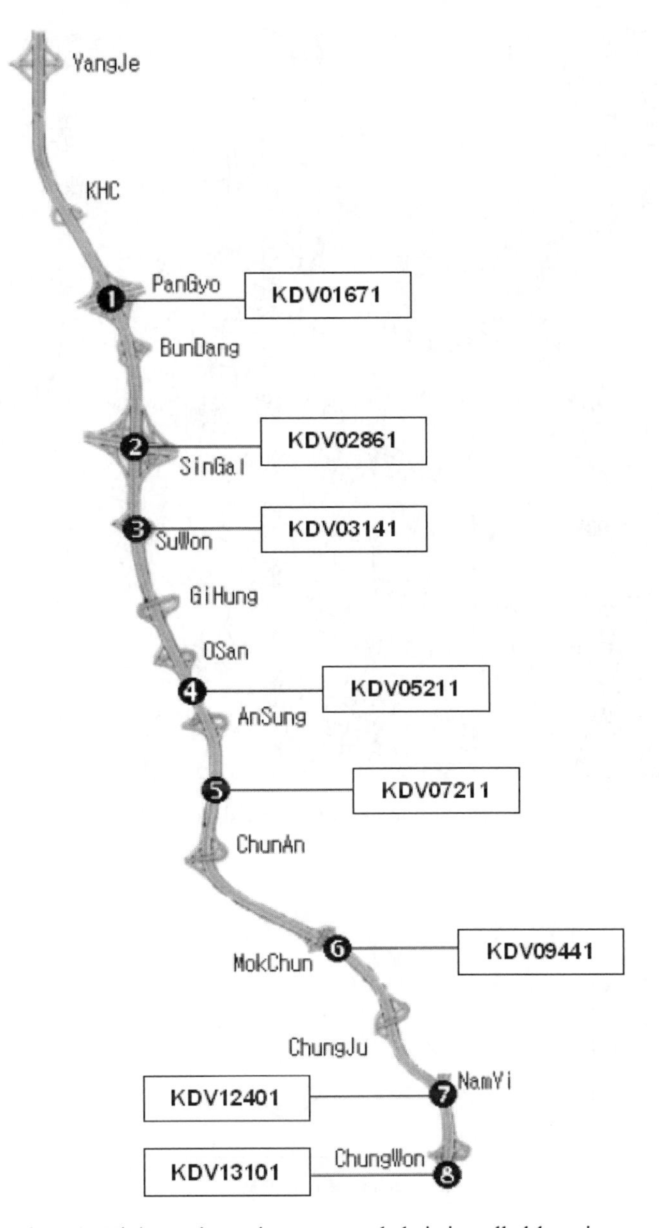

Fig. 11. Highway loop detectors and their installed locations.

$$\rho = \frac{\sum_{i=1}^{n} \left(\hat{x}_i - \hat{x}_{\text{mean}}\right)(x_i - x_{\text{mean}})}{n\hat{\sigma}\sigma},$$

where observed data is represented by x_i, predicted data by $\hat{x}_i$, mean of observed data x_i by x_{mean}, mean of forecasted data $\hat{x}_i$ by $\hat{x}_{\text{mean}}$, correlation coefficient by ρ, standard deviation of observed data x_i by σ, and standard deviation of forecasted data $\hat{x}_i$ by $\hat{\sigma}$.

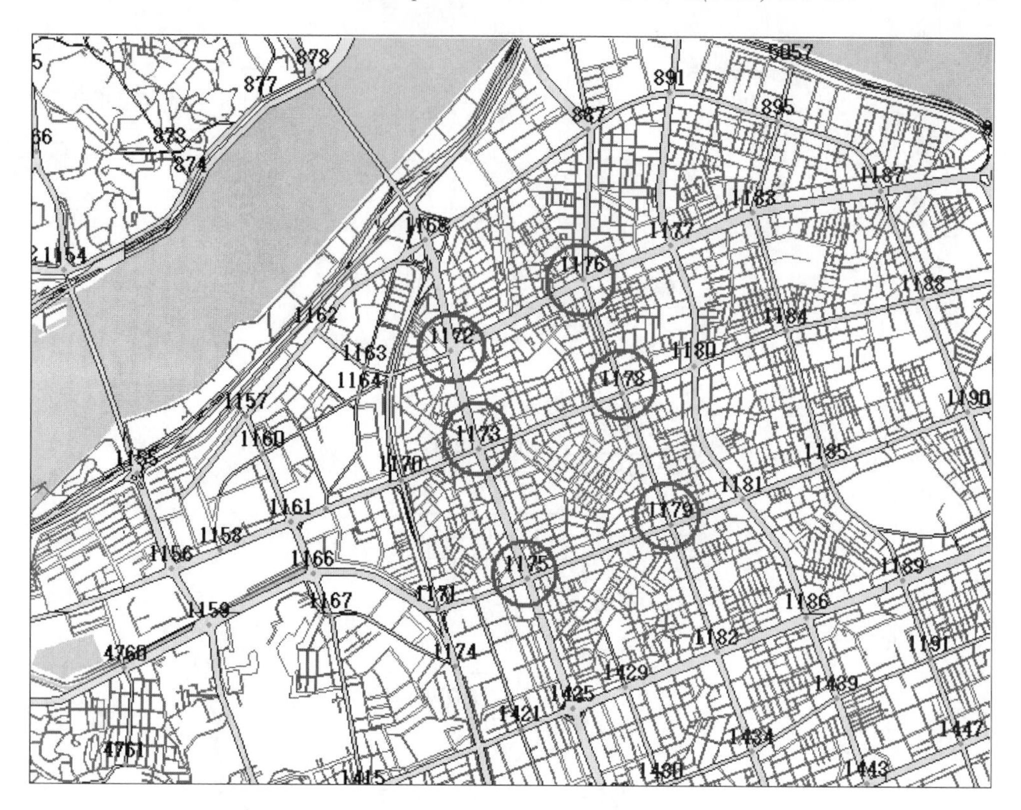

Fig. 12. Six roadside beacons located in south of Seoul.

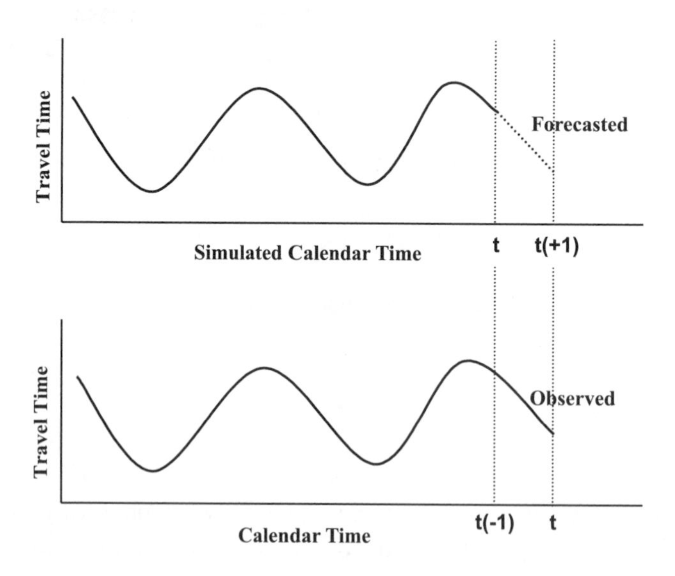

Fig. 13. Simulated forecasting using historical database.

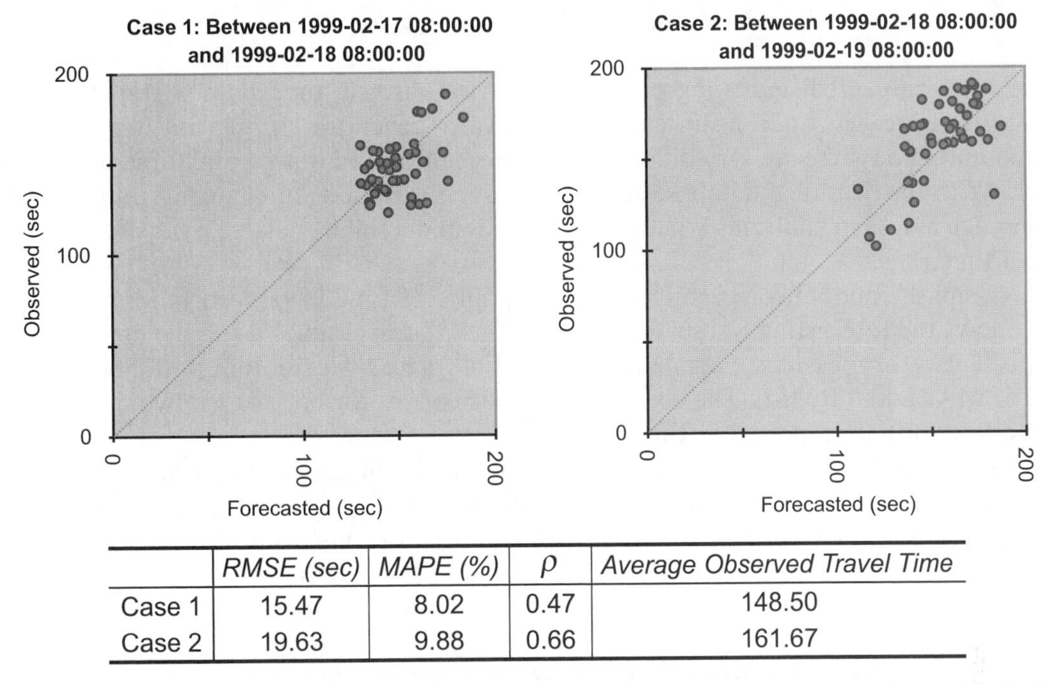

	RMSE (sec)	*MAPE (%)*	*ρ*	*Average Observed Travel Time*
Case 1	15.47	8.02	0.47	148.50
Case 2	19.63	9.88	0.66	161.67

Fig. 14. Forecasting based on randomly selected discrete time points with arterial data (origin: B1173; destination: B1179).

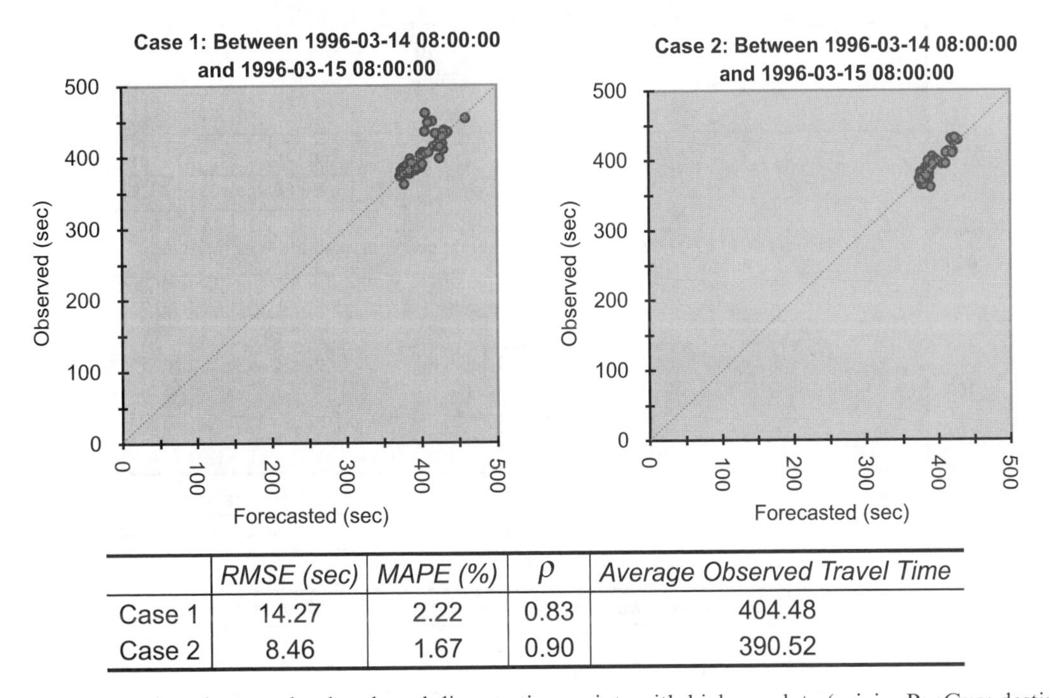

	RMSE (sec)	*MAPE (%)*	*ρ*	*Average Observed Travel Time*
Case 1	14.27	2.22	0.83	404.48
Case 2	8.46	1.67	0.90	390.52

Fig. 15. Forecasting based on randomly selected discrete time points with highway data (origin: PanGyo; destination: SinGal).

4.6. Forecasting based on randomly selected discrete time points

Travel times are predicted using randomly selected discrete time points from the historical databases that are generated using a random number generator. In general, travel conditions measured during daytime are very different from those measured during nighttime on a particular network. With the historical databases that reflect such variations of changing traffic conditions, this approach assists in analyzing whether the hybrid model fulfills reliable predictions for various points of time.

To accomplish random forecasting tests, 200 points of time have been randomly generated. Fig. 14 shows the forecasting results for "15-min ahead" travel times using the arterial historical database. The error measures indicate that the model forecasts travel time with about 10% error margin of MAPE and RMSE. The correlation coefficient (ρ) for both cases (two different time at the same location) are reported in Fig. 14.

This has resulted in the application of the model to the highway historical database for predicting "15-min ahead" travel times, as shown in Fig. 15. The error measures indicate that the hybrid forecasting model performs better with the highway historical database than with the

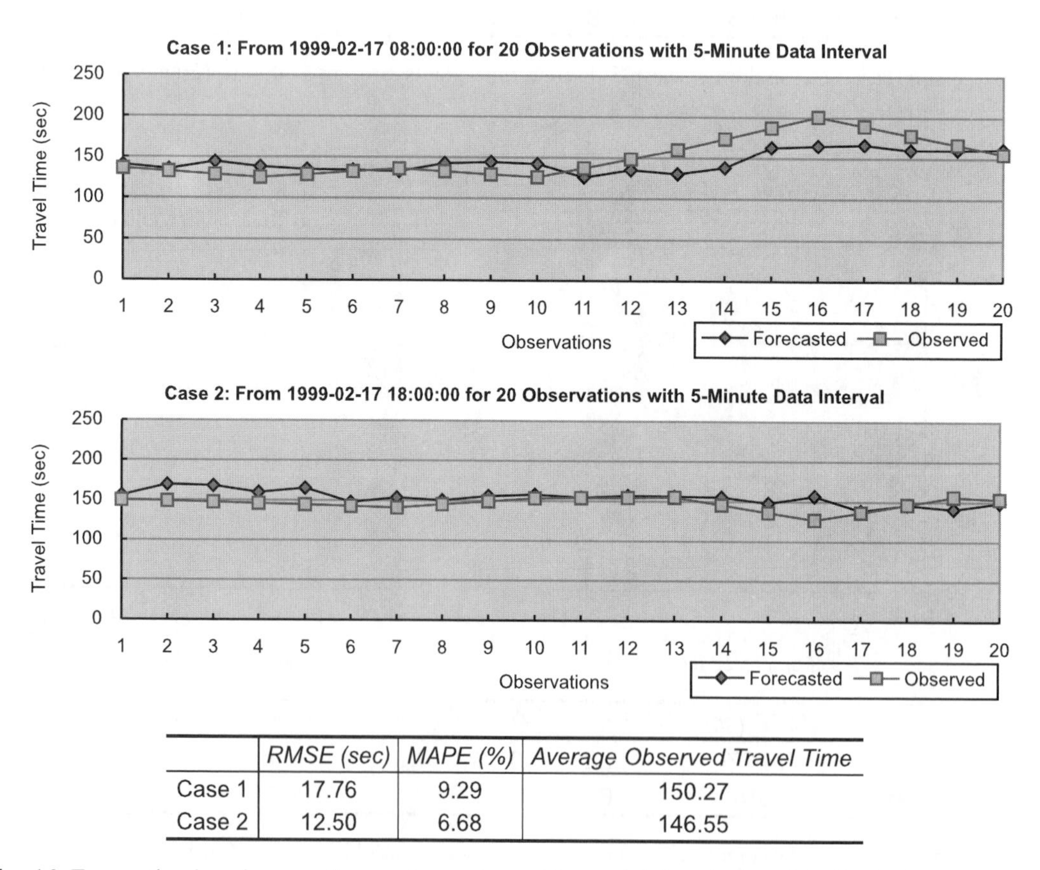

	RMSE (sec)	MAPE (%)	Average Observed Travel Time
Case 1	17.76	9.29	150.27
Case 2	12.50	6.68	146.55

Fig. 16. Forecasting based on continuous time points with arterial data (origin: B1173; destination: B1179).

arterial historical database. For both cases of the highway data (applying the model for the two different times at the same location), MAPE shows less than 3%, and the correlation coefficients are close to 1.

4.7. Forecasting based on continuous time points

In this simulation, travel times are predicted using continuous travel time data from a given point of time. This is to verify if the hybrid model predicts any biased results. This analysis is particularly useful in understanding whether the hybrid model corresponds to continuously changing traffic conditions on highways or arterial road networks.

Inspecting Figs. 16 and 17, it is evident that there is no specific tendency of underestimation or overestimation for "15-min ahead" travel time forecasting. As in the previous simulation, this analysis indicates that the hybrid model produces a relatively high accuracy of less than 10% error in MAPE with the arterial historical data and a very high accuracy of less than 3% error in MAPE with the highway historical data.

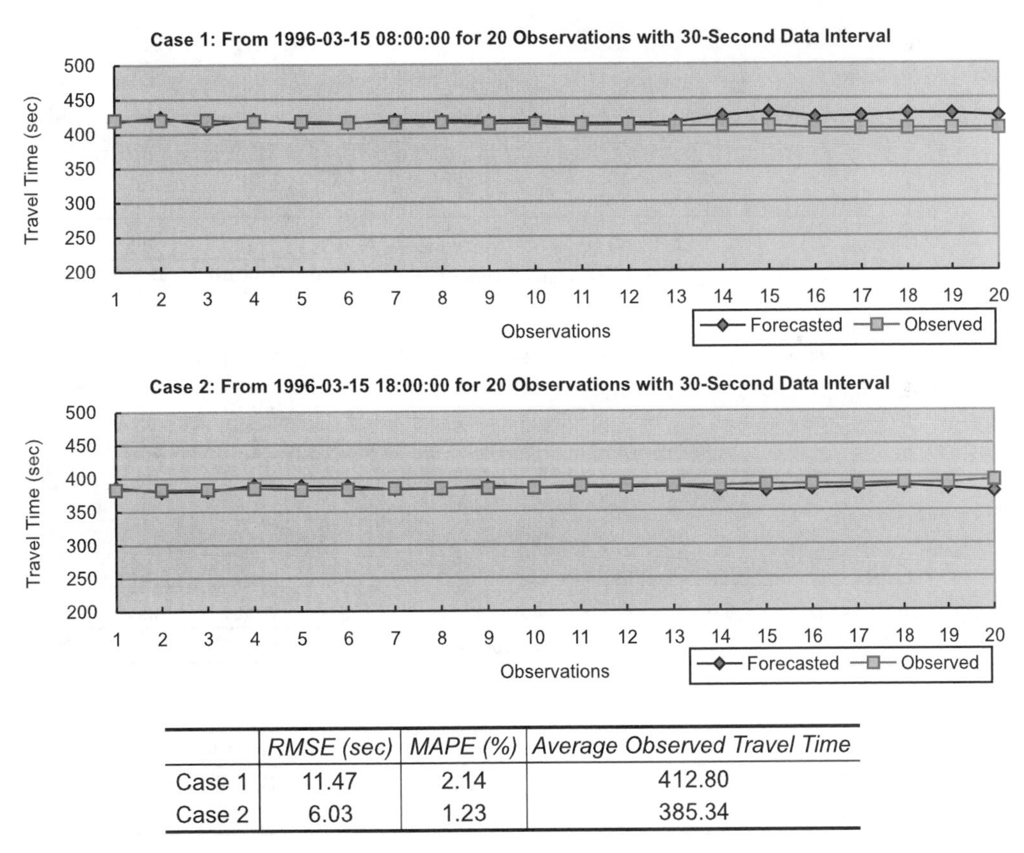

	RMSE (sec)	MAPE (%)	Average Observed Travel Time
Case 1	11.47	2.14	412.80
Case 2	6.03	1.23	385.34

Fig. 17. Forecasting based on continuous time points with highway data (origin: PanGyo; destination: SinGal).

4.8. Evaluation on the performance of the ML module

For the performance evaluation of the ML module, we compared the model's performance with and without the ML module between the forecasted travel time and observed travel time for 50 min from the arterial historical database, as shown in Fig. 18. In the figure, the forecast with the ML module performs better than the forecast without the ML module, which is evident from the values of RMSE (9.16 vs 20.90) and MAPE (10.81 vs 25.44).

4.9. Computation time

For the performance evaluation, we used the following number of parameter values: forecasting range (15, 30, 45, and 60 min), search data segment length (15, 30, 45, and 60 min), day of the week (yes or no), search range (± 1, ± 2, and $\pm$ 3 h), K (1–10), k (1–10), local estimation methods (the average or the fitting), and data preprocessing (the wavelet or the outlier detection analysis). The total possible combination of mutation from the input parameter values is 21,120 $(4 \times 4 \times 2 \times 3 \times 55 \times 2 \times 2)$. Among these possibilities, we have implemented the model 10,560

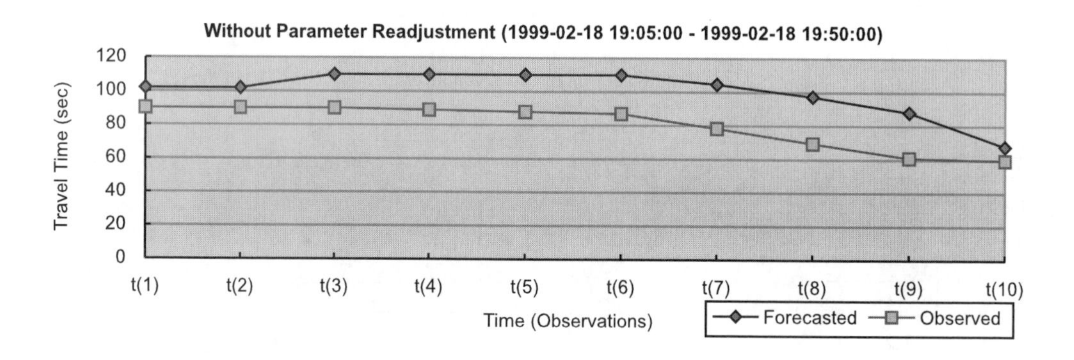

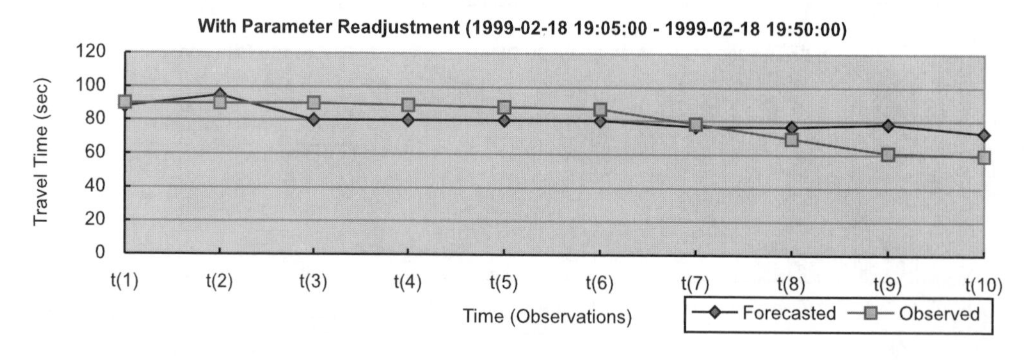

	RMSE (sec)	MAPE (%)	Average Observed Travel Time
Without Readjustment	20.90	25.44	80.27
With Readjustment	9.16	10.81	80.27

Fig. 18. Performance evaluation for the ML module.

times. Fig. 19 shows the minimum and the maximum computation time for implementing the model for these experiments using a single link. In dealing with arterial data, the minimum computation time is 4 s and the maximum is 8 s. On the other hand, for highway data the minimum computation time is 12 s and the maximum is 39 s. We emphasized achieving the

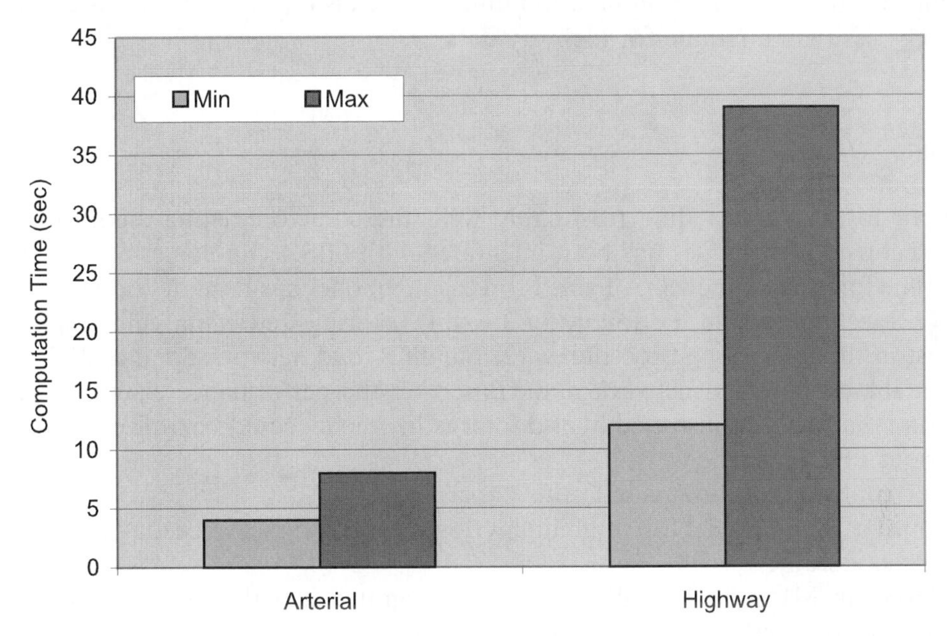

Fig. 19. Minimum and maximum computation times to predict a single link.

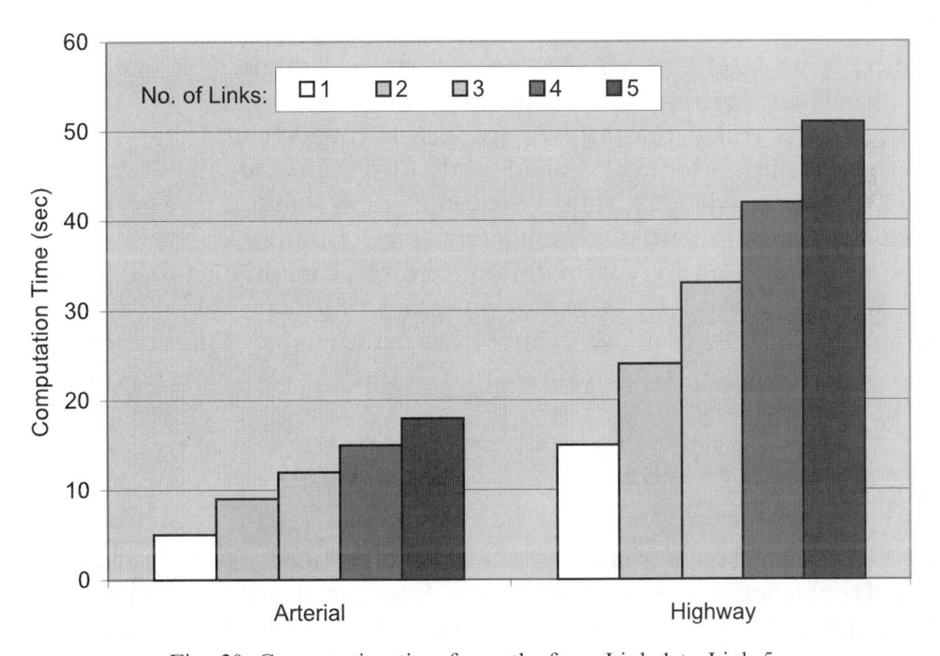

Fig. 20. Computation time for paths from Link 1 to Link 5.

minimum computation time in implementing the forecasting model, believing that fast computation times are very important in designing a hybrid forecasting model.

To evaluate the changes in computation times when more links are added, we added five links, one by one. As shown in Fig. 20, each additional link imposes relatively small amounts of time. For predictions with 5 links, the computation time required is less than 20 s for arterial data, and approximately less than a minute for highway data.

5. Conclusion

In order to forecast travel time reasonably well, the core forecasting algorithm with non-parametric regression techniques has been integrated with GIS technologies to implement a hybrid travel time forecasting model. A hybrid forecasting model has been developed and tested by deploying GIS technologies in the following areas: (1) storing, retrieving, and displaying traffic data to assist in the forecasting procedures, (2) building road network data, and (3) integrating historical databases and road network data. Based on the performance evaluation results, we strongly assure that the demonstrated hybrid forecasting model could be utilized as an important tool for various ITS applications.

Two major contributions are made in this paper:

1. we developed a Type I model for combining GIS and travel time forecasting model and implemented it successfully;
2. we developed an ML module and successfully integrated it with the travel time forecasting model with positive performance evaluation results.

Software vendors mentioned in this paper

Caliper Corp., Newton, MA (http://www.caliper.com/gisdk.htm).
ESRI, Inc., Redlands, CA (http://www.esri.com/software/).
Inprise Corp., Scotts Valley, CA (http://www.borland.com/delphi/).
INRO Consultants, Inc., Montreal, Canada (http://www.inro.ca/).
Intergraph Corp., Huntsville, AL (http://www.intergraph.com/software/gis/).
MapInfo Corp., Troy, NY (http://www.mapinfo.com/software/).
Microsoft Corp., Redmond, WA (http://msdn.microsoft.com/products/default.asp).
MVA Ltd. Woking, Surrey, UK (http://www.trips.co.uk/).
RST International Inc., Bellevue, WA (http://www.rstii.com/pages/product.htm).
The Urban Analysis Group (UAG), Hayward, CA (http://www.uagworld.com/).

Acknowledgements

We would like to express our sincere gratitude to the Ministry of Construction and Transportation, the Republic of Korea and the Campus Research Board, the University of Illinois at Urbana-Champaign for the support of this research. We also thank the Korea Highway Corporation and the LG Traffic Information Systems (now Road Traffic Information Systems:

ROTIS), Seoul, Korea, for providing traffic data. We also wish to extend our thanks to Mr. Sathyamoorthy Ponnuswamy for editing this paper. We take, however, the full responsibility for the contents of this paper.

References

Abkowitz, M., Walsh, S., Hauser, E., 1990. Adaptation of geographic information systems to highway management. Journal of Transportation Engineering 116, 310–327.

Azad, B., Cook, P., 1993. The management/organizational challenges of the "Server-Net" model of GIS-T as recommended by the NCHRP 20-27. In: Proceedings of the GIS-T'93: Geographical Information Systems for Transportation Symposium, pp. 327–342.

Ben-Akiva, M., Cascetta, E., Gunn, H., 1995. An on-line dynamic traffic prediction model for an inter-urban motorway network. In: Gartner, N.H., Improta, G. (Eds.), Urban Traffic Networks, Dynamic Flow Modeling and Control. Springer, New York, pp. 83–122.

Ben-Akiva, M., Koutsopoulos, H.N., Mukundan, A., 1994. A dynamic traffic model system for ATMS/ATIS operations. IVHS Journal 2 (1), 1–19.

Berkbigler, K.P., Bush, B.W., Davis, J.F., 1997. TRANSIMS software architecture for IOC-1. Report No. LA-UR-97-1242, Los Alamos National Laboratory.

Bush, B.W., 1999. The TRANSIMS framework. TRANSIMS Opportunity Forum, Santa Fe, NM, 28 June.

Choi, K., Kim, T.J., 1994. Integrating transportation planning models with GIS: issues and prospects. Journal of Planning Education and Research 13, 199–207.

Choi, K., Kim, T.J., 1996. A hybrid travel demand model with GIS and expert systems. Computers, Environment and Urban Systems 20 (4/5), 247–259.

Davis, G.A., Nihan, N.L., Hamed, M.M., Jacobson, L.N., 1990. Adaptive forecasting of traffic congestion. Transportation Research Record 1287, 29–33.

Gillespie, S., 1993. The benefits of GIS use for transportation. In: Proceedings of the GIS-T'93: Geographical Information Systems for Transportation Symposium, pp. 34–41.

Gilmore, J.F., Abe, N., 1995. Neural network models for traffic control and congestion prediction. IVHS Journal 2 (3), 231–252.

Holm, T., Stavanger, A.V., 1997. Using GIS in mobility and accessibility analysis. Paper Presented at the 17th Annual ESRI User Conference, 8–11 July, San Diego, CA.

Kriger, D., Schlosser, M., 1992. Integration of GIS with a travel demand forecasting model for transportation planning. Transportation Forum 4, 106–115.

Lockfeld, F., Speed, V., 1993. GIS links county to other transportation planners. Public Works 124, 43–44.

Loukes, D., 1992. Geographic information systems in transportation (GIS-T): an infrastructure management information systems tool. In: Proceedings of the TAC Annual Conference 3, pp. B27–B42.

Lussier, R., Wu, J.H., 1997. Development of a data exchange protocol between EMME/2 and ARC/INFO. Paper Presented at the 17th Annual ESRI User Conference, 8–11 July, San Diego, CA.

Mahmassani, H.S., Peeta, S., Chang, G.-L., Junchaya, T., 1991. A review of dynamic assignment and traffic simulation models for ADIS/ATMS applications. Technical Report DTFH61-90-R-00074. Center for Transportation Research, The University of Texas, Austin.

Nagel, K., Rickert, M., Simon, P.M., 1998. The dynamics of iterated transportation simulations. Paper presented at TRISTAN-III. Report No. LA-UR 98-2168. Los Alamos National Laboratory.

Nygard, K., Vellanki, R., Xie, T., 1995. Issues in GIS for transportation. MPC Report No. 95-43. Mountain-Plains Consortium, Fargo, ND.

Peeta, S., Mahmassani, H.S., 1995. Multiple user classes real-time traffic assignment for on-line ATIS/ATMS operations: a rolling horizon solution framework. Preprints of papers at the Transportation Research Board 74th Annual meeting, Washington, p. 27.

Ries, T., 1993. Design Requirements for Location as a Foundation for Transportation Information Systems. In: Proceedings of the GIS-T'93: Geographical Information Systems for Transportation Symposium, pp. 48–66.

Sen, A., Sööt, S., Ligas, J., Tian, X., 1997. Arterial link travel time estimation: probes, detectors and assignment-type models. Preprints of papers at the Transportation Research Board 76th Annual meeting, Washington, p. 21.

You, J., Kim, T.J., 1999a. An integrated urban systems model with GIS. The Journal of Geographical Systems 1 (4), 305–321.

You, J., Kim, T.J., 1999b. Implementation of a hybrid travel time forecasting model with GIS-T. In: Proceedings of the Third Bi-Annual Conference of the Eastern Asia Society for Transportation Studies, 14–17 September, Taipei, Taiwan.

PERGAMON

Transportation Research Part C 8 (2000) 257–285

TRANSPORTATION
RESEARCH
PART C

www.elsevier.com/locate/trc

Integration of the global positioning system and geographical information systems for traffic congestion studies

Michael A.P. Taylor *, Jeremy E. Woolley, Rocco Zito

Transport Systems Centre, School of Geoinformatics, Planning and Building, University of South Australia, North Terrace, Adelaide 5000, Australia

Abstract

The Transport Systems Centre (TSC) has developed an integrated Global Positioning System (GPS) – Geographical Information System (GIS) for collecting on-road traffic data from a probe vehicle. This system has been further integrated with the engine management system of a vehicle to provide time-tagged data on GPS position and speed, distance travelled, acceleration, fuel consumption, engine performance, and air pollutant emissions on a second-by-second basis. These data are handled within a GIS and can be processed and queried during the data collection (from a notebook PC in the vehicle) or saved to a file for later analysis. The database so generated provides a rich source of information for studies of travel times and delays, congestion levels, and energy and emissions. A case study application of the system is described focusing on studies of congestion levels on two parallel routes in a major arterial corridor in metropolitan Adelaide, South Australia. As part of these investigations, a discussion of the nature of traffic congestion is given. This provides both a general definition of traffic congestion and the discussion of a number of parametric measures of congestion. The computation of these parameters for the study corridor on the basis of data collected from the integrated GPS–GIS system is described. The GIS provides a database management platform for the integration, display, and analysis of the data collected from GPS and the in-vehicle instrumentation.

Keywords: Moving observer traffic studies; Traffic congestion; Global Positioning System; Geographic Information System; Traffic data analysis

1. Introduction

Transportation data, in common with many other data sets in civil engineering and the social sciences, often have spatial attributes. For example, traffic counts come from specific sites, travel

* Corresponding author. Tel.: +61-8-8302-1861; fax: 61-8-8302-1880.
E-mail address: map.taylor@unisa.ed.au (M.A.P. Taylor).

0968-090X/00/$ - see front matter
PII: S0968-090X(00)00015-2

time data refer to particular routes, and origin–destination data apply to a given area. Conventional database systems cannot make much use of the spatial or locational attributes of a data set, other than hold reference details for it. Geographical Information Systems (GIS), on the other hand, can absorb a database, relate its spatial attributes to maps of the region, and offer spatial integration with other pertinent databases for that region.

This paper describes the use of GIS techniques in the collection and analysis of travel time, delay, and congestion data for urban road corridors using data collected by a moving observer. The techniques have applications in both planning and design and in real-time monitoring of traffic systems. A key element of the GIS-based system is the use of Global Positioning Systems (GPS) data to determine locations, for both static observations and dynamic recording of vehicle positions over time. The GIS takes on the central role in data management, in terms of data entry and integration, data management, and some aspects of data analysis and display. Quiroga and Bullock (1998a) provide a prime example of the capabilities of GPS as a tool in moving observer traffic studies.

The Transport Systems Centre (TSC) at the University of South Australia has developed a GIS-based system for collecting on-road traffic data from an instrumented probe vehicle driven in a traffic stream. The probe vehicle can record travel time, distance covered, location, speed, fuel consumed, air pollutant emissions, engine performance, operating state variables, and delay and queuing data over time (second-by-second). Table 1 lists the data items recorded by the vehicle. The integration of the various data collection modules in the vehicle is a major task facilitated by the use of GIS. The data collection modules include differential GPS (for time-based location, speed, and direction of travel), and on-board vehicle data (speed, acceleration, fuel consumption, engine performance, and emission rates).[1] The GIS system makes strong use of GPS for real-time data collection, and integration with a radio communications system allows for the real-time tracking of a fleet of probe vehicles, as well as post-processing of the on-road data. The TSC test vehicle has a customised interface with its engine management system and Global Positioning Receivers (GPS) (Zito and Taylor, 1999). It can provide the data shown in Table 1 on a second-by-second basis. The data can be viewed in real time using a notebook PC in the vehicle or logged to a file for post analysis. A typical output string is of the form:

> time (s), distance (m), speed (km/h), fuel (ml), engine RPM, MAF, A/C, gear, P/E, eng temp, Tpos, lat, long 123.45, 1876.456, 62.1, 534.21, 1700, 111, 0, 4, E, 80, 45, −349197536, +1386064077.

The paper outlines the elements of the data recording system and their integration through GIS. It then illustrates the use of the system with a recent study of the southern road corridor in the Adelaide metropolitan area where the first stage of a new 'high-tech' reversible flow Expressway was recently opened. At the time of the study, the engine maps for pollutant emissions were not available. Therefore, emission data are not reported here. The maps were not available

[1] All variables except emission rates are measured directly by the on-board instrumentation. Pollutant emission rates are measured indirectly, using the engine performance variables engine revolutions (RPM), manifold pressure and engine temperature, and then applying a calibrated engine maps for each pollutant determined from dynamometer testing of the vehicle (Zito and Taylor, 1999).

Table 1
Vehicle parameters logged in real time by the TSC probe vehicle

Variable	Measurement units	Variable	Measurement units
Time	s	Air conditioning	on/off
Distance	m	Power/economy mode	on/off
Speed	km/h	Engine gear	gear (1–4)
Fuel consumption	l	Hydrocarbons (HC)	ppm
Engine revolutions	rpm	Nitrogen oxides (NO_X)	ppm
Manifold pressure	Pa	Carbon monoxide (CO)	ppm
Throttle position	ratio	Carbon dioxide (CO_2)	ppm
Engine temperature	°C	Oxygen (O_2)	ppm
GPS position	Latitude + Longitude		

until February 2000. However, the recorded data can be re-analysed using these maps to provide the emissions, and this will be done in subsequent stages of the southern road corridor study.

2. GIS–GPS integration

The use of GIS as a database integrator for a transport study area is illustrated in Fig. 1. This shows a schematic diagram representing a set of individual databases for the study area, which comprise a mixture of spatial, numerical, and perhaps textual data. These are integrated within the GIS and displayed by superposition of a separate map layer for each database. The layers shown in Fig. 1 include topographical and land use data, transport network infrastructure, socio-economic and demographic data, traffic flow data, and pollution and environment impact data. Areas of modelling and analysis in transport planning and transport systems management involving the use of different subsets layers are also indicated. An example of a GIS map showing a combination of data layers is given in Fig. 2. This figure includes probe vehicle locations over time along a route together with street centreline data and street map data layers. It is a snapshot of the data display available in the TSC instrumented vehicle, as described in this paper. A useful feature of the display is the use of a GIS 'info tool', providing a display window on the map of Fig. 2, showing the data relating to the journey and the vehicle's performance for the point highlighted by the arrow. The info tool window is in fact displaying at that point on the journey all of the recorded variables for the probe vehicle, as listed in Table 1. This display is available at any stage whilst the data are being logged (on the notebook PC in the probe vehicle) or subsequently in post-processing and analysis of the recorded data. The system thus allows the entire history of the journey to be studied at any time.

GPS receivers provide a fast and convenient method for obtaining position information that can be collected in real time and is easily employed within a GIS. This is because the basic GPS position information is provided in the form of latitude and longitude on the surface of the Earth, which is compatible with common GIS location specifications. Differential GPS corrections can be used to enhance positional accuracy to about ±5 m (with 95% confidence) under almost all conditions. In addition, instantaneous speed accuracy to ±2 km/h (with 95% confidence) can be

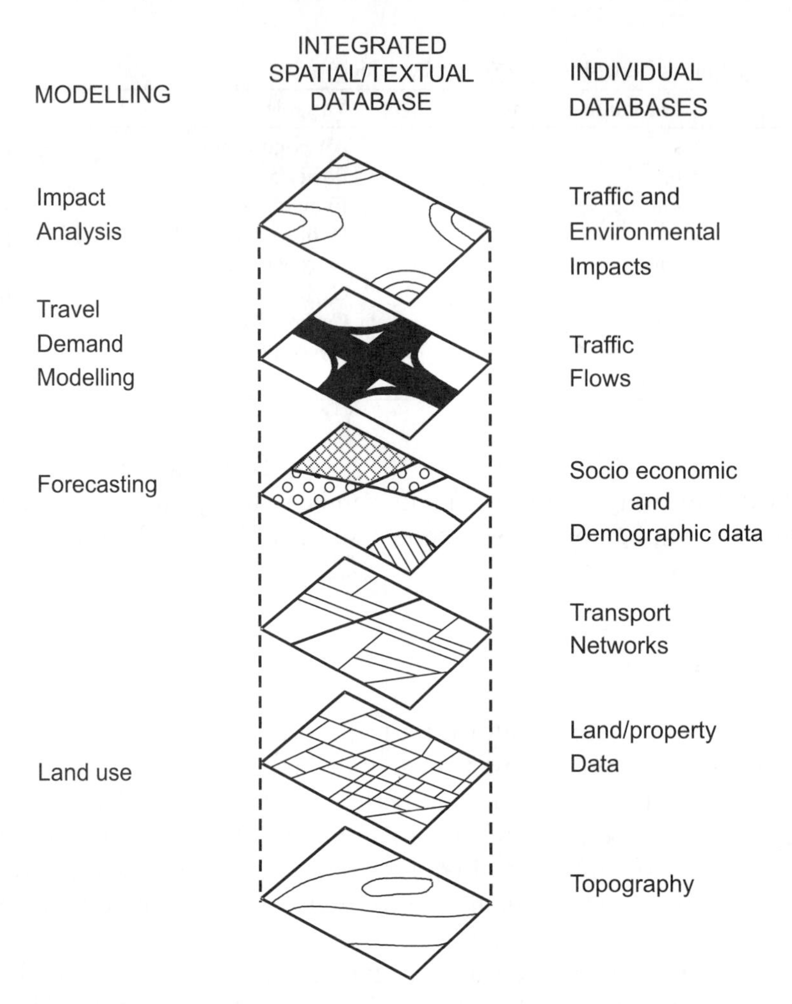

Fig. 1. The concept of superposition of data layers in GIS.

achieved, with vehicle speed being observed directly and recorded *independently* of the GPS position data, using commonly available GPS receivers. Zito et al. (1995), Quiroga and Bullock (1998a), and D'Este et al. (1999) explain the use of GPS in traffic studies and the theory of and methods for instantaneous speed observation using GPS and describe a series of experiments which validated the above results for both position and the independent speed observations.

A critical factor in the use of a GIS is the data that are available with it. Rigorous spatial analysis using GIS is only possible when the appropriate map databases are available. For Adelaide, South Australia the only road centreline database available until very recently was that derived from the Digital Cadastral Database (DCDB) held by the South Australian Department of Lands. This database is used for land titling in the state and shows the boundaries and gives the coordinates of the points that make up all parcels of land in South Australia. The street centrelines were then formulated by taking the centreline between the land boundaries to estimate the road centreline position. However, this method did not yield an exhaustive coverage of roads in

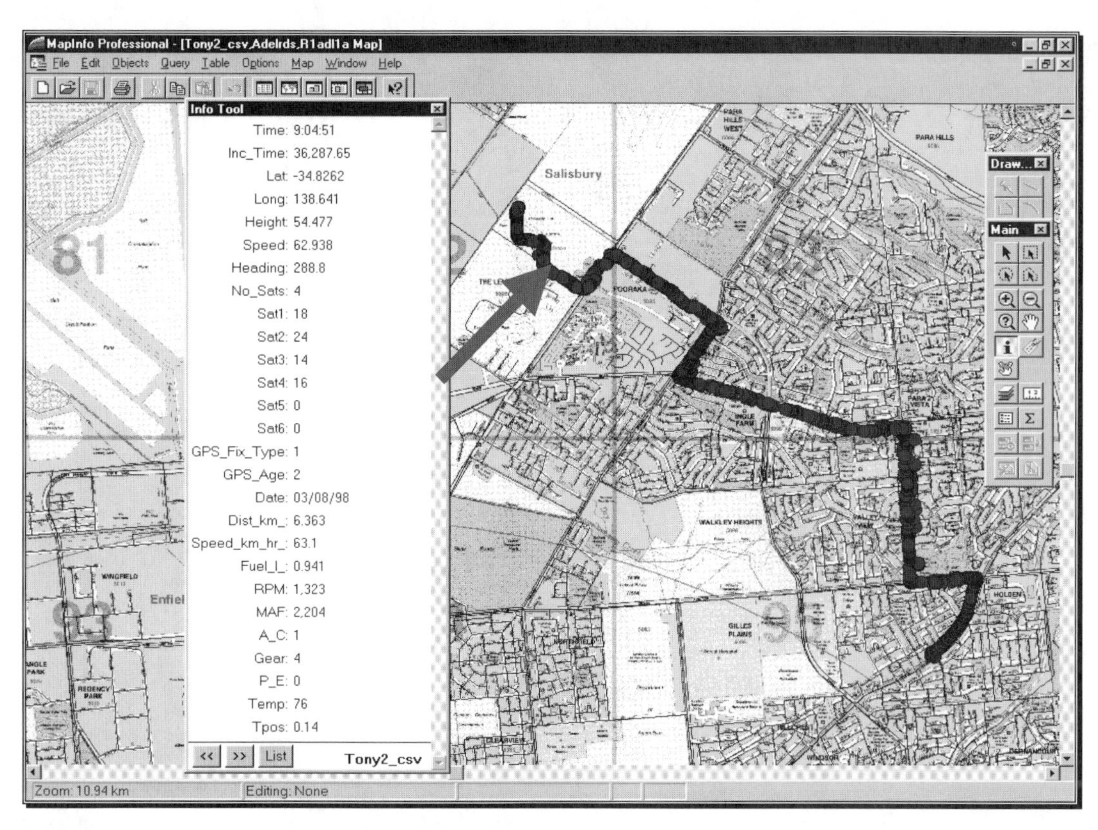

Fig. 2. GIS map of probe vehicle journey through street network, showing different data layers and 'info tool' display of journey parameters at the indicated location along the route.

Adelaide so the database was complemented with other digitised data from various maps, as well as the use of GPS to fill in data on missing road segments. This database has proved to be more than adequate for application in traffic studies. The spatial coverage of the database, shown in Fig. 3, is some 100-km east-west and almost 170-km north-south. The basic attributes associated with each link in the raw street centreline database, besides the start and endpoints of each link, are shown in Table 2. These attributes are basic, being little more than street address references and a broad road type code, but they open the door to more detailed descriptions through linkages with other databases.

Whilst the database provides a comprehensive coverage of the roads in the region, it does have some deficiencies. Firstly, Fig. 3 shows an inset of a blow up of a sample non-linear road segment centreline. The inset clearly shows that the road curve is made up of four independent links. This coding of road curves by a sequence of straight-line segments was adopted to facilitate the digitising of the road database. However, this greatly increases the numbers of links and nodes in the overall road network. Nor are 'arc' or 'polyline' objects a part of this database, which further decreases its efficiency. The current database is made up of some 135,000 records. Maintenance of the database is also an issue, for the street centreline database only truly represents the road network at a given date. This is a significant issue in areas undergoing new development, as is the

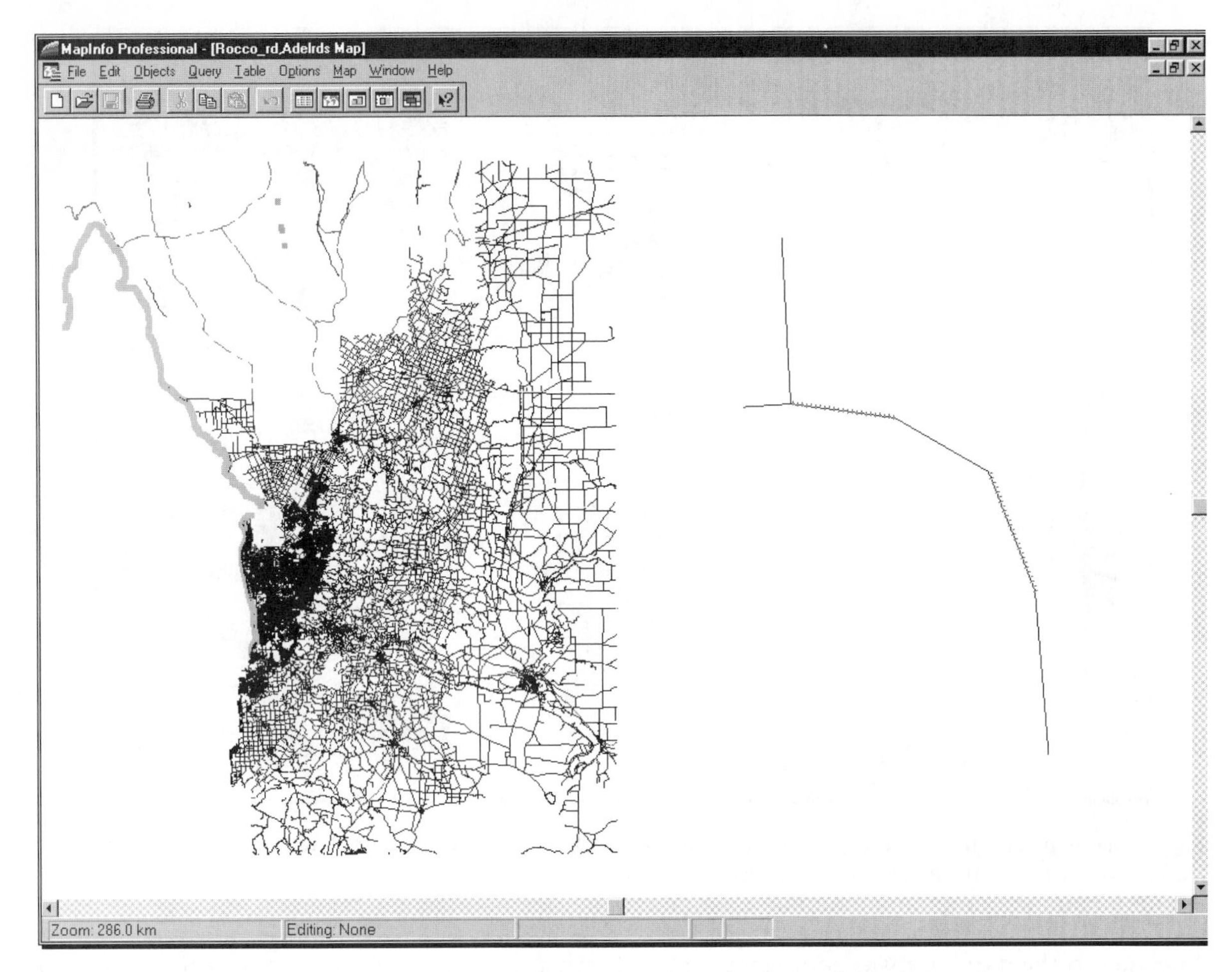

Fig. 3. Adelaide DCBD street centreline database, showing inset of a typical road segment.

Table 2
Sample road centreline attributes for link segments in the DCDB database

Road name	From left	To left+	From right	To right	Road type
ACACIA RD	29	51	26	48	1
REBECCA AV	3	3	4	4	0
WENTWORTH ST	160	154	159	153	0

case in the case study region considered in this paper. Updates must be released to include newly constructed roads and any modifications to existing roads through re-alignment or traffic management changes, such as the conversion of a two-way road into a one-way road. Until the late 1990s, updates were not available; now, updates to the database are released annually.

Thus, the database, while useful as an overlay to GPS data points, is not completely suitable for navigation purposes. Its limitations need to be understood and the data used accordingly. In the congestion and travel time studies described in this paper, the street centreline database is used as a backcloth for the (differentially corrected) GPS data points. The analysis of travel times, speed profiles, and other travel factors is based on the data pertaining to the GPS data points, stored within the GIS and subject to analysis using the GIS. This system has proved both reliable and pertinent for the moving observer traffic studies, as reported in this paper.

For moving observer traffic studies, the advantages of using GPS and GIS include:

- the ability to obtain second-by-second position data and simultaneous speed profile data from the GIS;
- the spatial display and analysis of data in a GIS for ease of interpretation, analysis, and integration with other data sets (such as that from the on-board instrumentation of the probe vehicle); and
- the ease of transferability of GPS equipment from vehicle to vehicle. Although this paper is concerned with data collected by a single probe vehicle specially equipped to record a variety of data items, the GPS system (which provides the principal data items of vehicle position and speed profile over time) can be moved easily between vehicles and has been employed on vehicles ranging from bicycles to large freight vehicles in a range of different traffic studies.

3. Traffic studies

Traffic studies are an essential part of traffic planning and engineering. They provide the basic inventory and performance data, as well as area-specific travel demand data, that are required for project planning and design and traffic systems management and monitoring (e.g., Taylor and Young, 1988; Liu, 1994; Robertson, 1994; Taylor et al., 1996; Taylor et al., 2000). Much of the data have spatial attributes, and many of the traffic study techniques involve data surveys based on observations from a variety of sites or collected by moving observers. Such data include traffic counts from road sections and turning movement counts at intersections, road crash data, speed, travel time and delay data, and origin–destination data. GIS techniques have a significant role in the storage, analysis, and reporting of these data sets and also allow for the integration of different databases relating to a study area, including both traffic-based data and related land use, demographic, topographic, and environmental data.

Moving observer methods are commonly used for travel time and delay surveys, including assessments of traffic congestion, and are also beginning to be used for studies of fuel consumption and pollutant emissions (for examples see Taylor et al., 2000). The basic method involves the surveyor ('observer') being in a vehicle in the traffic stream and noting, among other things, the time taken to travel between specified points. These techniques are particularly suitable for relatively long or complex journeys which do not have sufficient through-flow to support other travel time survey techniques, such as registration plate matching or input/output methods (Taylor et al., 2000). Another advantage of the moving observer methods is that they can yield information about travel times and traffic conditions for intermediate stretches of road and so identify the reasons for any abnormality in the overall journey time. The obvious problem with the basic moving observer method is that, unless repeated many times by different drivers, it is not likely to be representative of actual conditions and may be unduly influenced by the driver's driving style.[2] A general rule is that in order to overcome the sampling problem, each route should be driven at least 15 times with different drivers each time. This is seldom a practical proposition, and may introduce its own inaccuracies (e.g., on the shoulders of peak periods when traffic conditions may be changing rapidly, the different runs may not equate to random observations from the same population). Adopting the floating car method, whereby the drivers are instructed to attempt to 'float' in the traffic stream, overtaking as many vehicles as overtake them can reduce the extent of bias due to driving style. Overzealous attempts to comply with this instruction can, of course, have safety consequences, but even where it is safe to overtake or hang back it may not be obvious whether it is appropriate. It can, for example, be difficult to judge whether a given vehicle is a part of the main traffic stream in which the driver has to float or whether it is slowing down to join a queue of traffic waiting to exit onto a side road. The developers of the floating car method (Wardrop and Charlesworth, 1954) suggested a method of correcting for failure to float perfectly; see also O'Flaherty and Simons (1970) and Cowan and Erikson (1972) for independent assessments of the method. Wardrop and Charlesworth suggested a correction formula to calculate the mean travel time ($\bar{t}_{ab}$) from a to b as shown in Eq. (1):

$$\bar{t}_{ab} = t_{ab} + \frac{O}{q}, \tag{1}$$

where t_{ab} is the time taken by a survey car to travel the route from a to b, and O the net number of vehicles overtaken by the survey car vehicle (i.e. vehicles overtaken minus vehicles who overtake) while travelling from a to b, and q is the mean flow rate. The mean flow rate appears in the formula in order to reflect the fact that overtaking one 'too many' vehicles in a flow of 1000 veh/h is less significant than the same thing in a flow of 100 veh/h. Since q is rarely available from conventional sources (it is likely to differ on different parts of the route) an estimate can be made by having a vehicle travel in the opposite direction ($b \to a$) recording travel time (t_{ba}) and number of vehicles met (m) in the opposing flow (i.e. travelling in the $b \to a$ direction). Then an estimate of q is given by Eq. (2):

$$q = \frac{m - O}{t_{ab} + t_{ba}}. \tag{2}$$

The accuracy of the method is improved by taking several runs in the $a \to b$ direction (with matching $b \to a$ runs to estimate q), and it is convenient to utilise pairs of vehicles running back and forth between a and b for this purpose (note that, although common, the use of just one vehicle to provide the $a \to b$ and $b \to a$ times is theoretically flawed since the $a \to b$ and $b \to a$ run should be simultaneous). If travel times are required for subsections within the $a \to b$ route, it will be necessary to record data to allow separate correction factors to be calculated for each subsection. If executed correctly, this method can produce useful data, albeit at some cost. However, it is subject to error if conditions fluctuate markedly during the survey. This method for volume estimation is not recommended if alternative methods can be employed, e.g. the use of traffic

[2] Note that the problem is not overcome by instructing the driver to tail another vehicle – that merely replaces a bias due to the surveyor's driving style by one due to that of a randomly selected member of the driving population.

analysers to record volumes at sites along the route or volume data extracted from an urban traffic control system (as is the case in the study reported in this paper). Information derived from moving observer surveys can be enriched if data is simultaneously recorded on time spent in queues within each segment of the journey, and it is in this regard that use of an instrumented probe vehicle becomes important.

Modern moving observer surveys thus often require the use of specially equipped probe vehicles to collect data in real time, as described in D'Este et al. (1999), Quiroga and Bullock (1998a), Woolley and Taylor (1999), and Zito and Taylor (1999). Increasing use is being made of GPS for the recording of vehicle positions over time, see Quiroga and Bullock (1998a), and previous research at the TSC has indicated that pseudo-instantaneous travel speeds can also be recorded accurately by GPS, independently of the GPS position data. See Zito and Taylor (1994), Zito et al. (1995), and D'Este et al. (1999) for research results clearly indicating that the direct speed measurements from GPS correlate closely with 'actual' vehicle speeds (as recorded by on-board instrumentation). GPS location data along with other data recorded from the GPS or simultaneously from other on-board instrumentation on the probe vehicle are most conveniently handled through GIS.

3.1. Probe vehicle

The probe vehicle used in the research study reported in this paper is the TSC's instrument vehicle, currently a General Motors Holden VS Commodore sedan car. This vehicle has been instrumented so that it can log data continuously (at time intervals of one second) directly from its engine management system in synchronisation with GPS receivers. The GPS data provides both spatial and time/distance based data from which various traffic parameters can be derived, including travel time, stopped time, travel speeds (instantaneous and average), and various congestion indices. Differentially corrected GPS position data is recorded in real time using the differential corrections broadcast on the FM radio network 'JJJ' in Australia; see Zito and Taylor (1996) for an explanation of this system and a case study application concerned with residential area traffic calming. The engine management system module directly provides data, such as time, distance, speed, fuel consumption, engine revolutions (RPM), throttle position, engine temperature, engine gear, use of air conditioning, and economy/power mode (see Table 1). Air pollutant emissions are determined using calibrated engine maps for the vehicle and are based on the observed engine performance variables as noted above. The development and use of the engine maps is described in Zito and Taylor (1999). The elemental data provided by the vehicle and the GPS are stored in GIS software running on a notebook PC in the vehicle. Thus the recorded data may be displayed, interrogated, and analysed using GIS-specific functionality both during the data collection and afterwards, as indicated in Fig. 2.

Once the GPS and other vehicle-based data have been collected, they are imported into a GIS, where they can be displayed and analysed spatially, as well as analytically. Street centreline data are included as a layer in the GIS so that the exact route, speed profile, and time data can be determined on a link-by-link basis. In addition, other map layers, such as aerial photography, electronic street directory maps, topographical features and landmarks are included. Fig. 2 provides a sample plot of probe vehicle locations over time along a route superimposed on the street centreline data and electronic street map data layers (a raster image of street layout and land uses). Vehicle location at each point in time is represented on the map by a small coloured circle. The circles can be colour-coded to indicate relative values of a specific data item, such as instantaneous speed or fuel consumption rate. The user may select the specific data item to be displayed in this way by the GIS.

As well as indicating the ability of GIS to display the position of the probe vehicle in the street network, Fig. 2 also illustrates the use of GIS to associate the GPS time-tagged, vehicle-based data variables to each circle as shown by the 'Info Tool' dialogue box. In this way, data can be queried using standard database techniques (e.g. display those time-tagged observations, where the vehicle was stationary or where its speed exceeded 80 km/h, etc.) and also queried spatially (e.g. what was happening to the vehicle at a certain point along the route (as in Fig. 2), or what data were collected within 100 metres upstream and downstream of a given intersection, etc.). To give the data collected added value, GIS has the ability to overlay different databases. Fig. 2 shows how the collected data have been overlaid with a raster street directory image allowing the user to read directly off the map the street names. In addition, another vector layer of street centrelines has been added to the attributes associated with this database, thus including the geographic coordinates of the links, as well as the address ranges for each side of the road. This database allows the raster image to be georeferenced so that it is spatially correct, hence GPS positions can be overlaid on the map.

3.2. Socio-economic factors

A major advantage of the use of GIS in management and analysis of traffic data is the ability to integrate these data with other data sets, such as land use and socio-economic and demographic data relating to the region. This is especially important in environmental impact studies, where the possible juxtaposition of poor environmental conditions and residential land use in some locations is of concern. This issue and the use of GIS to determine such locations is described in Klungboonkrong and Taylor (1998), Affum and Brown (1999), and Taylor et al. (2000).

Fig. 4 shows some typical demographic maps of the Adelaide metropolitan area. These data were collected in the 1996 Australian National Census and are readily available in a GIS format and suitable for immediate mapping and analysis. The map on the left-hand side shows the total number of passenger cars per census collector district (the basic zoning system for the census data). The second map shows the dwelling densities in the same region. In both maps, the darker colours reflect the higher values, while the lighter colours represent the lower values. Kenworthy and Newman (1982) suggested that these parameters amongst others could be used to segment a larger region into various socio-demographic areas that could possibly represent different travel demand and driving patterns within them. Typically all of the demographic variables suggested by Kenworthy and Newman (1982) when applied to the Adelaide situation showed similar characteristics as those in Fig. 4. There are basically three distinct regions: northern, southern, and central. This result suggests the need to undertake separate analyses of travel characteristics and traffic conditions within these three regions. The Southern Expressway corridor studied in this paper lies in the southern region.

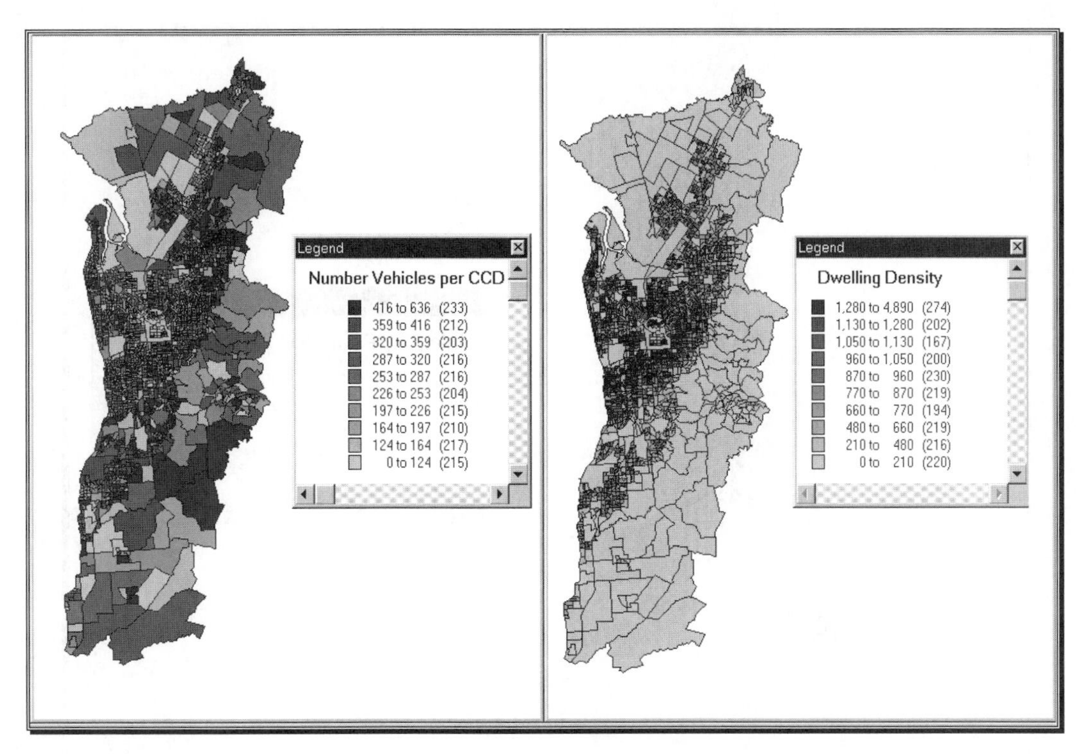

Fig. 4. Typical demographic features of the Adelaide Statistical Division in 1996, indicating the broad division of the metropolitan area into three social regions: northern, central and southern.

3.3. Traffic volumes

Given the regions that have been established, it is a matter of determining which roads in those regions should be driven on to obtain a representative sample of driving in that region. Fig. 5 shows a GIS map of the main arterial routes in the Adelaide metropolitan area, which displays average morning peak hour traffic volumes. Higher volumes are shown as darker shades and lower volumes in lighter shades. These network flow maps may be analysed by the GIS to indicate the proportions of driving (vehicle-kilometres of travel or vehicle-hours of travel) within the three regions (and by time of day). They may be further used to estimate total delay loads (e.g., vehicle-hours of delay) in the network, given delay time information. This can be obtained from the moving vehicle studies, as indicated in Section 4.

4. Congestion

Traffic congestion presents a common if not inevitable facet of traffic activity in a region, particularly in urban areas. The spread, duration, and intensity of congestion, the processes that lead to it, and the consequences of it are of special concern in urban policy making and transport planning. Traffic studies, such as travel time and delay and queuing studies provide

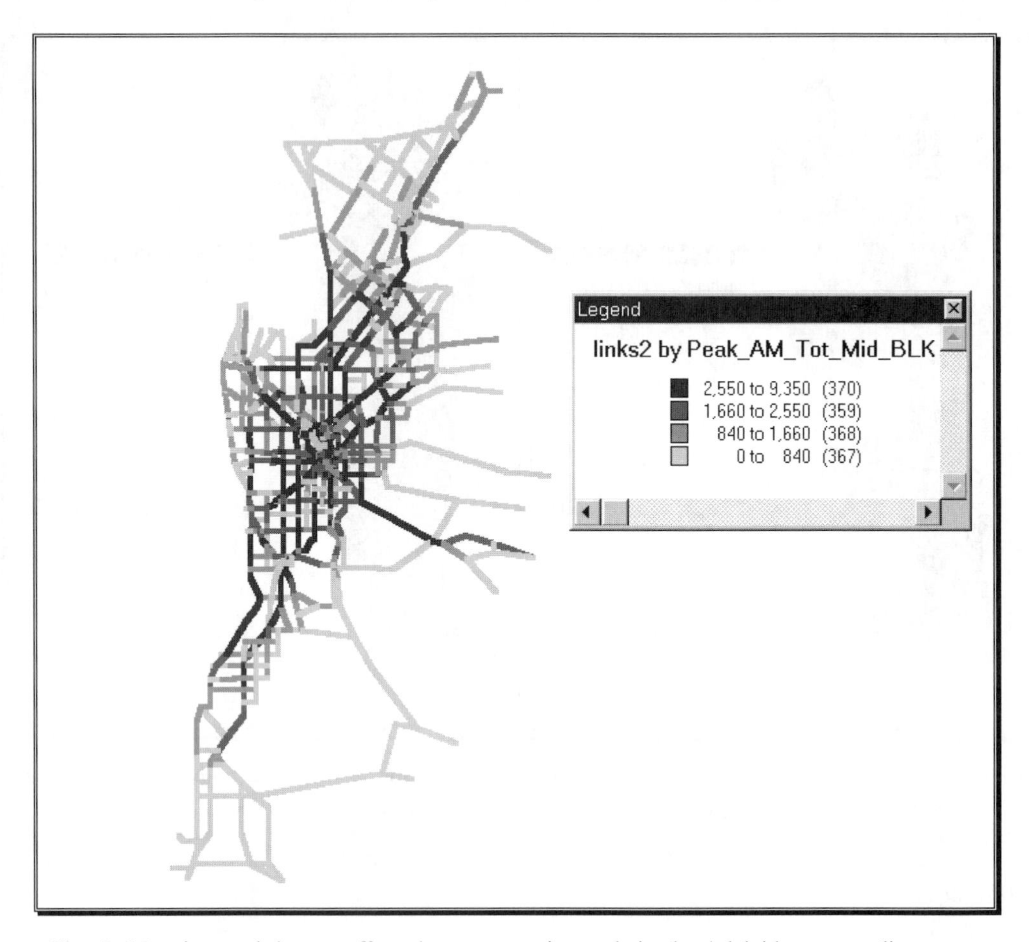

Fig. 5. Morning peak hour traffic volumes on main roads in the Adelaide metropolitan area.

basic information about the occurrence of congestion in a network. Some of this information can be collected from permanent monitoring sites, whilst other information can be collected using moving observer methods. The temporal and spatial distribution of congestion in a region is important, and the use of GIS software for database integration, data analysis, and data display is most advantageous (Affum and Taylor, 1999).

If knowledge about congestion and its extent and intensity is important, then the first consideration is to define just what congestion is. Congestion is an integral part of a transport system, but its specific definition and identification are not immediately obvious. Taylor (1992) reviewed a number of different definitions of traffic congestion and the observed phenomena associated with it. On the basis of this review, three recurrent ideas that occurred in the various definitions of congestion were identified:

- congestion involves the imposition of additional travel costs on all users of a transport facility by each user of that facility;
- transport facilities (e.g. road links, intersections, lanes and turning movements) have finite capacities to handle traffic, and congestion occurs when the demand to use a facility approaches

or exceeds the capacity;
- congestion occurs on a regular, cyclic basis, reflecting the levels and scheduling of social and economic activities in a given area; and
- in addition to the recurrent congestion, special episodes of congestion may occur at different points in a network due to irregular incidents, such as roadworks, breakdowns or accidents.

The following definition of congestion was subsequently proposed for use in traffic studies (Taylor, 1999):

> 'traffic congestion is the phenomenon of increased disruption of traffic movement on an element of the transport system, observed in terms of delays and queuing, that is generated by the interactions amongst the flow units in a traffic stream or in intersecting traffic streams. The phenomenon is most visible when the level of demand for movement approaches or exceeds the present capacity of the element and the best indicator of the occurrence of congestion is the presence of queues'.

This definition is an extension of that given in Taylor (1992). The more recent extension, as presented here, recognizes that the capacity of an element in a traffic systems may vary over time, e.g. when traffic incidents occur or for minor stream traffic movements where capacity may depend on the traffic volume in the major stream.

Thus, congestion may always be present in any part of a transport system, but the level of congestion may have to exceed some threshold value to be recognised. The threshold may be context-specific, for instance owing to the occurrence of incidents, such as breakdowns, road works, or road crashes. Peak periods are recognised as prone to congestion, but it must also be recognised that congestion can occur at other times due to different traffic management regimes in place off-peak or due to traffic incidents or unusual local traffic generating activity. For strategic transport planning purposes, a satisfactory definition of the level of congestion on a network component (e.g. a route, link or intersection turning movement) is the excess travel time incurred by a traveller when traversing that network component. Excess travel time is the additional travel time over and above the free flow travel time (T_0), which is the minimum amount of time required to cover the component. Thus the excess travel time corresponds to the 'system delay' pertaining to the component under the given traffic conditions (Taylor et al., 2000). It should be noted that system delay is a measure of the total delay experienced, including 'stopped delay', delays due to decelerations and accelerations (which may be due to the vehicle joining a queue or in negotiating road features, such as bends or traffic control devices), and delays incurred through interactions with other road users in a traffic stream. The GPS-equipped probe vehicle can measure some of these components of total delay, such as stopped time or acceleration noise (AN) (as described below).

Traffic movement along a link in a network may be seen as consisting of two components. The first component is cruising with traffic moving along the link largely uninterrupted (except for the possibility of side friction, say due to vehicle parking manoeuvres). Travel along the link may also be punctuated by points of interruption, say pedestrian crossings, bus stops and, most importantly, road junctions. For example, the junction at the downstream end of the link may dictate the traffic progression along the link. Movement through the interruption points can be handled using the methods for intersection analysis and queuing theory. What is also needed, particularly for urban areas or other places where congestion is expected, is a composite relationship that can include the two components simultaneously.

Having established concepts and principles that may be used to define and identify traffic congestion in a network, it is necessary to seek some measures or indices that may be used as indicators of the levels of congestion in the network.

4.1. Delay

Delay may be used as one measure of the extent of travel time. A commonly accepted definition of delay is the system delay (d), defined as the excess travel time above the minimum (free flow) travel time needed to traverse a network element (e.g. a link, road section, or intersection). If T is the actual travel time and T_0 is the free flow travel time, then the system delay (as described above) is

$$d = T - T_0. \tag{3}$$

Because it is a direct measure of the additional time costs imposed by one motorist on others, system delay is usually the most appropriate measure of delay for use in travel time studies in congested networks. There are alternative definitions of delay (e.g. stopped time (T_s)) as discussed in Taylor et al., 2000), and care is needed to ensure compatibility between definitions and computational procedures when making comparisons between results from different studies. The probe vehicle data collected in this study is used to estimate both system delay and stopped time delay. The cause of the delay may also be important, with the need to distinguish between recurrent delays due to cyclic variations in travel demand and incident-based delays. With knowledge of traffic volumes along the links in a route, the delay time based on Eq. (3) may be expressed in terms of vehicle-hours of delay (the product of link volume and delay on the link). Thus a measure of the total delay load along the route or on its component links can be obtained.

4.2. Congestion index

Delay time may be used as a measure of congestion for a specific road section but is not particularly useful when making comparisons between different road sections or routes. This is because delay time may depend on specific features (e.g. length) relating to the section. The units of time (e.g. hours, minutes or seconds) used to describe the delay may also mask the interpretation of the actual level of delay unless compared with some other parameter relating to the travel time on the section or route. A more general measure of delay may thus be sought. The level of delay as defined by the system delay can be expressed in terms of a dimensionless congestion index (CI), as described by Richardson and Taylor (1978). This index is given by

$$\mathrm{CI} = (T - T_0)/T_0. \tag{4}$$

This index takes values greater than or equal to zero. A CI value of zero means that the actual travel time is equal to the free flow travel time. A value of one means that the actual travel time is twice the free flow travel time (and that the system delay takes up 50% of the total travel time). The index allows for comparisons between routes, links, and sections because it is independent of route length, route geometry or intersection control and capacity factors that may distort comparisons of actual travel times and delays at different sites. Alternative dimensionless parameters that may be of use include the ratio of actual speed to free speed or the degree of saturation (e.g. see Akcelik, 1981).

4.3. Proportion stopped time

The proportion of stopped time (PST) on a trip segment is another useful measure of traffic congestion and journey quality that can be compared between different routes, links or network segments. It is the ratio of the stopped time to the total journey time, i.e.

$$\mathrm{PST} = T_s/T. \tag{5}$$

Stopped time is taken to be the time during which the vehicle is stationary or during which it is moving at less than some threshold speed value (e.g. Zito et al., 1995 used a threshold speed of 2 km/h). The total trip time T, thus consists of the stopped time and the running time (T_r), i.e.

$$T = T_r + T_s, \tag{6}$$

which may be compared to the definition of system delay (Eq. (3)). Stopped time is only one component of the total (system) delay as given by Eq. (3). PST is a useful parameter that provides a measure of the overall extent of queuing along a route as it measures the time spent in stationary queues. It is also a dimensionless parameter, and can therefore be used in comparisons of different routes and route segments. Given the definition of traffic congestion offered above, there is need for congestion parameters that reflect the extent of queuing. Note that, as suggested below, it is not feasible to directly observe queue lengths at many points in a network. Thus, PST has an important place in principle as a measure of the extent of congestion, along with the CI and other parameters such as AN. The use of multiple congestion parameters to describe an overall congestion state is warranted on the basis of the complexity of the traffic phenomena involved in studies of traffic congestion.

4.4. Acceleration noise

Another measure of congestion is the AN parameter defined by Underwood (1968). As a vehicle moves along a route, its speed varies continuously. The extent of the variation depends on the roadway and traffic conditions and on driver behaviour and vehicle characteristics. On an ideal road with free flowing traffic and steady driving, speed variations are relatively small. However, the variations increase as the roadway condition deteriorates, the influence of other vehicles increases, the levels of traffic and congestion increase, etc. Speed variations produce an acceleration pattern that varies with time. This pattern is defined by the speed-time profile for the journey. Examples of this are given later in this paper. Speed-time profile data may be measured using either on-board instrumentation or by GPS (as indicated in Zito et al., 1995; Zito and Taylor, 1996). This information is of great importance in modelling the fuel consumption and emission performance of vehicles (Biggs and Akcelik, 1986; Taylor and Young, 1996).

AN is a parameter derived from the acceleration pattern contained within the speed-time profile for the journey and provides a measure of the quality of traffic flow and thus of the level of congestion. It may be seen as a measure of the amount of acceleration and deceleration used by a driver when travelling along a road. A journey at a steady speed will have a low value of AN, a journey involving a considerable amount of stop-start driving with other periods of high-speed travel will have a large value of the AN index. AN is given by the equation

$$\mathrm{AN}^2 = \frac{1}{T_\mathrm{r}} \sum_{i=1}^{n} \frac{\Delta v_i^2}{\Delta t_i}, \qquad (7)$$

where Δt_i is the time interval taken for a speed change Δv_i. Underwood (1968) found that AN provides a useful parametric measure of the level of congestion when overall average travel speed ($\bar{v}$) exceeded 30 km/h and that it could be used as a measure of the quality of traffic progression along a route. More recently, it has been speculated that AN may be a useful secondary parameter in describing the level of congestion on different road sections when used in concert with the CI (see Eq. (4)) and the proportion stopped time (Eq. (5)). In this scheme

- the CI provides an overall measure of the congestion on a route,
- the proportion stopped time (PST) indicates the amount of time spent in queues during the journey, and
- the AN indicates the quality of traffic flow on the journey, in terms of the speed variations experienced on it.

An alternative to the AN parameter, believed by Underwood (1968) to be applicable to a wider range of traffic conditions, is the mean velocity gradient (MVG) defined as

$$\mathrm{MVG} = \mathrm{AN}/\bar{v} \qquad (8)$$

and, using the usual definition of a mean value,

$$\bar{v} = \frac{1}{T} \int_0^T v \, \mathrm{d}t, \qquad (9)$$

where T is the total travel time.

The parameters AN and MVG are 'data hungry' as they require knowledge of the speed-time profile for a given journey or route segment – which of course is the full vehicle trajectory over time for the journey or trip segment. Such data were seldom available in the past. This has limited the use of AN and MVG as measures of traffic congestion until now. The development of probe vehicles with instrumentation, such as the systems described in this paper, stands to change this situation, especially given the ease of use of GPS for collecting accurate speed-time profiles, allowing a reappraisal of the parameters for general use in traffic studies. The ability to plot the distribution of such parameters over a road network using GIS is a further enhancement for their use.

4.5. Queue length

If the presence of congestion is identified by the presence of queues, as suggested above, then knowledge of the incidence of queuing in a network is essential in any congestion monitoring system. Queue management is an important initiative in controlling the spread of congestion. One difficulty is that potential locations for queuing, except at intersections, are difficult to determine and require the use of sophisticated incident detection procedures (Govind et al., 1999). The PST parameter may be used as a measure of the extent of queuing along a route or route segment. This parameter is easily measured using GPS, and thus this experimental method enables account to be taken of queue lengths in congestion studies.

5. Application – Adelaide's Southern Expressway corridor

The Southern Expressway is a unique and novel approach towards traffic management in the southern part of metropolitan Adelaide in South Australia. It was built to provide additional peak period capacity for the southern suburbs with the aim of reducing travel times for commuters and alleviating congestion along the existing main arterial road (South Road). The Expressway effectively provides three additional lanes of capacity for the peak direction of flow.

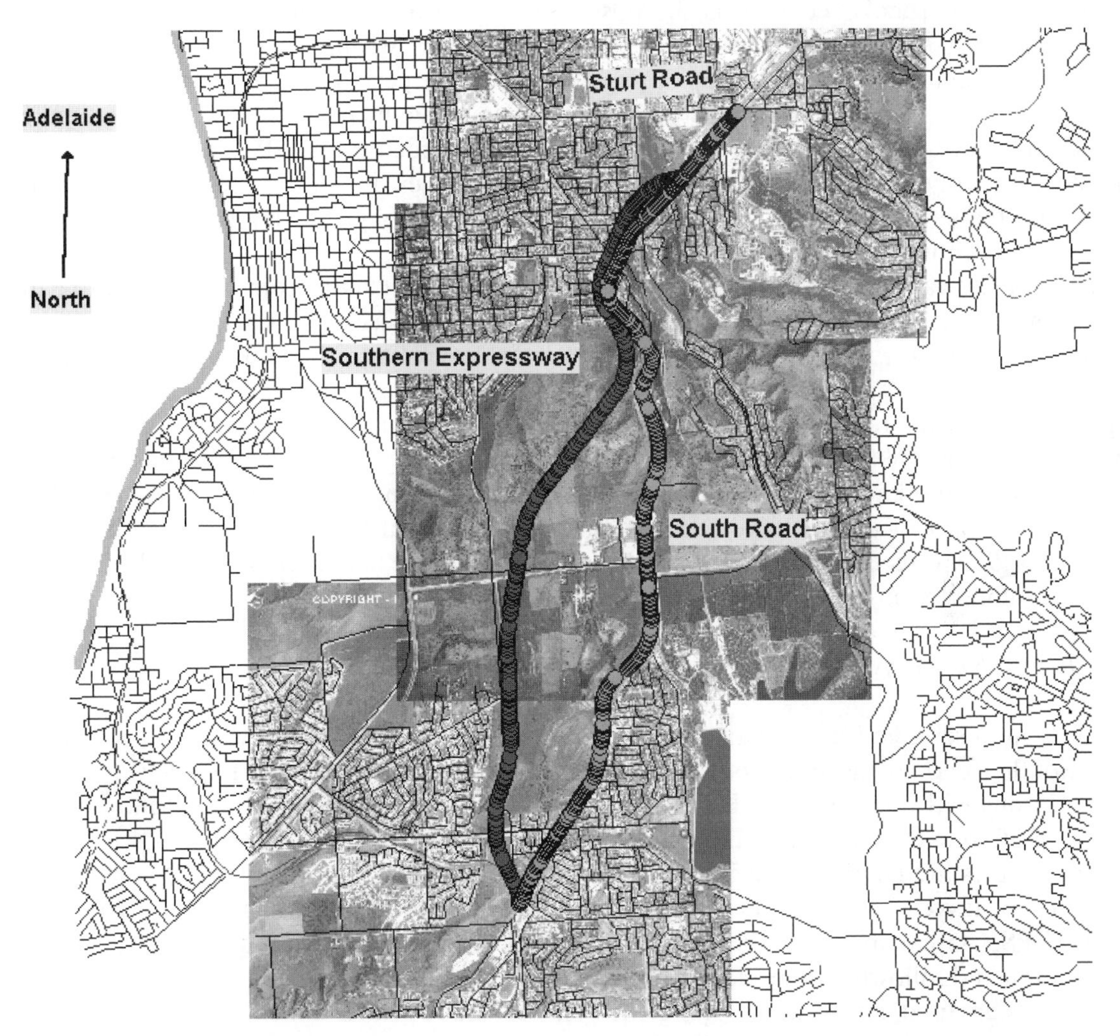

Fig. 6. Map showing GPS plotted outlines of the Southern Expressway and South Road superimposed on GIS layers of street centrelines and aerial photography.

These lanes also have the advantages of higher free flow speeds than those available on the existing road. The unique feature of the Southern Expressway is that it is a wholly tidal road. It allows traffic operations in one flow direction only at any point in time; the flow direction is switched in the middle of the day and late at night. Operationally, the Expressway is open to northbound traffic heading into the city during weekday mornings and southbound when traffic is heading out of the city on weekday afternoons. The directions of flow in these periods are reversed at weekends.

The Southern Expressway is characterised by variable message signs, pavement lights (for time-variant lane markings at the entry points), and a series of boom gates to manage the single direction of flow and restrict vehicles from entering in the wrong direction. Extensive sets of surveillance cameras and road detector loops allow remote operators to monitor the Expressway, rapidly detect any incidents which may arise, and invoke incident management plans and alarms as needed. A particular concern is the possibility of 'wrong way' traffic movements, either by errant motorists managing to enter the Expressway in the wrong direction or by vehicles that might be caught on the Expressway during the transition period. There is considerable care to ensure that such events do not occur. The locality and alignment of the Expressway and South Road are shown in Fig. 6. The road alignments were recorded using GPS position data collected by the TSC probe vehicle. This figure contains two GIS layers in addition to the road alignments. These are the street centreline database and aerial photography showing some details of the topography and urban development in the corridor.

The Expressway is being constructed in two stages. The first stage from Darlington to Reynella constituting some 8 km was opened in December 1997. The second 12 km stage, due for opening at the end of 2000, will extend a further 12 km from Reynella to the Onkaparinga River region. With the inaugural opening of the first stage of the Southern Expressway in December 1997, an opportunity existed to examine the effects of the new infrastructure in terms of a *before and after* type comparison. With the commissioning of the Expressway, direct comparisons could also be made between the Expressway route and the South Road route. The TSC's instrumented vehicle was used as part of an extensive series of data collection surveys between December 1997 and February 1999 to record data for the assessment of the before and after effects of the Expressway. The data collected in these surveys were stored and analysed using GIS.

5.1. Route characteristics

The routes selected for the data collection study extended from the intersection of Sturt Road and South Road to the intersection of South Road and the Southern Expressway. Hence all comparisons with the Southern Expressway share a common piece of South Road between the Northern Expressway entrance and the Sturt Road intersection (see Fig. 6).

Both routes are characterised by a 150 m climb into the hills and contain three lanes in each direction of travel. Fig. 7 shows the longitudinal height variations along the two routes logged using GPS. As can be seen, the height profiles are similar with South Road, showing more fluctuation and following more closely to the natural topography along its route. South Road also has nine signalised intersections along the selected study route, which cause interrupted flow traffic conditions along that route. The Expressway has signals at its entrance and exit, and traffic is free flowing along the route, apart from that section at the exit.

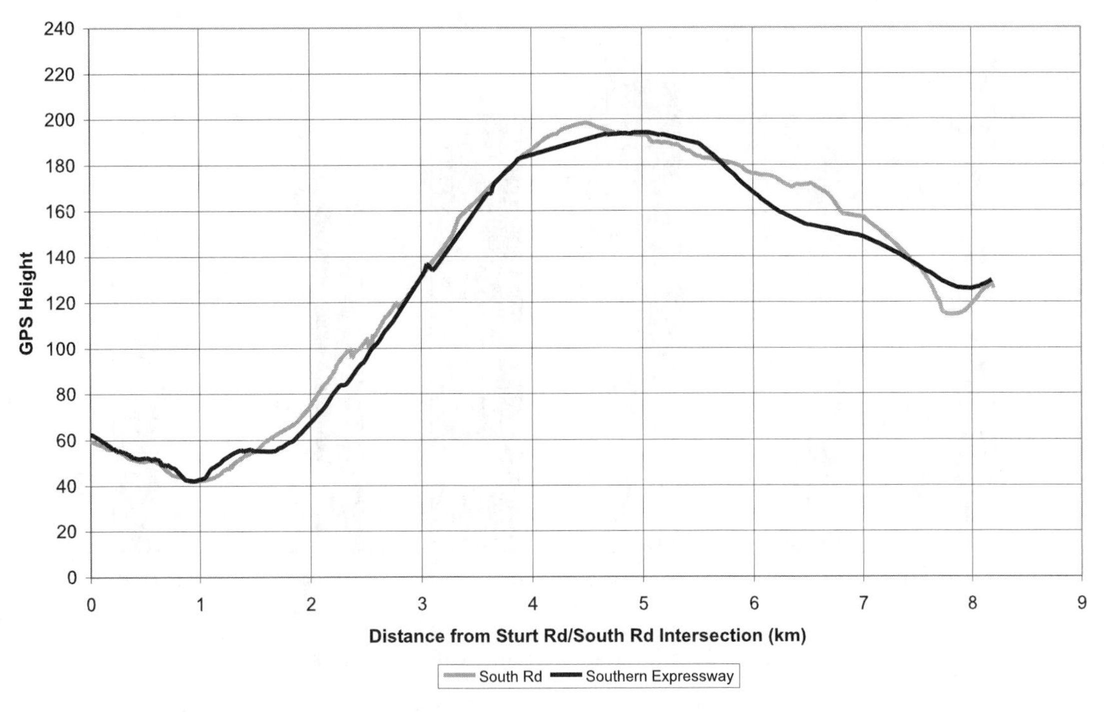

Fig. 7. GPS height profiles for the Southern Expressway and South Road.

5.2. Travel times

Travel times were recorded whenever the vehicle totally crossed the stop line at the selected intersection. In 1997, while the Expressway was being built, an estimation (using roadside furniture as a reference point) was made as to where the Expressway intersection stop lines would be located. During the initial data collection phases, construction speed limit zones operated at the intersection of the Expressway and South Road. At the Southern Entrance, a speed limit of 60 km/h was in place for the final 300 m of the route, instead of the usual 80 km/h. A time penalty of five seconds was therefore subtracted from all travel times measured before the Expressway was opened.

Average travel times of before and after along South Road are indicated in Fig. 8. In the figure, '*before nth bound*' refers to South Road in the northbound direction before the Expressway was opened and likewise for the other abbreviations. As can be seen, in terms of travel times on South Road, no visually significant difference can be detected. During the periods during which data were collected, no serious congestion problems were encountered, suggesting that in both cases (before and after) travel times are largely dependent on the physical characteristics of South Road rather than the delay caused by excess traffic. It should also be noted that the results presented represent average delays over the entire week, while the low-frequency occurrences of congested conditions do not greatly influence the data. This is consistent with a review of the system as a whole rather than isolated worst case scenarios. More detailed information is provided

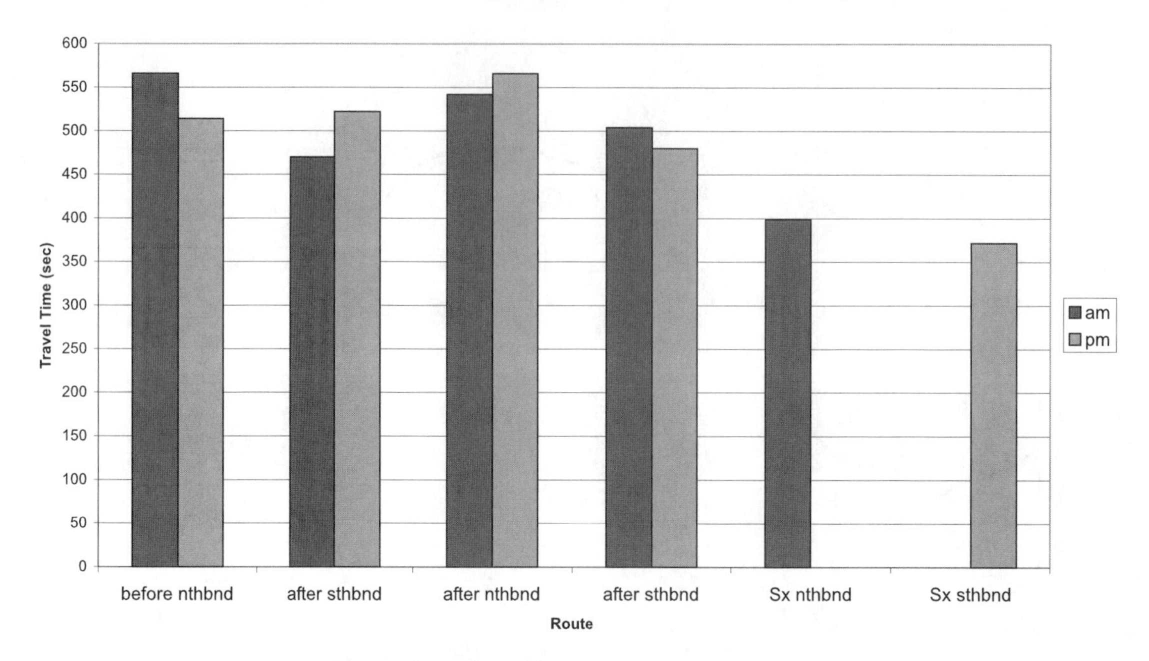

Fig. 8. Overall weekly average total travel times.

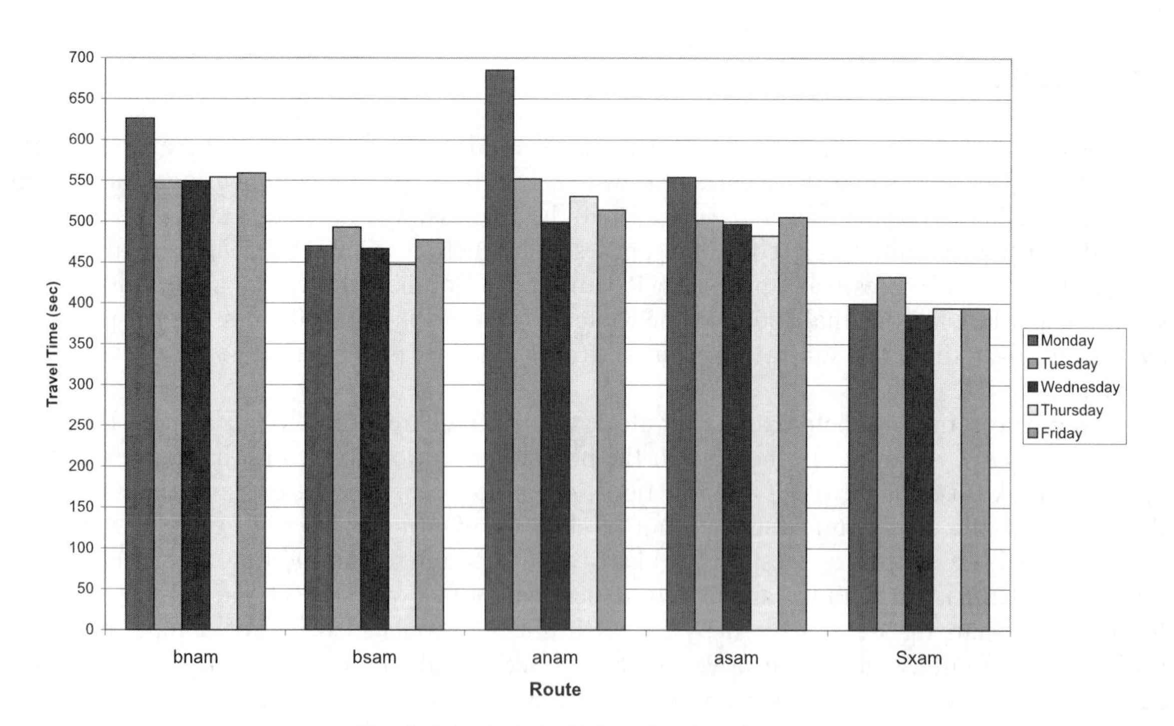

Fig. 9. Morning travel times by day of week.

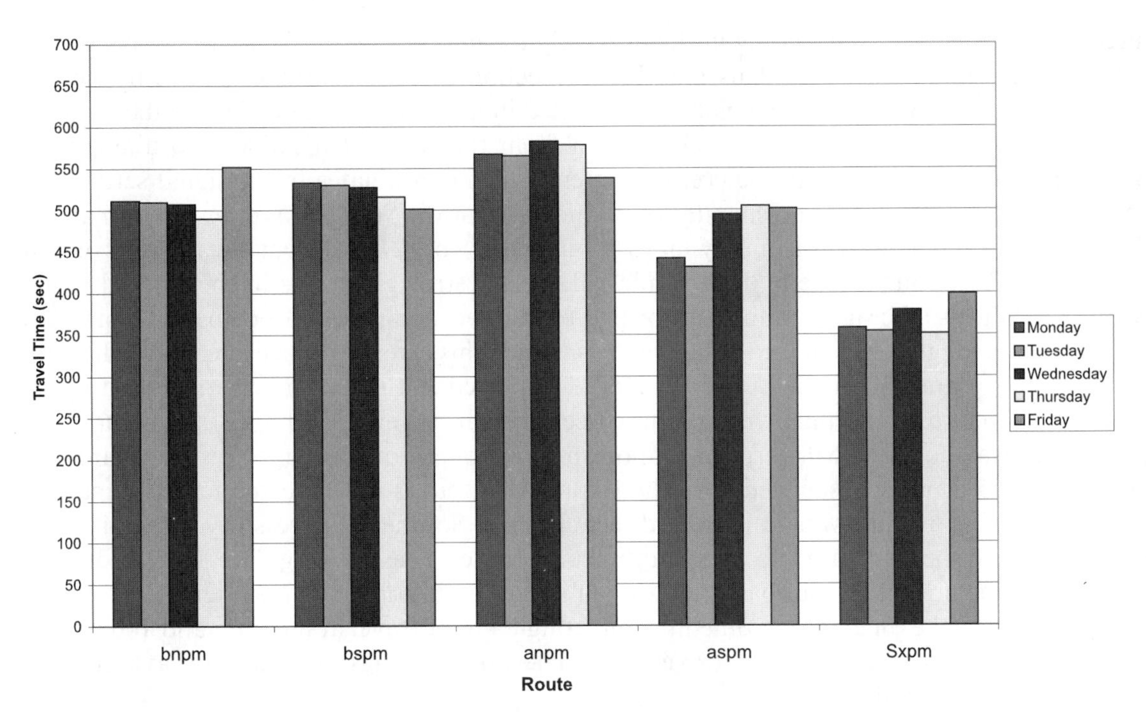

Fig. 10. Afternoon travel times by day of week.

by Figs. 9 and 10, which represent the morning and evening peaks (respectively) broken into days of the week. The codes used in the figures are summarised in Table 3.

The travel time savings from using the Expressway are again evident for the morning and evening cases, but some surprises emerge when looking at the travel times on South Road before and after the Expressway was operational. It would appear that in the morning, travel times are slightly reduced for the busier northbound direction. However, for the southbound direction travel times have increased slightly. This represents the traffic that cannot use the Expressway at this time and constitutes the minority flow in the opposing direction. Likewise in the afternoon period, travel times are reduced in the busy southerly direction but increased in the northerly

Table 3
Codes used for the comparison of routes

CODE	Description
Bnam	Before Expressway opened South Road northbound morning (a.m.)
Bsam	Before Expressway opened South Road southbound morning (a.m.)
Anam	After Expressway opened South Road northbound morning (a.m.)
Asam	After Expressway opened South Road southbound morning (a.m.)
Sxam	Southern Expressway morning regime (northbound)
Bnpm	Before Expressway opened South Road northbound afternoon (p.m.)
Bspm	Before Expressway opened South Road southbound afternoon (p.m.)
Anpm	After Expressway opened South Road northbound afternoon (p.m.)
Aspm	After Expressway opened South Road southbound afternoon (p.m.)
Sxpm	Southern Expressway afternoon regime (southbound)

direction. Once again, traffic going in the northerly direction is unable to use the Expressway and constitute the minority traffic. This may be a reflection of general traffic growth in the region. Data on traffic volumes over time is being collected in a parallel exercise. The traffic volumes on road links within the corridor are being taken from the *ACTS* (*A*delaide *C*oordinated *T*raffic *S*ystem) demand responsive traffic control system which coordinates traffic signal settings in the corridor. ACTS is the local implementation of the well-known *SCATS* (*S*ydney *C*oordinated *A*rea *T*raffic *S*ystem) widely used in many cities around the world. As a demand responsive system, ACTS/SCATS monitors lane-by-lane traffic volumes at strategic points in the arterial road network and adjust traffic signal timings according to the level and pattern of demand. It may also be used as a data logging system for traffic volumes. Given this extensive, accurate, and rich source of traffic volume data, the moving observer studies reported in this paper were not used to collect volume information in keeping with best practice for such surveys. Analysis of the volume data in the Expressway corridor will form an important component of the larger project searching for possible induced traffic associated with the Expressway, but is not reported in this paper.

In considering overall averages, it would seem that the Southern Expressway saves 140 s or 2:20 min in travel time compared to the South Road route. These savings are largely due to the physical characteristics of the roads in terms of speed limit, gradients, and signals. To place this into perspective, the difference in time in completing 8 km of travel at 80 km/h and 100 km/h is 72 s or approximately half the saved travel time. The other savings are made by avoiding delays caused by traffic signals.

5.3. Speed profiles

Plots of speeds were generated and stopped time proportions (PST) calculated. Speed-time profiles provide the basic information on trip quality for the calculation of the congestion in-

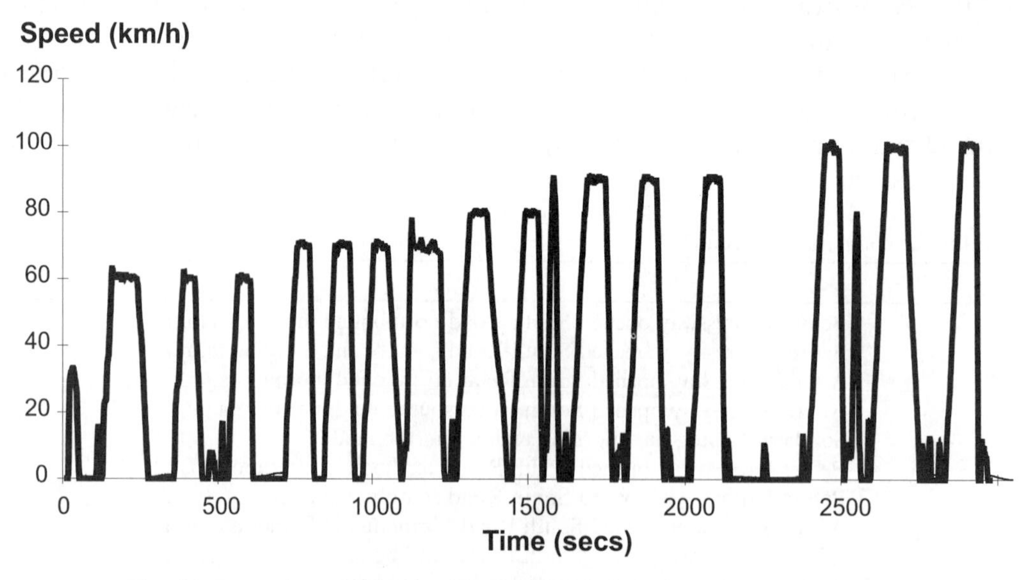

Fig. 11. Example speed-time profile from the Southern Expressway corridor.

dices, as previously described. Typical speed-time profiles are plotted in Fig. 11 for South Road and for the Expressway. These plots indicate the different nature of travel along the two roads, in keeping with the differences in design standard and traffic controls in place on each. The speed-time profile is essential data for studies of traffic flow quality and for fuel usage and emissions analysis. The speed-time graph, while data-rich, is, however, somewhat limited in its immediate information value to the planner or engineer. Knowledge of the spatial location corresponding to each speed-time observation is available from the GPS data, which thus allows the speed-time data to be displayed on a GIS map. The map display is a more useful resource for data interpretation.

Fig. 12 shows alternative GIS displays of the speed-time profiles and the differences in speed regimes for the Southern Expressway and South Road (note that the figure represents flow in opposite directions of travel). The GIS maps indicate the recorded speeds of the probe vehicle as is travelled along each road. The darker (or blue) regions indicate very low speeds, and the lighter (or red) regions indicate high speeds. Thus, the traces on the map indicate the

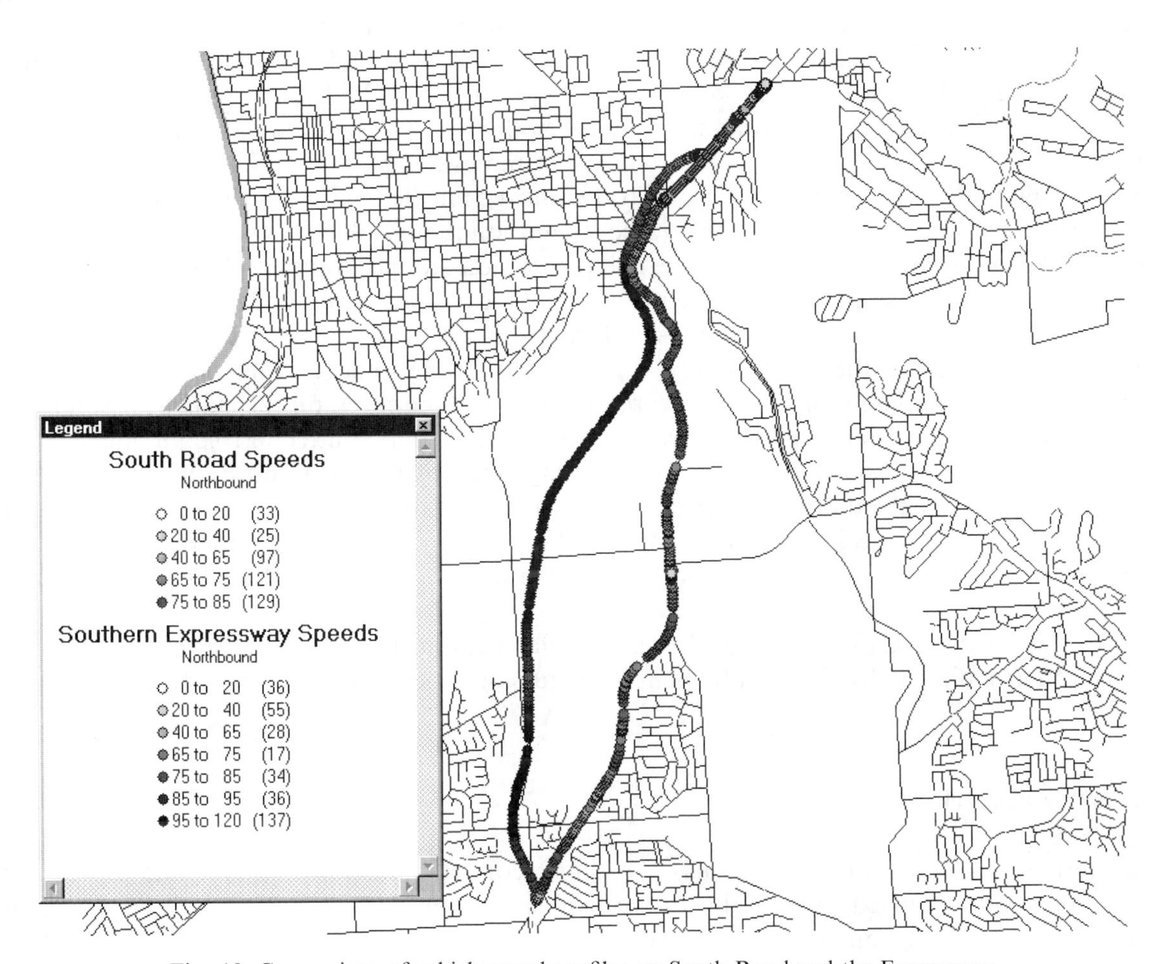

Fig. 12. Comparison of vehicle speed profiles on South Road and the Expressway.

fast and slow segments of each route. In general, serious congestion was not encountered during data collection. Taking the worst case, during the before data collection runs on a Monday morning, the proportions of time of the journey spent at speeds 0–2, <20, and <40 km/h were 9%, 33%, and 58%, respectively, for a total travel time of 884 s. This compares to a typical run on a Friday morning where the total travel time was 580 s with the proportion of time of the journey spent at speeds 0–2, <20, and <40 km/h were 1%, 5%, and 7%, respectively.

5.4. Congestion indices

The congestion indices defined earlier in this paper were calculated for travel time runs during morning and evening peaks on each route, thus yielding mean values and standard deviations of the indices for both routes. Tables 4 and 5 show these results for the Southern Expressway and South Road, respectively. Note that these tables show distance travelled during each data collection run. These distances are included as a check on the individual runs, to ensure that the same route sections were studied, and therefore, allowing the observed data to be compared. The observed data and computed congestion parameters in Tables 4 and 5 clearly indicate the flow advantages gained by use of the Expressway. Mean journey time, PST, and CI are all significantly lower on the Expressway than on South Road for both morning and evening peaks, although the differences are perhaps slightly less pronounced in the evening peak. AN is also lower on the Expressway, indicating a higher quality of traffic movement (less variations in speeds and hence less acceleration and deceleration in the traffic stream) on that road. The results of statistical tests of the differences in mean values of the CI, proportion-stopped time, and AN are shown in Table 6. All mean values from the two roads in both time periods were significantly different at the 1% level of significance, except for the AN during the p.m. peak, where the mean values were significantly different at the 5% level only. These results strongly suggest that there are significant differences in congestion levels on the two roads. As a further check on the statistical validity of these results, minimum required sample sizes for specified permitted errors in mean travel speed were computed using the 'hybrid' method suggested by Quiroga and Bullock (1998b) and then compared to the actual sample sizes (as indicated in Tables 4 and 5). The comparisons are shown for a 95% statistical confidence limit in Table 7. This table suggests that the survey results are consistent with a permitted error in average journey speed of about four km/h at 95% confidence.

Fig. 12 provides a GIS representation of the differences in travel conditions on the two alternative routes indicated by the variations in instantaneous speeds along each of them.

6. Further research

A fertile area for further research is the direct comparison of congestion and travel time data from GPS-equipped probe vehicles with the data collected by the conventional 'floating car' technique (Wardrop and Charlesworth, 1954; Robertson, 1994; Taylor et al., 2000), involving:

Table 4
Congestion indices for the Southern Expressway

Run code	Total distance (m)	Travel time (s)	Stopped time (s)	Mean journey speed (km/h)	Proportion stopped time	Acceleration noise	Mean velocity gradient	Congestion index
Morning peak direction		*Southern Expressway, free travel time* = 326.0 s, *all data collected in period* 07:00–09:00						
301198amn1	8154.7	424.0	22.0	69.2	0.052	0.459	0.024	0.301
3011981mn3	8149.1	374.0	0.0	78.4	0.000	0.420	0.019	0.147
011298amn2	8141.8	469.0	61.0	62.5	0.130	0.548	0.032	0.439
011298amn4	8166.4	396.0	11.0	74.2	0.028	0.574	0.028	0.215
021298amn1	8142.8	365.0	0.0	80.3	0.000	0.481	0.022	0.120
021298amn3	8164.9	416.0	29.0	70.7	0.070	0.515	0.026	0.276
021298amn5	8167.6	376.0	18.0	78.2	0.048	0.440	0.020	0.153
031298amn1	8165.3	350.0	0.0	84.0	0.000	0.494	0.021	0.074
031298amn3	8153.3	378.0	0.0	77.7	0.000	0.400	0.019	0.160
031298amn5	8150.7	454.0	70.0	64.6	0.154	0.581	0.032	0.393
041298amn2	8144.2	379.0	5.0	66.4	0.013	0.473	0.022	0.163
041298amn4	8155.8	421.0	57.0	69.7	0.135	0.562	0.029	0.291
041298amn5	8158.9	382.0	24.0	76.9	0.063	0.503	0.024	0.172
Mean	8155.0	398.8	22.8	74.14	0.053	0.496	0.024	0.223
S.D.	9.1	35.6	24.9	6.34	0.055	0.058	0.005	0.109
Evening peak direction		*Southern Expressway, free travel time* = 326.0 s, *all data collected in period* 16:00–18:00						
301198pms2	8200.8	403.0	26.0	73.3	0.645	0.627	0.031	0.236
011298pms2	8198.8	355.0	0.0	83.1	0.000	0.363	0.016	0.089
011298pms4	8199.0	353.0	0.0	83.6	0.000	0.343	0.015	0.083
021298pms1	8216.6	373.0	0.0	79.3	0.000	0.385	0.018	0.144
021298pms3	8196.0	375.0	17.0	78.7	0.045	0.498	0.023	0.150
021298pms5	8217.7	390.0	15.0	75.9	0.039	0.551	0.026	0.196
031298pms1	8208.6	354.0	5.0	83.5	0.014	0.482	0.021	0.086
031298pms3	8200.4	360.0	0.0	82.0	0.000	0.348	0.015	0.0104
031298pms5	8201.8	348.0	1.0	84.9	0.003	0.461	0.020	0.068
041298pms1	8213.1	399.0	32.0	74.1	0.080	0.467	0.023	0.224
041298pms3	8199.2	428.0	35.0	69.0	0.082	0.581	0.030	0.313
041299pms5	8197.7	370.0	0.0	79.8	0.000	0.349	0.016	0.135
Mean	8204.1	375.7	10.9	78.92	0.027	0.455	0.021	0.152
S.D.	7.7	24.7	13.6	4.95	0.033	0.099	0.057	0.076

- field studies for GPS data, such as those described in this paper, run in parallel with data collection using the conventional techniques. [The integrated GPS-vehicle performance data collection system will need to be extended to collect information on net overtakings (O) and number of vehicles met in the opposing flow (m) as described in Eqs. (1) and (2), and
- analytical studies comparing the basis of the two methods, possibly using the previous analytical studies by Cowan and Eriksson (1972) for the floating car method and Quiroga and Bullock (1998a) for the GPS probe vehicle data as starting points.

Table 5
Congestion indices for South Road

Run code	Total distance (m)	Travel time (s)	Stopped time (s)	Mean journey speed (km/h)	Proportion stopped time	Acceleration noise	Mean velocity gradient	Congestion index
Morning peak direction		*South Road, north-bound free travel time* = 384.0 s, *all data collected in period* 07:00–09:00						
301198amn2	8202.9	685.0	113.0	43.1	0.165	0.620	0.052	0.784
011298amn3	8213.7	564.0	85.0	52.4	0.151	0.673	0.046	0.469
011298amn5	8192.7	540.0	113.0	54.6	0.209	0.684	0.045	0.406
021298amn2	8195.5	496.0	4.0	59.5	0.008	0.569	0.034	0.292
021298amn4	8216.6	500.0	57.0	59.2	0.114	0.682	0.042	0.302
031298amn2	8216.7	558.0	86.0	53.0	0.154	0.653	0.044	0.453
031298amn4	8242.8	503.0	76.0	59.0	0.151	0.716	0.044	0.310
041298amn1	8243.5	505.0	38.0	58.8	0.075	0.586	0.036	0.315
041298amn3	8210.9	523.0	111.0	56.5	0.212	0.657	0.042	0.362
Mean	8215.1	541.6	75.9	55.12	0.138	0.649	0.043	0.410
S.D.	18.1	59.5	37.3	5.27	0.065	0.048	0.005	0.155
Evening peak direction		*South Road, south-bound free travel time* = 383.0 s, *all data collected in period* 16:00–18:00						
301198pms1	8205.4	452.0	12.0	65.4	0.027	0.427	0.024	0.180
301198pms3	8217.3	431.0	17.0	68.6	0.039	0.430	0.023	0.125
011298pms3	8211.2	431.0	21.0	68.6	0.049	0.465	0.024	0.125
021298pms2	8218.5	497.0	42.0	59.5	0.085	0.664	0.040	0.298
021298pms4	8230.1	492.0	37.0	60.2	0.075	0.586	0.035	0.285
031298pms2	8225.3	491.0	37.0	60.3	0.075	0.542	0.032	0.282
031298pms4	8217.1	518.0	65.0	57.1	0.126	0.587	0.037	0.353
041298pms2	8239.2	517.0	86.0	57.4	0.166	0.651	0.041	0.350
041298pms4	8224.8	486.0	51.0	60.9	0.105	0.545	0.032	0.269
Mean	8221.0	479.4	40.9	62.00	0.083	0.544	0.032	0.252
S.D.	10.1	33.5	23.9	4.43	0.044	0.089	0.007	0.088

Table 6
Statistical differences in mean values of the congestion parameters on the two routes (Southern Expressway – South Road)[a]

	Congestion index (CI)	Proportion stopped time (PST)	Acceleration noise (AN)
a.m. peak			
t-Statistic	−3.340	−3.311	−6.511
Degrees of freedom	20	20	20
Result	*	*	*
p.m. peak			
t-Statistic	−6.328	−3.337	−2.219
Degrees of freedom	19	19	19
Result	*	*	**

[a] All a.m. and p.m. peak data were collected in similar time periods and under similar traffic conditions.
* Significantly different at the 1% level.
** Significantly different at the 5% level.

Table 7
Comparison of computed minimum sample sizes required for travel time runs at specified permitted error ε (km/h) and actual sample sizes[a]

Route	Time/direction	Minimum sample size for permitted error ε (km/h)				
		ε = 3		4		5
Southern Expressway	a.m. peak, north bound	22	*13*	12		8
Southern Expressway	p.m. peak, south bound	14	*12*	8		5
South Road	a.m. peak, north bound	17		10	*9*	6
South Road	p.m. peak, south bound	12	*9*	7		5

[a] 1. Values in italics are the actual sample sizes (see also Tables 4 and 5). 2. Computed minimum sample sizes are based on the 'hybrid method' described by Quiroga and Bullock (1998b).

This research will serve to determine the full effectiveness and/or efficiency of the GIS–GPS technique.

Further research is also needed to extend the usage of the congestion parameters (e.g. CI, PST and AN, as defined in this paper) by considering these parameters over a wide range of traffic conditions, and perhaps for different vehicle and road user classes within a given traffic stream. The new ability to collect emission data in real time using the probe vehicle will add a new capability for future studies and will open the door to the use of fuel and emission-related parameters in future congestion studies.

7. Conclusions

GIS have long been seen to have an important role in transport planning, although the universal implementation of GIS methods in planning practice has yet to occur. Much of the focus on GIS in transport has been directed at higher-level planning studies and decision support systems. This paper has argued for the use of GIS techniques in more detailed work, especially the collection and analysis of traffic data. The paper described a specific area of traffic studies methodology, where GIS has a crucial role – the integration of GIS and GPS in the collection of data in real time from probe vehicles. In this area, accurate position information is essential, and this information then allows the tagging of other data attributes concerning the characteristics of a vehicle journey (e.g., instantaneous speed and fuel consumption (and emissions)) to specific locations along the vehicle's route. The identification of zones of traffic congestion or sites of environmental sensitivity can then be made. The GIS facilitates such analysis through the integration of the probe vehicle data with other databases (e.g. land use, socio-economic characteristics, and topography) pertaining to the area under investigation. Areas for further research aimed at confirming and extending the range of application of the GIS–GPS method were also identified.

The use of the integrated GIS–GPS system was illustrated with a case study application to a major road corridor in the southern part of the Adelaide metropolitan region, which involved comparisons of travel conditions along two parallel routes in the corridor.

References

Affum, J.K., Brown, A.L., 1999. Estimating urban air pollution levels from road traffic in TRAEMS. Journal of the Eastern Asia Society for Transportation Studies 3 (1), 139–150.

Affum, J.K., Taylor, M.A.P., 1999. Integration of geographic information systems and models for transport planning and analysis. In: Meersman, H., Van Der Voorde, E. Winkelmans, W. (Eds.), World Transport Research. Selected Proceedings from the Eighth World Conference on Transport Research, vol. 3, Transport Modelling/Assessment. Elsevier–Pergamon, Oxford, pp. 295–308.

Akcelik, R., 1981. Traffic signals, capacity and timing analysis. Research Report ARR123. Australian Road Research Board, Melbourne.

Biggs, D.C., Akcelik, R., 1986. Estimation of car fuel consumption in urban traffic. Proceedings of the Australian Road Research Board 13 (7), 124–132.

Cowan, R.J., Erikson, S.C.B., 1972. On the floating vehicle problem. Proceedings of the Australian Road Research Board 6 (3), 137–143.

D'Este, G.M., Zito, R., Taylor, M.A.P., 1999. Using GPS to measure traffic systems performance. Computer-Aided Civil and Infrastructure Engineering 14, 255–265.

Govind, S., Ardekani, S.A., Kazmi, A., 1999. A PC-based decision tool for roadway incident management. Computer-Aided Civil and Infrastructure Engineering 14, 299–307.

Kenworthy, J.R., Newman, P.W.G., 1982. A driving cycle for Perth: Methodology and preliminary results. In: Proceedings of the Second International Conference on Traffic, Energy and Emissions. Society of Automative Engineers and Australian Road Research Board, Melbourne.

Klungboonkrong, P., Taylor, M.A.P., 1998. A microcomputer-based system for multicriteria environmental impacts evaluation of urban road networks. Computers, Environment and Urban Systems 22 (5), 425–446.

Liu, T.K., 1994. Field tests of travel time survey methodologies and development of a standardized data processing and reporting system. In: Proceedings of the National Traffic Data Acquisition Conference. Connecticut Department of Transportation, Hartford, CN.

O'Flaherty, C.A., Simons, F., 1970. An evaluation of the moving observer method of measuring traffic speeds and flows. Proceedings of the Australian Road Research Board 5 (3), 40–54.

Quiroga, C.A., Bullock, D., 1998a. Travel time studies with global positioning and geographic information systems: an integrated methodology. Transportation Research C 6C (1), 101–127.

Quiroga, C.A., Bullock, D., 1998b. Determination of sample sizes for travel time studies. ITE Journal on the Web, August 1998, URL:http://ww.ite.org.

Richardson, A.J., Taylor, M.A.P., 1978. Travel time variability on commuter journeys. Journal of Advanced Transportation 12 (1), 77–99.

Robertson, H.D., (Ed.), 1994. ITE Manual of Traffic Engineering Studies, fourth ed. Prentice-Hall, Englewood Cliffs, NJ.

Taylor, M.A.P., 1992. Exploring the nature of urban traffic congestion: concepts, parameters, theories and models. Proceedings of the Australian Road Research Board 16 (5), 83–106.

Taylor, M.A.P., 1999. An extended family of traffic network equilibria and its implications for land use and transport policies. In: Meersman, H., Van Der Voorde, E., Winkelmans, W. (Eds.), World Transport Research. Selected Proceedings from the Eighth World Conference on Transport Research, vol. 4, Transport Policy. Elsevier–Pergamon, Oxford, pp. 29–42.

Taylor, M.A.P., Bonsall, P.W., Young, W., 2000. Understanding Traffic Systems: Data, Analysis and Presentation, second ed. Ashgate, Aldershot.

Taylor, M.A.P., Young, T.M., 1996. Developing of a set of fuel consumption and emissions models for use in traffic network modelling. In: Lesort, J.-B. (Ed.), Transportation and Traffic Theory. Pergamon–Elsevier, Oxford, pp. 289–314.

Taylor, M.A.P., Young, W., 1988. Traffic Analysis: New Technology and New Solutions. Hargreen Publishing Company, Melbourne.

Taylor, M.A.P., Young, W., Bonsall, P.W., 1996. Understanding Traffic Systems: Data, Analysis and Presentation. Avebury Technical Books, Aldershot.

Underwood, R.T., 1968. Acceleration noise and traffic congestion. Traffic Engineering and Control 10 (3), 120–123.

Wardrop, J.G., Charlesworth, G., 1954. A method of estimating speed and flow of traffic from a moving vehicle. Proceedings of the Institution of Civil Engineers 3 (2), 154–171.

Woolley, J.E., Taylor, M.A.P., 1999. Investigations of travel time and fuel consumption characteristics of the Southern Expressway. In: Proceedings of the Third International Conference of ITS Australia. ITS Australia Inc., Canberra, CD-ROM.

Zito, R., D'Este, G.M., Taylor, M.A.P., 1995. Global positioning systems in the time domain: how useful a tool for intelligent vehicle-highway systems. Transportation Research C 3 (4), 193–209.

Zito, R., Taylor, M.A.P., 1994. The use of GPS in travel time surveys. Traffic Engineering and Control 35 (12), 685–690.

Zito, R., Taylor, M.A.P., 1996. Speed profiles and vehicle fuel consumption at LATM devices. Proceedings of the Australian Road Research Board 18 (7), 391–406.

Zito, R., Taylor, M.A.P., 1999. An approach to fuel consumption and emissions modelling of the South Australian vehicle fleet. Journal of the Eastern Asia Society for Transportation Studies 3 (1), 115–122.

PERGAMON

Transportation Research Part C 8 (2000) 287–306

TRANSPORTATION
RESEARCH
PART C

www.elsevier.com/locate/trc

Performance measures and data requirements for congestion management systems

Cesar A. Quiroga *

Texas Transportation Institute, Texas A&M University System, 3500 N.W. Loop 410, Suite 315, San Antonio, TX 78229, USA

Abstract

Many metropolitan areas have started programs to monitor the performance of their transportation network and to develop systems to measure and manage congestion. This paper presents a review of issues, procedures, and examples of application of geographic information system (GIS) technology to the development of congestion management systems (CMSs). The paper examines transportation network performance measures and discusses the benefit of using travel time as a robust, easy to understand performance measure. The paper addresses data needs and examines the use of global positioning system (GPS) technology for the collection of travel time and speed data. The paper also describes GIS platforms and sample user interfaces to process the data collected in the field, data attribute requirements and database schemas, and examples of application of GIS technology for the production of maps and tabular reports. © 2000 Elsevier Science Ltd. All rights reserved.

Keywords: Congestion management systems; Travel time; GPS; GIS; Dynamic segmentation; Performance measures

1. Introduction

Traffic congestion is a critical problem in urban areas. Several indicators confirm this trend. For example, between 1976 and 1996, the number of vehicle-miles traveled (VMT) in the United States increased by 77%, while the mileage of roads and streets increased only by 2% (FHWA, 1998). Over the years, the percentage of the peak-hour VMT that occurs under congested conditions has steadily increased, although at a slower pace in recent years. In 1996, that percentage was 54% for the urban interstate system and 45% for the urban national highway system. Congestion usually results in time delays, increased fuel consumption, pollution, stress, health hazards, and added

* Tel.: +1-210-731-9938; fax: +1-210-731-8904.
E-mail address: c-quiroga@tamu.edu (C.A. Quiroga).

PII: S0968-090X(00)00008-5

vehicle wear. The associated cost is huge. For example, in 1997 congestion cost travelers in 68 urban areas in the United States 4.3 billion hours of delay, 6.6 billion gallons of wasted fuel consumed, and $72 billion of time and fuel cost (Schrank and Lomax, 1999).

Travel delay is perceived by many as the most noticeable impact of congestion. Not surprisingly, numerous efforts have been made to eliminate it or, at least, to alleviate its effects. In the past, adding capacity was considered the main solution to eliminate or reduce travel delays. However, this approach has frequently proved to be insufficient. Faced with this reality, many urban areas have opted for implementing alternative management measures. Examples of these alternative management measures include improved traffic surveillance and control systems, dedicated high-occupancy vehicle (HOV) lanes, improved transit service, and congestion and parking pricing. The objective of these measures is to manage and reduce congestion by improving traffic flow, enhancing mobility and safety, and reducing demand for car use. Many of these management measures are the result of federal, state or local legislation. Such is the case, for example, of the measures that had to be implemented in urban areas designated as non-attainment for ozone or carbon monoxide (CAAA, 1990; ISTEA, 1991) and that continue to be eligible for federal funding under TEA 21 (1998).

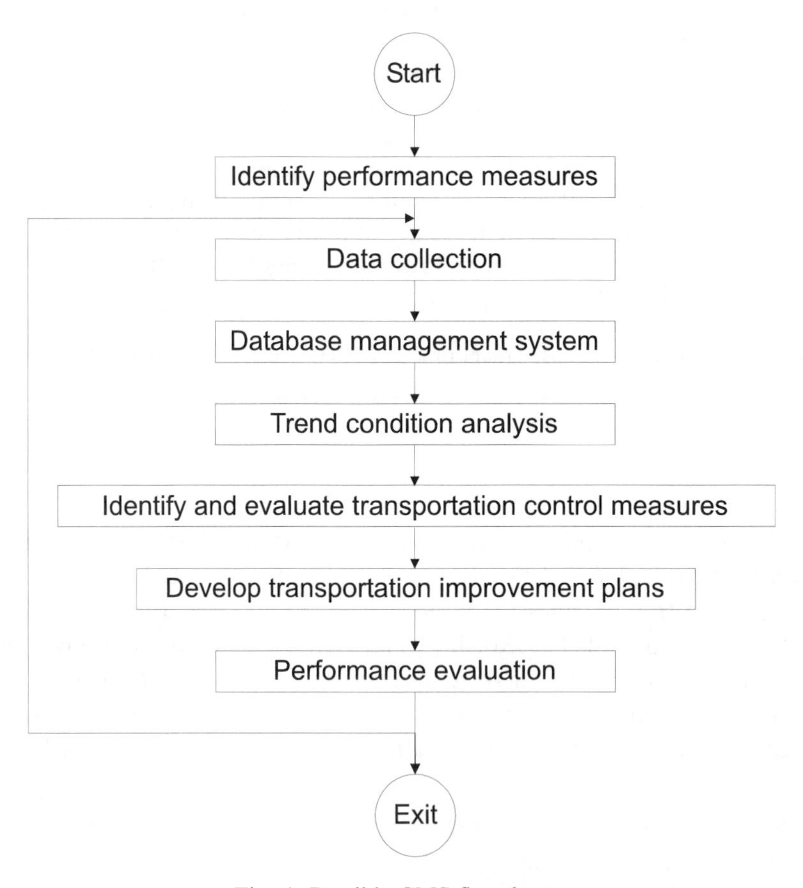

Fig. 1. Possible CMS flowchart.

A congestion management system (CMS) is intended to provide information on transportation system performance and identify alternative actions to alleviate congested roadway conditions (Lindquist, 1999). Although a CMS can use data obtained from a traffic operation surveillance and control system, in general a CMS is considered to be more a planning tool than an operations tool. This characteristic determines to a large degree the type of data collected and the types of data collection procedures implemented. In general, a CMS follows a decision making process that involves the identification of performance measures, a system-wide program of data collection and system monitoring, identification and evaluation of possible transportation control measures (TCMs), development of transportation improvement plans, and evaluation of the effectiveness of implemented strategies. Fig. 1 summarizes this process.

This paper focuses on those steps in Fig. 1 that have a strong geographic information system (GIS) component: identification of performance measures, data collection, data management, and identification and evaluation of transportation control measures. However, rather than focusing on GIS per se, this paper attempts to provide a review of technical and non-technical issues that influence the development of CMSs from a transportation engineering perspective and describes the role that GIS techniques can play in that development. It summarizes recent work in the area of travel time-based performance measures and describes the results of a recent project that involved the use of global positioning system (GPS) technology for the collection of travel time and speed data.

2. Performance measures

Numerous performance measures have been suggested or used for quantifying congestion (Francois and Willis, 1995; Schwartz et al., 1995; Lomax et al., 1997). Following Lomax et al., (1997), congestion measures can be grouped into highway capacity manual (HCM) measures, queuing-related measures, and travel time-based measures. Three commonly used HCM measures are volume to capacity (V/C) ratio, average intersection delay, and level of service (LOS). V/C ratios are frequently used because of the relative ease of traffic volume data collection and because surrogate measures such as LOS can be derived from V/C values. Average intersection delays are normally used on arterial streets and also provide a foundation for surrogate measures such as LOS. LOS measures normally range from "A" – best service to "F" – worst service. While conceptually simple and easy to understand by the professional transportation community, HCM performance measures tend to be somewhat abstract for the traveling public. They usually require detailed, location-specific input data, which makes them more appropriate for localized analyses and design than for area-wide planning. Finally, HCM measures are difficult to use for long-range comparisons because concepts such as capacity and speed-flow rate relationships tend to change over time (HRB, 1950; HRB, 1965; TRB, 1985; TRB, 1994; TRB, 1997).

Two commonly used queuing-related measures are queue length and lane occupancy. Queue length and duration can be determined by direct observation. Lane occupancy (or percentage of time a traffic lane is occupied by traffic) can be measured from vehicle detectors that are part of roadway surveillance and control systems. Queuing-related measures are increasingly being used to quantify roadway congestion because of the increasing availability of vehicle detectors and other sensors. However, although queues best reflect the public's perception of congestion, measuring queues remain laborious, site-specific, and time-specific. Because it is usually impractical to measure queues on a broader spatial scale, queuing-related measures tend to be inappropriate for planning and policy-related analyses.

Three commonly used travel time-based measures are travel time, travel speed, and delay. Travel time data collection is an integral component of traffic engineering studies. Travel speed and delay data can be derived from travel time data by using a reference desired/acceptable travel time or speed. Travel time-based measures are easy to understand by both the professional transportation community and the traveling public. They are flexible enough to describe traffic conditions at various levels of resolution in both space and time. This makes travel time-based measures appropriate for handling specific locations as well as entire corridors. It also allows analysts to perform comparisons over long periods of time, e.g., years or decades. Travel time-based measures translate easily into other measures like user costs, and can be used directly to validate planning models such as travel demand forecasting models (Laird, 1996). Travel time-based measures are applicable across modes. So important is travel time in this regard that the year 2000 edition of the HCM is being structured around travel time as a common measure of effectiveness for all modes (JHK, 1996). All these reasons make travel time-based measures extremely powerful, versatile, and desirable. Not surprisingly, an increasing number of transportation agencies are switching to travel time measures to monitor and manage congestion.

Unfortunately, budgetary limitations usually impose severe restrictions on the number and coverage of travel time studies. Recent technological advances, however, are assisting transportation officials in providing the necessary tools to make travel time data collection more affordable and reliable. Examples of those technological advances include GPS, GIS and a variety of subsystems and components normally associated with intelligent transportation system (ITS) deployments.

A number of travel time-based measures could be used (Lomax et al., 1997; Quiroga and Bullock, 1999). A sample of measures follows.

2.1. Travel time

Travel time (t_L) is the total time required to traverse a roadway segment of length L. It can be measured directly using field studies, although it could also be derived using simulation models or empirical relationships between volume and roadway characteristics.

2.2. Acceptable travel time

Acceptable travel time (t_{L_0}) is the travel time associated with a performance goal established for the transportation facility. The acceptable travel time should be influenced by community input and should, explicitly or implicitly, provide a balance between transportation quality, economic activity, land use patterns, environmental issues, and political concerns. In the absence of a more detailed analysis, a number of transportation agencies define the acceptable travel time as that associated with free flow conditions (or sometimes posted speed limits). A more detailed analysis should provide a differentiation by time period (i.e., peak vs off-peak), by functional class (e.g., freeway vs major arterial), and by geographic location (e.g., central business district vs suburban area) (Lomax et al, 1997). For example, based on a consensus involving technical and non-technical groups, an acceptable travel time per km (or travel rate, as defined below) during peak periods for a major arterial located in the central business could be, say, 3 min. For the same arterial, an acceptable travel time per km during off-peak periods could be, say, 2 min.

2.3. Segment speed

Segment speed (u) is the result of dividing the length (L) of the segment along which travel time data are collected by the corresponding travel time t_L:

$$u = L/t_L. \tag{1}$$

Length is an established item in a roadway inventory database and is normally measured in the field with a distance measuring instrument (DMI). GIS packages can also be used to provide estimates of distance, but it is clear that the accuracy of these estimates depends on the accuracy of the underlying digital base map. Digital base map accuracy has dramatically increased in recent years (submeter accuracy is now commonplace) and, as a result, measuring distances with GIS packages has become a feasible alternative. Readers should realize, however, that GIS packages usually measure distances on the ellipsoid, i.e., they "simulate" ground distances by using a mathematical model of the surface of the Earth. Some GIS packages take into consideration the eccentricity of the ellipsoid (i.e., explicitly account for the fact the Earth is not a perfect sphere), while other GIS packages simply assume a spherical model of the surface of the Earth. GIS packages that do consider the eccentricity effect provide more accurate estimates of distance than GIS packages that assume a simple spherical model of the surface of the Earth. Differences between the two models can be quite significant. For example, at 30° of latitude, differences between the two models could be up to 3 m/km on the N–S direction and up to 2 m/km on the E–W direction (Quiroga, 1999).

In general, travel time studies involve several runs and more than one segment. In this case, it may also be of interest to compute representative speeds and travel times for each segment and/or for all segments combined. In general, as shown in Fig. 2, if there are interchanges or intersections along the route, the number of runs per segment may be different.

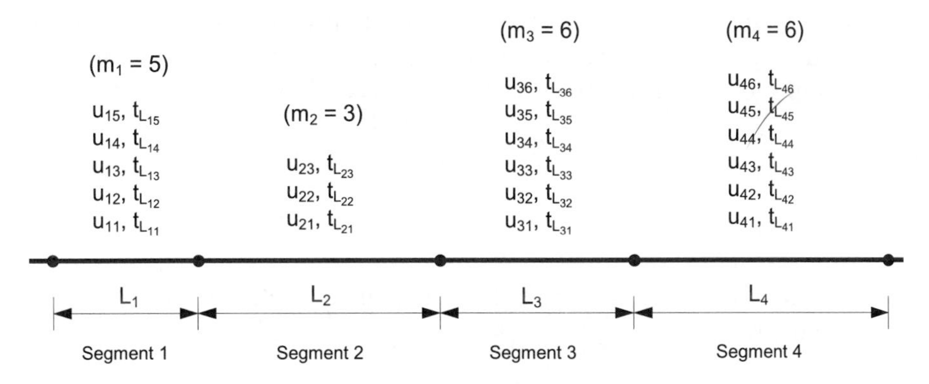

Fig. 2. Sample segment speeds and travel times.

Assume a representative segment travel time is given by the arithmetic average of all travel time values associated with a segment. Following Quiroga and Bullock (1999), the total representative travel time and speed over all segments can be expressed as

$$t_{T_L} = \sum_{i=1}^{n} \bar{t}_i = \sum_{i=1}^{n} \left[L_i \frac{1}{m_i} \sum_{j=1}^{m_i} \frac{1}{u_{ij}} \right], \tag{2}$$

$$\bar{u}_L = \frac{L_T}{t_{T_L}} = \frac{\sum_{i=1}^{n} L_i}{\sum_{i=1}^{n} \bar{t}_i} = \frac{1}{\sum_{i=1}^{n} \left[\frac{L_i}{L_T} \frac{1}{m_i} \sum_{j=1}^{m_i} \frac{1}{u_{ij}} \right]}, \tag{3}$$

where t_{T_L} is the total representative travel time over all segments, n the number of contiguous segments, $\bar{t}_i$ the average (arithmetic mean) of all travel time values associated with segment i, L_i the length of segment i, m_i the number of travel time and speed records per segment, u_{ij} the jth speed record associated with segment i, L_T the total length considered, and $\bar{u}_L$ is the overall average speed.

Eq. (3) represents a "weighted" harmonic mean of segment speeds, where the weight is the ratio of the length of each segment to the total length considered. One disadvantage of this equation is its sensitivity to outlying low speeds (which tend to occur on atypically adverse traffic conditions), resulting sometimes in very small average speeds. A more solid, robust estimator of central tendency can be obtained by using median segment travel times instead of arithmetic mean segment travel times. The median speed formulation is

$$\bar{u}_L = \frac{L_T}{\sum_{i=1}^{n} t_{m_i}} = \frac{1}{\sum_{i=1}^{n} \left[\frac{L_i}{L_T} \frac{1}{u_{m_i}} \right]}, \tag{4}$$

where t_{m_i} is the median travel time associated with segment i and, u_{m_i} is the median speed associated with segment i.

2.4. Travel rate

Travel rate (t_r) is the inverse of average speed and is usually expressed in min/km (or min/mile). While not readily understood by all audiences, travel rate provides a useful measure that can be averaged for a facility, geographic area, or mode. It can also be used to compare performance among transportation facilities more effectively than speed. Travel rate can be expressed as

$$t_r = t_L / L. \tag{5}$$

2.5. Delay

Delay (d_L) is the difference between travel time and the acceptable travel time on a road segment. Delay can be expressed as

$$d_L = t_L - t_{L_0}. \tag{6}$$

2.6. Total delay

Total delay ($D_{L_{\rm tp}}$) is the sum of delays for all vehicles traversing the segment during the time period for which travel time data are available and is normally expressed in vehicle-minutes or vehicle-hours. Total delay can be expressed as

$$D_{L_{\rm tp}} = V_{\rm tp} d_{L_{\rm tp}}, \tag{7}$$

where tp is the subscript associated with the time period for which data are available (e.g., 15 min), $V_{\rm tp}$ is the number of vehicles traversing the segment during the time period for which travel time data are available. Traffic volumes are established items in a roadway inventory database.

2.7. Delay rate

Delay rate ($d_{r_{\rm L}}$) is the rate of time loss for a specified roadway segment. It is calculated as the difference between the travel rate and the acceptable travel rate. Delay rate can be expressed as

$$d_{r_{\rm L}} = \frac{t_L - t_{L_0}}{L} = \frac{d_L}{L}. \tag{8}$$

2.8. Relative delay rate

Relative delay rate ($d_{r_{\rm R}}$) is a dimensionless index that can be used to compare congestion on facilities, modes, or systems in relation to different mobility standards. It is calculated as the delay rate divided by the acceptable travel rate. Relative delay rate can be expressed as

$$d_{r_{\rm R}} = \frac{t_L - t_{L_0}}{t_{L_0}} = \frac{d_L}{t_{L_0}}. \tag{9}$$

3. Data collection

There are essentially two groups of travel time data collection techniques: roadside techniques and vehicle techniques. Roadside techniques are based on the use of detecting devices physically located along the study routes at pre-specified intervals. They obtain travel time data from vehicles traversing the route by recording passing times at predefined checkpoints. Examples of these techniques include license plate matching and automatic vehicle identification (AVI). License plate matching is based on recording of the license plate number of individual vehicles and the corresponding time stamps as they pass checkpoints. Travel times are determined as differences in time stamps between checkpoints. An assumption of this technique is that each individual vehicle does not make intermediate stops. This may be limiting, particularly if there are intersections, on-ramps, off-ramps, or interchanges between checkpoints.

AVI is an example of a data collection technique included as part of traffic surveillance and control system deployments at traffic management centers. AVI systems are based on the used of in-vehicle transponders (or tags), roadside reading units, a communication network, and a central computer system. The roadside reading units detect individual vehicles equipped with transponders as they pass nearby and transmit the corresponding transponder data to the central computer system. Travel times between consecutive checkpoints are computed in a similar manner as with the license plate technique, except that transponder identification numbers are used to compare time stamps instead of vehicle license plate numbers. One advantage of AVI technology is that area-wide real-time travel time data collection and dissemination are possible. With GIS-based Internet tools, for example, cities like Houston, Chicago, and Seattle are using AVI technology to disseminate up-to-date geo-referenced travel time and speed data to the traveling public.

Vehicle techniques are based on the use of detection devices carried inside the vehicle. Examples of these techniques include the traditional stopwatch and clipboard technique and automatic vehicle location (AVL). In the stopwatch and clipboard technique, travel time and passage of specific landmarks are manually recorded along the route. Two technicians are required in the vehicle: one of them to drive and the other one to manually record items such as the location and time of individual checkpoints and the length and time spent in queues. Unfortunately, this process tends to be labor intensive during the data collection and data reduction phases, and spatial resolution and coverage are limited. In addition, problems such as missing checkpoints or inaccurately marked checkpoints are common. To avoid some of these problems, many transportation agencies use distance measuring instruments (DMIs) in their probe vehicles. With a DMI, only one technician is needed in the vehicle. In some cases, it is even possible to log route and checkpoint locations. However, DMIs require frequent calibrations to avoid inaccurate speed and distance readings (Benz and Ogden, 1996). In addition, DMI data (which by definition are linearly referenced) are not always compatible with geographic databases because of the difficulty to ensure those critical checkpoints on the survey, mainly the beginning and ending points, have been properly geo-referenced in the field.

AVL is a generic term that groups several techniques that use receivers or transmitters on board to determine vehicle location (in latitude and longitude) and speed. Examples of these techniques are ground-based radio navigational system techniques and GPS techniques. GPS techniques are particularly advantageous because they do not need receiving towers on the ground as traditional radio navigational systems do. Several GPS-based techniques have been developed in recent years (Guo and Poling, 1995; Laird, 1996; Quiroga and Bullock, 1996; Quiroga and Bullock, 1999; Zito et al., 1995) and the number of implementations in urban areas is constantly increasing.

One of the significant advantages of AVL techniques compared to other techniques is that traffic monitoring is roadway network and driver independent. This makes AVL suitable for many applications, including tracking the motion of special-purpose probe vehicles and entire fleets. When used with single probe vehicles, AVL systems are usually configured so that data are collected and stored on board, and then post-processed in the office. When used with entire fleets, AVL systems are usually configured so that data are collected and transmitted via radio or cellular phone to a central location where they can be automatically processed.

Table 1 is a summary of characteristics and applicability of the travel time data collection techniques described previously. Roadside techniques are obviously infrastructure dependent, as opposed to vehicle techniques. Roadside techniques have lower levels of resolution and accuracy than vehicle techniques. However, vehicle techniques are generally based on a limited number of probe vehicles, which means that area wide coverage is limited. This makes roadside technique (specifically AVI) better suited for daily or real-time monitoring. In contrast, vehicle techniques are best for determining initial conditions and for annual monitoring.

Table 1
Comparison of travel time data collection techniques (adapted from Liu and Haines, 1996; Turner, 1996)

Criteria	Roadside techniques		Vehicle techniques	
	License plate matching	AVI	DMI	AVL
Characteristics				
Infrastructure dependent	Yes	Yes	No	No
Travel time/speed resolution	Low	Low	High	High
Travel time/speed accuracy	Good	Good	Good	Very good
Area wide coverage	Low	Very good	Low	Low
Technology status	Proven	Proven	Proven	Proven[a]
Capital costs	Low	High	Low	Low to moderate
Operating costs per unit	Moderate	Low	High	Low to moderate
Applicability				
Annual monitoring	Yes	Yes	Yes	Yes
Daily monitoring	Limited	Yes	Limited	Limited
Real-time travel information	Limited	Yes	No	Yes
Incident detection	Limited	Yes	Limited	Limited

[a] GPS is a proven technology. However, its applicability to travel time studies has been limited until recently.

4. Data management

Regardless of the data collection technique used to collect travel time data, the data management component of a CMS is critical. In general, that component should be built using a geographic relational database model and provide all the necessary interfaces and procedures to provide the capability to measure congestion accurately, reliably, and efficiently. Obviously, the structure of the data management component depends on the data collection techniques used and the performance measures chosen.

As an illustration, this section summarizes the data management component of a travel time data collection application developed recently in support of a congestion management system for Baton Rouge, Louisiana (Quiroga and Bullock, 1998; Quiroga and Bullock, 1999). This application, called travel time with GPS (TTG), is based on the use of GPS receivers to collect travel time data and GIS-based procedures to manage the data collected in the field. TTG includes a spatial model, a geographic relational database, a procedure for linearly referencing GPS data using dynamic segmentation tools, and data reporting procedures. TTG was used to process 2.4 million GPS records on 40,000 km (25,000 miles) of travel time runs on a 240-km (151 miles) highway network.

4.1. Spatial model

The spatial model is based on highway links, where highway links are defined as directional centerlines delimited by physical discontinuities like signalized intersections, ramps, and interchanges (Fig. 3). Fig. 3 also shows sample GPS data being mapped to highway links. Linearly referencing these GPS points involves computing cumulative linear distances for GPS points located along the route of interest.

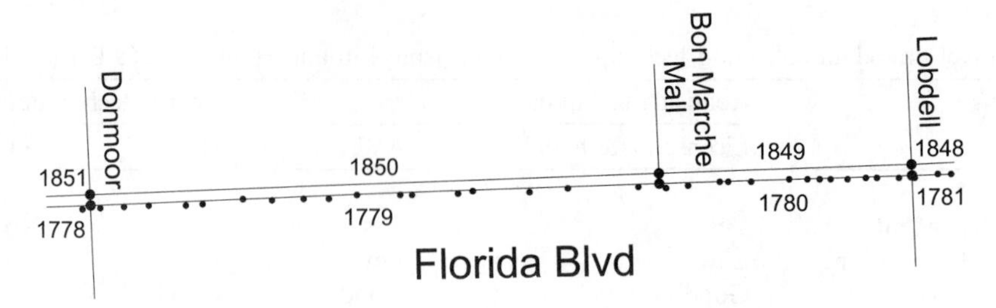

Fig. 3. Highway links on Florida Boulevard in Baton Rouge, Louisiana, with overlaid GPS data collected during a travel time run.

Fig. 4 shows a generic time–distance diagram for a probe vehicle that traverses a segment of length L (Note: a segment does not necessarily have to be same spatial entity as a link). The segment travel time, t_L, could be estimated by interpolating the time stamps of the two GPS points located immediately before and after the segment entrance, and the time stamps of the two GPS points located immediately before and after the segment exit. If instantaneous speed values are recorded along with the GPS positional data, a representative speed value for the segment could

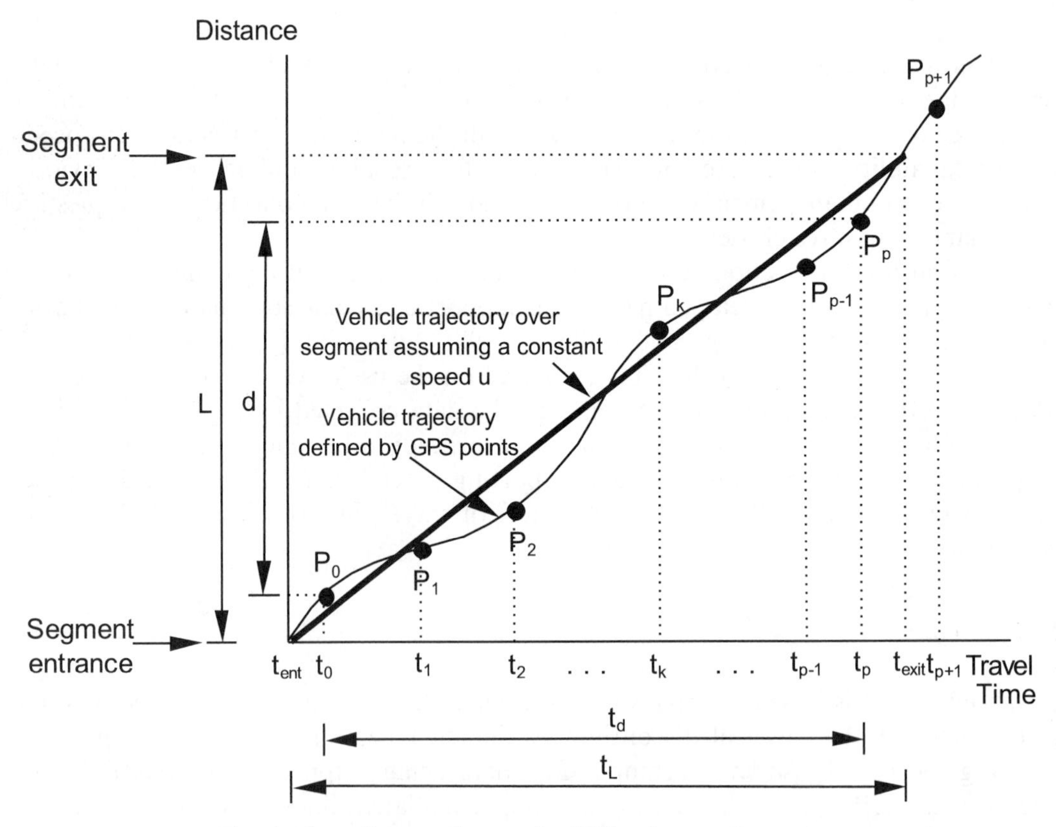

Fig. 4. Time–distance diagram for GPS points on a segment.

be computed. In this case, the segment travel time, t_L, could be estimated by dividing the segment length by the representative segment speed value.

4.2. Geographic database

Each GPS data file contains data such as time stamps, speed, latitude, longitude, and satellite navigational data at regular time intervals, say every 1 s. These data need to be linearly referenced so that GPS data can be associated with routes on the highway network. To manage all this information efficiently, a geographic relational database was developed. For illustration purposes, Fig. 5 shows the database schema (or database structure) and includes tables, field names and relationships (both one-to-one and one-to-many) among tables. To assist readers in the process of understanding the database structure, Fig. 6 shows a few sample records of each table included in the database.

The database structure assumes that a link code is explicitly associated with each GPS point (LinkCode attribute in table GPS_DATA). This link code results from the linear referencing process and is the same as that associated with links in the highway network map (e.g., Link code = 1779 in Fig. 3 and table GPS_DATA of Figs. 5 and 6). This way, users can easily build queries based on the same segmentation scheme as that used for generating the highway network map. Strictly speaking, however, all that is required from the linear referencing process is a route

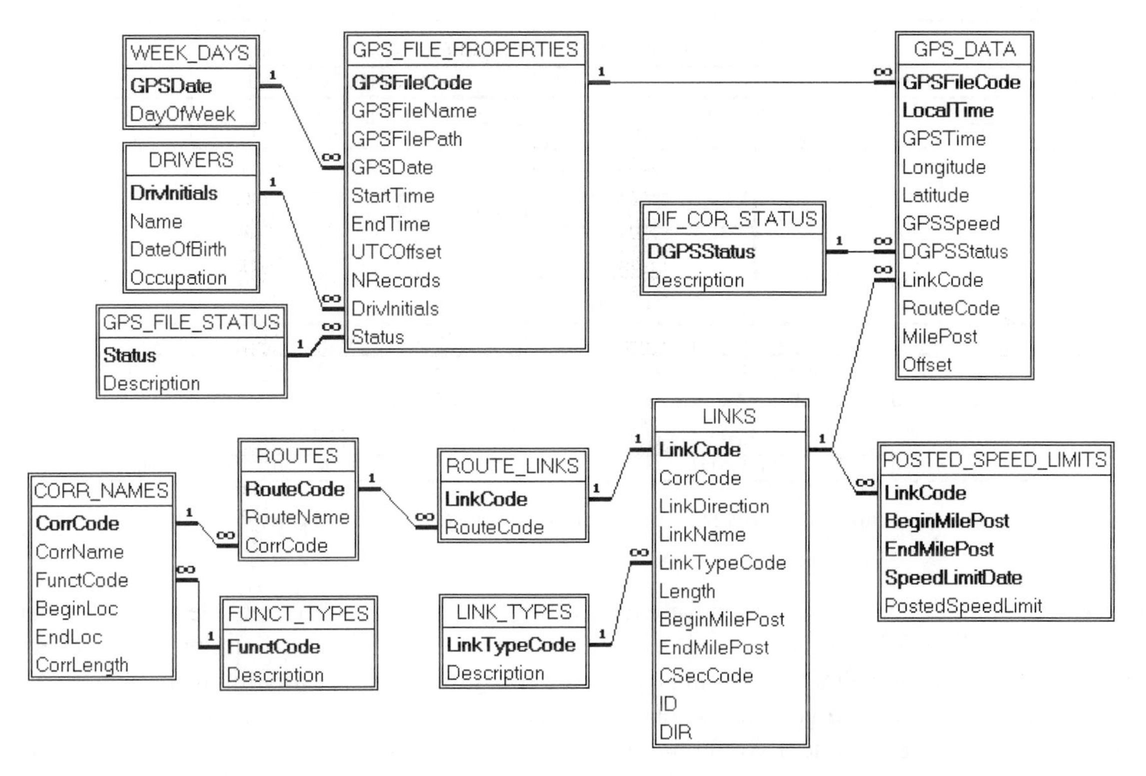

Fig. 5. Geographic database schema showing tables and relationships among tables (primary keys are indicated in bold).

Table CORR_NAMES

CorrCode	CorrName	FunctCode	BeginLoc	EndLoc	CorrLength
6	Airline Hwy	2	La 1145	2 mi east of Highland Rd	22.87
7	Florida Blvd	2	River Front	Range Ave	14.69

Table FUNCT_TYPES

Funct Code	Description
1	Interstate
2	Principal arterial

Table ROUTES

RouteCode	RouteName	CorrCode
07FLBDE	Florida Blvd EB	7
07FLBDW	Florida Blvd WB	7

Table ROUTE_LINKS

LinkCode	RouteCode
1779	07FLBDE
1849	07FLBDW

Table LINKS

LinkCode	CorrCode	LinkDirection	LinkName	LinkTypeCode	Length (mi)	BeginMilePost	EndMilePost	CSecCode	ID	DIR
1779	7	EB	Florida Blvd	1	0.2613	4.2880	4.5493	13-04	28035	1
1780	7	EB	Florida Blvd	1	0.1153	4.5493	4.6645	13-04	28063	1

Table LINK_TYPES

LinkTypeCode	Description
1	main
2	interchange

Table POSTED_SPEED_LIMITS

LinkCode	**BeginMilePost**	**EndMilePost**	**SpeedLimitDate**	PostedSpeedLimit (mph)
1779	4.2880	4.5493	07/31/97	50
1780	4.5493	4.6645	07/31/97	50

Table WEEK_DAYS

GPSDate	DayOfWeek
10/19/95	TH
10/20/95	FR
10/21/95	SA
10/22/95	SU
10/23/95	MO

Table DRIVERS

DrivInitials	Name	DateOfBirth	Occupation
MS	Driver No. 1	01/01/73	LSU Student
BP	Driver No. 2	01/01/73	LSU Student

Table GPS_FILE_STATUS

Status	Description
0	GPS file entry generated
1	Point geographic file generated
2	MI/MO operation completed
3	Linear referencing completed
4	GPS data imported into Access

Table GPS_FILE_PROPERTIES

GPSFileCode	GPSFileName	GPSFilePath	GPSDate	StartTime	EndTime	UTCOffset	NRecords	DrivInitials	Status
178	10191129.txt	G:ttg\gpsdata\95fall\10191129	10/19/95	41,695.75	49,111.75	-5	6140	MS	4
179	10192210.txt	G:ttg\gpsdata\95fall\10192210	10/19/95	80,184.25	82,286.25	-5	1,719	BP	4

Table DIF_COR_STATUS

DGPSStatus	Description
2	2-D differential correction
3	3-D differential correction

Table GPS_DATA

GPSFileCode	**LocalTime**	GPSTime	Longitude	Latitude	GPSSpeed	DGPSStatus	LinkCode	RouteCode	MilePost	Offset
178	27,173.75	45,173.75	-91.1183809	30.4514496	44.2	3	1779	07FLDBE	4.2910	8.0
178	27,174.75	45,174.75	-91.1181741	30.4514556	44.4	3	1779	07FLDBE	4.3033	8.8
:	:	:	:	:	:	:	:	:	:	:
178	27,190.75	45,190.75	-91.1147336	30.4515751	41.6	3	1779	07FLDBE	4.5084	9.5
178	27,193.75	45,193.75	30.4515929	-91.1141925	37.1	3	1779	07FLDBE	4.5407	9.1

Fig. 6. Sample of records from the database (primary keys are indicated in bold).

code and a cumulative distance value for each GPS point (attributes RouteCode and MilePost in table GPS_DATA). With this information and any table containing cumulative distances associated with links or segments along routes, generic GPS data tables for any highway segmentation scheme can be produced to generate segment aggregated travel time and speed data.

TransCAD was used to linearly reference GPS data and to display results in a map format. Like other desktop GIS packages, TransCAD's architecture is based on a fairly large number of associated files. For example, a geographic file can easily involve 10–20 associated files including graphical elements, tables, indexes, and data dictionaries. Most tables are in dBase IV format (.dbf extension), which means that each table is stored in a separate dBase file. This kind of architecture is intended to provide flexibility to typical TransCAD users. However, it can also complicate data management problems in an environment where tens or hundreds of GPS data files are being generated and processed. With GPS data scattered in several dBase files, the process of enforcing data integrity constraints, building queries, and producing reports that involve aggregating or summarizing travel time and speed data by, say, time period or corridor, could be quite challenging. To address this issue, all database tables were stored in a single access file. Tables LINKS, ROUTE_LINKS, and GPS_DATA contain records imported from TransCAD files. Table LINKS contains the same records as file links.dbf, which is a file generated by TransCAD for viewing attribute data associated with each link in the highway network. Table

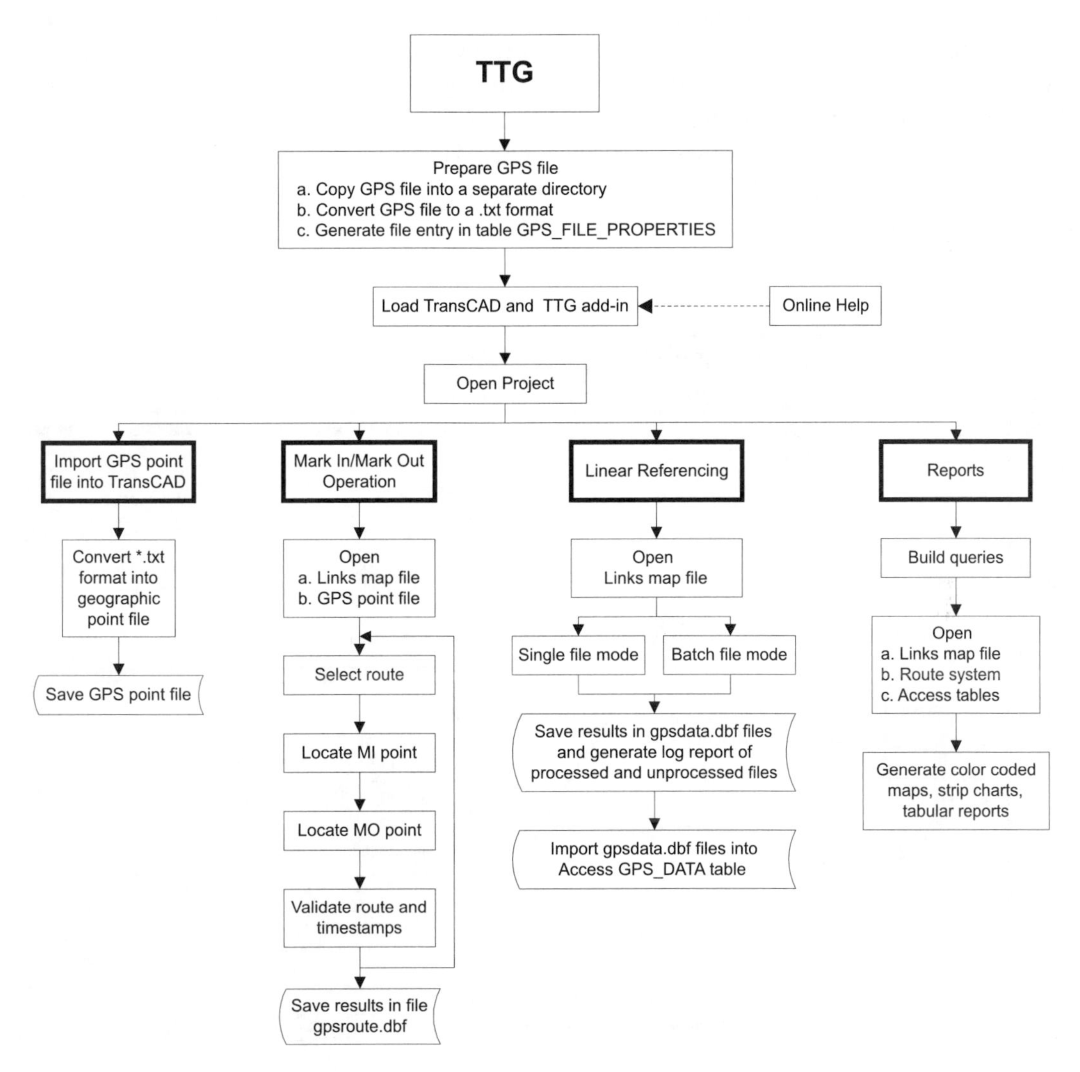

Fig. 7. Data reduction work flow.

ROUTE_LINKS is the result of joining file links.dbf and all route link.dbf files in TransCAD. Table GPS_DATA contains linearly referenced GPS data that results from the linear referencing process in TransCAD.

4.3. Data reduction procedure

Fig. 7 shows a generic view of a typical work flow using TTG. For completeness, Fig. 7 shows both data reduction steps and data reporting steps. In this section, only the data reduction steps are discussed. In summary, the data reduction procedure allows users to

- Import GPS data file into the GIS to generate a point geographic file that can be overlaid on the highway network vector map;
- Specify when and where the vehicle enters and exits a study route using an animated GPS playback utility (Fig. 8). Of particular interest in Fig. 8 are the mark-in (MI) and mark-out (MO) buttons. The MI button is used to specify when the vehicle enters a route by marking the first GPS point associated with that route. Similarly, the MO button is used to specify when the vehicle exits a route by marking the last GPS point associated with that route. For added flexibility, the GPS player utility allows users to define MI–MO pairs anywhere along a route and define more than one MI–MO pair per route. This technique is useful for filtering out spurious GPS points and for partial route analysis;

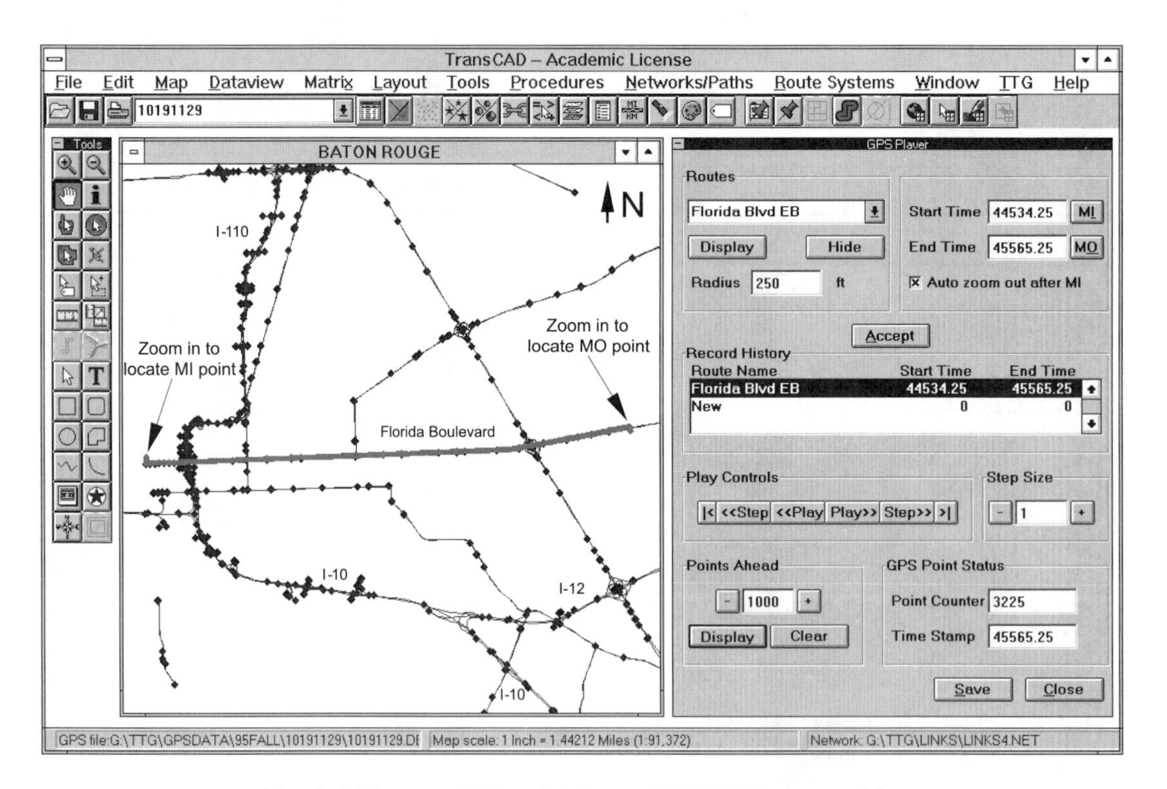

Fig. 8. Links map, GPS point file, and TTG GPS player utility.

- Linearly reference GPS points to the highway network. TTG also measures the transversal distance from each GPS point to the mapped link on the network (Offset field in table GPS_DATA of Fig. 6). This offset provides a verification of GPS positional accuracy and allows users to flag GPS points that may have been referenced to incorrect routes;
- Import the linearly referenced GPS data into a repository database (Microsoft Access).

4.4. Data reporting procedure

After storing the linearly referenced GPS data in a repository database, the next step involves constructing queries and reports. For the sake of brevity, only a sample of databases querying and data reporting options are included here. For additional data reporting examples, readers are referred to other sources (Quiroga, 1997; Quiroga and Bullock, 1998; Turner et al., 1998). Suppose it is of interest to produce reports showing average link speeds on a corridor. The procedure to do this can be summarized as follows:

- Build a query to retrieve the GPS records associated with the time period of interest (e.g., 7:00–7:15 am).
- Calculate link travel time and speed for each travel time run conducted during the time period of interest. Because relational database and GIS packages do not have tools to readily perform numerical interpolations, a special purpose utility was developed. This utility automatically calculates link speeds based on a table such as GPS_DATA and outputs the results to a table called SEGMENT_SPEEDS_x (where x represents the number associated with the procedure to calculate link or segment speed chosen). For example, by selecting a time interpolation procedure (Fig. 4), the output table would be called SEGMENT_SPEEDS_1. The utility provides users with the capability to compute segment speeds and travel times for any highway segmentation scheme.

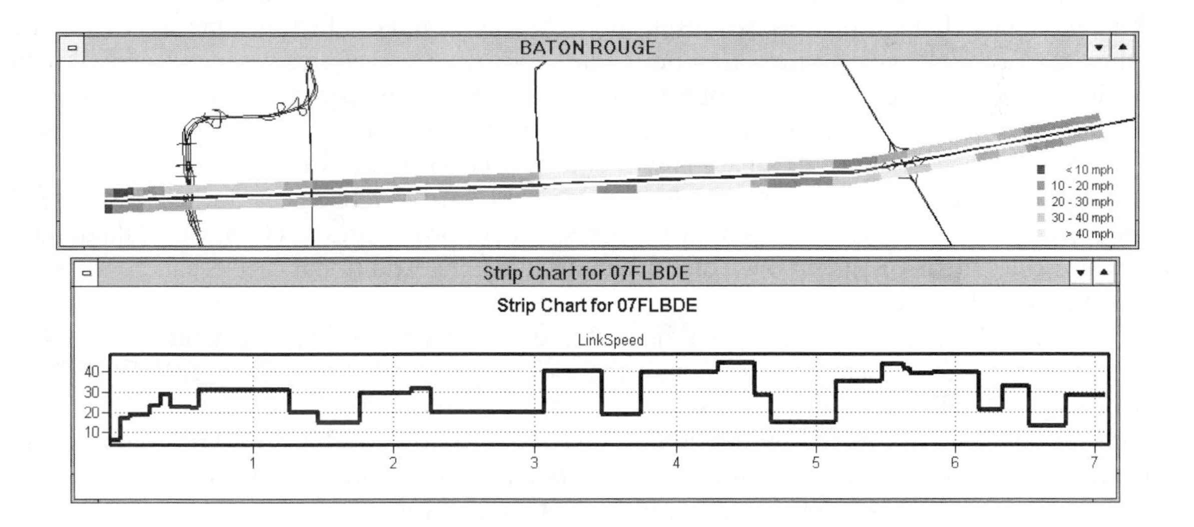

Fig. 9. Link speeds on Florida Boulevard in Baton Rouge, Louisiana.

- Calculate representative segment speeds (Eqs. (2)–(4)). Off-the-shelf database functions give users the capability to compute minimum, average, and maximum speeds, but not median speeds. As a result, a utility to compute median speeds was developed.
- Produce reports documenting average speeds on the corridor of interest. Examples of reports include maps and strip charts (Fig. 9) and tabular reports.

4.5. Discussion of results

TTG was tested using GPS data from 428 files collected in Baton Rouge, Louisiana. The 428 files included 2.4 million GPS records on 40,000 km (25,000 miles) of travel time runs on a 240-km (151 miles) highway network. Of the 2.4 million GPS records collected, 1.8 million GPS records were located on the main routes. The remaining 0.6 million GPS records were located on the other parts of the network including service roads, on-ramps, off-ramps, and intersecting streets. The 1.8 million GPS records located on the main routes were stored in table GPS_DATA in the access database file. The resulting size of this database file, including the other tables shown in Figs. 5 and 6, was 238 Mbytes. This number translates to approximately 132 bytes per linearly referenced GPS record.

Processing and storing 1.8 million linearly referenced GPS records in a relational database provided the capability to improve and optimize the data reduction application, develop generalized data quality control checks, and detect limitations and bugs of the developing platform. In general, the centralized database approach requires some extra effort on the part of users when setting up projects and at the beginning and end of the data reduction process (mainly to generate entries for the GPS data file in Access and to import and append gpsdata.dbf files to Access). However, it appears the extra effort is worth the benefits the system provides, particularly with respect to the ability to build comprehensive queries to generate travel time reports, the ability to conduct generalized data quality control checks, and the ability to process data much faster than with manual data collection methods. An analysis of data reduction speeds indicates that the automated data reduction process can result in 15–17 min of data reduction time for two hours worth of GPS data. By comparison, traditional data reduction procedures based on manual data collection procedures require about one hour of data reduction per hour of data collection (Turner et al., 1998). In other words, the GPS/GIS data reduction procedure described here is about 7.5 times faster than traditional manual data collection procedures.

During the processing of the GPS data files, a number of areas were detected where errors tend to occur frequently. To assist readers in this process, a set of procedures or checks for data quality control were developed. Some of these checks are listed below. Additional checks can be found in Quiroga and Bullock (1998).

- *File system.* Make sure each GPS data file is stored in a separate subdirectory under the GPSDATA subdirectory. Likewise, make sure the file entry in table GPS_FILE_PROPERTIES, particularly the file name and path is correct. The need to check for the location of the GPS data file could be eliminated by developing a script to automatically store GPS data files in the appropriate subdirectories as they are being uploaded from the data collection equipment. The script would also generate an entry in the database automatically.
- *MI/MO operation.* Verify that route assignments and beginning and ending time stamps are correct. An effective way of doing this is by checking the contents of the output file from the

data reduction process. As an aid to users, this file is automatically displayed at the end of every MI/MO session.

- *Route system files*. Before beginning with the formal linear referencing process, verify that all route system files are correct, i.e., that routes contain only valid links and that the beginning and ending mileposts of individual links are correct. TransCAD automatically calculate mileposts, however, if links must be edited, the GIS does not always recalculate distances correctly.
- *Linear referencing*. Verify that link codes, mileposts, and offsets are correct and meaningful. Offsets are transverse distances from the GPS points to the highway links and provide an indirect measurement of either GPS data positional errors (if the underlying highway network base map is more accurate than the GPS data) or GPS data mapping errors (unusually high offsets could be an indication that the route assignment was incorrect). Table GPS_DATA can be used to check for large offset values.
- *Average link/segment speeds*. Verify that representative link/segment speeds are meaningful. TTG includes two formulations for the computation of representative link/segment speeds: using a harmonic mean formulation (Eq. (3)), and using a median formulation (Eq. (4)). As discussed previously, harmonic mean speeds are based on arithmetic mean travel times. However, harmonic mean speeds are very sensitive to low outlying speeds which to occur under atypically adverse traffic conditions. By comparison, median speeds are not sensitive to outliers and, therefore, they tend to provide more robust estimates of central tendency than harmonic mean speeds.

5. Conclusions

This paper presented a summary of procedures, collectively called travel time with GPS (TTG), and examples of application of GPS and GIS technologies for the collection of travel time data needed for monitoring and managing congestion. The paper examined transportation network performance measures and discussed the benefit of using travel time as a robust, easy to understand performance measure. The paper addressed data needs and examined the use of new technologies such as global positioning system (GPS) technology for the collection of travel time and speed data.

TTG is built using a general data model that includes a spatial model, a geographic relational database, and a procedure for linearly referencing GPS data. The spatial model uses a GPS-based directional vector representation of the network. In this vector representation of the network, routes are partitioned into links and links are assigned unique identification numbers. The geographic relational database is composed of a series of tables that store information about links, routes, posted speed limits, GPS file descriptors, and linearly referenced data. The procedure for linearly reference GPS data uses GIS dynamic segmentation tools. To automate this process, an application that allows users to determine when a vehicle enters and exits a route and to automatically calculate mileposts for all GPS points along the routes of interest was developed.

TTG was implemented using a PC-based TransCAD-Access environment. This environment is relatively inexpensive and allows users to process vast amounts of GPS data and produce reports quickly and cost-effectively. This environment works well in most cases although it was found that the software had some deficiencies that could produce erroneous results if care is not taken in processing the data.

Following an analysis of the travel time and speed data and an evaluation of the congestion situation, the next step would to identify and evaluate potential TCMs. If appropriate and properly implemented, TCMs could result in reductions of congestion levels. A sample of TCMs is included in Table 2. Notice that some of the TCMs can be evaluated using GIS techniques, particularly those that involve spatial analyses such as accessibility analysis, routing analysis, and demand/market program analysis. As an illustration, consider the case of defining appropriate locations for park-and-ride lots. Each park-and-ride location has an associated cost to locate, build, and operate. In addition, each location is expected to serve a number of drivers. For each driver using the park-and-ride lot, there is an associated cost of service, e.g., the travel time between the driver's home and the park-and-ride location. A goal for the park-and-ride location could be to minimize the cost of service for all drivers. Cost of service values is stored in cost matrices. In a cost matrix, each row represents the location of each alternative park-and-ride lot and each column represents a driver. Each cell, therefore, represents the cost of service for a single location-driver combination. For the GIS analysis, the following layers of data are needed: a network map layer with travel time and/or speed data, a park-and-ride location layer, and a driver location layer. Using GIS routing and dynamic segmentation functions, it is possible to construct the corresponding cost matrix. The minimum total cost of service could be obtained by adding the cost of service for all drivers who are expected to use each park-and-ride location and by comparing the total cost of service among locations. The locations with the lowest total cost of service are then retained for further analysis.

Another example could be that of development of para-transit operation improvements. The objective would be to service all origins and destinations based on a specified fleet of vehicles, while minimizing the total travel time for their customers. Each vehicle has a fixed capacity and each destination has a demand given by the number of passengers that must be transported there. This is a typical routing problem that requires the construction of a vehicle routing matrix. The vehicle routing matrix contains the distance and travel time between each origin (i.e., each customer's home) and each destination and between every pair of origins. It is

Table 2
Transportation control measures likely to have a positive impact on congestion

Transportation control measure group	Examples
High occupancy vehicle (HOV) lanes	Entrance ramp priority, dedicated HOV lanes
Traffic flow improvements	Traffic signal optimization, incident management systems
Parking management	Preferential parking for HOVs, parking zoning regulations, park-and-ride facilities, shuttle services
Vehicle use restrictions	Route diversion, downtown vehicle restrictions, no-drive days, truck movement control
Special event and activity center programs	Remote parking with shuttle service, parking management
Improved public transit	Service expansion, operational improvements, rail expansion
Employer-based transportation management programs	Carpooling, transit, financial incentives to employees, telecommuting, flextime, compressed work weeks
Trip reduction ordinances	Special use permits, mandated ridersharing
Rideshare incentives	Commute management organizations, tax incentives
Bicycle and pedestrian programs	Bicycle routes and storage facilities, sidewalks and walkways

possible to build the vehicle routing matrix by using GIS routing functions applied to a network map layer, a customer home layer, and a destination layer. Once the vehicle routing matrix is developed, the system attempts to find efficient routes that service as many customers and destinations as possible while trying to minimize the total travel time. The output from the procedure is an itinerary for each vehicle summarizing the route and all stops and destinations associated with that route.

References

Benz, R.J., Ogden, M.A., 1996. Development and benefits of computer-aided travel time data collection. Transportation Research Record 1551, TRB, National Research Council, Washington, DC, pp. 1–7.

CAAA, 1990. Clean Air Act Amendment. US Code, Title 1, Section 103.

FHWA, 1998. Our Nation's Highways – Selected Facts and Figures. Publication No. FHWA-PL-98-015, US Department of Transportation, Washington, DC, p. 28.

Francois, M.I., Willis, A., 1995. Developing effective congestion management systems. Federal Highway Administration, Technical Report No. 8, p. 22.

Guo, P., Poling, A.D., 1995. Geographic information Systems/Global positioning systems design for network travel time study. Transportation Research Record 1497, TRB, National Research Council, Washington, DC, pp. 135–139.

HRB, 1950. Highway Capacity Manual; Practical Applications of Research. National Research Council, Washington, DC, p. 147.

HRB, 1965. Highway Capacity Manual, second ed. Special Report 87, National Research Council, Washington, DC, p. 397.

ISTEA, 1991. Intermodal Surface Transportation Efficiency Act. US Code, Title 23, Chapter 3, Section 303.

JHK, 1996. Performance Measures and Levels of Service in the Year 2000 Highway Capacity Manual. NCHRP Project 3-55(4), TRB, National Research Council, Washington, DC, p. 25.

Laird, D., 1996. Emerging Issues in the Use of GPS for Travel Time Data Collection. National Traffic Data Acquisition Conference, Albuquerque, NM, 6–9 May, 1, pp. 117–123.

Lindquist, E., 1999. Assessing effectiveness measures in the ISTEA management systems. Southwest Region University Transportation Center, Texas Transportation Institute, College Station, TX, p. 103.

Liu, T.K., Haines, M., 1996. Travel Time Data Collection Field Tests – Lessons Learned. Report FHWA A-PL-96-010, US Department of Transportation, p. 116.

Lomax, T., Turner, S., Shunk, G., Levinson, H.S., Pratt, R.H., Bay, P.N., Douglas, G.B., 1997. Quantifying congestion. Final Report, National Cooperative Highway Research Program, Transportation Research Board, p. 184.

Quiroga, C.A., 1997. An integrated GPS-GIS methodology for performing travel time studies. Ph.D. Dissertation, Louisiana State University, Baton Rouge, LA, p. 171.

Quiroga, C.A., 1999. Accuracy of linearly referenced data using GIS. Transportation Research Record 1660, TRB, National Research Council, Washington, DC, pp. 100–107.

Quiroga, C.A., Bullock, D., 1996. Architecture of a congestion management system for controlled-access facilities. Transportation Research Record 1551, TRB, National Research Council, Washington, DC, pp. 105–113.

Quiroga, C.A., Bullock, D., 1998. Development of CMS Monitoring Procedures. Report No. FHWA/LA-314, Louisiana Transportation Research Center, p. 87.

Quiroga, C.A., Bullock, D., 1999. Travel time information using GPS and dynamic segmentation techniques. Transportation Research Record 1660, TRB, National Research Council, Washington, DC, pp. 48–57.

Schrank, D.L., Lomax, T.J., 1999. The 1999 Annual Mobility Report. Texas Transportation Institute, p. 123.

Schwartz, W.L., Suhrbier, J.H., Gardner, B.J., 1995. Data collection and analysis methods to support congestion management systems. ASCE Transportation Congress, Proceedings V. 2, San Diego, CA, pp. 2012–2023.

TEA 21, 1998. Transportation Equity Act for the 21st Century. US Code, Title 23, Section 149.

TRB, 1985. Highway Capacity Manual, third ed. Special Report 209, National Research Council, Washington, DC, p. 504.

TRB, 1994. Highway Capacity Manual, third ed. Update, Special Report 209, National Research Council, Washington, DC.

TRB, 1997. Highway Capacity Manual, third ed. Update, Special Report 209, National Research Council, Washington, DC.

Turner, S.M., 1996. Advanced techniques for travel time data collection. Transportation Research Record 1551, TRB, National Research Council, Washington, DC, pp. 51–58.

Turner, S.M., Eisele, W.L., Benz, R.J., Holdener, D.J., 1998. Travel Time Data Collection Handbook. Report No. FHWA-PL-98-035, Texas Transportation Institute, College Station, TX, p. 346.

Zito, R., D'Este, G., Taylor, M.A.P., 1995. Global positioning systems in the time domain: how useful a tool for intelligent vehicle-highway systems? Transportation Research Part C 3 (4), 193–209.

PERGAMON

Transportation Research Part C 8 (2000) 307–320

TRANSPORTATION
RESEARCH
PART C

www.elsevier.com/locate/trc

The effects of highway transportation corridors on wildlife: a case study of Banff National Park

Shelley M. Alexander *, Nigel M. Waters

Department of Geography, University of Calgary, 2500 University Dr. NW, Calgary, Alta., Canada T2N 1N4

Abstract

Road fragmentation is a concern for wildlife viability in and adjacent to protected areas in the Rocky Mountains. Roads create a barrier to wildlife movement and have documented demographic effects, including the alteration of animal communities, the reduction of biological diversity, and the increased threat of extinction. Wildlife movement across and adjacent to the Trans-Canada Highway (TCH) (14,000 annual average daily traffic, AADT) and Highway 1A (3000 AADT) was studied in Banff National Park, Alberta. Animal tracks were observed crossing roadways and on transects adjacent to roads for wolves, cougar, lynx, wolverine, marten, elk, deer, sheep, hare, and red squirrel relative to road types. Data were analyzed to assess the barrier effect and a geographical information system (GIS) was used to identify landscape attributes associated with species movement. The TCH was found to be a barrier to movement for all species. In less perturbed environments, it was observed that movement patterns for the wildlife communities were spatially continuous and that individual species movement was complex. This movement was not observed across the TCH. An interpolation of point data showed sites of high crossing frequency within the continuum of crossing points. These sites ranged from 250 to 2000 m in diameter. General predictors for movement by aspect were found to be the south, southwest and west facing slopes. Flat slopes, areas of low topographic complexity, and slopes lower than 5° were also effective predictors of animal movements. The data suggest that maintaining contiguous tracts of habitat with the above attributes facilitate normal wildlife movement most effectively. Mitigation that approximates previous patterns can be achieved only by elevating and/or burying extensive sections of highway. © 2000 Elsevier Science Ltd. All rights reserved.

Keywords: Habitat fragmentation; Wildlife movement corridors; Multi-species; Mitigation

* Corresponding author. Tel.: +1-403-220-6398; fax: +1-403-282-6561.
E-mail address: nwaters@ucalgary.ca (N.M. Waters).

PII: S0968-090X(00)00014-0

1. Introduction

Transportation oriented geographic information systems (GIS-T) have traditionally involved little consideration of the environment. While it is true that both early and more recent applications of GIS have been used to select transportation routes which minimize the route's impact on the environment (see the pioneering work of McHarg, 1969, and also Goodchild, 1977; Lee and Tomlin, 1997), it is also the case that contemporary academic and technical reviews of GIS-T (Waters, 1992, 1999; Nyerges, 1995; Noronha, 1999) have not considered how GIS might be used to investigate the environmental impact of a transportation route once it has been selected. It is time to redress this deficiency.

This paper begins with a brief review of environmentally-oriented papers presented at a recent American Association of State Highway and Transportation Officials (AASHTO) sponsored GIS-T conference, papers presented at the International Conference on Wildlife Ecology and Transportation (ICOWET) Conferences, and papers published by the Transportation Research Board, Washington, DC. In the past, these organizations have been concerned primarily with research dealing with GIS and transportation, research on GIS and the environment, and research into the impact of transportation on the environment, respectively.

This paper presents a case study that demonstrates how GIS-T can be used to conduct research into the destructive impacts of transportation developments on wildlife habitat. The hope is that this will lead to further studies that will expand the usefulness of GIS-T for mitigating habitat destruction and the degradation of the environment.

2. Current GIS-T research and the environment

One of the richest sources of original GIS-T research has been the AASHTO sponsored GIS-T conferences held annually since 1988. Cooke et al. (1998) note that many state departments of transportation (DOT) use GIS to support transportation decision-making processes. However, they argue that their case study is novel in that it uses GIS to integrate environmental issues into the transportation systems planning process in the North Carolina Department of Transportation (NCDOT). The previous lack of environmental data had, according to the authors, led to a feeling of frustration with the decision-making process and confrontation rather than cooperation among transportation planners, engineers, and environmental planners. One of the main benefits of GIS was found to be the ease with which consensus could be reached on environmentally preferred corridors among a variety of agencies with widely differing perspectives. The use of GIS by the NCDOT has allowed site-specific environmental information to be incorporated in the earliest phase of the planning process. Later in this paper, similar site-specific information is used to determine the best locations for road mitigation measures, such as under- and over-passes on wildlife movement patterns. The NCDOT used their GIS to develop a series of environmental sensitivity maps that would allow the least sensitive areas to be set aside for future transportation corridors.

If the AASHTO sponsored GIS-T symposia have traditionally neglected the environment, then the ICOWET conferences have conversely paid scant attention to GIS as they address how the environment is impacted by transportation infrastructure. That this is beginning to change is shown in the proceedings of the 1998 ICOWET conference, where one explicitly GIS-oriented paper was presented by Carr et al. (1998). Carr and co-authors made use of ESRI's Arc/Info GIS (ESRI 1995) to identify alternative transportation corridors, to determine horizontal and vertical alignments, to design environmentally sensitive new roadway construction, and to design the best mitigation strategies. The trend towards the use of GIS to study such environmental issues as transportation corridors and their impacts on wildlife (the specific focus of the present research) was even more evident in the recently published 1999 ICOWET Proceedings (see, for example, Callaghan et al., 1999; Klein, 1999, among others). Several recent articles concerning the environmental impact of transportation infrastructure that have appeared in the Transportation Research Record might have benefited from a GIS-based methodology (Adler et al., 1997; Bardman, 1997; Lackey, 1997).

With an ever increasing public awareness of the impact of transportation infrastructure on the environment and the continuing impact of legislation, this paper presents a detailed case study from Banff National Park, Alberta, Canada, showing how GIS may be used to mitigate the impact of transportation corridors on wildlife habitat use.

3. Impact of transportation corridors on wildlife habitat

Habitat fragmentation contributes to habitat decline and species loss (Wilcox and Murphy, 1985; Hobbs 1993; Reed et al., 1996). "Roads precipitate fragmentation by dissecting previously large patches into smaller ones" and by creating a barrier to movement and dispersal between adjacent habitat patches (Reed et al., 1996; Forman and Alexander, 1998). The barrier effect is manifested through direct mortality of wildlife and indirect psychological effects that may cause wildlife to avoid roads (Paquet and Callaghan, 1996; Forman and Alexander, 1998). The consequences of barrier effects include the alteration of animal communities, the reduction of biological diversity and the increased threat of extinction (Wilcox and Murphy, 1985; Oehler and Litvaitis, 1995; Bascompte and Solé, 1996; Forman and Alexander, 1998). Despite knowledge of these consequences, "the barrier effect remains little studied with regards to roads" (Forman and Alexander, 1998).

Road mortality can be a serious threat to species with low population levels (i.e., rare or endangered species) or poor resilience, which results from life history traits, such as late age of first reproduction, and predisposes populations to extirpation (Weaver et al., 1996). For example, in the Bow Valley of Banff National Park (BNP), high rates of road and rail mortality of wolves have resulted in a localized extinction of their population (Callaghan, 1999).

Wide-ranging species are especially vulnerable to road effects. Wolves, bears, and elk have sizeable home ranges that typically require them to cross roads frequently. Moreover, road right-of-ways often attract vertebrate species because they provide good habitat for hunting, grazing, and movement (Oehler and Litvaitis, 1995; Forman and Alexander, 1998). Consequently, the chance of vehicular-caused mortality of large mammals increases.

Habitat patches between roads may encompass the home ranges of smaller ranging species, such as the marten. In the previous case, "meta-population processes, such as patch-specific colonization and extinction rates, may be directly affected" by road kill, increased predation, and impaired migration (Mader et al., 1990; Ims et al., 1993).

The creation of linkages between isolated patches of habitat has been identified as one method for improving biotic exchange across road barriers (Forman and Alexander, 1998). Linkages across barriers facilitate dispersal and migration processes, which are critical to species persistence (Weaver et al., 1996). Dispersal refers to movements by juvenile animals when leaving their natal range, while migration is the movement of individuals or groups between two areas (Weaver et al., 1996).

4. Study area

Traffic on the Trans-Canada Highway (TCH) in BNP acts as a barrier to wildlife movement and causes wildlife injury and mortality. It is argued here that an effective highway mitigation design must approximate normal wildlife movement patterns. Moreover, these patterns will likely vary by species. Thus, it is critical for park managers to understand the placement and site requisites for mitigation structures that meet the needs of wildlife communities. GIS provides an excellent medium for addressing this question.

This research was conducted in BNP, Alberta. BNP is approximately 6640 km^2 in area and is the most heavily visited National Park in Canada with over 5 million visitors per year (Banff-Bow Valley Study, 1996). The study region is characterized by rugged mountainous terrain, steep valleys, and narrow (2–5 km), flat valley bottoms. Roads and other human development primarily occur along the valley floors.

Conclusions of this paper are based on two years of data collected on the non-twinned (2 lane), unfenced section of the TCH from Castle Junction to the British Columbia border (Phase IIIB) and along the Bow Valley Parkway (1A), as shown in Fig. 1. Data were collected from November

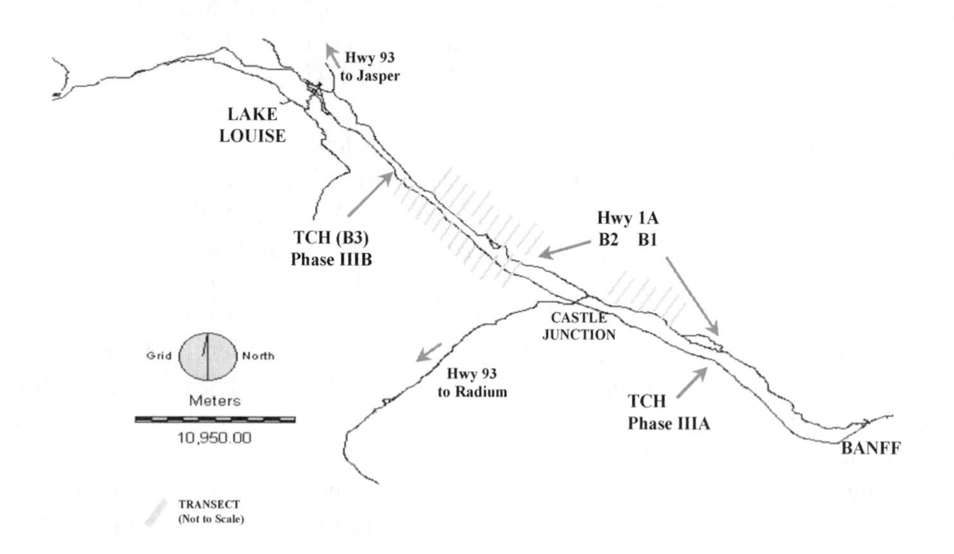

Fig. 1. BNP study area.

to April (1997/98 to 1998/99). Except for wildlife warning signs, none of the study roads are currently mitigated.

Three road sections were surveyed; each is approximately 30 km in length. Annual average daily traffic (AADT) volume on the TCH is about 14,000 vehicles and volume is about 3000 vehicles on the 1A (Banff-Bow Valley Study, 1996). The BNP road survey sections used in this study are classified as:

B1: Bow Valley Parkway (Hwy 1A) from 5 Mile Bridge to Castle Junction,

B2: Bow Valley Parkway (Hwy 1A) from Castle Junction to Lake Louise, and

B3: TCH from Castle Junction to the British Columbia Border (unfenced, non-twinned section of the TCH).

4.1. Methodology

Roads were surveyed no less than 18 h and no more than 48 h after the end of a snowfall. Tracks entering or exiting the road right of way were recorded for coyote (*Canis latrans*), fox (*Vulpes vulpes*), wolf (*Canis lupus*), cougar (*Felis concolor*), bobcat (*Lynx rufus*), lynx (*Felis lynx*), marten (*Martes americana*), fisher (*Martes pennanti*), wolverine (*Gulo gulo*), elk (*Cervus elaphus*), moose (*Alces alces*), sheep (*Ovis canadensis*), and deer *(Odocoileus virginianus and Odocoileus hemionus*).

Repeat surveys were conducted, as conditions permitted, approximately 3–4 days after initial surveys until the next new snowfall. On repeat surveys, only observations of large carnivore crossings were recorded, because the frequency of crossings by other species would prohibit completing the surveys. No attempt was made to differentiate between the various deer species tracks because of the degree of overlap in track characteristics. Tracks were observed from a field vehicle while driving slowly (15–20 km/h) and verified on foot.

Data collected at crossing sites included: species type, GPS location (UTM-Nad27), direction of travel, travel direction relative to vegetation and road, distance traveled on the road, vegetation type and closure estimate, and whether the animal was crossing the road multiple times or had aborted a crossing attempt. If no obvious tracks existed on the road surface, it was assumed that tracks entering/exiting the road surface were crossing attempts. If no companion tracks exiting/entering the road were found within 300 m, then the crossing was marked as unconfirmed. Same species tracks recorded past that interval were identified as separate crossing attempts. The effect of traffic volume on landscape permeability was examined statistically by comparing track frequencies for each species by road type. Landscape permeability is defined as the ease of movement between adjacent habitat patches that are bisected by roads. Higher crossing frequencies would suggest ease of movement and greater permeability.

One kilometer transects consisting of twenty 50-m long sub-transects were fixed perpendicular to each road type. Forty transects were surveyed in BNP. Transects were surveyed randomly between 18 and 108 h after snow, immediately following each road survey. The randomization involved selecting a starting pair of transects for each road section and sampling without replacement. These pairs were used to start the survey, and the researchers chose the order by which other transects were surveyed. Staffing, weather, snow condition, avalanche hazards, and logistical limitations prohibit transects being surveyed in a completely random fashion as researchers must be flexible to these previous limitations. Transects are surveyed on foot and require an extended survey period relative to the road survey. Within daylight hours, a researcher is physically capable of surveying no more than five to seven 1-km transects per day. Tracks were recorded on transects for all road crossing species plus squirrel, weasel, and hare. The additional species were recorded for use as explanatory variables in the presence of other species. These three species were not included on road surveys because their abundance would prohibit timely completion of the surveys. Transect survey data provide an estimate of available migrants and are critical to explaining variances in movement at different traffic volumes.

Landscape attributes have been used to predict movement of species across the landscape. Research on wolves in BNP has indicated that movement can be predicted by presence or use of landscape features, such as slope, aspect, elevation, and vegetation type and closure (Paquet, 1993). A disproportionate use of certain physiographic attributes is assumed to reflect a preference (Alexander and Waters, 1999).

The Idrisi GIS (Eastman, 1997) was used to analyze landscape attributes coincident with crossings observed in 1998/99 for marten, wolf, lynx, cougar, and elk. Slope and aspect coverages were derived using the SURFACE module and a digital elevation model (DEM-Alberta Environment, Provincial Base Map) with a 30-m resolution. Topographic complexity was calculated from the DEM using the PATTERN module within Idrisi (Eastman, 1997). One function in this module, which determines the number of different classes around a centroid, was used to specify a 3×3-moving window for the complexity analysis. For details on this operation refer to the Idrisi instruction manual (Eastman, 1997). A combination of the Idrisi CROSSTAB module and the Minitab statistical package was used to calculate χ^2 and related statistics (Loether and McTavish, 1988; Fosnight and Fowler, 1996) to examine the various landscape associations discussed below.

To visualize the spatial distribution of points, a map overlay operation was used to render a composite surface of all crossing points. An Idrisi SURFACE interpolation was then run on the previous road coverage to identify crossing "hot spots" for multiple species. These hot spots may define suitable locations and spatial extent required for successful mitigation.

5. Results

5.1. *Traffic volume and landscape permeability*

Tables 1 and 2 summarize crossing counts for all species surveyed during the winter seasons 1998/99 and 1997/98. Each of the three road sections is approximately 30 km in length.

Crossing frequencies were compared by highway section using the χ^2 statistic at the 99% confidence interval. Observed crossings were tested against a uniform distribution of crossings. Results of this analysis are presented in Tables 3 and 4.

5.2. *Predictive landscape attributes*

Observed crossings were compared with expected counts using output from the Idrisi CROSSTAB module, and the χ^2 statistics were calculated using Minitab (Kitchin and Tate, 2000), while Cramer's V measures of association were calculated by the authors (Loether and McTavish, 1988). Observations were tested against a uniform distribution of crossings by attribute class

Table 1
Banff road crossing summary: total for 1998/1999 (17 surveys)

Species	1A: East (B1) (unmitigated)	1A: West (B2) (unmitigated)	Phase IIIB (B3) (unmitigated)
Marten	219	106	65
Coyote	76	36	28
Wolf	30	5	1
Lynx	0	2	6
Cougar	18	0	0
Wolverine	0	0	0
Elk	69	8	2
Moose	0	1	0
Sheep	6	0	0
Deer	57	1	0
Fox	1	0	0
Fisher	0	1	0

Table 2
Banff road crossing summary: total for 1997/1998 (12 surveys)

Species	1A: East (B1) (unmitigated)	1A: West (B2) (unmitigated)	Phase IIIB (B3) (unmitigated)
Marten	68	15	16
Coyote	77	9	23
Wolf	14	7	1
Lynx	3	6	5
Cougar	12	0	0
Wolverine	6	0	1
Elk	50	3	2
Moose	0	0	0
Sheep	3	0	0
Deer	7	0	0
Fox	1	0	0
Fisher	0	0	0

Table 3
1997/98 summary statistics

Sections compared	d*f*	χ^2	Cramer's V	Significance level (%)
B1 vs B3	9	27.268	0.31	99
B2 vs B3	5	11.314	0.35	95
IIIA vs B3	6	393.775	0.66	99
IIIA vs B2	8	455.355	0.44	99
B1 vs B3 vs B3	18	224.585	0.54	99

Table 4
1998/99 summary statistics

Sections compared	d*f*	χ^2	Cramer's V	Significance level (%)
B1 vs B3	8	72.276	0.35	99
B2 vs B3	7	9.734	0.19	Under 95
B1 vs B2 vs B3	20	120.603	0.29	99

Table 5
Marten ($N = 489$)

Attribute	df	χ^2	Significance (%)	Cramer's V
Aspect	8	272.528	99	0.0290
Slope	21	375.096	99	0.0350
Topographic complexity	8	343.693	99	0.0314

Table 6
Elk ($N = 134$)

Attribute	df	χ^2	Significance (%)	Cramer's V
Aspect	8	83.282	99	0.016
Slope	11	149.232	99	0.021
Topographic complexity	8	109.716	99	0.019

Table 7
Wolf ($N = 19$ – points surveyed opportunistically not included)

Attribute	df	χ^2	Significance (%)	Cramer's V
Aspect	8	32.810	99	0.010
Slope	11	30.466	99	0.010

Table 8
Lynx ($N = 3$ – points surveyed opportunistically not included)

Attribute	df	χ^2	Significance (%)	Cramer's V
Aspect	8	19.475	99	0.008
Slope	11	20.357	95	0.007

Table 9
Cougar ($N = 7$ – points surveyed opportunistically not included)

Attribute	df	χ^2	Significance (%)	Cramer's V
Aspect	8	11.716	Under 95	N/A
Slope	11	10.166	Under 95	N/A

(aspect, slope, and complexity). The relationships between crossings and landscape attributes are listed by species in Tables 5–9. Details on the variability explained by specific attribute classes are presented in the next section.

6. Discussion of results

6.1. Landscape permeability

Traffic on the TCH impedes movement and dispersal in the Bow River Valley, as shown by statistically lower crossing frequencies along the TCH (B3). A three-way comparison of frequencies for B1, B2 and B3 indicated a significant difference at 99% confidence between highway crossing frequencies in both 1997/98 and 1998/99 ($\chi^2 = 224.585$, $df = 18$; $\chi^2 = 120.603$, $df = 20$, respectively). This difference is explained primarily by variation between B1 (1A East) and B3 (TCH). A high crossing frequency by marten, coyote, cougar, wolf, wolverine, and elk on B1 (1A East) contributed most to the differences between highway segments in both data sets.

Pairwise comparisons showed a significant difference at 99% confidence between the TCH (B3) and the 1A East (B1) in 1997/98 and 1998/99 ($\chi^2 = 27.268$, $df = 9$; $\chi^2 = 72.276$, $df = 8$, respectively). In 1997/98, a significant difference was observed between B2 and B3 at the 95% confidence level but did not reach the 99% level of confidence ($\chi^2 = 11.314$, $df = 5$). In 1998/99, no significant difference was observed between B2 and B3, at either the 99% or even the 95% confidence level ($\chi^2 = 9.734$, $df = 7$).

A preliminary analysis of transect data shows that species richness and abundance were comparable in habitat adjacent to the B3 (TCH) and B1 (1A East) but lower in neighboring B2 (1A West). This finding further supports the conclusion of a barrier effect along B3 (TCH) because the lower ratio of actual crossings over potential crossings increases the severity of the existing statistical differences. The crossing frequencies on B2 are low because of the presence of less suitable habitat adjacent to the road. In contrast to the habitat bordering sections B1 and B3, that adjacent to B2 is more steep and rugged, which reduces the suitability of the habitat for multiple species and thereby reduces the number of available migrants. Subsequent formal analysis of transect data is planned and will provide an index of habitat suitability in these regions.

It was expected that all species would have lower crossing frequencies on the TCH compared with lower traffic volume roads (e.g., 1A). However, lynx crossings were higher on the TCH (B3) compared to other highway sections and increases in movement appear to coincide with the offspring dispersal period.

6.2. *Predictive landscape attributes for individual species*

6.2.1. *Marten*

The distribution of marten crossings was significantly different from uniform (99% confidence) for the landscape attributes aspect, slope, and topographic complexity.

Marten tracks were observed more often than expected in aspect classes south (157.5–202.5°), southwest (202.5–247.5°), west (247.5–292.5°), and flat (slope = 0°). Marten were observed less than expected in classes north (337.5–22.5°), northwest (292.5–337.5°), northeast (22.5–67.5°), east (67.5–112.5°), and southeast (112.5–157.5°). These findings are consistent with expected occurrence of marten by vegetation type, the latter being correlated by aspect (Buskirk and Ruggiero, 1994).

A statistically non-uniform distribution of crossings by slope angle (99% confidence) was shown for marten. The higher frequency of observations over expected values at 0°, 1°, and 2° contributed most to the statistical variability. All slopes above 5° showed lower crossing frequencies than expected. No crossings occurred in areas over 21°. The latter is related to sampling on roads, which rarely occur on steep slopes.

Marten observations coincided statistically (99% confidence) with areas of low topographic complexity, as represented by the number of different elevation classes around a centroid (3×3 moving window). Substantially higher observed versus expected values for cells with one, two, and three different neighboring classes and lower observed versus expected for cells with eight and nine different neighbors contributed most to the non-uniform distribution of points.

6.2.2. Elk

It was found that elk distribution was statistically non-uniform by aspect category (99% confidence). Most of the variance can be explained by the higher use than expected rates on flat and southern slopes and the disproportionately lower use than expected on north and northeast slopes.

Use of slopes by elk was significantly non-uniform (99% confidence). Slopes below 5° were used more frequently than expected. Much of the variance from a uniform expectation can be explained by the relatively higher observed compared to expected values on flat and 1° slopes, and lower observed versus expected values for slopes above 11°.

Elk use of habitat of varying complexity was shown to be statistically non-uniform (99% confidence). Overuse of areas of low complexity (one, two, and three different types of neighboring cells) explained this statistical difference. Under-utilization of areas of the highest complexity class (nine different neighboring cells) also made a substantive contribution to the non-uniformity.

6.2.3. Wolf

Wolf crossings were statistically non-uniform (99% confidence) in distribution by aspect class. Seven crossings were observed in the "flat" aspect class, explaining most of the variance in distribution. The small sample size weakens the validity of these results and additional research will be necessary to confirm these observations.

The occurrence of wolf points was non-uniform by slope class (99% significant). The majority of wolf points were observed in flat aspects.

6.2.4. Lynx and cougar

The small sample size for wolves was problematic when using the χ^2 statistic, even considering arguments for more liberal assumptions in the use of χ^2 analysis (see Baker and Lee, 1975, for a discussion). Consequently, the analysis of lynx and cougar crossings was modified and only aspect and slope relationships were examined.

Lynx showed a significantly non-uniform distribution of points by aspect class as higher use was confined to the northeast and southwest aspects. It was also determined that lynx distribution occurred only on slopes of 7° or less, but the sample size was too small for continued analysis. Cougars did not show a distribution different from uniform for aspect. Future analyses will combine annual observations of species with low distribution from successive field seasons to enhance sample size.

6.3. Data visualization and surface interpolation

Qualitative observations indicated that wildlife crossings on a community level are not spatially clustered but are spatially continuous. See Fig. 2 for an example. In contrast, the existing TCH wildlife mitigation constrains movement to narrow and infrequent sites. On lower traffic volume roads, such as B1 and B2, crossings for many survey species are characterized by multiple

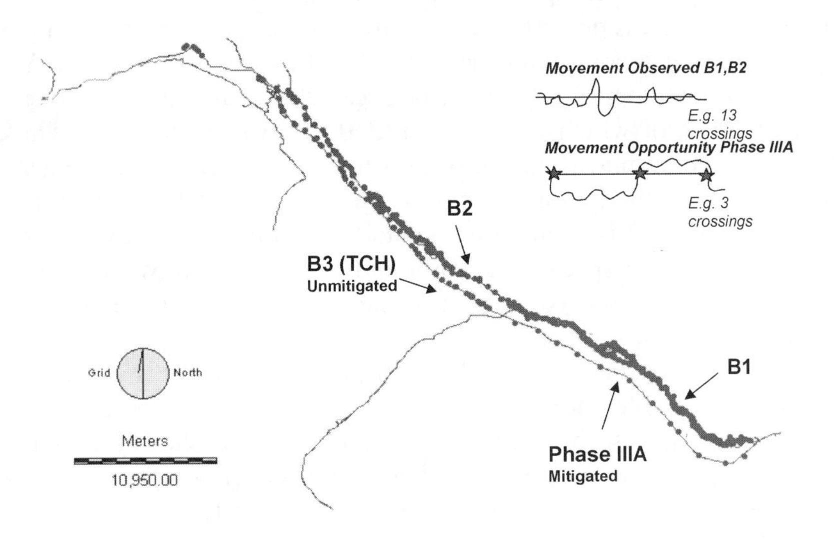

Fig. 2. Multiple species movements vs mitigation placement.

crossings in one movement session (e.g., in one day). For example, one wolf pack was observed to cross the road 13 times in 24 h (12 February 1999). This type of movement is not possible on the mitigated portion of the TCH (Phase IIIA) because of inadequate mitigation design, such as narrow culverts that wolves refuse to cross (Paquet, 1993).

The Idrisi surface interpolation module was run on the continuum of points shown in Fig. 2. This procedure identified nodes of high crossing frequency and indicated that these hot spots range from 250 to 2000 m in diameter.

7. Conclusions

The present findings yield strong evidence that traffic on the TCH, Phase IIIB creates a barrier to movement for most species.

All the statistical analyses of crossings by landscape attributes showed low strength of relationship (using Cramer's V). This is a common occurrence in raster-based GIS analysis because of the overwhelming presence of zero values in the crossing image.

It may be concluded that general predictors for movement for aspect are south, southwest, west, and flat slopes. In addition, lynx were found on northeastern aspects. Road mitigation should be designed to maintain connectivity of wildlife populations between these suitable aspects. Low topographic complexity and slopes lower than 5° were shown to be optimal areas for movement. These findings are consistent with predictive models for wolf movement that were constructed using radio-telemetry observations (Paquet, 1993). It is suggested that areas exhibiting the previous qualities may be useful to delineate movement corridors and to locate sites for mitigation that have high movement potential along roadways.

These results indicate that the mitigation approach employed in BNP at present does not facilitate natural movement. In less perturbed environments, the spatial movement pattern observed for individual species and at a community level was continuous and complex. Within this continuum of points shown on the multi-species crossing surface, sites of high crossing frequencies were observed, the average of which ranged from 250 to 1000 m in width. Thus, in designing mitigation that meets the continuous nature of movement, the frequent placement of mitigation structures that have considerable spatial extent (e.g., greater than 250 m wide) is recommended. It is hypothesized that the best mitigation design should simultaneously allow use of the mitigated area as a home range for smaller species, such as a marten, and as a movement corridor for larger species, such as the wolf. The proposed design is achieved most appropriately by elevating and/or burying substantial portions of the highway.

The present research has used a traditional GIS framework and tools of analysis to determine the spatial requisites of wildlife movement patterns, and compare these to existing styles of mitigation. Only through the use of such tools is it possible to show the limitations of these measures and suggest appropriate remedies. To date such research integrating GIS environmental concerns and transportation has been relatively rare in the literature.

Acknowledgements

This project was made possible by the generosity of the following financial contributors: the Natural Sciences and Engineering Research Council of Canada, the University of Alberta and Alberta Conservation Association Biodiversity Fund, the Province of Alberta Graduate Fellowship, Alberta Environmental Protection, Canadian Pacific Corporation, Banff National Park Wildlife and Highways Divisions, Edward Alexander, Dr. Margaret P. Hess, the University of Calgary, the Alberta Sport-Recreation-Parks and Wildlife Foundation, Employment Canada, the Western Forest Carnivore Society and the Paquet Fund for Wildlife Biology.

References

Adler, J.L., Franks, T., Nelson, D., Benware, T., Ivery, M., McVoy, G., 1997. Expert system architecture for computer-aided environmental analysis. Transportation Research Record No. 1601, Environmental Issues in Transportation, Transportation Research Board, National Research Council, National Academy Press, Washington, DC, pp. 29–34.

Alexander, S., Waters, N., 1999. Decision support methods for assessing placement and efficacy of road crossing structures for wildlife. In: Evink, G.L., Garrett, P., Zeigler, D. (Eds.), Proceedings of the Third International Conference on Wildlife Ecology and Transportation. FL-ER-73-99, Florida Department of Transportation, Tallahassee, FL, pp. 237–246.

Baker, E.J., Lee, Y., 1975. Alternative analyses of geographical contingency tables. Professional Geographer 27, 179–188.

Banff-Bow Valley Study, 1996. Banff-bow valley: at the crossroads. Technical report of the Banff-Bow Valley Task Force (Robert Page, Suzanne Bayley, J. Douglas Cook, Jeffrey E. Green, J.R. Brent Ritchie). Prepared for the Honourable Sheila Copps, Minister of Canadian Heritage, Ottawa, ON.

Bardman, C.A., 1997. Applicability of biodiversity impact assessment methodologies to transportation projects. Transportation Research Record No. 1601, Environmental Issues in Transportation, Transportation Research Board, National Research Council, National Academy Press, Washington, pp. 35–41.

Bascompte, J., Solé, R., 1996. Habitat fragmentation and extinction thresholds in spatially explicit models. Journal of Animal Ecology 65, 465–473.

Buskirk, S.W., Ruggiero, L.F., 1994. American marten. In: The Scientific Basis for Conserving Forest Carnivores: American Marten, Fisher, Lynx and Wolverine in the Western United States. US Dept. of Agriculture, Forest Service, Gen. Tech. Rep. RM-254, Fort Collins, Colorado, USA 80526. p. 184.

Callaghan, C., 1999. Personal communication.

Callaghan, C., Paquet, P., Wierzchowski, J., 1999. Highway effects on gray wolves within the golden canyon, British Columbia. In: Proceedings of the Third International Conference on Wildlife Ecology and Transportation. Evink, G.L., Garrett, P., Zeigler, D. (Eds.). FL-ER-73-99, Florida Department of Transportation, Tallahassee, FL, pp. 39–51.

Carr, M.H., Zwick, P.D., Hoctor, T., Harrell, W., Goethals, A., Benedict, M., 1998. Using GIS for identifying the interface between ecological greenways and roadway systems at the state and sub-state levels. In: Proceedings of the International Conference on Wildlife Ecology and Transportation. Evink, G.L., Garrett, P., Berry, J.B. (Eds.), FL-ER-69-98, Florida Department of Transportation, Tallahassee, FL, pp. 68–77.

Cooke, P.D., Foster, D., Schuller, E.R., 1998. Geographic information systems (GIS) implementation in the phased environmental process for systems planning at the North Carolina department of transportation (NCDOT). In: Proceedings of the Geographic Information Systems for Transportation Symposium, American Association of State Highway and Transportation Officials, Washington, pp. 230–253.

Eastman, J.R., 1997. Idrisi for Windows. User's Guide. Version 2. Clark Labs for Cartographic Technology and Geographic Analysis, Clark University, Worcester, MA.

ESRI, 1995. Understanding GIS: The Arc/Info Method. Environmental Systems Research Institute, Geoinformation International, Cambridge, UK.

Forman, R.T.T., Alexander, L.E., 1998. Roads and their major ecological effects. Annual Review of Ecological Systems 29, 207–231.

Fosnight, E.A., Fowler, G.W., 1996. Measures of association and agreement for describing land cover characterization classes. In: Proceedings of the Spatial Accuracy Assessment in Natural Resources and Environmental Sciences Symposium, USDA Forest Service, General Technical report RM-GTR-277, pp. 425–433.

Goodchild, M.F., 1977. An evaluation of lattice solutions to the problem of corridor location. Environment and Planning A 9, 727–738.

Hobbs, R.J., 1993. Effects of landscape fragmentation on ecosystem processes in the western Australian wheatbelt. Biological Conservation 64, 193–201.

Ims, R.A., Rolstad, J., Wegge, P., 1993. Predicting space use responses to habitat fragmentation: Can voles (Microtus oeconomus) serve as an experimental model system (ems) for capercaillie grouse (Tetrao urogallus) in Boreal forest? Biological Conservation 63, 261–268.

Kitchin, R., Tate, N.J., 2000. Conducting Research into Human Geography. Prentice-Hall, Englewood Cliffs, NJ.

Klein, L., 1999. Usage of GIS in wildlife passage planning in estonia. In: Evink, G.L., Garrett, P., Zeigler, D. (Eds.), Proceedings of the Third International Conference on Wildlife Ecology and Transportation. FL-ER-73-99, Florida Department of Transportation, Tallahassee, FL, pp. 179–184.

Lackey, A.E., 1997. Reconciling transportation corridor preservation and national environmental poicy act process: evaluation of north carolina phased environmental approach. Transportation Research Record No. 1601, Environmental Issues in Transportation, Transportation Research Board, National Research Council, National Academy Press, Washington, pp. 21–28.

Lee, B.D., Tomlin, C.D., 1997. Automate transportation corridor location. GIS World 10 (1), 56–60.

Loether, H.J., McTavish, D.G., 1988. Descriptive and Inferential Statistics: an Introduction, third ed. Allyn & Bacon, Newton, MA.

Mader, H.J., Schell, C., Kornacker, P., 1990. Linear barriers to arthropod movements in the landscape. Biological Conservation 54, 209–222.

McHarg, I., 1969. Design with nature. Doubleday, Natural History Press, New York.

Noronha, V., 1999. Unit 183 Transportation Networks. Core Curriculum in Geographic Information Science, National Center for Geographic Information and Analysis, Department of Geography, University of California, Santa Barbara, California, URL:http://www.ncgia.ucsb.edu/giscc/units/u183/.

Nyerges, T., 1995. Geographic information system support for urban/regional transportation analysis. In: Susan Hanson (Ed.), The Geography of Urban Transportation, second ed. Guilford Press, New York, pp. 240–265.

Oehler, J.D., Litvaitis, J.A., 1995. The role of spatial scale in understanding responses of medium-sized carnivores to forest fragmentation. Canadian Journal of Zoology 74, 2070–2079.

Paquet, P.C., 1993. Summary reference document-ecological studies of recolonizing wolves in the Central Canadian Rocky Mountains. BNP Warden Service, Banff, AB.

Paquet, P.C., Callaghan, C., 1996. Effects of linear developments on winter movements of gray wolves in the bow river valley of Banff National Park, Alberta. In: Evink, G.L., Garrett, P., Zeigler, D., Berry, J. (Eds.), Trends in Addressing Transportation Related Wildlife Mortality, Proceedings of the Transportation Related Wildlife Mortality Seminar. FL, USA. June 1996.

Reed, R.A., Johnson-Barnard, J., Baker, W.L., 1996. Contribution of roads to forest fragmentation in the rocky mountains. Conservation Biology 10 (4), 1098–1106.

Waters, N. (Ed.), 1992. Geographic information systems in transportation. In: Proceedings of the Transportation Research Forum 6, 112–29.

Waters, N., 1999. Transportation GIS: GIS-T. In: Longley, P., Goodchild, M., Maguire, D., Rhind D. (Eds.), Geographical Information Systems: Principles, Techniques, Applications and Management, second ed. Wiley, New York.

Weaver, J.L., Paquet, P.C., Ruggiero, L.F., 1996. Resilience and conservation of large carnivores in the rocky mountains. Conservation Biology 10 (4), 964–976.

Wilcox, B.A., Murphy, D.D., 1985. Conservation strategy: the effects of fragmentation on extinction. American Naturalist 125, 879–887.

PERGAMON

Transportation Research Part C 8 (2000) 321–336

TRANSPORTATION
RESEARCH
PART C

www.elsevier.com/locate/trc

Mapping evacuation risk on transportation networks using a spatial optimization model

Richard L. Church [a,*], Thomas J. Cova [b]

[a] *Department of Geography, University of California at Santa Barbara, National Centre for Geographic Information and Analysis, Santa Barbara, CA 93106, USA*

[b] *Department of Geography, University of Utah, 260 S. Central Campus Drive, Room 270, Salt Lake City, UT 84112, USA*

Abstract

The focus of this paper is on the development of a methodology to identify network and demographic characteristics on real transportation networks which may lead to significant problems in evacuation during some extreme event, like a wildfire or hazardous material spill. We present an optimization model, called the critical cluster model, that can be used to identify small areas or neighborhoods which have high ratios of population to exit capacity. Although this model in its simplest form is a nonlinear, constrained optimization problem, a special integer-linear programming equivalent can be formulated. Special contiguity constraints are needed to keep identified clusters spatially connected. We present details on how this model can be solved optimally as well as discuss computational experience for several example transportation networks. We describe how this model can be integrated within a GIS system to produce maps of evacuation risk or vulnerability. This model is now being utilized in several research projects, in Europe and the US.

Keywords: Emergency evacuation; Network capacity; Optimization; Integer programming; Geographical information systems

1. Introduction

Disasters that require some type of evacuation are relatively common (Perry, 1985). Buildings, industrial plants, airplanes, flood zones are but a few examples of areas or structures, which may require immediate and quick evacuation. The risk involved may be well understood. For example, buildings are often modeled for their evacuation or egress potential (Owen et al., 1996). Room

* Corresponding author. Tel.: +1-805-893-4217; fax: 1-805-893-3146.
E-mail address: church@geog.ucsb.edu (R.L. Church).

0968-090X/00/$ - see front matter
PII: S0968-090X(00)00019-X

capacities are established to prevent too many people from crowding an exit in times of an emergency, exit signs are placed at high locations on ceilings and walls above doorways and are easy to identify, and fire drills may be practised.

Evacuation, per se, requires a mode of transit and infers some type of movement. A number of factors can affect the safety of an evacuation. These factors include the number of people needing evacuation (i.e. demand), the transportation capacity provided for evacuation, the rate at which the demand is exerted, the rate at which capacity is actually provided, the differences between these rates, human behavior (Lindell and Perry, 1991), and accidents. A problem exists during an evacuation whenever the demand exerted during a given time interval exceeds the capacity provided for the evacuation. For example, let us say an area needs to be evacuated due to a hurricane. Even though the bulk capacity of all vehicles may be large enough to carry all people out of the area, the maximum flow rate of the transportation network to carry vehicles out of the area becomes a defining condition for the time it will take to accomplish a complete evacuation, assuming no accidents. Human behavior and accidents can also have a great effect on the outcome. As such, one should look at an evacuation involving a number of stochastic elements, which may have a significant effect on the outcome. Further, the best methodology to analyse such stochastic elements is simulation (Sheffi et al., 1982; Hobeika and Jamei, 1985; Stern and Sinuary-Stern, 1989; Southworth, 1991; Tufekci and Kisko, 1991; Pidd et al., 1997). Pidd et al. (1997) integrate their simulation model with a GIS.

There are several scales of concern within the context of evacuation. They can be roughly divided into the following general categories: regions, neighborhoods, buildings, ships, and airplanes. Large area evacuation analysis begins with setting the boundaries of the area that might be evacuated: e.g. within 10 miles of a nuclear power plant, a 100 year flood plain within the urbanized area, a potential landfall zone for a hurricane, etc. Such bounded regions are called emergency planning zones (EPZ) (Sorensen et al., 1987). EPZs are generally defined in advance and may represent the basis for a long-term planning project and the development of an evacuation plan. The same can be said for the evacuation analysis of buildings, ships, and airplanes. Namely, an implied EPZ exists, which represents the outside shell of the building, ship or airplane. The evacuation problem at the neighborhood scale is a bit problematic, less studied, and less understood. One of the main reasons for this is that an easily recognized EPZ may not exist. We know, however, that the evacuation of a neighborhood can be a problem. For example, the Oakland/Berkeley Hills fire in 1991 left 25 people dead, 150 injured, and destroyed 3354 single-family dwellings and 456 apartments in a matter of hours (Shough et al., 1992; OES, 1992). But, what made that neighborhood and disaster such a lethal combination? As research and analysis for evacuation at this level of scale matures, it is expected that such phenomena will be better understood and better advanced planning can be accomplished. Within a transportation focus, we might ask whether lethal combinations of population and transportation networks exist at the neighborhood scale which hinder evacuation and increase the risk of loss of life. For example, would it be possible to identify and recognize the potential for an evacuation problem in an area such as the Oakland Hills neighborhood before a disaster and then mitigate the risks by better advanced planning? This is the subject of this paper, evacuation at the neighborhood scale.

The remainder of this paper is devoted to describing a network-based model which has been developed to identify small areas or neighborhoods that may be difficult to evacuate. The basic idea is that the model can be used to search in a locally defined area of the network for a neighborhood that might be difficult to evacuate. Then, that area can be classified as to the degree to which evacuation difficulty exists. By applying the model numerous times across the network and classifying each local area as to evacuation difficulty, a map of evacuation vulnerability emerges. With such a model it is possible to map "evacuation risk" just as flood risk can be mapped. In the next section we describe one possible approach to estimating evacuation risk. In subsequent sections we integrate the approach to estimating evacuation risk into a model that can be used in conjunction with GIS to map evacuation risk.

2. Clearing time as an evacuation risk factor

Little is known about small area evacuation as it is nearly impossible to measure accurately during an emergency. Further, few models have been developed to simulate such an emergency for a transportation network at the scale of a neighborhood or relatively small urban area (Sinuany-Stern and Stern, 1993). To use a simulation model, the neighborhood needs to be defined in advance. Since the difficulty involved in evacuation is inextricably tied to the definition of the neighborhood, then the problem of neighborhood evacuation analysis includes identifying the critical size and shape of the neighborhood in question. Thus, defining the exact boundaries of the neighborhood is part of the evacuation risk mapping problem. Consequently, we begin our discussion here with two factors that are indisputably important: (1) total demand, and (2) transport capacity. Consider the following terms:

pop_k the population of neighborhood k, which may be estimated as the product of the number of houses times the number of people per household

cpp_k people per vehicle during a sudden evacuation of neighborhood k

cap_k capacity of lanes leading out of neighborhood k in vehicles per minute

cte_k clearing time estimate for neighborhood k.

Using this notation, we can now define the following ratio:

$$\mathrm{cte}_k = \frac{\mathrm{pop}_k * \frac{1}{\mathrm{cpp}_k}}{\mathrm{cap}_k} = \text{the clearing time in minutes for neighborhood } k.$$

The above cte value for a neighborhood is a simple estimate on how much time it would take to clear a neighborhood of its inhabitants. It assumes no accidents, that all inhabitants would either drive or ride a vehicle, and that the critical transportation element is the outbound road capacity of the neighborhood. Let us suppose that it takes time to notify and to marshal people in terms of the urgency of evacuation. Technically the time for notification and marshalling needs to be added to the cte value. In the event of an accident, then outbound or exiting capacity will likely be impaired and therefore clearing time increased. Thus, the above ratio is a lower bound estimate on the actual time to clear a neighborhood of its inhabitants. If the cte is low in value, then it is theoretically possible to make a speedy clearance of the neighborhood. If the cte is large in value, then it might signal a potential difficulty, especially if the time is too large in comparison to the time available due to the hazard (e.g. encroaching toxic cloud or a moving firestorm).

If exit capacity is measured in terms of the number of exit lanes, rather than a flow rate of vehicles per minute, then the above ratio is a measure of bulk demand per lane and has units of vehicles per lane that must evacuate. We can designate this ratio as bulk lane demand (bld). The higher the bulk lane demand, the higher the risk in leaving a neighborhood. Both bld and cte values are estimates of evacuation risk. The larger these values are, the greater the time and volume of traffic per lane that is required to clear a neighborhood.

Although the cte and bld values are not the only possible measures of evacuation risk, they are both easy to compute and estimate. These factors (i.e. clearing time and bulk demand), undoubtedly, are directly related to evacuation vulnerability. An alternate approach would be to employ a micro-simulation model of a neighborhood and estimate an expected clearing time. The difficulty that we face here is that we do not have a neighborhood defined in advance. Since cte and bld values can be easily calculated (as compared to using a simulation model) for a very large number of possible neighborhood definitions, it will allow us the capability to search for the critical neighborhood definition. For the remainder of this paper, we will assume that the cte and the related bld value are reasonable surrogate measures for evacuation risk. In the following section, we discuss the definition of an "evacuation neighborhood".

3. Defining an evacuation neighborhood

In order to calculate a cte or a bld value, it is necessary to delineate a neighborhood on a network. An example calculation for a bld value of a predefined neighborhood, designated as an EPZ, is given in Fig. 1(a). Demand in population (as estimated by # houses x # people per household) is multiplied by the number of vehicles per person (i.e. 1/2) and divided by the number of exiting lanes (i.e. 2) to estimate bld as 12.5 vehicles per lane. In many geographical contexts, it is possible to specify an EPZ in advance as done in Fig. 1(a), but for neighborhood evacuation

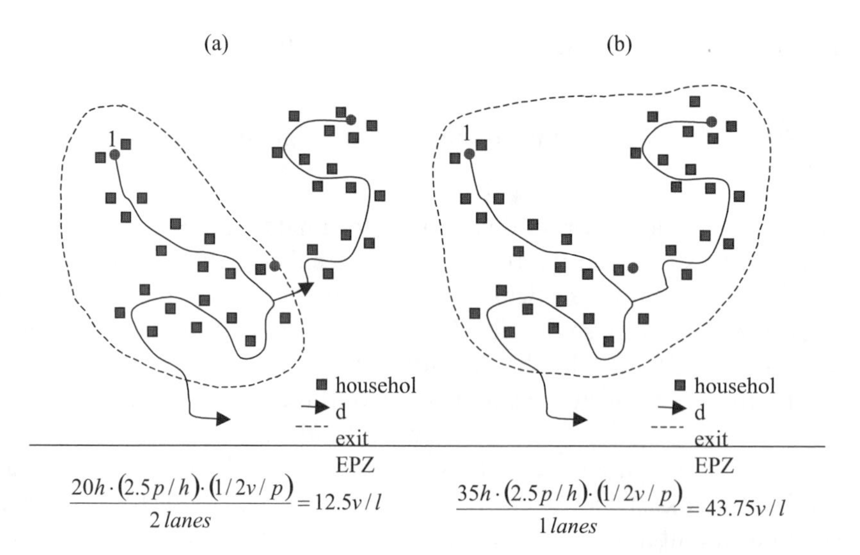

Fig. 1. (a) A bld calculation for a specific neighborhood. (b) A bld calculation for a larger neighborhood.

analysis, this may not be appropriate. For example, in Fig. 1(b), the bld is computed for a slightly larger neighborhood. The bld value increases since demand increases and the number of exit lanes decreases to one. Thus, the bld and cte values will be quite sensitive to the definition of the neighborhood. Taken from a more personal perspective, let us say you lived at node 1 which is indicated by the numeral 1 in Fig. 1. Your interest in evacuation might involve the question: what is the most difficult (worst-case) evacuation scenario within which I might participate (given that I am at node 1)? Clearly, this dictates finding the neighborhood which surrounds you at node 1, which has the highest bld or cte values. We will call the neighborhood defined about a given node, which maximizes bld or cte, as the "critical neighborhood". The nodes and street segments of which the "critical neighborhood" is comprised will be called the *critical cluster*. For each node of the network, there exists a critical cluster. Identifying the critical cluster for a given node is the critical cluster problem (CCP). In the following section, we present a model which can be used to solve an instance of the CCP.

4. The critical cluster model

Given a node of interest, called an anchor node, the critical cluster problem involves identifying a critical neighborhood of connected nodes and arcs about that anchor node which has the highest cte or bld value. As our focus is on delineating a small area or neighborhood EPZ, an upper limit will be used to restrict the critical cluster size. That is, since the area of the hazard that is faced is small and concentrated like the Oakland Hills fire, the immediate area of evacuation will be relatively small. Consequently, for the work presented here, we assume that each critical cluster will be limited and not exceed some maximum possible size.

There are many possible clusters that can be defined about a given anchor node. The critical cluster for that node is the one that maximizes some measure of risk or vulnerability like cte or bld. An example of a critical cluster is given in Fig. 2 for node 1. In this example each node has a population of 1 and each arc has one lane in each direction. Fig. 2 also presents a plot of the

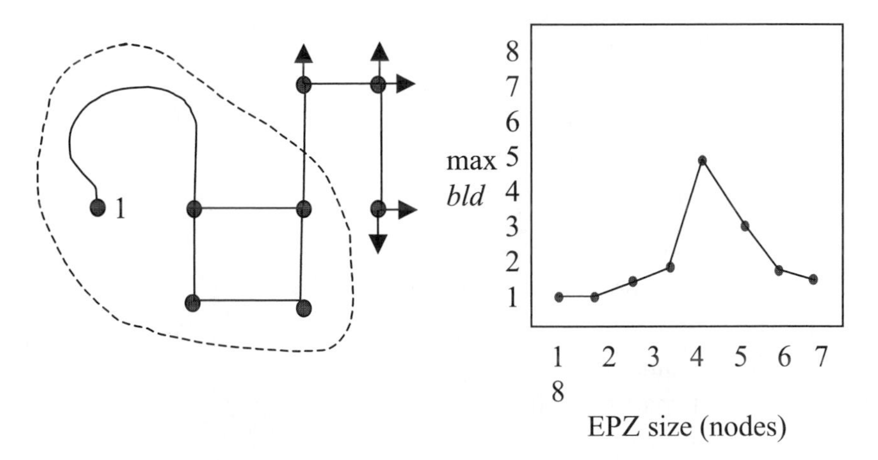

Fig. 2. A network and its corresponding max bld plot for node 1.

relationship between cluster size and the bld value. It is quite likely that one of the critical points in evacuation occurs when the neighborhood is relatively small as depicted in Fig. 2. Note that for the example anchor node, the critical cluster occurs at a cluster size of five. The focus in spatial evacuation analysis is on the local spatial variation in critical cluster values from anchor node to anchor node.

The problem of finding an anchor node's critical cluster falls within the broad category of graph or network partitioning problems (Cova and Church, 1997). A partition is a subset of nodes within a larger graph, and an optimal partitioning maximizes or minimizes some specified criteria related to the partition. In this context, we are only interested in contiguous partitions which will be referred to as clusters. The CCP is related in concept to the Graph Partitioning problem in the operations research literature (see, for example, Kernighan and Lin, 1970; Johnson et al., 1989; Jin and Chan, 1992; Laguna et al., 1994; Pirkul and Rolland, 1994). The critical cluster problem for node r can be stated formally as:

> Given a graph G with capacities on its edges and population at its nodes (set V), find a contiguous partition less than a given maximum size that contains anchor node r (call it partition V_r) so as to maximize some measure of evacuation vulnerability/risk such as cte or bld. We can formulate the CCM with respect to maximizing bld in the following manner:

Objective:

$$\text{Maximize}: \quad \frac{\sum_i a_i x_i}{\sum_i \sum_j c_{ij} y_{ij}} \tag{1}$$

Subject to:

$$\begin{aligned} x_i - x_j &\leqslant y_{ij} \quad \forall i,\ j \in V, \\ x_j - x_i &\leqslant y_{ij} \quad \forall i,\ j \in V, \end{aligned} \tag{2}$$

$$\sum_i x_i \leqslant s, \tag{3}$$

$$x_r = 1. \tag{4}$$

Contiguity constraint set (5)

$$x_i, y_{ij} \in \{0, 1\} \quad \forall i,\ j \in V, \tag{6}$$

where

$$x_i = \begin{cases} 1 & \text{if node } i \text{ is in } V_r, \\ 0 & \text{otherwise,} \end{cases}$$

$$y_{ij} = \begin{cases} 1 & \text{if node } i \text{ is in } V_r \text{ and node } j \text{ is not in} V_r, \\ 0 & \text{otherwise,} \end{cases}$$

a_i = weight of node i (population),
c_{ij} = number of lanes of road segment i, j,
s = maximum size of V_r,
r = index of anchor node $\in V_r$.

The objective function (1) maximizes the bld ratio of the partition's total population to the number of lanes leaving the partition associated with arcs having one node in the set V_r and the other end-node is not in the set V_r. It could be as easily cast in the form which maximizes cte. Constraint (2) ensures that if node i is in the partition V_r and node j is not, then arc y_{ij} must be equal to 1, as it is a connection between the partition containing node r (i.e. V_r) and the rest of the network. If an arc's y_{ij} value is equal to 1, then the arc's capacity as measured in terms of the number of lanes (heading from i to j) is included in the denominator of the objective function and the total exit capacity of the partition. Constraint (3) limits the search to partitions less than a specified size as measured in terms of the number of nodes within the partition. Such a condition limits the size of the search neighborhood, since the objective is to find small geographical areas that may have difficulty in a sudden evacuation scenario. The size of the search neighborhood must always be less than the size of the entire network. Otherwise, it may be possible to select the entire network where no exit capacity exists. Constraint (4) maintains that anchor node r must be in the partition node set V_r. Constraint set (5) specifies that the partition must be contiguous, which implies that it is a connected cluster. This constraint set will be specified in more detail in a subsequent section. Constraint set (6) ensures that all y_{ij} and x_i are binary integer variables. The above formulation is related to the partition model of Pirkul and Rolland (1994), with the exception that it employs a nonlinear objective function.

The objective (1) in this formulation can be stated in the following equivalent form:

$$\frac{\sum_i a_i x_i}{\sum_i \sum_j c_{ij} y_{ij}} \geqslant M \quad \text{or} \quad \sum_i a_i x_i \geqslant M \sum_i \sum_j c_{ij} y_{ij},$$

where the objective is to maximize M. On the right-hand side of the equation, it is evident that this is a nonlinear objective as we have the variable M multiplied by other variable terms $c_{ij} y_{ij}$. To transform the formulation into a linear integer programming (LP/IP) problem, we can fix M as a constant, add the above right-hand expression as a constraint, and append a new objective: minimize the size of the cluster. In other words, find the smallest cluster (in nodes) that includes node r, such that the ratio of the cluster's population to lane-capacity is greater than M. This alternative LP/IP formulation is given as follows:

Objective:

$$\text{Minimize} \sum_i x_i \tag{1}$$

Subject to:

$$\sum_i a_i x_i \geqslant M \sum_i \sum_j c_{ij} y_{ij} \tag{2}$$

$$x_i - x_j \leqslant y_{ij} \quad \forall i,\ j \in V \tag{3}$$

$$\sum_{i=1}^{n} x_i \leqslant s \tag{4}$$

$$x_r = 1 \tag{5}$$

Contiguity constraint set. (6)

$$x_i, y_{ij} \in \{0,1\} \quad \forall i,\ j \in V \tag{7}$$

where M is the minimum partition threshold, and where all other notation is as defined previously.

This formulation can be used to solve for critical clusters which maximize cte or bld values depending upon which values are used in the definition of the objective (7). Technically, it can be used to search for the smallest cluster which has a specified minimum cte or bld value. It may be necessary to use several constraint values, M, in order to identify the highest feasible M value and its associated critical cluster for a given anchor node r.

Because a cluster can alternatively be viewed as either a set of nodes or a set of arcs, the y_{ij} variables are not seen as important in finding a critical cluster. In the model only the x_i variables must be restricted to integer values with an upper bound of 1. The y_{ij} variables will be integer as long as the x_i are integer in value. For large data sets, this greatly reduces the number of branches that occurs during the branch-and-bound procedure.

5. Ensuring that the critical cluster is contiguous

The contiguity constraint set (6) was not specified mathematically above in order to keep the CCM formulation as simple as possible. In application it is worthwhile to attempt to solve a problem without such constraints. If the optimal solution to the above model is connected without such constraints, then these constraints are not necessary. We will describe how we can make selective use of such constraints later. However, it is worth specifying these constraints unambiguously. To structure such a constraint set consider a node l that is different from the anchor node, but has been selected as a part of the cluster. If the cluster is contiguous, then a path must exist that connects the anchor node to node l and which lies entirely within the cluster. We can ensure that a path exists within the cluster between the anchor node r and node l in the following manner:

Contiguity constraint set for node l, (6)

$$\sum_j P^l_{rj} = X_l, \tag{6a}$$

$$\sum_k P^l_{ik} - \sum_k P^l_{ki} \quad \text{for each } i \text{ where } i \neq r, l, \tag{6b}$$

$$\sum_k P^l_{kl} = X_l \quad \text{for each } l, \tag{6c}$$

$$\sum_j P^l_{ij} \leqslant x_i \quad \text{for each } i, j \in V, \tag{6d}$$

$$\sum_j P^l_{ij} \leqslant x_j \quad \text{for each } i, j \in V, \tag{6e}$$

where additionally,

$$P_{ij}^{l} = \begin{cases} 1 & \text{if arc } ij \text{ is in path leading from node } l \text{ to node } r, \\ 0 & \text{otherwise,} \end{cases}$$

r = index of anchor node $\in V_r$.

The above contiguity constraint set ensures that a path will exist wholly within the cluster between node l of the cluster and the anchor node. The first constraint (6a) specifies that an arc in a path leading to node l from the anchor node r must be chosen to depart from the anchor node r if node l is chosen for the cluster. If node l is not chosen for the cluster then no such path requirement is enforced. This constraint ensures that if node l is in the cluster, then there is one arc that originates from the anchor node that is an element of the path between the anchor node and node l. The next constraint (6b) specifies that for any node i that is not the anchor node or the destination node l, the number of arcs chosen in the path to l that terminate at node i must be equal to the number of arcs chosen in the path l that originate from node i. This assures that a path does not terminate at an intermediate node on its way to node l. Constraint (6c) specifies that the number of arcs chosen in the path l that terminate at node l is equal to 1 if node l is in the cluster and 0 if it is not. This constraint ensures that if node l is in the cluster, then there is one arc terminating at node l that is an element of the path between the anchor node and node l. The last two constraints specify that for any arc used in the path from the anchor node to node l must have both endpoints within the cluster.

The set of constraints (6a)–(6e) given above for node l, can be replicated for all nodes of the network other than the anchor node. This would then represent a complete contiguity constraint set (6). The size of the model without contiguity constraints, in terms of the numbers of variables and constraints is: n x_i variables; $2m$ y_{ij} variables; $2m + 3$ constraints, where n is the number of nodes and m is the number of arcs. A full complement of contiguity constraints adds: $2mn$ P_{ij}^{l} variables and $n^2 + 4mn + 2n - 1$ constraints. It is easy to see that the above model is a manageable size without the contiguity constraints, but is far too large to be practical when all possible contiguity constraints are employed. Our approach in using selected contiguity constraints is described in the next section.

The contiguity constraints structured above use a set of path variables for each potential node l that could be selected for the cluster. An alternative approach could be to use a network flow approach, where flow originates at the anchor node and must flow across arcs wholly within the cluster and where one unit of flow must arrive at each node selected for the cluster. Such an approach requires fewer constraints and variables, however, constraints that ensure that flow must be on arcs wholly within a cluster cannot be written in an integer friendly form (in contrast to the integer friendly form of constraints (6d) and (6e)).

6. Generating optimal and heuristic solutions for the CCM

The general approach to deriving optimal solutions for the CCM involved writing a software driver to generate a mathematical programming system (MPS) file. In this context, a driver is a program that generates an instance of an optimization model (e.g. critical cluster model) in the

MPS file format. CPLEX™ was used for solving all integer programming problems. Fig. 3 shows the procedure for deriving an optimal solution to the CCM. The driver accepts a textual representation of the network and a set of parameters that specify the anchor node and scale limit. This driver outputs an MPS file for input into CPLEX, which solves the problem optimally. The arrow running from the solution back to the driver implies that this procedure is iterative.

In many cases, the initial solution to the model (without contiguity constraints) for a given anchor node and scale limit was not contiguous. Because the objective of the modified linear formulation is to minimize the size of the cluster, the model consistently selected the anchor node and a non-contiguous set of heavily weighted satellite clusters. In these cases, an optimal contiguous solution could be "teased out" by fixing one of the heavily weighted nodes from a satellite cluster out of the solution with an additional constraint. However, there needed to be a basis for determining whether or not the node to fix out was clearly not an element of the optimal solution.

To accomplish this, the node was first "fixed-in" to the solution, along with the required contiguity constraints for that particular node only. This usually generated a solution that was clearly inferior to a simple heuristically derived solution about the anchor node. Hence, one can, without loss of generality, fix that node out in subsequent model runs. Deriving heuristic solutions to the CCM is the subject of a companion paper (Cova and Church, 1997). The best known contiguous cluster value that was derived from a heuristic algorithm described in Cova and Church was used as the initial M value. If CPLEX failed to find a cluster that contained the "fixed-in node" and matched or exceeded the M value, then that node could not be part of the optimal solution, and it could justifiably be fixed out. If a new "higher" cluster was found that included the fixed-in node, then that cluster would become the new best M value, and the search for a greater cluster could be continued.

The M parameter that was introduced in order to make the model linear can be viewed as a form of threshold. Fig. 4 shows a "clustergram" that depicts a graphic interpretation of the M value. In a sense, the M value changes the query to, "Is there an evacuation starting scenario within which residents at this node might participate, within the scale limit, that exceeds M"? In short, the size limit (s) (see model formulation) bounds the size of the cluster to search for, and the M value puts a lower bound on the cluster difficulty value to search for. This means that it is possible that no solution may exist for a given anchor node and scale limit if there is no cluster with a difficulty value that exceeds M. For this reason, this approach to solving the model is useful for performing an optimality test on a known cluster. In other words, if a known cluster has a difficulty value m and a size s, we can test whether it is the optimal cluster for any of its nodes by substituting m for M.

Optimal solutions were derived for 40 randomly selected nodes from 4 real world street networks (10 each) at 3 scale limits 10, 25, and 50 (120 problems). The networks ranged in size from 200 to

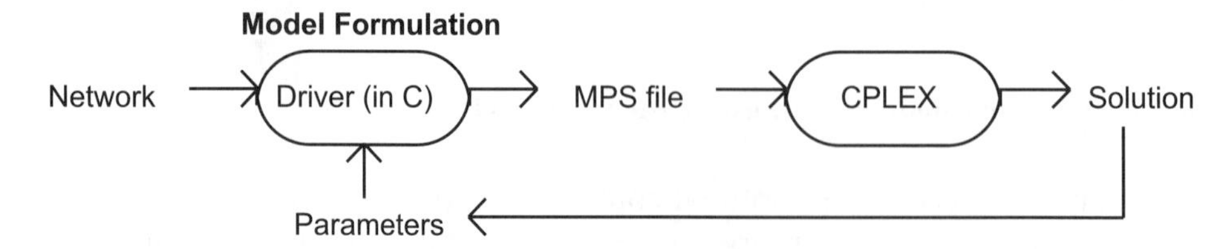

Fig. 3. The process for deriving optimal solutions to the CCM.

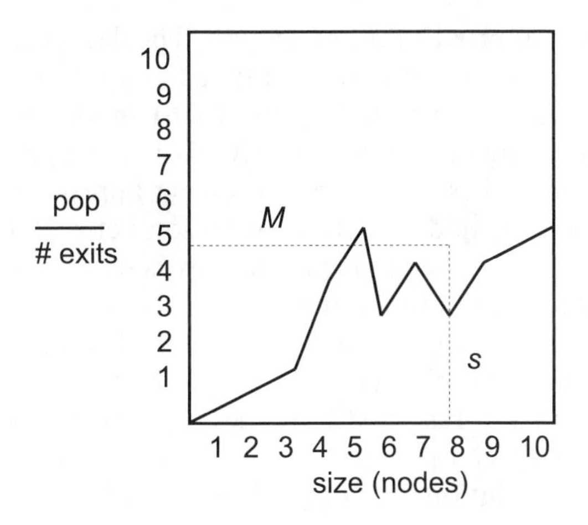

Fig. 4. A graphical interpretation of the M parameter.

300 nodes, and the overall process was labor intensive due to the fact that the decision process as to which nodes to fix-in and fix-out was not completely automated. An average problem took anywhere on the order of a half-hour to an hour to completely solve and resulted in approximately 30 runs of CPLEX to systematically fix-out as many as 15 nodes to arrive at the optimal contiguous cluster. Such a process could be automated to reduce operator involvement, but still would be computationally intensive in terms of the use of the CPLEX software. Obviously, such a model relying solely on general purpose software for generating solutions is of limited use. Research on alternate optimal approaches could be of considerable value for application.

The real value of the above optimization model and general purpose solution software is in its use to test heuristic approaches to the CCM. Without a competitive solution algorithm, it is

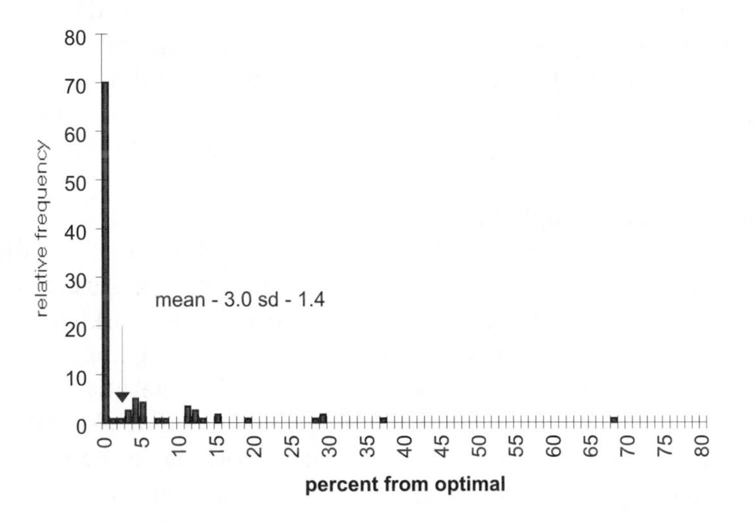

Fig. 5. An example of the heuristic's solution quality.

necessary to rely on a heuristic approach for application. The design and acceptance of a heuristic can be made with confidence, only when known optimal solutions can be used in testing a heuristic's performance. The solutions generated by the CCM model and general purpose optimization software were used in testing the efficacy of a CCM heuristic presented in Cova and Church (1997). The heuristic algorithm relies on a region-growing approach to construct a contiguous cluster from each potential anchor node in a transportation network by iteratively adding nodes on the fringe of the current cluster. A semi-greedy strategy is used to add a node to the cluster for each iteration, where the next node is randomly selected from a list of nodes within a specified percent of the node that most increases the objective value. The heuristic was found to be relatively robust at identifying optimal critical clusters. Fig. 5 shows a histogram of the mean percent from optimal when the heuristic was restarted 128 times from each node and any node within 77% of the greedy choice at each step is a potential candidate to add to the cluster. In this case, the heuristic identified the optimal solution 70% of the time and the mean percent from optimal was 3%.

7. Producing risk maps using the CCM and GIS

Maps are often used in depicting risk which varies spatially, like flood plains and seismic zones subject to liquefaction. The designation of a flood plain, for example, involves identifying which areas will be flooded with some annual frequency (e.g. once in 50 yr). A flood plain map then depicts those areas which are subject to a reasonable risk of flooding (i.e. probability of an event occurring on a given land unit). Such maps depict "event risk occurrence", but not the risk the occupants face when an emergency evacuation is made. The most common map associated with evacuation involves the depiction of evacuation routes and safe zones (like shelters). Such maps depict an evacuation plan for a designated area, but not the risks. A critique of such maps can be found in Dymon and Winter (1993).

What is proposed in this work and in Cova and Church (1997) is the development of risk maps of potential evacuation difficulty, which could be used in conjunction with event-risk maps to develop better evacuation planning maps. The major objective would be to identify places of high event-risk and high evacuation-risk. For example, hind-sight shows that the Oakland Hills area was such an area. Advanced knowledge and public recognition could have possibly averted the tragedy of 1991.

The CCM represents one possible model that can be used to estimate small neighborhood evacuation risk difficulty. The higher the bld or cte, the greater the possible problems that may be encountered in an emergency evacuation. If each node of the network is used as a anchor node for the CCM, it is then possible to label each node on the transport network with a risk measure like bld or cte. Given a spatial depiction of node locations and risk values, it is possible to perform two types of mapping functions: (1) interpolate lines of equal risk, like elevation contours, and develop a risk contour map; or (2) classify each node according to its "relative risk value" and map nodes and arcs using some type of color scheme or gray scale according to the risk.

To develop such a map would be realistically out of the question for a large area, unless many of the functions were automated. It is only natural to select a GIS platform to supply much of this functionality. For our project, we selected the ARC INFO GIS system, although other systems would support such an application as well. Essentially, data in the GIS was represented by a set of coverages, including network and population information. An export function using the ARC INFO macro language was developed which produced a forward star data structure for the associated road network. This data structure was used by the MPS setup program and the heuristic.

All network data was stored within the GIS and was exported in the special form for a given selected anchor node. The CCM model was then solved (using either heuristic or CPLEX general purpose LP/IP software system) and the result was imported back into the GIS. The application was automated so that each node was systematically selected as an anchor node and solved by the

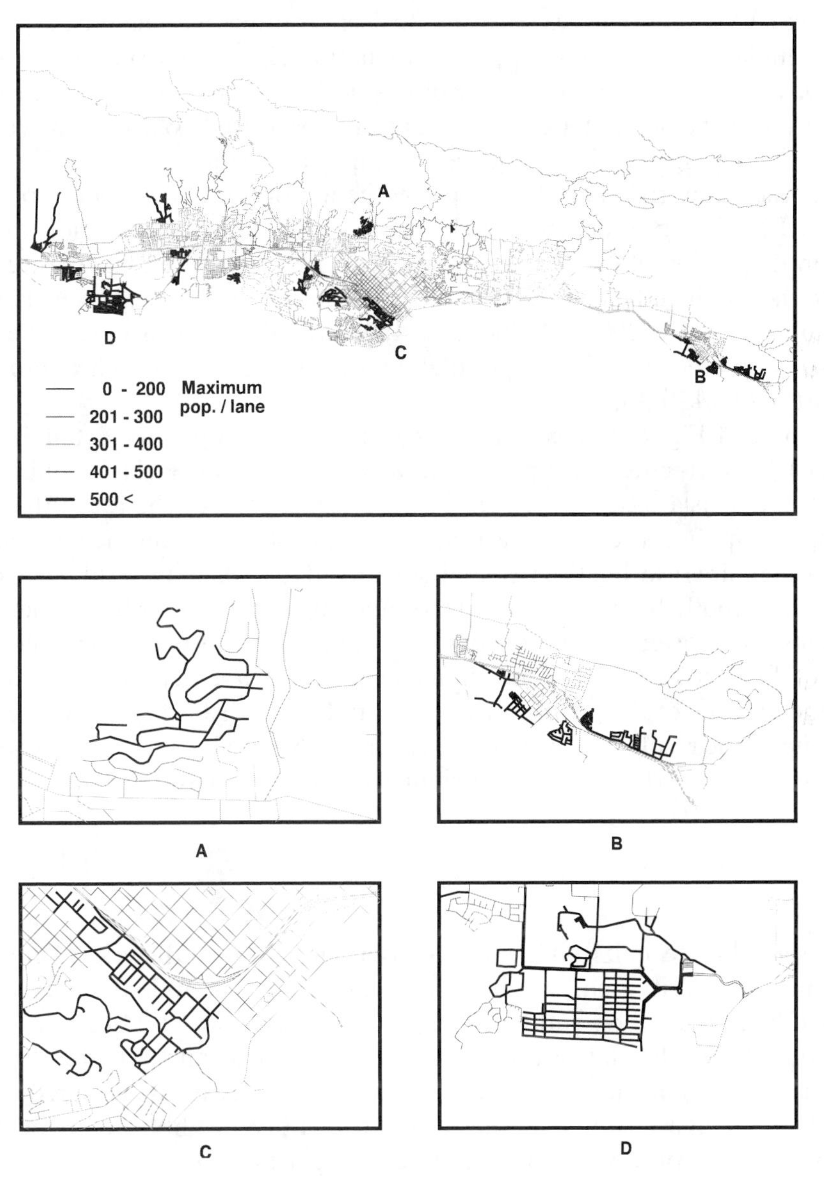

Fig. 6. An evacuation vulnerability map of Santa Barbara.

heuristic. The risk values, e.g. bld, were then imported into the GIS and assigned as attributes to the nodes. Each identified critical cluster represented a set of nodes and a bld value. After solution for a specified anchor node was determined, each node within the critical cluster was tested to see if the new bld value was higher than the current bld value in the data layer. If the new value was higher, then the node attribute for the critical value was set at that new value. If the new value was lower, then the node attribute for the critical value was left unchanged. Essentially, for each node the critical value is the highest value found associated with all critical clusters found which included that node.

After all nodes were considered as possible anchor nodes for solving a CCM, then several Arc Macro Language (AML) routines were executed. These routines assigned the evacuation difficulty of each arc the higher of the nodal endpoint evacuation risk values (i.e. bld values). Then each node and arc was categorized by the relative bld value. These categories were then assigned either a color or a gray scale value, and a complete map was produced. Without the capabilities of the GIS, this type of mapping exercise would be too time consuming.

An example evacuation risk map developed by the use of the CCM model coupled with ARC/info is given in Fig. 6. Fig. 6 depicts a risk map for the south coast region of Santa Barbara County. The upper portion of Fig. 6 depicts the entire south coast region. This region is bounded by the ocean to the south and a mountain range to the north. A gray scale was used to depict bulk lane demand in ranges of 0–200; 201–300; 301–400; 401–500; >500 people per exit lane. Census population data was used to estimate population values and a NavTech database was used to depict transport network links.

In the lower part of Fig. 6, four separate areas are shown in greater detail. They are Mission Canyon (upper left), Carpenteria (upper right), downtown Santa Barbara southwest of highway 101 (lower right), and Isla Vista (lower left). These depict four of the areas that appear as very dark in the upper map. Because of the natural foliage and steep terrain the mission canyon area is a region of very high fire risk. If a fire risk map were available, then it would be possible to identify Mission Canyon as both high evacuation risk and high fire risk. Most other Santa Barbara foothill locations have lower evacuation risk, thus planning efforts can be concentrated in such special areas of high risk. As an aside, homeowners and fire department officials have been convinced of the urgency of the problem by this map. Even a district supervisor called a planning meeting based upon the results of this mapping exercise. Members of the homeowners association are now suggesting the need for a detailed simulation and evacuation plan.

8. Conclusion

We have presented a specialized network partition model called the critical cluster model. This model can be used to identify small neighborhoods about a given node that have potentially risky combinations of high population and low exit road capacity. The CCM can be used to identify a contiguous nodal cluster that maximizes bulk lane demand or an estimate of network clearing time. Both measures, while not exact, are assumed to be reasonable surrogate measures of evacuation risk. Although it would be desirable to identify neighborhoods at risk by a micro simulation model, such a process would require defining the neighborhood in advance. The CCM model can be used to perform this task.

We have presented details on the solution of the CCM using general purpose optimization software. Although the model does take considerable time to solve optimally, optimal solutions have been used to test the efficacy of a heuristic process presented in a companion paper (Cova and Church, 1997). Further research is needed in testing alternative approaches for solving the CCM.

The general issue of mapping event-risk (e.g. flooding) was discussed along with how the CCM model can be used to map evacuation risk. Details on the integration of the CCM with GIS were also presented. Finally, results of a loosely coupled model system using ARC/info GIS and the CCM were presented and discussed. Results from this model as presented in map form have affected the perception of such risk in the Santa Barbara area. Local fire department officials, homeowners, and public officials are currently working to address this problem in one of the areas identified as higher than average risk (in terms of both event-risk and evacuation risk).

Acknowledgements

The authors appreciate the helpful comments provided by the reviewers of the original draft of the manuscript. Network data was supplied by Navigational Technologies, under an agreement to the National Center for Geographic Information and Analysis. Support by the National Science Foundation (NSF SBR96-00465) is gratefully acknowledged.

References

Cova, T.J., Church, R.L., 1997. Modeling community evacuation vulnerability using GIS. International Journal of Geographic Information Science 11, 763–784.

Dymon, U.J., Winter, N.L., 1993. Evacuation mapping: the utility of guidelines. Disasters 17, 12–24.

Hobeika, A.G., Jamei, B., 1985. MASSVAC: a model for calculating evacuation times under natural disasters. Emergency Planning, Simulations Series 15, 23–28.

Jin, L.M., Chan, S.P., 1992. A genetic approach for network partitioning. International Journal of Computer Mathematics 42, 47–60.

Johnson, D.S., Aragon, C.R., McGeoch, L.A., Schevon, C., 1989. Optimization by simulated annealing: an experimental evaluation; part I, graph partitioning. Operations Research 37, 865–892.

Kernighan, B.W., Lin, S., 1970. An efficient heuristic procedure for partitioning graphs. Bell Systems Technical Journal 49, 291–307.

Laguna, M., Feo, T.A., Elrod, H.C., 1994. A greedy randomized adaptive search procedure for the two-partition problem. Operations Research 42, 677–687.

Lindell, M.K., Perry, R.W., 1991. Understanding evacuation behavior: an editorial introduction. International Journal of Mass Emergencies and Disasters 9, 133–136.

Office of Emergency Services, 1992. The East Bay Hills Fire – A Multi-agency Review of the October 1991 Fire in the Oakland/Berkeley Hills. East Bay Hills Fire Operations Review Group, Governor's Office, Sacramento, CA.

Owen, M., Galea, E.R., Lawrence, P.J., 1996. The EXODUS evacuation model applied to building evacuation scenarios. Fire Engineers Journal 56, 26–30.

Perry, R., 1985. Comprehensive Emergency Management: Evacuating Threatened Populations. JAI Press, London.

Pidd, M., Eglese, R., de Silva, F.N., 1997. CEMPS: a prototype spatial decision support system to aid in planning emergency evacuations. Transactions in GIS 1, 321–334.

Pirkul, H., Rolland, E., 1994. New heuristic solution procedures for the uniform graph partitioning problem: extensions and evaluation. Computers and Operations Research 21, 895–907.

Sheffi, Y., Mahmassani, H., Powell, W.B., 1982. A transportation network evacuation model. Transportation Research 16A, 209–218.

Shough, W.H., Magdalena, A.T., Stalberg, C.E., 1992. Hazard mitigation report for the East Bay fire in the Oakland-Berkeley hills. FEMA, FEMA-919-DR-CA.

Sinuany-Stern, Z., Stern, E., 1993. Simulating the evacuation of a small city: the effects of traffic factors. Socio-Economic Planning Sciences 27, 97–108.

Sorensen, J.H., Vogt, B.M., Mileti, D., 1987. Evacuation: An Assessment of Planning and Research. Oak Ridge National Laboratory ORNL-6376, Tennessee.

Southworth, F., 1991. Regional Evacuation Modeling: A State-of-the-Art Review. Oak Ridge National Laboratory ORNL-11740, Tennessee.

Stern, E., Sinuany-Stern, Z., 1989. A behavioral-based simulation model for urban evacuation. Papers of the Regional Science Association 66, 87–103.

Tufekci, S., Kisko, T.M., 1991. Regional evacuation modelling system REMS: a decision support system for emergency area evacuations. Computers and Industrial Engineering 21, 89–93.

PERGAMON

Transportation Research Part C 8 (2000) 337–359

TRANSPORTATION RESEARCH PART C

www.elsevier.com/locate/trc

Spatial decision support system for hazardous material truck routing

William C. Frank [a], Jean-Claude Thill [b,*], Rajan Batta [a]

[a] *Department of Industrial Engineering and National Center for Geographic Information and Analysis, State University of New York at Buffalo, Buffalo, NY, USA*

[b] *Department of Geography and National Center for Geographic Information and Analysis, State University of New York at Buffalo, Buffalo, NY 14261, USA*

Abstract

Shipping hazardous material (hazmat) places the public at risk. People who live or work near roads commonly traveled by hazmat trucks endure the greatest risk. Careful selection of roads used for a hazmat shipment can reduce the population at risk. On the other hand, a least time route will often consist of urban interstate, thus placing many people in harms way. Route selection is therefore the process of resolving the conflict between population at risk and efficiency considerations. To assist in resolving this conflict, a working spatial decision support system (SDSS) called Hazmat Path is developed. The proposed hazmat routing SDSS overcomes three significant challenges, namely handling a realistic network, offering sophisticated route generating heuristics and functioning on a desktop personal computer. The paper discusses creative approaches to data manipulation, data and solution visualization, user interfaces, and optimization heuristics implemented in Hazmat Path to meet these challenges. © 2000 Elsevier Science Ltd. All rights reserved.

Keywords: Hazardous material; Routing; Spatial decision support system; GIS; Highway network

1. Introduction

Transportation of hazardous materials (hazmat) is necessary for an industrial economy to exist. For example, gasoline is consumed by millions of cars, trucks and buses in the US, but needs to be transported from refineries/shipment points to gas stations. Gasoline is clearly

* Tel.: +1-716-645-2722 ext. 24; fax: +1-716-645-2329.
E-mail address: jcthill@acsu.buffalo.edu (J.-C. Thill).

PII: S0968-090X(00)00007-3

hazardous in the sense that a truck carrying gasoline, if involved in a collision, can result in personal casualty and serious damage to other vehicles and property. However, it begs us to the question: what is hazardous material? The Federal Hazardous Materials Transportation Law (October '94) states that "The secretary of transportation shall designate material (including an explosive, radioactive material, etiologic agent, flammable or combustible liquid or solid, poison, oxidizing or corrosive material, and compressed gas) or a group or class of materials as hazardous when the secretary decides that transporting the material in commerce in a particular amount and form may pose an unreasonable risk to health and safety or property".

To understand the magnitude of the problem, one simply has to note that there are approximately 500,000 daily hazmat shipments (Nozick et al., 1997). Hazmat is transported by air, highway, railway, and water. In a perfect world, incidents would not occur when transporting hazmat. However, due to the risks associated with transportation, hazmat incident costs to health and property are quite substantial. Highway incidents alone were responsible for 15 deaths, 511 injuries and $19,800,000 of property damage in 1993 (US DoT, 1994).

Often, a hazmat accident results in the spillage of a dangerous chemical that necessitates an evacuation of the population in the surrounding area. Evacuations are costly and are often used as a surrogate measure of unreasonable risk to health and safety. US DoT (1994) statistics indicate that the average number of people evacuated per highway incident is above one hundred, which is substantial since people evacuated during an incident are placed at an unreasonable risk.

There are several strategies that come to mind to mitigate the risk associated with such a situation. First, by judiciously selecting the route we can reduce the probability of an accident. Also, choosing a route through less populated areas could also be used to reduce the number of people placed at unreasonable risk. Next, vehicle and container design could be altered to reduce the severity of a release once an accident has occurred. Finally, accident probability could also be reduced by improved driver training.

In this paper, we focus on risk mitigation via route selection. We present a spatial decision support system (SDSS) that recommends one or more routes compatible with stated objectives and constraints. While previous hazmat route solutions have often been advanced for small illustration networks, our goal is to operationalize our SDSS on a large-scale transportation network. For this purpose, the implementation makes use of a realistic road network of about 57,000 intersections maintained by the Bureau of Transportation Statistics. The sheer size of the network calls for innovative design solutions discussed in the rest of the paper.

In Section 2, the problem of hazmat routing is discussed in greater detail and our general approach to its solution is presented. A brief literature review follows in Section 3. The remaining sections discuss in detail the components of the proposed DSS called Hazmat Path. The first part of Section 4 describes the user interface, which includes the routing parameters, network intersections and display options. The second part of Section 4 discusses the data structures that include the map objects and link attribute costs. The methods used to generate link attributes that are SDSS inputs (time-dependent population at risk, accident rates and link travel speeds) are presented in Section 5. Section 6 is devoted to the route generation algorithm of the SDSS: the mathematical formulation of the problem is presented and solution methods are outlined. The paper concludes in Section 7 with an overview of the main properties of the proposed routing system and of future extensions that it can accommodate.

2. Route selection strategies: problem statement

The goal of this paper is to develop a working SDSS capable of handling a realistic transportation network covering a multi-state region or an entire country and their numerous loading and delivery points. DSSs are computer-based systems that share several key characteristics, including (Sprague and Carlson, 1982; Geoffrion, 1983; Turban, 1995):

- Assisting users in their decision making in a flexible and interactive manner.
- Solving all classes of problems, including ill-structured ones.
- Having a powerful and user-friendly interface.
- Having a data analysis and modeling engine.

Conceptually, SDSSs are special cases of DSSs. Densham (1991) effectively argues that they differ markedly from general DSSs in some key respects. They need spatial capabilities for data input, display of complex relations and structures, analysis, and cartographic output. The architecture of our Hazmat Path SDSS is depicted in Fig. 1. Hazmat Path is a full-featured SDSS allowing for interactive problem editing, comparison of solutions, and evaluation of decision criteria.

There is a difference between conceptualizing and actually developing a working SDSS. When developing a working SDSS, numerous trade-offs exist between level of effort and SDSS quality. A conceptual SDSS does not have this real world trade-off. Commercial geographic information systems (GIS) software applications, which often provide the spatial information processing engine of SDSSs, are designed with tremendous flexibility but at a cost to time for producing results. Displaying results in a timely manner is imperative for a SDSS. Therefore, the choice has been

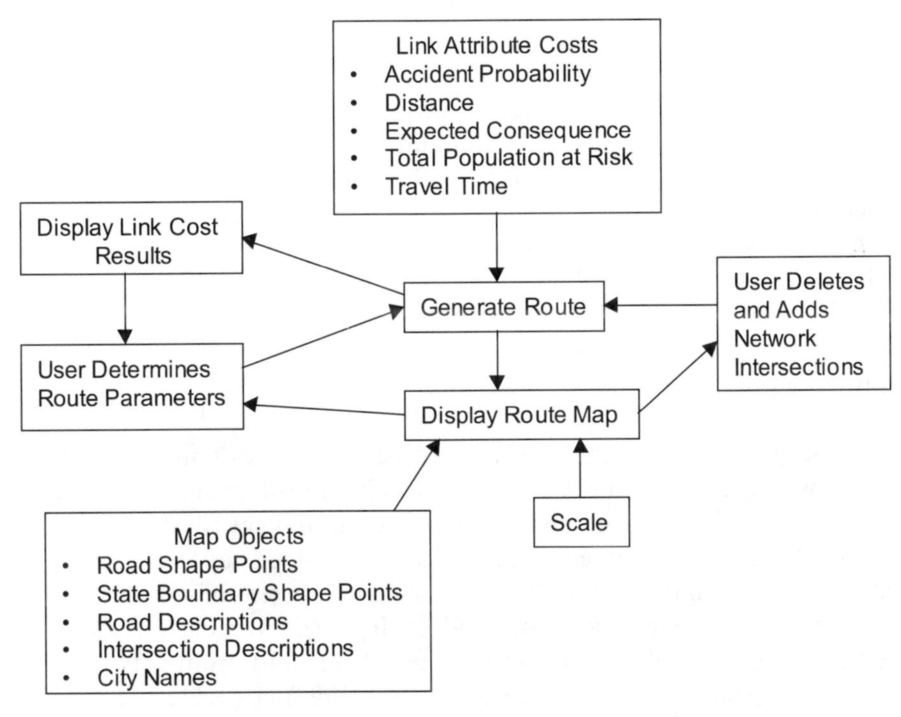

Fig. 1. Flow of information in Hazmat Path.

made to design a Windows-based software application tailored to the SDSS and running on mid-range desktop computers instead of using an existing commercial application.

Since most long-distance travel occurs on highways, they compose the network on which travel time is minimized. They primarily connect and transverse large population centers. Therefore, minimizing travel time puts a large population at risk. Population centers can be avoided by using slower and less direct non-interstate roads. Thus, the strategy of selecting a route to minimize travel time and the strategy of minimizing population at risk are conflicting in nature.

Arbitration between these conflicting costs makes use of the capability of the SDSS to generate several alternative truck routing solutions based on single optimization criteria. A widely used approach is to combine several attribute costs into a single cost. The new cost is often taken to be a linear function of population at risk, distance, time and accident probability. With a single link cost, a simple solution method (e.g., Dijkstra's shortest path algorithm) can be used to determine a vehicle route. By varying the weights of the attribute costs, different routes can be generated. The process of varying the weights indicates the sensitivity that the attribute costs have on route selection.

Another approach to route selection is having multiple objectives. Minimizing travel time and total population at risk is an example of multiple objectives. Many routes can be in the solution set because it contains all of the non-dominated routes. The number of non-dominated paths can become very large in networks typical of real-world applications, thus rendering the approach unpractical.

Still another approach is to minimize one cost attribute while limiting the sums of other cost attributes. This type of problem is called a constrained shortest path (CSP). It is used in this research as follows. Travel time is minimized while four other criteria – total population at risk, distance, accident probability, and consequence – are constrained. Travel time is chosen for the objective function because, it best represents financial cost. As noted earlier, minimizing time for a long route often produces a path, which places large numbers of people at risk. Consequently, the CSP problem is a method that resolves the minimum time and minimum population conflict.

An added complication to the CSP problem is that some link attributes are assumed in this research to be a function of the time of day. Link travel time is one of the temporal link attributes. The level of traffic congestion influences link travel time and, congestion is affected by time of day. This problem is called the time-dependent constrained shortest path (TCSP).

3. Literature review

The literature relevant to our hazmat routing problem can be organized into studies that implement a solution within a GIS and those that use a SDSS. Both groups use GIS techniques for data storage, data manipulation (for instance, to generate link attribute costs) and to display solutions on a map. SDSS-based implementations have the added capability of allowing the user to easily specify shipment origins and destinations, along with removing intersections and roads from route consideration. GIS provides an ideal environment for design and management of hazmat routes because of its ability to integrate multi-theme and multi-source data into an operational information system. Souleyrette and Sathisan (1994) advocate the use of GIS for the comparative study of pre-defined, alternative routes on selected characteristics. This type of analysis is illustrated by a case study of Nevada highway and rail routes for shipment of high-level radioactive materials.

Abkowitz et al. (1990) envision GIS to fulfill functions in hazmat transportation that reach beyond those of input data storage, data manipulation and output map display. They propose a GIS application of hazmat routing on a large-scale network of size similar to the one used in this research. The routing algorithm handles a single routing criterion, but compromise or negotiated solutions can be achieved by ex-post comparison of solutions generated on different routing criteria. In their implementation, the authors use criteria of distance (a measure of efficiency) and populations at risk (a measure of safety). The latter is measured by the tally of people within a given bandwidth (0.25, 0.5, 1, 3, 10, or 25 miles) along highway segments. A gradient method is used in conjunction with Thiessen polygons to allocate enumeration district population to predefined buffers.

In their study of transportation of aqueous hazardous waste in the London, UK, area, Brainard et al. (1996) apply weighting schemes to identify routing solutions that compromise between criteria. A single link attribute is calculated from a weighted combination of link attributes (population at risk, groundwater vulnerability, and accident likelihood computed from historical records). A labeling algorithm is then used to minimize this new attribute combination. Solutions associated with alternative weights can be compared visually (map display) and statistically (for instance, travel time, expected number of accidents, highway mileage versus mileage on local roads) for risk assessment purposes. A similar approach is suggested by Lepofsky et al. (1993). These authors stress that the network data model used by GIS to represent individual highway segments allows for their detailed attribute characterization and for the efficient modeling of segment-specific risk of hazmat shipping. Following established practices in the matter, they define risk by a combination of accident likelihood, probability of a release, consequence of an incident measured in terms of population exposed, and risk preference of affected interest groups. Generation of the accident likelihood and consequence factors is considerably enhanced in a GIS environment. Lepofsky et al. (1993) present the case study of a shipment from the California/Arizona border to Vandenberg, CA, with a rather small-scale network representing the highway system of Southern California. A route through Los Angeles is produced when travel time was minimized. On the down side, this route has the highest population exposure. A compromise solution is retained with weights of 25% travel time and 75% accident likelihood, allowing for an acceptable trade-off between efficiency and safety.

Useful components of a hazmat SDSS are outlined by Baaj et al. (1990) and discussed in greater detail by Erkut (1996). They include using different routing solution methods. Also, interactive post-editing of generated and displayed routes gives control of the process to the analyst: a user may wish to create a detour around a sensitive location, or it may be deemed desirable to remove some links from the network and have a new route generated. Usually there is no "best" route. Minimizing one criterion typically conflicts with minimizing another. By an iterative process of displaying routes, using different solution methods and creating detours, a compromise route can be developed.

A working SDSS is developed by Lassarre et al. (1993). Their network covers 600 km^2 in the region of Haute-Normandie, France. The SDSS has the capability of loading geographical overlays that include hydrology, railways and population densities. Dijkstra's algorithm is used to compute a route with the lowest risk. Risk is defined to be the product of accident rate and people affected in an accident. The latter is composed of the day population in adjacent polygons (buffers of given width around highway segments), on links (traffic of road segments), and nearby points (children at school). Their SDSS also has the ability of removing links and nodes or areas with particular characteristics (i.e., single-lane roadways) from the network.

Coutinho-Rodrigues et al. (1997) propose a personal computer-based SDSS for multi-objective hazmat location and routing problems. The routing solution generator offers several alternative multi-objective optimization techniques, including the weighted method, the constraint method, and goal programming. The user interacts with the SDSS through multiple graphical and numerical solution displays. Time-dependence of attributes and solutions are not accommodated by the SDSS in its current form. Geographic visualization of attributes and solution paths is limited to graph windows that hardly allow for geographic reasoning on solutions and other geo-referenced information.

4. User interface and data structures

4.1. Map objects display

The user may elect to display several relevant map objects. Fig. 2 shows the options available to the user. In this dialogue box, truck route selection is gray because no route has been generated so far and therefore, cannot be displayed. The user can also specify the size and color of many of these objects.

Displaying individual road attributes is useful when evaluating truck routes. Questions such as "Are there any alternative routes available with a lower total population at risk without greatly increasing travel time"? can be explored and answered by visual examination of suitably attributed map objects. The geo-visualization method used here to display link costs is by road width.

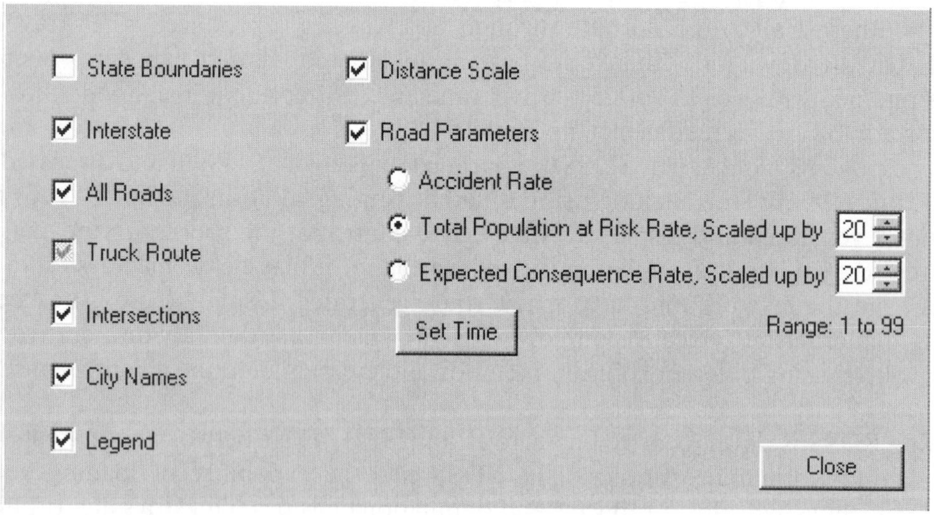

Fig. 2. Map objects selected dialogue box.

In Fig. 2, the attribute "total population at risk" is selected for display and the resulting map is shown in Fig. 3. The truck route is shown with intersection arrival times.

A linear relationship between attribute cost and width of the representation of the road feature is used. The maximum road width is seven pixels. Each incremental pixel width corresponds to a range of the attribute cost, with the requirement that the range of the single pixel start at zero. Therefore, the area of a link on the monitor is a measure of its cost. Furthermore, the origin to destination path displayed with the least area (smallest number of colored pixels) is also the least cost path for that attribute.

The population at risk road parameters (total population at risk and "expected consequence" in Fig. 2) exhibit heavily skewed frequency distributions, where very large values appear in a few large urban areas. A strict application of the rule of proportional symbolic representation leads to the depiction of most roads outside these areas with a width of one pixel. While this poses no problem when the area of interest includes these few large urban areas, for proper visual differentiation, the pixel width should be scaled up as shown in Fig. 2 when they are not included. With default scaling of 20, roads with 1/20 to 20/20 of the true maximum population at risk are displayed with the same maximum width. Roads with zero to 1/20 of the true maximum population at risk have pixel width proportional to population at risk.

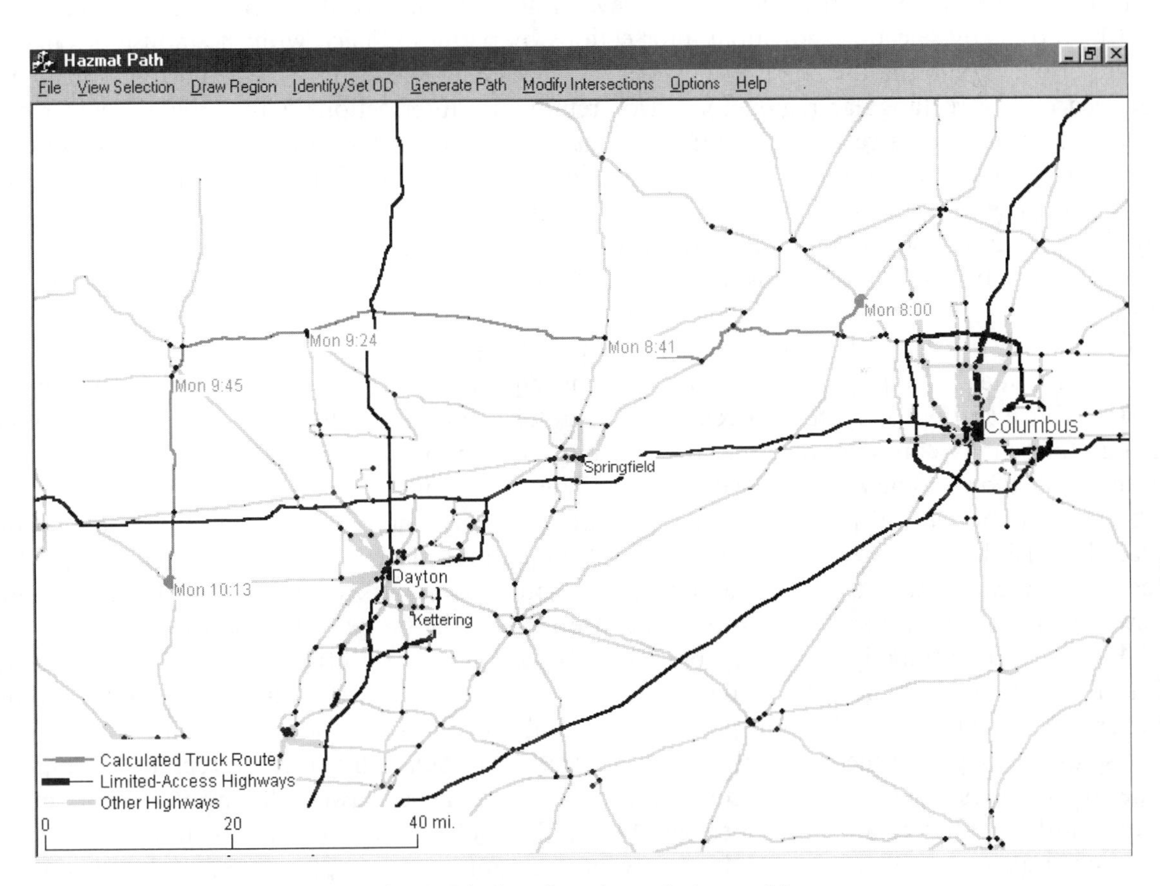

Fig. 3. Display of total population at risk.

Since the area of a link is linearly proportional to one of its attribute costs, the link length times the attribute cost needs to represent the cost to travel that link. The probability of an accident pixel width is a linear function of probability of an accident per mile. Expected consequence width pixel width is also a rate, expected consequence per mile. The same is true for total population. Average population at risk has no equivalent rate and therefore, cannot be displayed.

4.2. Deleting and adding network intersections

Hazmat Path uses the national highway planning network (NHPN) maintained by the Federal Highway Administration (Bureau of Transportation Statistics, 1999). The NHPN has 95,000 nodes. Pre-processing of the original network brings it down to 57,000 nodes. This operation involved removing nodes outside the continental US as well as most of the nodes with only two incident links.

The user has the option of temporarily removing intersections from the network. Intersections temporarily removed from the network cannot be part of the truck route being planned. There are two reasons for removing intersections, namely to reduce solution run-time and to prevent a solution from traversing a particular region for equity reasons or to comply with local or state hazmat traffic regulations.

In our SDSS, three different methods are available to exclude or deactivate intersections. The first method is by selecting individual intersections by point and click with the mouse or cursor. Intersections can be added to the network by the same operation. Also, the cursor can be dragged over a rectangular region to add or remove all intersections within the circumscribed area. Fig. 4 displays a rectangle of intersections being deleted: the intersections represented by clear circles have been removed from the network. The last method is associated with a constraint imposed by the user. Only intersections that are reachable while traveling from the specified origin to the specified destination without violating the said constraint are included in the network. In the current implementation, the constraint is based either on travel time or distance.

Often many hazmat shipments take place between the same origin–destination pairs. Following the same route for multiple shipments could concentrate the risk among a localized group of people above and beyond what is deemed acceptable. Therefore, with such equity considerations in mind, one may consider spreading the risk more evenly over a large population. This can be accomplished by finding D differentiated routes. The metric for total population at risk would be a set of D numbers (total population placed at risk by only one of the routes, total population placed at risk by any two of the routes... total population placed at risk by all D routes). This requires the calculation of the combined total population at risk of any two links along with developing a solution method, both of which are beyond the scope of this paper.

A less computationally intensive approach can be applied by utilizing the network editing features of the SDSS. The user can develop differentiated routes by reducing the overlap of route buffers in populated areas. For this purpose, the total population at risk along each road segment is displayed along with the mile scale so as to reveal the population density in regions of interest. The mile scale is needed for the user to estimate the distance between highway links or more importantly the degree of overlap between link buffers. The approach consists in manually subtracting and adding intersections and routes generated until an appropriate level of differentiation is obtained.

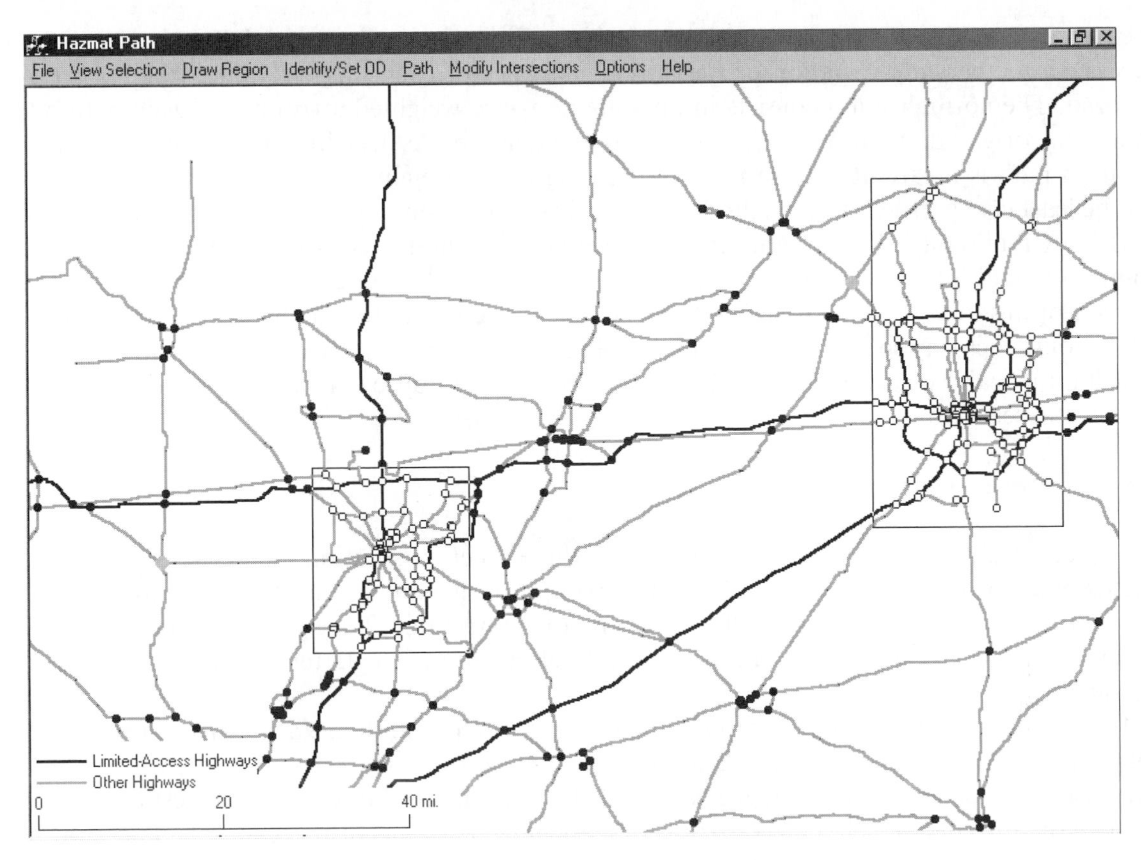

Fig. 4. Display of removal of intersections.

4.3. Route generation parameters

The SDSS can generate two different categories of routes, namely those along the least cost path and those along the least time path with limits on the other attributes. In both instances, the user is prompted to input the distance from a road within which an accident is expected to have harmful impacts on human populations. The area affected by the release of toxic materials and population exposure depends heavily on the properties of the material being shipped and the spill characteristics (Lepofsky et al., 1993). The user-supplied distance is therefore case-specific. It defines the radius of the moving circular buffer used to identify average population at risk when calculating the expected consequence of an accident as well as the bandwidth of buffers around highway segments used when determining total population at risk. More details on the calculation of the population at risk attributes is found in Section 5.

4.4. Least cost path

With the least cost path option, each of the five link attributes is weighted to produce a single composite link cost. As in Lepofsky et al. (1993) and Brainard et al. (1996), Dijkstra's algorithm is then applied to this cost to generate a single path. The user inputs the weights using a dialogue box. Due to the wide variation of measurement scales of each link attribute, weights are normalized. The normalization consists in dividing non-zero weighted attributes of each path by the minimum origin–destination path cost for that attribute. As weights are applied linearly, the solution path is invariant with respect to a scaling of the weight vector.

The least cost path solution method can conceivably use non-temporal accident and population at risk attributes only. If temporal attributes were used without a constraint on time, then a likely solution would involve excessive parking to avoid links with high cost during specific time windows. For instance, a vehicle assigned to a 500 mile ride could optimally be scheduled to travel for a few hours each day so as to take advantage of lower cost temporal attributes, thus unrealistically and impractically delaying delivery for several days. The approach proposed here avoids such unrealistic scenario.

4.5. Constrained least time path

The constrained least time path problem has three different solution methods, all of which are extensions of Handler and Zang (1980). The first approach assumes non-temporal attributes. The remaining two solution methods allow for temporal attributes. They are the intermediate node method and the weight-guided solution method. Discussion of the latter methods is conducted in Section 6.

If a solution method with temporal link attributes is selected, then start time influences the route selected. Start time can be input manually in a dialogue box. The month of the year can also be supplied to account for seasonal fluctuations in the level of congestion. The dialogue box through which the constrained least time path problem is specified is shown in Fig. 5(a).

The lower left panel is labeled shortest paths without constraints. With the exception of the minimum time, non-temporal attributes are used to calculate the minimum link attributes. Fig. 5(a) displays the results describing the solution path in terms of each of the five attributes (accident probability, distance, expected consequence, travel time, and total population) when travel time is minimized, while Fig. 5(b) displays the corresponding results when total population is minimized. By clicking on the appropriate radio button, the numerical results can be displayed for different attribute minima. This feature assists the user in determining appropriate constraint bounds.

Solution paths can be depicted in the decision space by displaying the appropriate map object in the map window. As an aid to decision making, the performance of solutions vis-a-vis the objectives and constraints of the routing problem can be visualized by displaying them in the objective space. Spider webs are used in this research (see also Coutinho-Rodrigues et al., 1997). Routing performance results with travel time minimization are depicted in Fig. 6. In this graph, the corners of the dark gray polygon represent the minima of the five attributes. The corners of the light gray polygon represent the attribute sums when travel time is minimized. Fig. 7 shows the solution diagram for a scenario where total population is minimized. It should be pointed out that the dark gray polygon is the same as in Fig. 6 but the light gray polygon is very different. These graphs illustrate the conflict that typically exists between minimizing travel time and minimizing population at risk.

Constrained Least Time Path

Path Parameters
Population at Risk, Diameter/Bandwidth
Start Time

Stopping Parameters
200 Maximum number of K-shortest paths in duality gap closure, Range 1 to 1000
Stop procedure if (lower / upper bound) is greater than 95 percent, Range 60 to 99
Stop procedure if run time is greater than 4 minutes, Range 1 to 999

Feasible solution found; (lower bound / upper bound) was 0.859 when procedure was stopped

Shortest Paths w/o Constraints		Attributes Constrained	RHS Values	Results	Non-Zero Multilpier
Accident Probability	0.000431	Accident Probability	0	0.000847	
Distance	272	Distance	0	298	
Expected Consequence	0.0170	Expected Consequence	0	0.0118	
Time	4.4	Time	0	5.7	
Total Population	38364	Total Population	20000.	16456	✓

Calculate Shortest Paths | Shortest Paths Graph | Solution Method | Calculate Constrained Path | Constrained Path Graph | Duality Gap Graph | Cancel | OK

(a)

Shortest Paths w/o Constraints		Attribute
Accident Probability	0.001481	Acc Pro
Distance	446	Dis
Expected Consequence	0.00837	Exp Cor
Time	8.8	Tim
Total Population	12615	Tot Pop

Calculate | Shortest | Solution

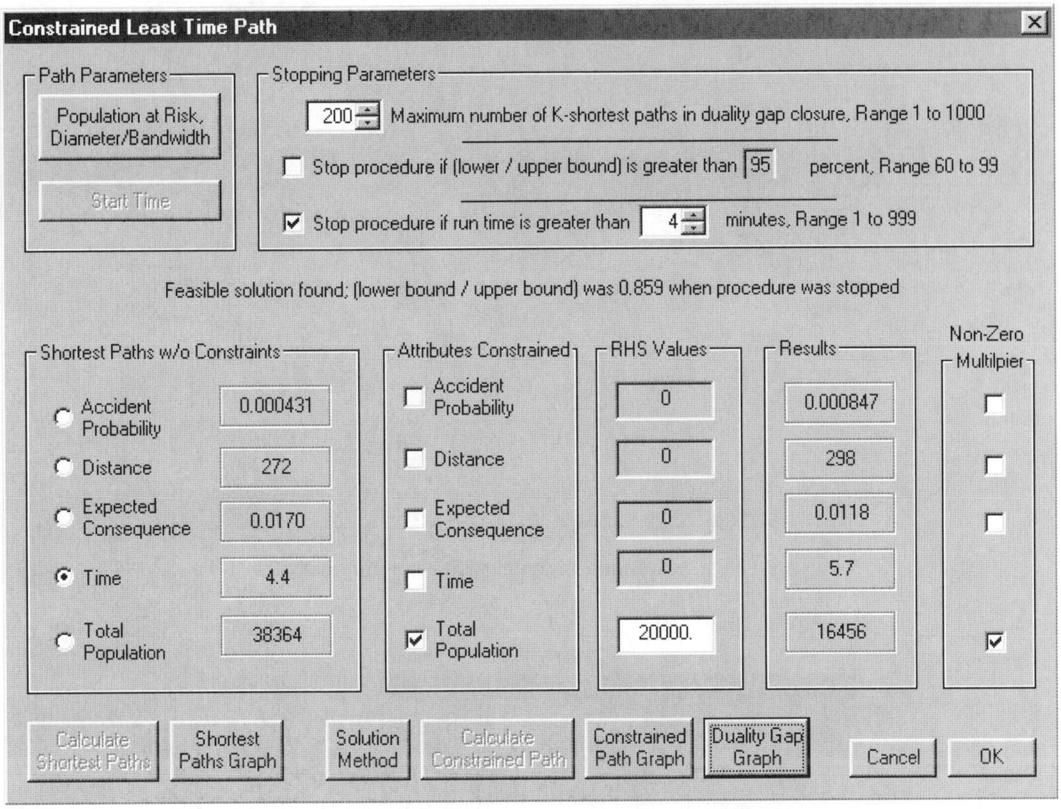

(b)

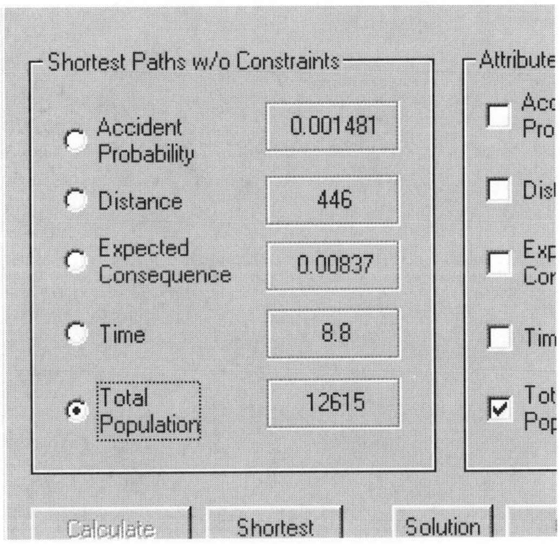

Fig. 5. (a) Constrained least time path dialogue box. (b) Total population minimized.

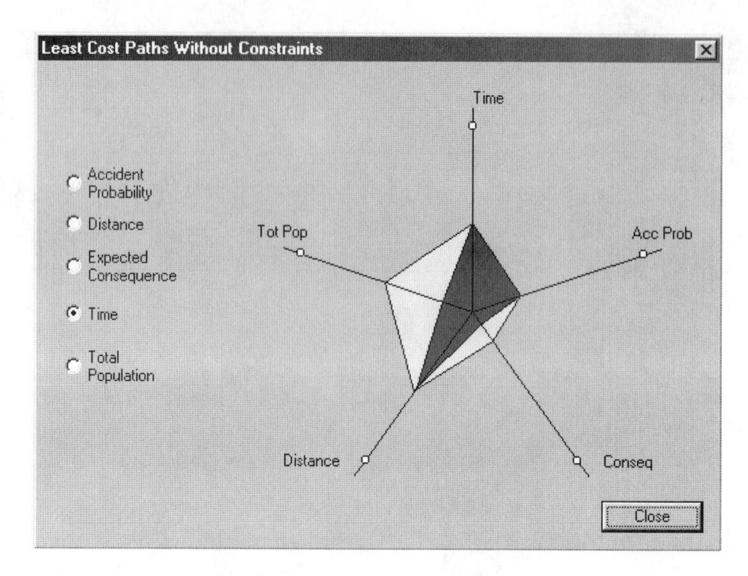

Fig. 6. Solution diagram with time minimized.

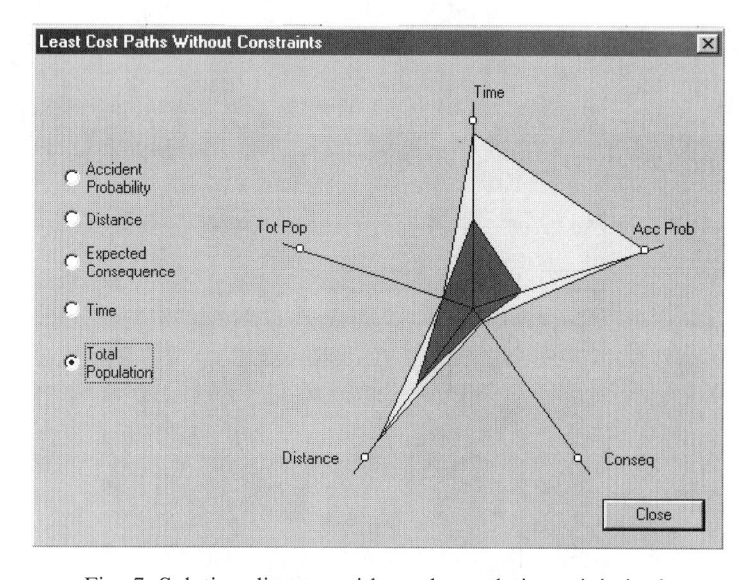

Fig. 7. Solution diagram with total population minimized.

The lower center panel of Fig. 5(a) is used to supply upper bounds on attribute sums. The numerical results are displayed in the lower right side of Fig. 5(a). Fig. 8 displays the results graphically. This figure is similar to Fig. 6 except that the constraint is added in the form of a dot on the total population axis. The attribute sums are shown in the form of a polygon without a colored interior. The differences between the light gray polygon and the uncolored polygon are caused by constraints.

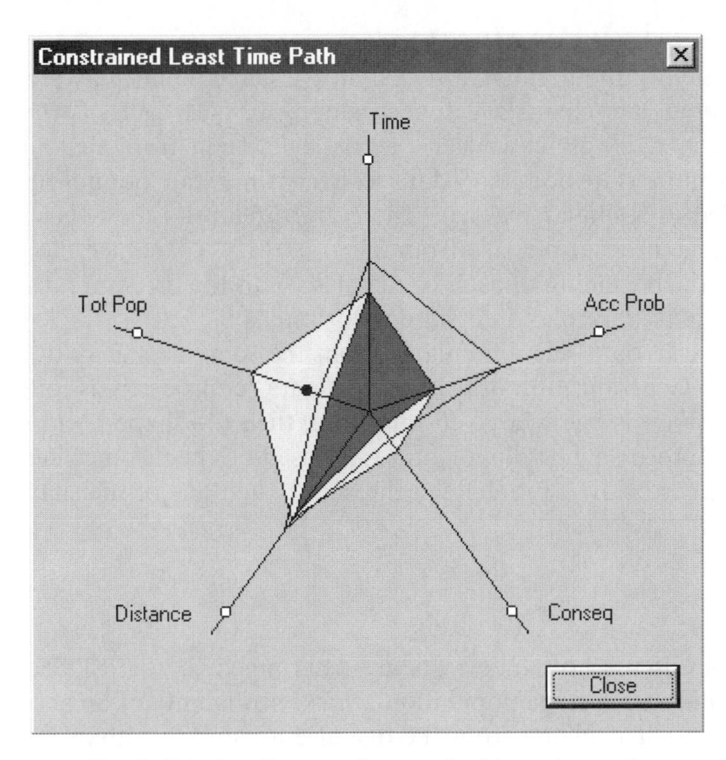

Fig. 8. Solution diagram of constrained least time path.

Parameters controlling the computer runtime can be specified by any of three different approaches in the upper center right panel of Fig. 5(a). The first method limits the number of paths generated when using Yen's K least cost algorithm (Yen, 1971). The second method stops the heuristic if the present solution is close enough to the best solution. The third method limits the total time in minutes.

4.6. Data structures

The Hazmat Path SDSS utilizes two data structure sets, namely link costs and map objects. The link costs set is used to determine the truck route. The development of these costs is presented in Section 5. To produce one TCSP route, Dijkstra's algorithm might be performed thousands of times. Therefore, to reduce computer runtime, this data is stored in RAM in a binary tree data structure. The tree is constructed only once when the program is initially started. The user has the option of choosing different solution methods. This option dictates the assumptions applied to the link attribute costs.

The second data structure set consists of information needed to draw a map. Most of this data are in the form of shape points used to draw roads and state boundaries. This data is stored on disk because of its large size. There are three complete sets of state boundaries. Each set has the same number of chains but the number of shape points varies dramatically. As the user zooms in or out, the resolution is adjusted by displaying a set with more or less shape points. The sets range in total file size from 300 KB to 8 MB. This decreases map drawing time with little noticeable decrease in map quality. Without different sets, multiple shape points may represent one pixel on a monitor. The state boundary chains are divided among 10 files, each of which covers a different region of the country. A file is read when at least one chain from that file is displayed. This prevents reading an entire state boundary data set every time state boundaries are redrawn. Road data sets are handled in a similar manner. There are three road data sets each divided into 45 regions. The sets range in total file size from 23 to 67 Mb. To reduce file read time, interstate chains are placed at the beginning of each of the files. With this structure, the entire file need not be read if the user wants to display interstate roads only.

The resolution-specific data sets are created from the same master data set by selectively removing shape points. Generalization of features in a data set proceeds as follows. If the change in direction at a shape point is below a given threshold, then the shape point is removed. Another approach for the pre-processing of chain features is based on the distance between adjacent shape points. If this distance is below a pre-defined threshold, then one of the shape points is removed.

5. Link attribute costs

In this section, we discuss how link costs are determined for the SDSS. While distance is a primary link cost, total and average population at risk, probability of an accident, and link travel time are secondary costs calculated from the distance attribute combined with appropriate domain-specific knowledge. The expected consequence link attribute is derived from average population at risk and probability of an accident. All secondary costs are time dependent.

5.1. Population at risk

There are two different approaches to determining population at risk:

1. Calculate total population at risk for a given bandwidth and link (Abkowitz et al., 1990).
2. Calculate average population at risk for a given radius of affected area and link (Saccomanno and Chan, 1985; Erkut and Verter, 1995).

With the first method, total population at risk, the population count within a buffer of given bandwidth is calculated for each link. The individual link values of total population at risk are summed to determine the total population at risk along a route from end to end. This method of calculating total population at risk has a flaw: people are double counted if they are located within half the bandwidth of the link end points. To avoid this double counting, a more elaborate method is developed as follows. A standard buffer is divided into three regions as shown in Fig. 9: regions A and C are the buffer drawn around the end points of the link, while region B is the remaining part of the standard buffer.

The total population at risk for a link is

$$\frac{1}{2}\text{ population in region A} + \text{ population in region B} + \frac{1}{2}\text{ population in region C.}$$

This method greatly reduces double counting but does not eliminate it. Indeed, if the B region of two links overlap, double counting occurs.

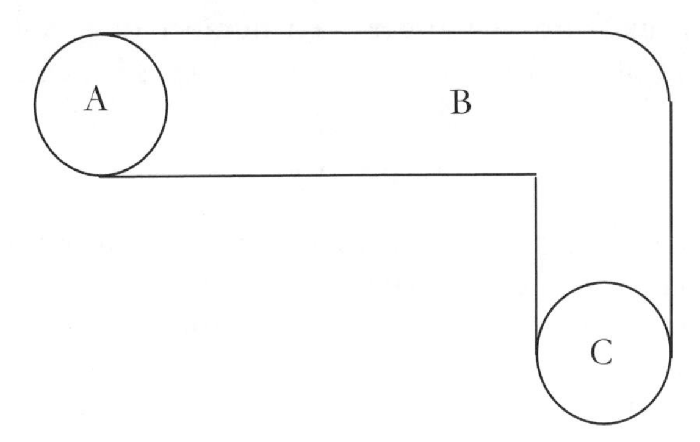

Fig. 9. Partitioned buffer.

The other population at risk metric is average population. It measures the average population at risk in case of accident on a link, under the assumption that the probability of an accident is constant along the length of the link. If an accident occurs while traveling a link, then the expected number of people exposed is the population within a given radius λ of the accident location. The value of the radius λ depends on the type of hazmat and the characteristics of the accident. Given that the location of future accidents is not known and that they are expected to occur with a probability that is uniformly distributed over the link, the expected number of people at risk can be approximated as follows. A series of circles are centered and equally spaced along the length of the link as shown in Fig. 10. The circle radius λ is taken to be constant in this implementation, but the methodology proposed here can accommodate a distribution of radii associated with a distribution of hazmat accidents drawn from historical records. The number of people within each circle is calculated by overlay with the layer of block group polygons produced by the Bureau of the Census and incorporated in the TIGER/Lines files. Their statistical mean produces the average population at risk. By design, the average population at risk calculation assigns more weight to people the closer they live to the link. A person nearer the link is more likely to be counted multiple times than a person further from the link. A similar discussion is held in Erkut and Verter (1995).

Average population at risk is of little use by itself. Once multiplied by the probability that an accident occurs on a link, it produces the expected consequence of a single truck accident.

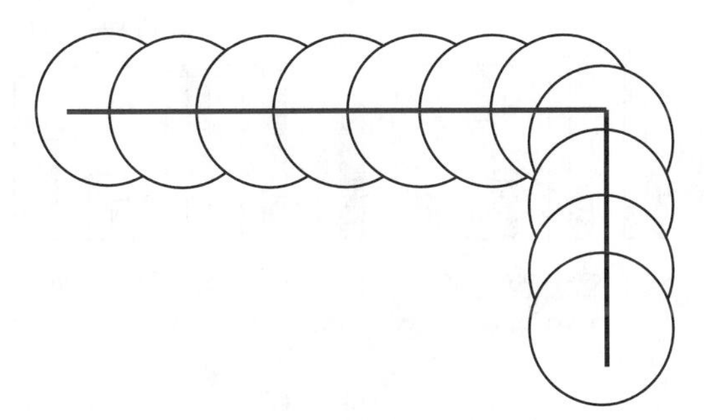

Fig. 10. Average population at risk.

Expected consequence is a measure of the expected number of people exposed on an origin–destination trip. Calculation of accident probabilities is discussed next.

5.2. Accident rates

Considerable research has been devoted to truck accidents, their forecast and measurement. Research usually focuses on a single geographic region, ranging in size from a metropolitan area to a state. Seldom is there any emphasis placed on the variation of the truck accident phenomenon by the hour, the day, or the month.

Mohamedshah et al. (1993) are interested in the non-temporal relationship between highway geometry and accident rates. Using data from Utah, they have identified horizontal curvature and vertical gradient as significant factors of truck accident rates. A highway safety information system summary report (US DoT, 1994a) determines truck accident rates using 1985–1987 Utah and Illinois accident data. The study also concludes that rates are greatly dependent on road type and state.

Hazmat routing in the Toronto area is analyzed by Saccomanno and Chan (1985) on the basis of 1981 Metropolitan Toronto police records. In this study, it is estimated that the probability of an accident on a 100 km/h dry urban expressway to be 2.379×10^{-6} per mile for unrestricted visibility and 4.054×10^{-6} for restricted visibility. Another study by Jovanis and Delleur (1983) also provides statistical estimates of accident rates. From accident records on the Indiana Tollway, they calculate the probability of an accident for a large truck with good weather to be 1.44×10^{-6} per mile for day travel and 1.47×10^{-6} night travel. Other temporal accident rates are derived by Lyles et al. (1991) for the state of Michigan. In addition, a few state agencies have collected, organized and analyzed accident data from their own state. Results of these limited studies exhibit tremendous inconsistencies and cannot reliably be transformed into national accident rates. Deficiencies in truck accident and truck usage data have previously been pointed out by other authors, including Lyles et al. (1991), Lepofsky et al. (1993), and Mohamedshah et al. (1993). Therefore temporal accident rates used in this study are derived from national databases, which offer great consistency and comprehensive geographic coverage at the expense of attribute specificity and detail.

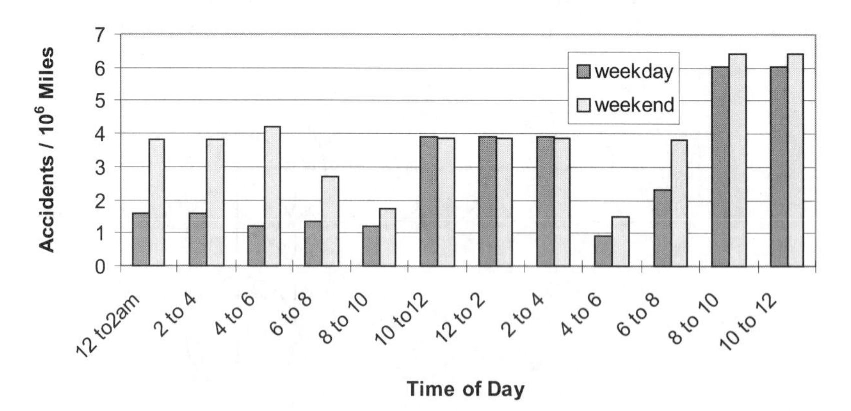

Fig. 11. Truck accident rates.

Three nationwide data sources are integrated to establish temporal accident rates:

1. Annual weighted accident police reports with type of roadway, time of day and day of week (National Highway Traffic Safety Administration, 1993).
2. Truck mileage by type of roadway for a one year period (FHWA, 1996).
3. Annual truck mileage by type of roadway, time of day and day of week (US DoT, 1992).

The second and third source are used to develop a nationally weighted truck mileage by type of roadway, time of day and day of week for a one year period. This statistic is then combined with the first source to determine accident rates by type of roadway, time of day and weekday/weekend. All data are relative to the 1993 calendar year. The results are shown in Fig. 11 for all types of roadways combined.

5.3. Link travel speeds

On most urban and all rural roads, the travel speed is not significantly affected by congestion. Therefore, on these roads, the travel speed can be assumed to be constant. Link travel time is calculated as the link length divided by the speed limit inferred from the road type. On the remaining urban roads where congestion is a problem, travel speed or time delay is time dependent. Ideally, a four-step travel demand methodology should be used to forecast link travel speeds in each metropolitan area afflicted with significant congestion. Given the enormity of such task in the context of the present research, an alternative method incorporating simulations is followed.

The approach to estimating travel time starts with a bivariate classification of links into groups with similar characteristics, namely average annual daily traffic (AADT) and lane width. A simulation is performed for each group to determine temporal travel speeds under expected recurrent and non-recurrent congestion conditions. The simulation follows the procedure outlined in US DoT (1986). Inputs to the model include freeway capacity reduction under incident conditions, incident frequency and average incident duration. There are five different lane widths and five AADT values for a total of 25 simulations. Linear interpolation is used for values between the five average AADT values. The average AADT on any link is estimated by a series of simple proportionality rules given the following link attributes: city name, month of the year, weekday or weekend, time of day, and link orientation with respect to the center of the metropolitan. AADT per lane for the 50 most congested metropolitan areas comes from the US DoT (1997). National averages are used to distribute the latter by month, weekday or weekend and, time of day.

Traffic directionality varies dramatically with the time of the day. More vehicles are heading towards a city in the morning, while more vehicles exit a city during the evening peak hours. These traffic patterns are modeled as a function of link orientation with respect to the city center by means of a directional flow. Equal flow in both directions is associated to an even 50% directional flow. Different directional flows are applied to links for morning and evening peak periods to capture the directionality of commuting patterns.

The directional flow is determined in two steps. The first step is determining the spatial relation between the center of an urban area and the direction of flow. Traffic flow angle is used as the metric in the first step. The traffic flow angle is converted into a directional flow in the second step.

The following algorithm calculates the traffic flow:

1. Draw a line through the link enter node to the link exit node. These are points A and C from the example in Fig. 12.

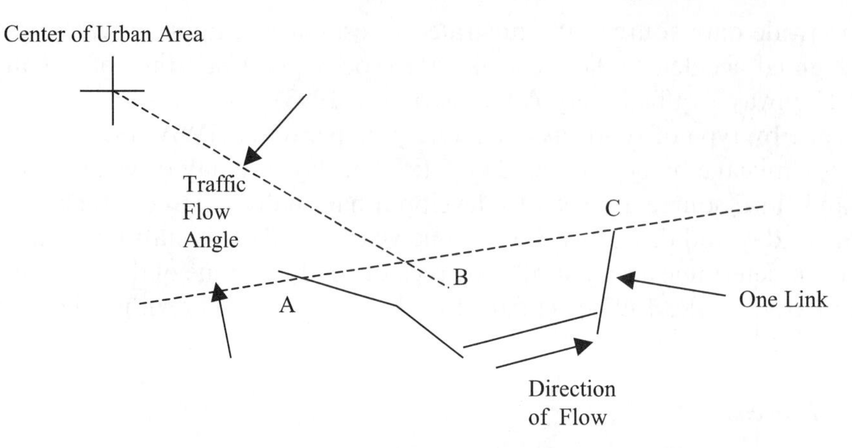

Fig. 12. Traffic flow angle diagram.

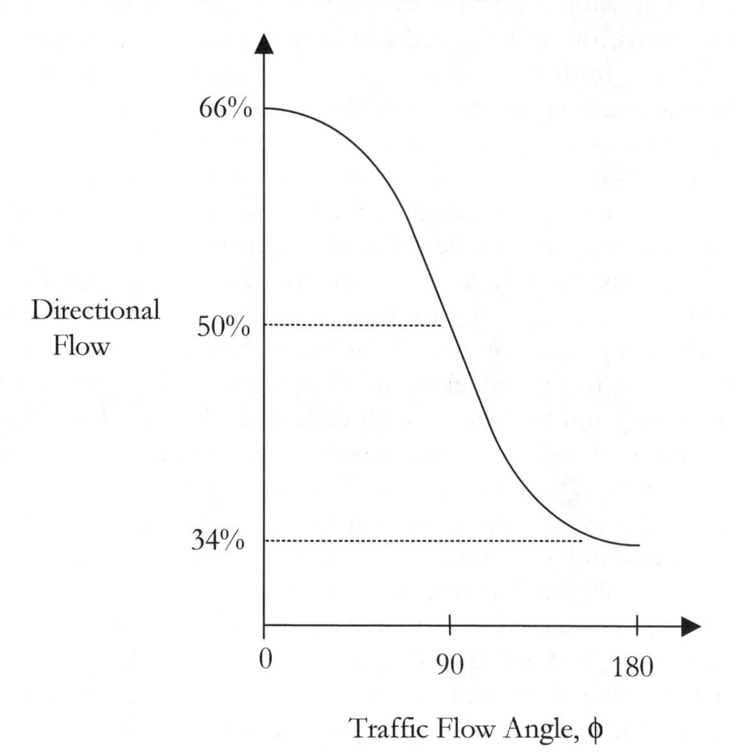

Fig. 13. Evening directional flow vs traffic flow angle.

2. Draw a line from the center of the urban area to a point midway between the entrance and exit nodes, point B.
3. The angle made by these two lines is the traffic flow angle.

For morning traffic, traffic varies directly with the traffic flow angle. The opposite is true for the evening. A sine function is used to convert traffic flow angle to directional factor. Proper scaling of the transform generates directional flows in the 34–66% range (Robinson et al., 1992). The traffic flow angle-directional factor relationship is given by

$$50 + 16\sin(\phi - 90^\circ)\% \quad \text{for mornings,}$$
$$50 + 16\sin(\phi + 90^\circ)\% \quad \text{for evenings,}$$

where ϕ is traffic flow angle, and depicted by Fig. 13. Notice the morning and evening functions are shifted 180°.

6. The temporal constrained shortest path

Costs (approximated by travel distance, travel time and risk exposure metrics) occur when traveling from the origin to the destination of the shipment. This section describes the methods implemented to evaluate these costs once the solution path has been identified. The mathematical formulation of the TCSP problem is also presented along with solution methods to solve it.

6.1. Problem formulation

Let path Π be an origin to destination path. Also, let us denote by δ_{ijt} the travel time on link *ij* when departing node *i* at time *t*. The departure time from node *i* is denoted by γ_i.

If parking is not allowed, then departure time is: $\gamma_j = \gamma_i + \delta_{ij\gamma_i}$. On the other hand, if parking is utilized, then departure time is $\gamma_j = \gamma_i + \delta_{ij\gamma_i} + D_j$, where D_j is the time parked on node *j*.

Let the probability of an accident on link (*i,j*) at time *t* be denoted by a_{ijt}. Summing up the costs for all the links on a path, the probability of an accident on Π is $\sum_{\forall(i,j)\in\Pi} a_{ij\gamma_i}$.

The average population at risk along link (*i,j*) at time *t* is denoted by s_{ijt}. The expected consequence on link (*i,j*) is $a_{ijt}s_{ijt}$. This assumes that an accident anywhere along link (*i,j*) is equally likely. The expected consequence on Π becomes $\sum_{\forall(i,j)\in\Pi} a_{ij\gamma_i}s_{ij\gamma_i}$.

The average population at risk along link (*i,j*) at time *t* is denoted by b_{ijt}, and the total population attribute associated with path Π is $\sum_{\forall(i,j)\in\Pi} b_{ij\gamma_i}$.

Distance is the only non-temporal link attribute. The total distance on Π is $\sum_{\forall(i,j)\in\Pi} d_{ij} \cdot d_{ij}$ is the distance when traveling link (*i,j*).

Total travel time is equal to the destination departure time (γ_D) with no parking on the destination.

Therefore, the TCSP takes the following formulation:

$$\text{Minimize } \gamma_D$$

$$\begin{aligned}
\text{Subject to: } & \sum_{\forall(i,j)\in\Pi} d_{ij} \leqslant \Lambda_{\text{distance}} \\
& \sum_{\forall(i,j)\in\Pi} a_{ij\gamma_i} \leqslant \Lambda_{\text{acc prob}} \\
& \sum_{\forall(i,j)\in\Pi} b_{ij\gamma_i} \leqslant \Lambda_{\text{tot pop}} \\
& \sum_{\forall(i,j)\in\Pi} a_{ij\gamma_i}s_{ij\gamma_i} \leqslant \Lambda_{\text{conseq}} \\
& \gamma_j = \gamma_i + \delta_{ij\gamma_i} + D_j \qquad \text{for link } i,j \in \Pi,
\end{aligned}$$

where $\Lambda_{\text{distance}}$, $\Lambda_{\text{acc prob}}$, $\Lambda_{\text{tot pop}}$ and Λ_{conseq} are attribute bounds.

6.2. Solution methods

The TCSP incorporates a number of link attributes whose valuation varies with time of the day, and day of the week. In addition, parking or stopping on a node and continuance of the trip at a later time is allowed in the temporal network. Waiting to enter a link may be advantageous when the cost of traversing this link is expected to decrease in the future. Parking can be thought of as a cycle where the only non-zero link attribute is time. Cycles may exist in the optimum solution if parking is not allowed. If neither cycles nor parking is allowed, then a less obvious method of time consumption may exist. An example is given to illustrate this problem. Consider the network in Fig. 14.

Assume the travel time of subpath 1–3–2 is greater than the travel time from node 1 to node 2. Also assume all other link attribute costs for subpath 1–3–2 are greater than subpath 1–2 attribute costs. Subpath 1–2 will be chosen over subpath 1–3–2 if parking is allowed that subpath 1–2 dominates subpath 1–3–2. The temporal costs of the link from node 2 to the destination have no effect on which subpath is chosen.

Now, consider the case where parking is not allowed and the costs to travel link 2D decrease with time. Traveling subpath 1–3–2 will delay entering link 2D. This delay will decrease the costs of traveling link 2D. However, this delay comes at the increased costs of traveling subpath 1–3–2. Subpath 1–3–2 will be chosen if the total cost is less.

By its sheer size, the 57,000 node network derived from FHWA's NHPN network entails a major challenge to solving the TCSP. Even under the optimistic scenario of a 90% reduction in network size obtained by applying the constraints discussed in Section 4, the network is still too large to apply a temporal dynamic programming solution method. A solution method with exponential complexity would not be practical with such a large network, so a major concern of this research is the development of efficient heuristic procedures.

A workable strategy to get around the network size problem is to break up the problem into manageable parts. The primary difficulty of this approach is relating information between the manageable parts. Two solution methods are implemented in Hazmat Path, namely the so-called weight-guided solution method, and the intermediate nodes solution method. Both approaches are an extension of Handler and Zang (1980) and are now outlined.

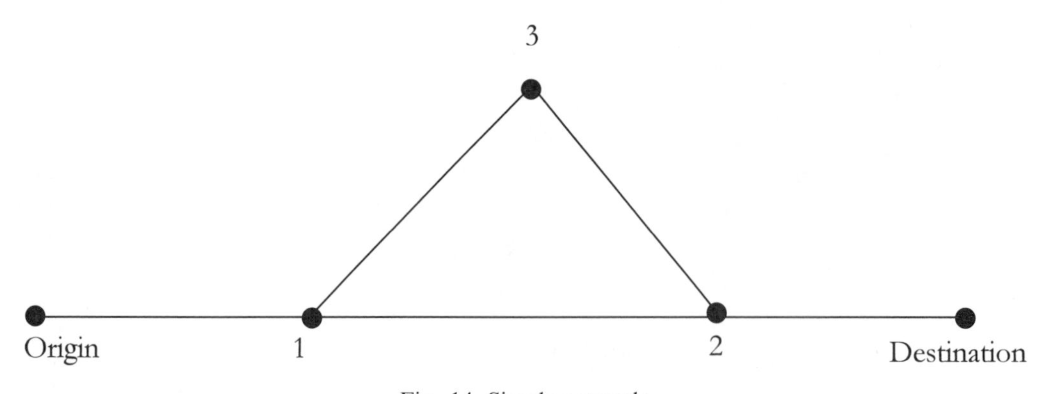

Fig. 14. Simple network.

The weight-guided solution method disentangles the original formulation into a master problem and a sub-problem. The master problem produces a path using Dijkstra's algorithm with a single link cost. This link cost is a linear combination of the five link attributes. The sub-problem determines the parking times at nodes. The attribute sums from this path along with the upper bounds on these attribute sums are inputs in determining the new single link cost. The master problem then produces another origin–destination path. This is repeated until there is no change in results between iterations.

The intermediate nodes solution method transforms the TCSP into many time-independent CSP problems. All link attributes are defined over discrete time interval, with the exception of travel time in congested cities which is a continuous function of time. The discrete time intervals range from 1 to 6 h. To implement this method, travel time in large urban areas is converted to a function of discrete time. The algorithm proceeds as follows. A vehicle starts at the origin and reaches a pre-determined intermediate node before any attributes change values. It stays at this parking position until attributes change value. Since no attributes change value while traveling from the origin to the intermediate node, a time-independent solution method can be used in this subpath route selection. The vehicle resumes its trip to another intermediate node or to the destination. This process is repeated as needed until the destination is reached.

7. Conclusions and future extensions

In this paper, we presented a working and easy to use hazmat routing SDSS that overcomes three significant challenges, namely handling a realistic network, offering sophisticated route generating heuristics and functioning on a desktop personal computer. Although many parts of this work can individually be found in previous work, never before have they been combined into one single working system.

A successful SDSS necessitates the development of custom software. Decision making is rendered considerable less cumbersome for several reasons. First, the user follows a logical procedure when developing a route. On the contrary, off-the-shelf software adapted to hazmat routing requires learning the general syntax of the software prior to delving into hazmat routing. Custom software also produces a route in a more timely manner before it incorporates efficient data structures and solution algorithms. The navigational simplicity and efficiency advantages help the decision-maker focus on creating solutions, negotiating trade-offs, and evaluating scenarios.

This paper outlines two solution methods to the TCSP problem implemented in Hazmat Path. Another possible solution method could be developed by combining the work of Handler and Zang (1980) and Lombard and Church (1993) approach to solving the gateway shortest path problem. This method can be outlined as follows. The best route generated from the gateway procedure is used as input to the Handler–Zang procedure, which calculates a set of link weights. These link weights are then applied to the network. The gateway procedure is run again and the best route is determined. This process is repeated until there is no improvement between iterations.

In Section 4, we discussed producing differentiated routes to spread risk over a larger population. In our approach, routes were constructed interactively by the decision-maker on the map window. A possible enhancement of the SDSS could involve adding a route generator to produce differentiated routes. Akgun et al. (1999) evaluated several methods for creating differentiated paths. The computational effort required to generate these paths is very low for some of these methods.[1] However, evaluating the quality of the set of differentiated paths would require considerable computational effort during route selection. This calculation consists of creating polygon overlays, which are used for determining the overlap of link buffers.

The display of temporal link attributes is another area where the current system could benefit from future enhancements. The SDSS presently displays attributes for one user-defined time period whereas a shipment may take a considerable amount of time, possibly several days. Displaying multiple maps of temporal attributes in a time loop would prove to be a useful decision support tool. Such capability would require creating, storing and retrieving multiple bitmaps.

Acknowledgements

The financial support of the National Science Foundation to the National Center for Geographic Information and Analysis (award number SBR-9600465) is gratefully acknowledged. Peer review of this manuscript was handled by Professor Hugh Calkins, Department of Geography, SUNY at Buffalo.

References

Abkowitz, M., Cheng, P.D.-M., Lepofsky, M., 1990. Use of geographic information systems in managing hazardous materials shipments. Transportation Research Record 1261, 35–43.

Akgun, V., Erkut, E., Batta, R., 2000. On finding dissimilar paths. European Journal of Operational Research 121, 232–246.

Baaj, M.H., Ashur, S.A., Chaparrofarina, M., Pijawka, K.D., 1990. Design of routing networks using geographic information systems: applications to solid and hazardous waste transportation planning. Transportation Research Record 1497, 140–144.

Brainard, J., Lovett, A., Parfitt, J., 1996. Assessing hazardous waste transport risks using a GIS. International Journal of Geographic Information Systems 10, 831–849.

Bureau of Transportation Statistics, 1999. National Transportation Atlas Databases, US Department of Transportation, Washington, DC.

Coutinho-Rodriques, J., Current, J., Climaco, J., Ratick, S., 1997. Interactive spatial decision-support system for multiobjective hazardous materials location-routing problems. Transportation Research Record 1602, 101–109.

Densham, P., 1991. Spatial decision Support Systems. In: Maguire, D.J., Goodchild, M.F., Rhind, D.W. (Eds.), Geographic Information Systems: Principles and Applications, vol. 1. Longman, London, pp. 403–412.

Erkut, E., 1996. The road not taken. OR/MS Today 23 (6), 22–28.

Erkut, E., Verter, V., 1995. A framework for hazardous materials transport risk assessment. Risk Analysis 15, 589–601.

FHWA, 1996. Annual Vehicle Miles of Travel and Related Data, US Department of Transportation, Publication No. FHWA-PL-96-024.

Geoffrion, A.M., 1983. Can OR/MS evolve fast enough. Interfaces 13, 10–25.

Handler, G.Y., Zang, I., 1980. A dual algorithm for the constrained shortest path problem. Networks 10, 293–310.

Jovanis, P.P., Delleur, J., 1983. Exposure-based analysis of motor vehicle accidents. Transportation Research Record 910, 1–7.

[1] Lombard and Church (1993) report similar experience.

Lassarre, S., Fedra, K., Weigkricht, E., 1993. Computer-assisted routing of dangerous goods for haute-normandie. Journal of Transportation Engineering 119, 200–210.

Lepofsky, M., Abkowitz, M., Cheng, P.D.-M., 1993. Transportation hazard analysis in integrated GIS environment. Journal of Transportation Engineering 119, 239–254.

Lombard, K., Church, R.L., 1993. The gateway shortest path problem: generating alternative routes for a corridor location problem. Geographical Systems 1 (1), 25–45.

Lyles, R.W., Campbell, K.L., Blower, D.F., Stamatiadis, P., 1991. Differential truck accident rates for Michigan. Transportation Research Record 1322, 62–69.

Mohamedshah, Y.M., Paniati, J.F., Hoheika, A.G., 1993. Truck accident models for interstate and two-lane rural roads. Transportation Research Record 1407, 35–41.

National Highway Traffic Safety Administration, National Accident Sampling System General Estimates System, National Center for Statistics and Analysis, US Department of Transportation.

Nozick, L.K., List, G.F., Turnquist, M.A., 1997. Integrated routing and scheduling in hazardous materials transportation. Transportation Science 31 (3), 200–215.

Robinson, C.C., Levinson, H.S., Goodman, L., 1992. Capacity in transportation planning, In: Edwards, J.D. (Ed.), Transportation Planning Handbook. Prentice-Hall, Englewood Cliffs NJ, pp. 410–446.

Saccomanno, F.F., Chan, A.Y.-W., 1985. Economic evaluation of routing strategies for hazardous road shipments. Transportation Research Record 1020, 12–18.

Souleyrette, R.R., Sathisan, S.K., 1994. GIS for radioactive materials transportation. Microcomputers in Civil Engineering 9, 295–303.

Sprague, R.H., Carlson, E.D., 1982. Building Effective Decision Support Systems. Prentice-Hall, Englewood Cliffs, NJ.

Turban, E., 1995. Decision Support and Expert Systems: Management Support Systems. Prentice-Hall, Englewood Cliffs, NJ.

US Department of Transportation, 1986. Quantification of Urban Freeway Congestion and Analysis of Remedial Measures, Report No. FHWA/RD-87/052.

US Department of Transportation, 1992. 1990 Nationwide Truck Activity and Commodity Survey, ORNL, TM-12188.

US Department of Transportation, 1994a. Highway Safety Information System Summary Report, Truck Accident Models, FHWA-RD-94-022.

US Department of Transportation, 1994b. Research and Special Programs Administration, Biennial Report on Hazardous Materials Transportation Calendar Years 1992–1993.

US Department of Transportation, 1997. Estimates of Urban Roadway Congestion – 1990, DOT-T-94-01, March 1993 (update to 1993 data).

Yen, J., 1971. Finding the K shortest loopless paths in a network. Management Science 17, 712–716.

PERGAMON

Transportation Research Part C 8 (2000) 361–380

TRANSPORTATION RESEARCH PART C

www.elsevier.com/locate/trc

A spatial decision support system for retail plan generation and impact assessment

T.A. Arentze, H.J.P. Timmermans *

Urban Planning Group, Eindhoven University of Technology, P.O. Box 513, Mail Station 20, 5600 MB Eindhoven, Netherlands

Abstract

Current geographic information systems typically offer limited analytical capabilities and lack the flexibility to support spatial decision making effectively. Spatial decision support systems aim to fill this gap. Following this approach, this paper describes an operational system for integrated land-use and transportation planning called Location Planner. The system integrates a wide variety of spatial models in a flexible and easy-to-use problem solving environment. Users are able to construct a model out of available components and use the model for impact analysis and optimization. Thus, in contrast to existing spatial decision support systems, the proposed system allows users to address a wide range of problems. The paper describes the architecture of the system and an illustrative application. Furthermore, the potentials of the system for land-use and transportation planning are discussed. © 2000 Elsevier Science Ltd. All rights reserved.

Keywords: Integrated land-use and transportation planning; Spatial decision support systems; Spatial models; Multi-purpose trips models

1. Introduction

The interest in geographical information systems in transportation research has recently increased considerably, as evidenced by the number of papers at transportation conferences. Especially in the area of integrated land-use and transportation planning, geographical information systems allow users to manage their geo-coded data, perform particular analyses, and map the results of these analyses. While the mapping and database aspects of geographical information systems are generally well developed, this cannot be said about the opportunities for analyses that

* Corresponding author. Tel.: +31-40-247-3315; fax: +31-40-247-5882.
E-mail addresses: t.a.arentze@bwk.tue.nl (T.A. Arentze), eirass@bwk.tue.nl (H.J.P. Timmermans).

PII: S0968-090X(00)00010-3

geographical information systems offer. In most commercially available systems, such opportunities are rather limited. Moreover, to the extent that systems offer modeling opportunities, they tend to be quite dated and often restricted to the simplest versions of that model.

Under such circumstances, the user is left with the option to either apply more or less outdated technology, or develop and apply dedicated software. Several geographical information systems now also provide some script language that may be used, but in general, this is not particularly efficient. In many cases, we have therefore chosen to build dedicated software that has been optimized with a particular class of problems in mind, and that incorporates the latest models that have been developed for that problem. Communication with widely used geographical information systems is guaranteed through the use of particular data file formats.

The present article focuses on an example of such a spatial decision support system that was made operational over the last couple of years: Location Planner. It has been developed with retail planning problems in mind, and allows the user to address issues of transportation and land-use planning. A major problem is how the supply of retail facilities, given the properties of the transportation system, affects spatial shopping behavior and related trip patterns, and vice versa.

The article is organized as follows. First, we will describe the objectives of the system against a brief overview of spatial decision support systems. This is followed by a more detailed discussion of the architecture and various components of the decision support system. Next, we will illustrate the use of part of the system in a case study of Veldhoven, Netherlands. The article is concluded with a summary and evaluation of the proposed system.

2. Review of spatial DSS approaches

Decision support systems (DSS) were first introduced in business management. The aim was to improve the decision support capabilities of the management information systems that were used at the time. The first DSS applications began to appear in the early 1970s. Since the early 1980s, DSS efforts gained in strength under influence of the PC revolution, the increasing performance–price ratio of hardware and software, and the increasing availability of public databases and other sources of external data (Sprague, 1989). Although there is not a generally agreed upon definition, the term DSS commonly refers to "computer-based systems which help decision makers utilize data and models to solve unstructured problems" (Sprague, 1989). Spatial DSS is generally defined as a DSS which combines geographic information with appropriate algorithms to support locational decision making (Crossland et al., 1995; Keenan, 1998; Maniezzo et al., 1998). This section first reviews approaches in spatial DSS for transportation and location planning and then positions our approach in this context.

Spatial DSS approaches reported in the literature are typically centered on a single modeling technique. According to the technique used, we can distinguish spatial interaction/choice modeling, mathematical programming, or multi-criteria decision modeling approaches. Examples of systems based on the first approach are Roy and Anderson (1988), Borgers and Timmermans (1991), Grothe and Scholten (1992), Kohsaka (1993), Birkin (1994), Birkin et al. (1996) and Clarke and Clarke (1995). The core of these systems is a spatial interaction/choice model for predicting destination choice or interactions between zones dependent on travel distance, size and attributes of retail or service facilities. Typically, predictions are represented in an origin–destination matrix from which market shares of stores and travel demands of consumers are derived. Thus, the systems allow users to predict and analyze impacts of possible location or transportation plans. Few systems also actively support location selection. The system proposed by Kohsaka (1993) incorporates a steepest descend algorithm to search for locations on a continuous potential surface (predicted by the shopping model). Clarke and Clarke (1995) and Birkin et al. (1996) describe dedicated GIS applications that incorporate a location–allocation model for optimizing the spatial configuration of a network of facilities.

In contrast, location–allocation models make up the core of systems based on the second approach. Location–allocation models simultaneously optimize the choice of locations for facilities and the allocation of consumers to those facilities. The mathematical programming approach implies that an optimization problem can be defined by users in terms of a (single) objective function and one or more constraints on solutions. For given candidate locations, the system generates optimal configurations of the retail/service network and, in some cases, optional outlet formats. Densham (1994) reviews applications for location selection. Armstrong et al. (1990) and Densham and Rushton (1996) describe spatial DSS applications for reorganizing service delivery systems. As Densham (1991) argues, the objective of spatial DSS (in this approach) is to provide a 'flexible problem-solving environment in which decision makers can explore a given problem, evaluate the possible trade-off between conflicting objectives, and identify unanticipated, possibly undesirable characteristics of the problem'.

Multi-criteria or multi-objective DSS represents a third approach aimed at providing tools for analyzing the complex trade-off between candidate locations in choosing a suitable location for a new facility (e.g., highway, waste dumps, power plant, hospital, etc.). Multi-criteria evaluation (MCE) refers to a set of techniques for ranking a given set of choice alternatives on a given set of multiple, possibly, conflicting criteria. Multi-criteria or multi-objective spatial DSS is typically applied in a group setting. It assists the group in identifying location candidates, developing lists of criteria for evaluation, determining weights of criteria, performing sensitivity analysis and alternative ranking (Janssen, 1991). The systems typically integrate MCE or multi-objective programming (e.g., goal programming) with analytical and mapping tools of GIS. There are many examples of such systems reported in the GIS literature (Fedra and Reitsma, 1990; Carver, 1991; Pereira and Duckstein, 1993; Jankowski, 1995; Lin et al., 1997; Malczewski, 1996; Jankowski and Ewart, 1996; Crossland et al., 1995). Thill (1999) gives a review of the field. The spatial DSS for solid waste planning described in MacDonald (1997) combines various techniques including mathematical programming, impact analysis, and the Analytical Hierarchical Process. Carver (1991) describes a method for integrating GIS and MCE techniques.

The spatial DSS that we propose intends to improve the flexibility and interactive properties of current spatial DSS. Sprague and Carlson (1982) distinguish two aspects of flexibility. First-level flexibility is the ability of the system to adapt to a solution path preferred by the decision maker. This is important because location problems are often ill-structured implying that no standard solution procedure exists. Second-level flexibility is the ability to modify the configuration of the DSS so that it can handle a different set of problems. This is an essential feature of a generic DSS. Existing systems perform poorly on both first and second-level flexibility. With respect to the first level, spatial interaction/choice model (SIM) systems focus on what Densham and Armstrong (1993) call the intuitive mode. In this mode, users can specify scenarios in terms of planned or anticipated developments (e.g., opening a new facility, population forecasts) and the system gives feedback in terms of impacts on criterion variables (e.g., travel demands). On the other hand, DSS based on mathematical programming (MP) or multi-criteria evaluation (MCE) methods are designed to support the goal-seeking mode. That is, users can specify the location problem in terms of criterion variables (e.g., criteria weights) and the system generates optimal solutions in terms of decision variables (i.e., optimal configurations or rankings). Where SIM-systems lack the analytical capabilities or an appropriate user-interface for the goal-seeking mode, the MP/MCE systems are weak in supporting the intuitive mode. The goal of our approach is to design a system that supports both interaction modes in terms of modeling capabilities, as well as user interfaces.

With respect to second-level flexibility, SIM-systems also tend to be highly restrictive. They offer users limited possibilities to change the specification of a shopping model (selection of attributes), the utility function form (e.g., nested logit or MNL), and sometimes even the values of model parameters (weights of attributes). The reason for this is that most systems are designed for a specific application. For each new application, systems must be re-designed to suit the specific information needs. A generic DSS requires a more fundamental solution. To be able to support problems in both public-sector and private-sector planning, the system must cover a wide range of planning objectives. Moreover, users should be able to choose attributes, attribute parameters, and even the form of the spatial shopping model.

Interactive properties relate to two dimensions of the user-interface: the extent to which users can view and manipulate relevant conditions of a problem (openness of the system) and communicative properties. To identify potentially relevant information categories, we use a model of the planning cycle. The model is schematically shown in Fig. 1. The boxes in this scheme correspond to different information categories or sections of the problem domain and the arrows represent the dependency relationships between them. The user-interface of the proposed system

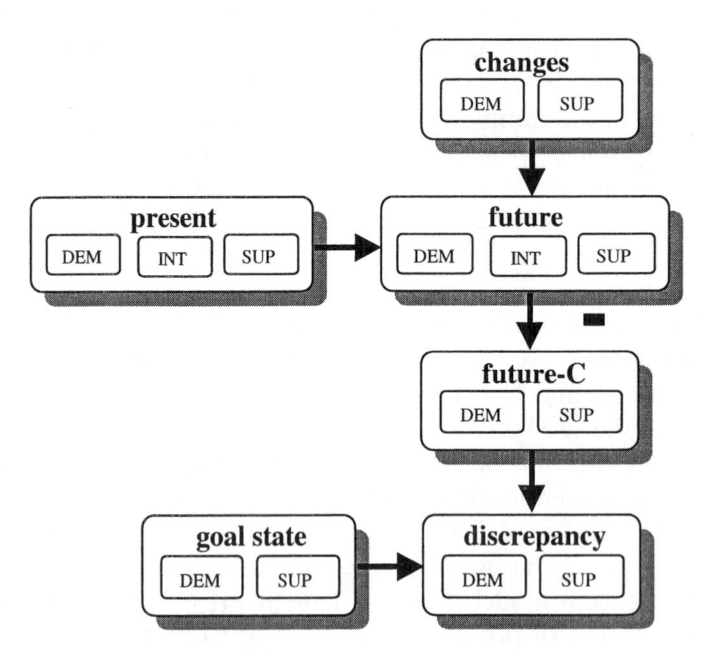

Fig. 1. Scheme of the planning cycle.

allows users to interact with each of the sections to view or change conditions of the system being planned. In particular, users should not only be able to evaluate alternative plans, but also demographic and economic scenarios of change in a study area.

To enhance the communicative properties, the second dimension of user-interfaces, we use dynamic-graphics techniques much in the same way as proposed by Densham and Armstrong (1993) and Densham (1994). That is, the proposed system supports multiple representation formats (views) of domain sections. These include map, graph and table format. The views are dynamically linked with each other so that changes in one view lead to automatically updating related views. For example, if the user selects a record in the table view, the system highlights the corresponding object in the map and graph views at the same time.

3. Architecture of the system

This section discusses the architecture of the system designed to achieve these objectives. The architecture is schematically shown in Fig. 2. The inference model supports the intuitive

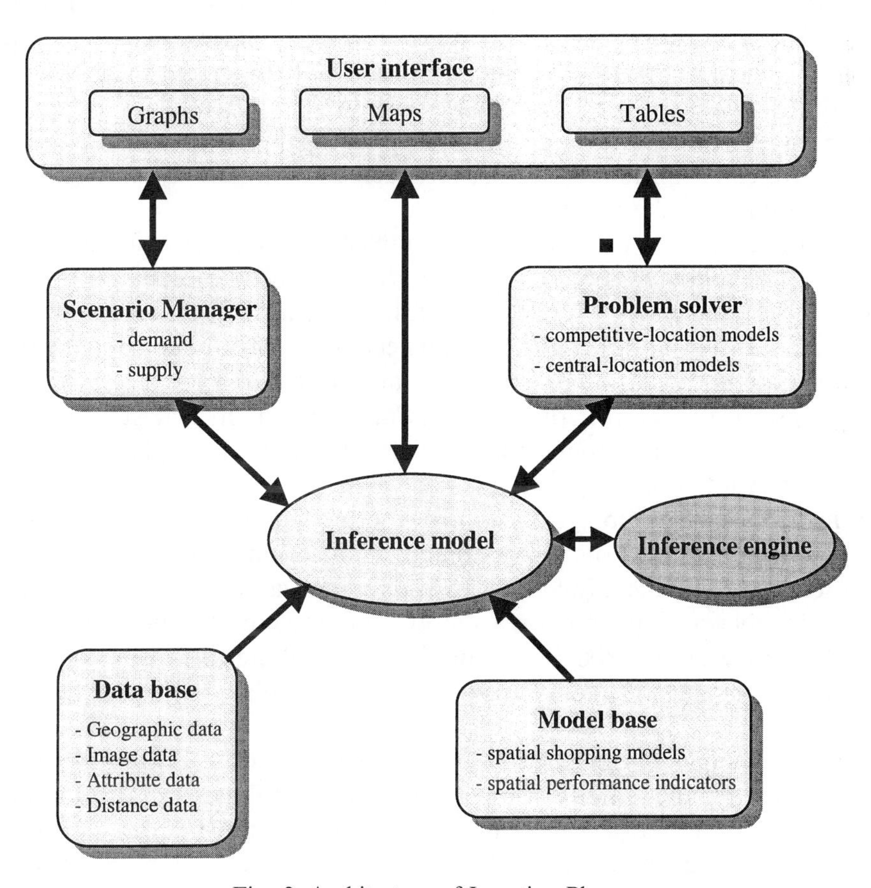

Fig. 2. Architecture of Location Planner.

interaction mode. Users can define scenarios, for example related to population or plans, and run models to predict and analyze impacts on planning objectives. To provide second-level flexibility, the inference model is not a fixed component of the system. Rather users are able to construct the model out of more elementary components that are available in the model base of the system. Thus, users can construct a model that suits information needs, available data and preference for model types. To provide first-level flexibility, the second major component called Problem Solver supports the goal-seeking interaction mode. Users can define a problem in terms of objectives and constraints and the system generates the optimum and near-optimum solutions. The system is implemented as a stand-alone Windows-95 application using the object-oriented programming tool C++-builder. The remainder of this section discusses each component in turn.

3.1. Constructing an inference model structure

Location Planner allows users to construct a model of the retail/service system out of elementary models available in the model base of the system. The internal database consists of several interconnected sections that correspond to the planning scheme mentioned before (Fig. 1). The *present state* and *goal state* represents the current and desired states of the system being planned respectively. The desired state refers to planning objectives to be attained at some future moment in time. The *future state* of the system is the result of anticipated or planned *changes* of the *present state*. The *future*-C state represents the future state in terms of the system performance characteristics that are used to describe the goal state. The *discrepancy* state is the result of comparing the future-C state and the goal state. Dependent on the specification of the goal state, discrepancies involve either an imbalance between demand and supply, or opportunities for expanding or otherwise improving a facility network.

Within each of these states of the system, the database stores information about different types of spatial objects. The following object types are distinguished: (i) demand objects (residential zones); (ii) supply objects (retail or service facilities); and (iii) interactions between demand and supply (trips or expenditure flows). Distances between locations are considered externally given through data files. Each combination of object type and state of the system constitutes a section of the internal database. For example, the combination demand–present state stores information about the present state of residential zones in the study area. For planning, not all possible combinations are relevant. For example, goals are usually not formulated at the level of interactions between demand and supply. Therefore, the goal-state-interaction section is not supported. Fig. 2 shows the sections that are available in Location Planner.

Internally, a section is an object holding a two- or three-way table. A demand section stores a demand × attribute table, a supply section a supply × attribute table, and an interaction section a demand × supply × attribute matrix. For example, the attribute dimension of interactions may concern trip type or consumer segment. Irrespective of the table type, attributes are dynamically defined.

Initially, a section is empty. Users construct the inference model by adding attributes to sections and defining for each attribute a model for evaluating that attribute. In Location Planner (and in object-oriented systems in general), a model related to an attribute is called a 'method'. Users can define a method by selecting a model from the model base. The following categories of models are available:

1. procedures for retrieving data from the external database (present-state, goal-state and changes attributes);
2. procedures for arithmetic summation (compute : future = present + changes);
3. spatial shopping models (future-state-interactions attributes);
4. demand-related performance analysis models (future-C-demand attributes);
5. supply-related performance analysis models (future-C-supply attributes); and
6. performance evaluation models (discrepancy attributes).

Having selected a model, users next specify model parameters, if any, and indicate which attributes in related sections (methods 2–6) or external databases (method 1) provide input data. Thus, methods establish connections between mutual attributes (2–6) or between attributes and the external database (1).

3.2. Inference model base

This part of the system contains the models that can be used as methods for evaluating attributes, as explained above. The data-retrieval and arithmetic models compute elementary functions and need no further explanation. Available performance-evaluation models simply calculate a ratio or difference between corresponding goal-state and future-state attributes. This section focuses on the available spatial shopping models and performance analysis models. Of each model type, different variants are available to allow users to choose the model variant considered appropriate for the application.

3.2.1. Shopping models

These models predict interaction probabilities between demand and supply locations (objects). The production-constrained spatial interaction model is available, as well as the MNL model and a nested-logit model. All models, including the spatial interaction model, assume multi-dimensional location-attractiveness terms, inelastic demand, rule-based definitions of choice sets, and an outflow measure per demand location (proportion of expenditure that leaves the study area). In the nested-logit model, location choice is embedded within the choice of trip type. The utility of each trip-type is calculated as the sum of a user-defined alternative-specific constant and the logsum of the lower nest. The logsum quantifies the expected maximum utility across the location alternatives within the nest. Thus, it represents the utility of the nest under optimal location choice. For each nest, users can specify a rule defining the choice set, utility function, and outflow attribute, just as in the case of a conventional MNL model. The nested-logit model can be specified as the multi-purpose-shopping model that we have developed and tested in earlier work (Arentze et al., 1997). When specified as a multi-purpose model, the trip-types refer to alternative trip purposes including possible combinations of purchases of different goods (multi-purpose).

As stated before, interaction sections support a demand × *supply* × attribute matrix. The attribute dimension can be specified in different ways. First, if the nested-logit model is used, Location Planner automatically generates a three-dimensional matrix of which the third dimension represents trip type. In a three-dimensional matrix, cell probabilities sum up to 1 for each demand location across the supply *and* attribute dimension. Second, users can use the same structure to define one or more two-dimensional matrices, i.e., matrices in which cell probabilities sum up to 1 across the supply dimension. A set of two-dimensional interaction matrices is relevant, for example, in cases where one wishes to use different specifications of an MNL model for predicting choice behavior of different consumer segments (e.g., socio-economic groups).

3.2.2. Supply related performance models

These models compute attributes in the future-C-supply section. Two model types are available. First, interaction-based models use the demand × supply × attribute matrix to calculate for each supply location a measure of allocated demand. Besides the interaction matrix, this method assumes input as a measure of demand per demand location and a measure of inflow of demand from outside the study area per supply location. Two variants are available. The first variant assumes a three-dimensional interaction matrix. Users can set the relative weight of each trip type (the third dimension). The weights should reflect differences between trip types in the amount of money spent per trip on the good or service of interest. The second variant assumes one or more two-dimensional interaction matrices. Users can set the origin demand for each matrix separately.

The second model type uses a deterministic rule, instead of an interaction matrix, for allocating demand to supply. The rule implemented assumes that consumers choose the nearest location that matches the trip purpose. Irrespective of the particular model used, interaction-based or rule-based, users have the option to choose between alternative ways of expressing allocated demand as turn over, floor productivity, or market share.

3.2.3. Demand-related performance models

These models compute performance measures related to demand locations. The following models are available:

1. distance to N nearest facilities or the number of facilities available within cost-band K (Breheney, 1978);
2. travel distance required to purchase a set of n goods while accounting for multi-purpose trips (Arentze et al., 1994a);
3. expected travel distance required for purchasing a good while accounting for multi-stop trips (Arentze et al., 1994b); and
4. the amount of allocated supply capacity for each demand location (Clarke and Wilson, 1994);
5. total distance traveled, average trip length and trip frequency while accounting for multi-purpose trips.

These models relate to different criteria for judging plans including accessibility/spatial choice range (models 1–3), service-provision efficiency (model 4), and travel demands (model 5). Models 2–3 can be viewed as more sophisticated alternatives for model 1 in that they account for multi-purpose, multi-stop trips. Models 4–5 assume a demand × supply × attribute matrix.

Deriving total travel distance from an interaction matrix is not straightforward if multi-purpose shopping is involved. The question is whether consumers keep the total purchase frequencies or the total trip frequency constant when they choose to make a multi-purpose trip (i.e., combining several purchases on the same trip). Model 5 allows users to specify, dependent on the assumption made, either a total trip frequency or purchase frequency and calculates the travel distance implied by the matrix accordingly.

3.3. External database

The external database consists of a collection of files specified by the user. The data files should contain attribute information of demand locations and supply locations, and distances between locations in table form. Geographic and image data about the study area are optional and allow the system to display a map of the area and an image background for the map.

Location Planner does not provide an editor for editing these files. Instead, the system supports the use of existing files, for example, generated by a GIS or other general-purpose software. Supported formats include DBase, TransCad-table and text files for attribute and distance data, BNA for geographical data, and bitmap for image data.

3.4. Inference engine

The inference engine controls the execution of methods for evaluating attributes and guards the internal consistency of the inference model. In this context, it is important to note that defining dependency relationships between attributes and methods constitutes a coherent model structure. Generally, the methods that need to be executed for evaluating an attribute at any position in the model constitute a tree structure of which the endpoints (leafs) consist of methods for reading external data. If the engine receives a command to evaluate an attribute, it identifies and executes the nodes (methods) of the tree starting from leaf nodes. On the other hand, if the definition of an attribute changes, the data of dependent attributes are affected as well. The attributes that are affected constitute a tree structure in the opposite direction, whereby leafs represent the endpoints of reasoning chains. If an action of the user leads to a change of the definition of an attribute, the engine identifies and resets the tree of affected attributes. Various user actions can lead to a change of variable definitions. These include re-specifying or deleting a method, changing the selection status of supply or demand objects, changing data files, editing attributes in present state, change or goal sections and adding or deleting attributes. In sum, the inference model works much like a spreadsheet where formulas need to be defined only once and are automatically executed if needed. Besides speeding up the process of evaluation, the advantage is that users are insulated from the technical details of the models.

3.5. Problem solver

This module incorporates models for solving well-defined location problems. There are two basic models available: the competitive location model and the central location model.

3.5.1. The competitive location model

This model allows users to optimize the configuration of a retail or service chain in terms of the number, attribute profile and location of outlets. Users can define the problem in terms of an objective function and a constraint function. For example, a typical problem is to maximize the market share for a chain as a whole under the constraint that every outlet of the chain meets a minimum floor productivity level. If users have specified different attribute profiles (maximally three) for new outlets, the model optimizes the format of each new outlet simultaneously. Furthermore, users can choose between two search algorithms for solving the problem: an exhaustive search algorithm and an interchange algorithm. The exhaustive search algorithm is suitable for problems of small size. Irrespective of the algorithm chosen, the form of the output depends on the chosen output option. Users can choose between displaying the N best solutions or the best solution per possible macro-strategy. A macro-strategy is a specific combination of the number and formats of outlets, which normally can be implemented in different ways in terms of locations (Ghosh and McLafferty, 1987).

3.5.2. The central location model

This model allows users to find the locations of p outlets that minimize the aggregated weighted travel distance from demand locations to nearest supply location. Users can specify the problem in terms of the size p of the network, the relevant existing supply locations (all or selected only), the attribute containing demand weights, and the yes/no possibility to relocate existing facilities (fixed or replaceable). Again, the exhaustive search and interchange algorithms are available for generating solutions.

3.5.3. Implementation of the models

The exhaustive, as well as interchange search algorithms, are not implemented as independent models. Instead, they use the inference model (defined by the user) to evaluate solutions in terms of both the objective function and constraint conditions. Since users can specify the inference model independent of the location model, a wide variety of model specifications are possible. For example, a complex multi-purpose shopping model can be used in combination with sophisticated performance models to evaluate solutions.

3.6. User interfaces

All settings defined by users, as well as the internal database, are stored in a single project file. When users open a new project file, every section of the database is empty and holds only one attribute. Sections can be viewed through a table, map and graph representation format. In the table view, users can add attributes (columns or sheets) and define methods for attributes.

The table and map views are bi-directionally linked. When activated, the map window displays demand objects as zones with centroids and supply objects as point locations. By clicking on a map object, the record in the table window is simultaneously highlighted. Conversely, by clicking on a table record, the corresponding map object receives focus. This functionality of the system is considered important, as it allows users to link attribute and geographical data of objects. Furthermore, the system automatically displays thematic information contained in the table view on the map. When users select a column of a table, the map is refreshed by the system and the attribute is displayed in the form of a circle diagram on the map. Thus, the spatial distribution of a quantity, such as population or floor space, can be easily assessed. Interaction data, on the other hand, are displayed in the form of lines connecting demand and supply locations. The width of a line represents the relative size of the flow. Finally, when activated, the graph view displays the selected attribute in the form of a bar diagram. This format gives a visual impression of the distribution of objects on an attribute (e.g., population). Table-graph links are uni-directional from table to graph. Specifically, the graph is refreshed each time the selection of a column changes.

Editing data is possible in the table view of change, future and goal sections. In Location Planner, the content of a demand-change or supply-change section is called a 'scenario'. A scenario can be implemented by editing in the table view or by retrieving data from the external database. By clicking on the update button, users give a command to the inference engine to update attributes. The discrepancy state shows consequences in terms of performance scores relative to the goal state (if any). The Scenario Manager module allows one to store and manage scenarios. Using the scenario manager, users can easily return to an old scenario or evaluate new combinations of demand and supply scenarios. Problem Solver supports, in a similar way, the management of generated solutions. Generated solutions are automatically stored. Users can display, remove, or re-generate solutions. A solution itself is a list generally consisting of *N* best solutions. The same linked table, graph and map views are available to view solutions.

Typically, the final result of this user-system interaction is a set of alternative scenarios and corresponding outcomes. Each scenario is based on specific assumptions about demographic and economic developments (demand scenarios), consumer choice behavior (shopping model) and objectives (optimization model or goal state). To facilitate political decision making, the results together with scenario assumptions should be compiled into a comprehensive report. Location Planner does not offer a facility for generating such a report. However, the views provide the required material in the form of tables, graphs and maps. Moreover, each attribute is explicitly defined in terms of the linked method specification.

4. Illustration

The case study conducted to illustrate the system considers a large-scale expansion of the major shopping center in Veldhoven, Netherlands. The multi-purpose trip model and complementary performance models are applied to predict the impacts of the expansion on travel demands of consumers and market shares of shopping centers. Thus, the case study focuses on the inference model component of Location Planner. This section discusses the definition of the study area, the estimation of a multi-purpose trip model, and the information that can be derived from the model.

4.1. The study area

Veldhoven is a city in the Southeast of Netherlands with 41,000 inhabitants in 1996. It is located nearby Eindhoven. The map in Fig. 3 shows the major neighborhoods of Veldhoven and the spatial distribution of the shopping centers in the area. As is typical for Dutch retail systems, the majority of retail facilities are concentrated in planned neighborhood centers. The major shopping center, called the City Center, is located in the central district (nr 1). The second center with an above local function is Kromstraat (nr 3). The only centers outside of Veldhoven that attract consumers from Veldhoven are the major shopping center of Eindhoven and some shopping centers in the eastern and central part of Eindhoven. These centers were identified based on a consumer survey and were also included in the analysis.

In 1997, a large-scale expansion of the City Center in Veldhoven was undertaken. Expansion involved an increase in floor space with approximately one half of the existing floor space in

Fig. 3. The neighborhoods and shopping centers in Veldhoven.

both the daily and the non-daily sector. To analyze the impacts of this development on the shopping and travel behavior of the Veldhoven population, a consumer survey was held before and after the expansion of the center. The after-survey provided suitable data for estimating a multi-purpose trip model. 498 households filled out a paper-and-pencil questionnaire. The stops, as well as purposes, were known for shopping trips made by the respondent in the last week.

4.2. Model estimation

Considering the general information needs of local government in retail impact studies, the purchased items reported by the respondents were exhaustively classified into daily goods (food and personal care) and non-daily goods (all other goods). Given this two-good system, three trip types based on trip purpose could be distinguished, namely (i) single purpose daily good trips, (ii) single purpose non-daily good trips, and (iii) multi-purpose trips where items of both categories are bought during the same visit to the center. Although data were available about multiple trip stops as well, the single stop variant of the multi-purpose model was estimated. Fig. 4 shows the structure of the model used in this case.

The reported shopping trips involved 1369 visits to identifiable shopping centers. These visits were used as observations for estimating the model. In this sample, 53% involved daily trips only, 27% non-daily only, and 20% were multi-purpose. Hence, a considerable portion of reported trips

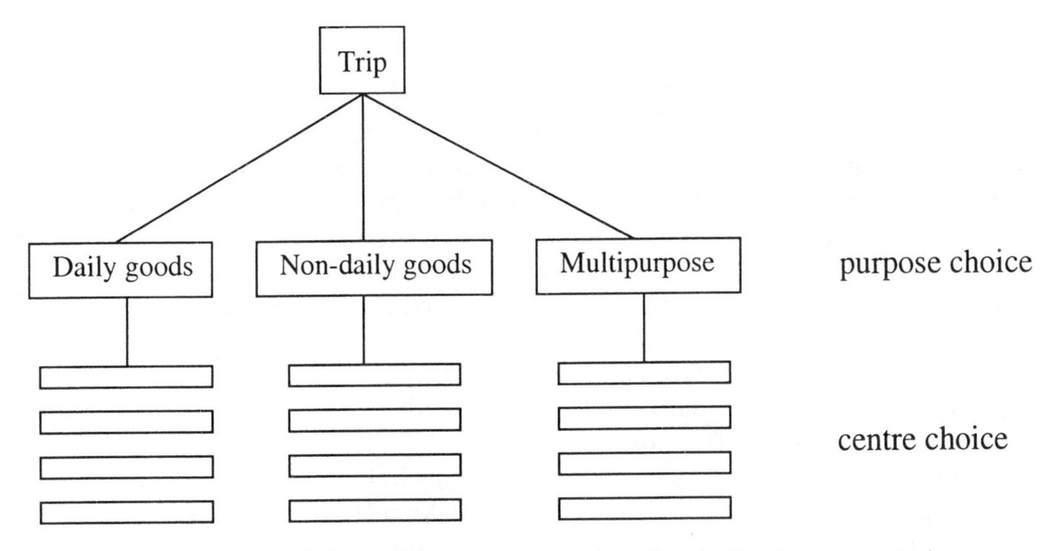

Fig. 4. Structure of the multipurpose trip model used in the Veldhoven case study.

was multi-purpose in terms of the two-good classification. The destination choice set for each trip was defined as the set of centers known to the individual, where stores required by the trip type under concern were available (daily, non-daily or both). Each destination alternative was described in terms of both travel distance from the home location and a set of attributes of the shopping center. Distance was measured as the length in meters of the shortest route across the road network. Respondents were assigned to the node of the network that was closest to the five-digit zip code of their home address. Similarly, the shopping centers were assigned to the nodes closest to the centroid of the shopping area (e.g., a street). A major road to the network linked the relevant shopping areas in Eindhoven. The GIS package TransCAD (Caliper Corporation, 1996) was used to digitize the geographic data and to generate a demand × supply distance matrix using a shortest path routine. The attributes used to describe shopping centers included the total floor space of stores in the daily and the non-daily sectors respectively, and a binary variable representing the presence of a low-price level image of the center. Center atmosphere and parking facilities are generally also influential factors, but were not included because the data was not available.

The software HieLow was used for full-information estimation of the hierarchical choice model (Bierlaire, 1995). The rho bar squared value of 0.163 indicates a satisfactory goodness-of-fit of the model considering the limited set of variables used to describe shopping centers. All parameter values were statistically significant and had values as expected. The logsum parameters have an interesting interpretation for impact analysis. For each trip type, a logsum parameter was estimated. The estimated values of these parameters indicate the extent to which the choice of a trip type is influenced by the attractiveness of locations. If this value is 0, trip type choice is not sensitive to supply. If this value is 1, the supply elasticity is maximal. The values found in this case study are 0.34 (daily), 0.64 (non-daily) and 0.53 (multi-purpose) suggesting that the higher-order trips are more sensitive to variation in supply than the lower-order trips.

4.3. Impact analysis

The model was implemented in Location Planner using the system facilities for model construction. For predicting shopping trips, the same travel distance matrix as in the estimation stage represented the transportation system. The travel distance matrix was based on a subdivision of the area into 89 zip code areas. However, demographic data was available only at a more aggregated level of 14 districts. The demographic data were disaggregated to the zip code level by assuming that populations within districts are evenly distributed across the zip code areas. Destination choice-sets per trip type and per zone were defined using deterministic rules available for that purpose in Location Planner. Specifically, for each trip type the choice-set was defined as the centers offering the required store types (daily or non-daily or both). Furthermore, an additional rule was used for single-purpose daily good trips to further reduce choice-sets to centers lying within a distance of 5000 meter from origin locations.

The multi-purpose trip model was used to predict trips in both the before and after situation. Since the expansion of the City Center was the only development that had taken place, differences found could be interpreted as impacts of the expansion. Impacts on the choice of multi-purpose trips, travel demands of consumers, and market shares of centers were considered using the demand and supply related performance models available in Location Planner.

4.3.1. Choice of multi-purpose trips

For each trip type, the average choice probability across residential zones weighted by the zone's population were determined. In the *before* situation, the probabilities are 51.8% (daily only), 30.4% (non-daily only) and 17.8% (multi-purpose trips) and in the *after* situation, 51.5%, 30.2% and 18.3%. This indicates a small increase of the probability of multi-purpose trips caused by the City Center's expansion.

Predicted trip type choice varies not only across time, but also across space. As an indication of this, predicted minimum and maximum probabilities across zones in the after situation are 45.9–53.0% (daily only), 28.3–33.6% (non-daily only), and 17.9–20.5% (multi-purpose trips). The spread in trip frequency is caused by the spatial differentiation of utilities of available destinations. Generally, zones near relatively large shopping centers are characterized by higher shares of single-purpose, non-daily, as well as, multi-purpose trips. This reflects the relatively big supply elasticity of higher-order trips compared to daily good trips. For example, relatively high shares of multi-purpose trips are predicted for zones nearby the City Center or the major shopping center of Eindhoven.

4.3.2. Travel demand

The expansion may lead to a change in destination choices as well as a change in the share of multi-purpose trips. The latter may have consequences for trip generation. As discussed in the previous section, the travel demand model allows a choice between two assumptions about how individuals adjust their travel behavior in response to changes in multi-purpose trips. Dependent on the assumption made, either the total number of trips across purposes or the total purchase frequency across goods is kept constant. If the first assumption is chosen, users need to specify the number of trips. If the second assumption is chosen, the total purchase frequency needs to be specified. In the present case, the latter option was chosen reflecting the assumption that

Table 1
Travel demand after expansion of the City Center expressed as a percentage of travel demand before expansion

District	Population	Total frequency	Average trip length	Total travel
1	1560	99.6	99.0	98.5
2	210	99.5	98.7	98.3
3	3225	99.5	97.9	97.6
4	160	99.6	99.6	99.1
5	3900	99.5	98.5	98.0
6	4690	99.6	99.7	99.3
7	3920	99.5	96.4	95.9
8	2430	99.5	96.4	95.9
9	5060	99.6	96.3	95.9
10	560	99.6	96.3	95.9
11	1500	99.5	98.1	97.6
12	4800	99.6	98.6	98.1
13	2975	99.6	98.1	97.6
14	5985	99.5	94.8	94.3
Average		99.5	97.5	97.1

individuals keep purchase frequencies constant and make multi-purpose trips in order to reduce the required number of trips.

The model was run for both the before and after situation. The output generated by Location Planner describes for each zone (i) the predicted trip frequency per capita, (ii) the average trip length, and (iii) the total distance traveled. Table 1 shows the after situation when the before predictions are set to 100. Average trip length is calculated as a weighted sum of trip lengths using probabilities of trip types as weights. Then, total distance traveled is simply calculated as the product of trip frequency, trip length and population weights. As the figures indicate, the multi-purpose model predicts a decrease in total travel across residential zones of 2.9%. The model predicts for each zone a small decrease in total trip frequency (on an average 0.5%). Assuming that the sum of purchases of daily and non-daily goods remains constant, the decrease is attributable to the predicted increase in the share of multi-purpose trips (on an average 3.6%). Also, the predicted average trip length has decreased for each zone (on an average 2.5%). A closer look at the destination choice probabilities reveals that this decrease is the result of two counter-acting effects. First, the City Center tends to attract trips which in the before situation went to local district centers inducing more travel inside Veldhoven. At the same time, the increased competitive strength of the center is responsible for a decrease of relatively long trips to the larger centers in Eindhoven. The net result of these two opposite effects is a decrease of average trip length.

4.3.3. Market shares

Location Planner further allows users to predict the market share for each shopping center and good. The market share of a center is defined as the share of the total amount of expenditure available in the study area attracted by that center. In this case, market shares were calculated for daily and non-daily stores in each shopping center. Recall that in the used market share model, the amount of expenditure per capita and per good is a given constant independent of available supply. However, the model takes into account that the amount of expenditure may differ between trip

Table 2
Attributes and market shares after expansion of the City Center[a]

	Center	Size daily (m^2)	Size non-daily (m^2)	Low price image	Market share daily (%)	Market share non-daily (%)
1	City Center	8000	10,800	0	137.9	131.6
2	Burg van Hoof	1198	1745	0	93.4	94.9
3	Kromstraat	1534	4148	0	94.1	95.1
4	Heikant	775	30	0	93.8	95.0
5	t Look	610	0	0	93.8	
6	Zonderwijk	1340	230	0	93.7	94.9
7	Mariaplein	115	745	0	93.3	95.0
8	Zeelst	657	1146	0	93.2	94.9
9	Oerle	100	0	0	93.3	
10	EH inner city	4273	88,273	0	92.6	94.6
11	EH Wonsoel	7780	12139	0	92.5	94.0
12	De Hurk	1225	3163	1	92.8	95.0
13	Kast. Plein	1653	2318	0	93.2	94.9
14	Trudoplein	207	2189	0	93.0	95.0

[a] Market shares are expressed as a percentage of the market share before expansion.

types and users can specify the relative weights of trip type. In the present case, the relative weight of multi-purpose trips was set to 0.5 for both daily and non-daily goods assuming that the amount of expenditure for each good is twice as much on single purpose trips than on multi-purpose trips. Furthermore, users can specify for each shopping center the amount of expenditure attracted from outside the study area. Because these data were not available in this case, inflows were assumed to be zero. Hence, the calculated market shares cannot be readily interpreted in terms of turnover, but they do give an indication of the competitive strength of centers in attracting consumers from within the study area, which was the primary concern in the present study.

Predicted market shares for the after situation when the before situation is set to 100 are shown in Table 2. As expected, the market share of the City Center has increased considerably in both the daily (38%) and non-daily sector (32%). The decrease in market share of competing centers is distributed almost evenly across the centers. Impacts range between 6.9% and 7.5% in the daily sector and between 4.9% and 6.0% in the non-daily sector. The competition with the major center of Eindhoven is also of interest. The loss of Veldhoven market share in Eindhoven inner city is 7.4% (daily sector) and 5.4% (non-daily sector).

4.4. Discussion

As the results indicate, the expansion of the center has reduced the total distance traveled by the population for shopping purposes (by approximately 3%). Multi-purpose trips have increased somewhat, but the effect is largely due to a substitution of long trips to the Eindhoven center by shorter trips to the Veldhoven center. The market share of the City Center has increased considerably (32–38%), but to a lesser extent than the increase in floor space (approximately 50%). The model did not take into account possible impacts on transport mode choice for the shopping trips. However, the same model structure could be used to define a nested model of transport-mode and location choice. In a mode-location-nested model, alternative destinations of trips are nested under mode choices (for example, car versus other modes), so that the impact of changes at destinations on mode choice can be predicted.

In this case study, the model was estimated on consumer data after the change had taken place. However, in most studies one would estimate the model on the before situation and use it to predict possible impacts in a planning stage. Several scenarios might be considered related to anticipated population developments or compensatory actions to reduce negative impacts, if any. For example, if parking facilities were included as additional attributes in the location utility function, the effects of simultaneous parking policies could be evaluated. The facilities offered by Location Planner may reduce the threshold for formulating scenarios. The inference engine automatically runs the entire inference model for evaluating impacts of each scenario and Scenario Manager can be used for generating and managing a scenario base.

The present case study focused on how a multi-purpose trip model can be specified, estimated and applied for impact analysis using Location Planner. The two-good system assumed in this case suits the data-availability and information needs of local governments. At least in Dutch retail planning, it is usual to analyze and collect floor space data at the level of daily and non-daily goods. Retail plans are also normally formulated at this level. The case study highlighted the extra information that can be derived from the multi-purpose model. Besides an origin–destination matrix, the model predicts trip-type probabilities and trip frequencies dependent on the choice of multi-purpose trips. Moreover, the model is sensitive to (i) settings of the relative weight of trip types in predicting market shares, (ii) assumptions about the trip-generation effects of multi-purpose trips, (iii) purpose-specific attractiveness of centers, and (iv) supply elasticity of trip choice.

5. Conclusions and discussion

This article described and illustrated with a case study the spatial decision support system, Location Planner, which we have developed. The primary objective of Location Planner is to provide a system that is easy to use and able to support a large variety of problems in retail/service planning. The system is relevant in addressing both issues of transportation planning and location decisions. It incorporates a wide range of spatial models, including spatial interaction/choice models, system performance models and location–allocation models (Fig. 5).

The system can be evaluated against general objectives of a decision support system. First, adaptability of the system to a wide range of problems is a strong point of the system. The structure of the inference model is not fixed but can be defined by users. Because the model base includes a wide range of model variants within each category a wide range of problems can be accommodated. Once the model structure is defined, the system has the flexibility to allow the choice between different modes of user–system interaction. The intuitive mode supports impact analysis of plans or market developments. The goal-seeking mode supports model-based optimization of the spatial configuration of retail or service networks.

Second, the system is strong on visual and interactive properties. The use of dynamic variable definitions strongly reduces the length of feedback loops. Users can manipulate a wide range of conditions and need only to click on an update button to see the implications. Using multiple active and linked views on data sections enhances the interactive properties. Users can view the

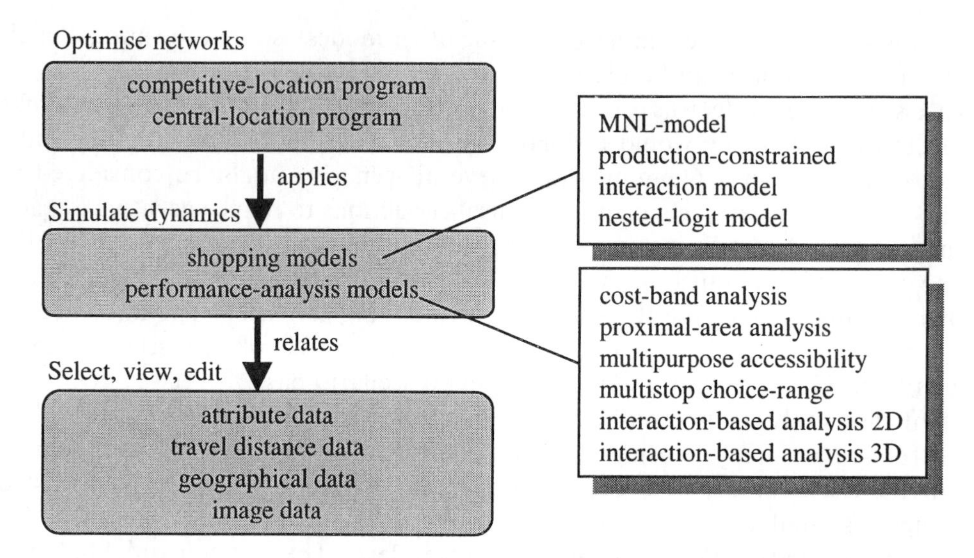

Fig. 5. Organization of data, models and knowledge in Location Planner.

same data set in table, map and graph views. The views are linked so that the selection of objects or attributes in one view is simultaneously implemented in linked views as well.

Third, it should be noted that the system has a limited focus. Not all stages of plan decision making are supported. The system emphasizes the stages of impact analysis and plan generating. To also support the preceding stage of monitoring developments and identifying problems, the system must be extended with a time dimension that make the representation and analysis of time series data possible. The identification of candidate locations and evaluation of plan alternatives also receive limited attention.

Given the emphasis on impact analysis and plan-generation, general-purpose GIS and MCE-software are considered complementary. Standard GIS tools support the elementary forms of spatial analysis required for identifying candidate locations. Commercially available MCE-software, such as Expert Choice (1995), can be used in addition for identifying criteria, deriving criterion weights, and ranking alternatives on the criteria. Hence, Location Planner is explicitly meant to be complementary to existing GIS and MCE (group) software. Communication is realized through data files.

Acknowledgements

This research is supported by the Technology Foundation (STW), Netherlands.

References

Arentze, T.A., Borgers, A.W.J., Timmermans, H.J.P., 1994a. Geographical information systems and the measurement of accessibility in the context of multipurpose travel: a new approach. Geographical Systems 1, 87–102.

Arentze, T.A., Borgers, A.W.J., Timmermans, H.J.P., 1994b. Multistop-based measurements of accessibility in a GIS environment. The International Journal of Geographical Information Systems 8, 343–356.

Arentze, T., Oppewal, H., Timmermans, H.J.P., 1997. A multipurpose destination choice model for shopping trips: some empirical results. In: Proceedings of the Paper presented at the Fourth Recent Advances in Retailing and Services Science Conference, 30 June–3 July, Scotsdale, Arizona.

Armstrong, M.P., De, S., Densham, P.J., Lolonis, P., Rushton, G., Tewari, V.K., 1990. A knowledge-based approach for supporting locational decisionmaking. Environment and Planning B: Planning and Design 17, 341–364.

Bierlaire, M., 1995. A robust algorithm for the simultaneous estimation of hierarchical logit models. GRT Report 95/3, Department of Mathematics, FUNDP, Namur, Belgium.

Birkin, M., 1994. Understanding retail interaction patterns: the case of the missing performance indicators. In: Bertuglia, C.S., Clarke, G.P., Wilson, A.G. (Eds.), Modelling the City: Performance, Policy and Planning. Routledge, London, UK, pp. 121–150.

Birkin, M., Clarke, G., Clarke, M., Wilson, A., 1996. Intelligent GIS: Location Decisions and Strategic Planning. Geoinformation International, Cambridge, UK.

Borgers, A.W.J., Timmermans, H.J.P., 1991. A decision support and expert system for retail planning. Computers Environment and Urban Systems 15, 179–188.

Breheney, M.J., 1978. The measurement of spatial opportunity in strategic planning. Regional Studies 12, 463–479.

Caliper Corporation, 1996. TransCAD: Transportation GIS software: User's Guide Version 3.0, Caliper Corporation, Newton, MA.

Carver, S.J., 1991. Integrating multicriteria evaluation with geographical information systems. International Journal of Geographical Information Systems 5, 321–339.

Clarke, C., Clarke, M., 1995. The development and benefits of customized spatial decision support systems. In: Longley, P., Clarke, G. (Eds.), GIS for Business and Service Planning. Geoinformation International, Cambridge, UK, pp. 227–254.

Clarke, G.P., Wilson, A.G., 1994. A new geography of performance indicators for urban planning. In: Bertuglia, C.S., Clarke, G.P., Wilson, A.G. (Eds.), Modelling the City: Performance Policy and Planning. Routledge, London, UK, pp. 55–81.

Crossland, M.D., Wynne, B.E., Perkins, W.C., 1995. Spatial decision support systems: an overview of technology and a test of efficacy. Decision Support Systems 14, 219–235.

Densham, P.J., 1991. Spatial decision support systems. In: Maguire, D.J., Goodchild, M.F., Rhind, D.W. (Eds.), Geographical Information Systems: Principles. Wiley, New York, pp. 403–412.

Densham, P.J., 1994. Integrating GIS and spatial modelling: visual interactive modelling and location selection. Geographical Systems 1, 203–221.

Densham, P.J., Armstrong, M.P., 1993. Supporting visual interactive locational analysis using multiple abstracted topological structures. In: Proceedings of AutoCarto 11, American Congress on Surveying and Mapping, Bethesda, MD, pp. 2–22.

Densham, P.J., Rushton, G., 1996. Providing spatial decision support for rural public service facilities that require a minimum work load. Environment and Planning B: Planning and Design 23, 553–574.

Expert Choice, 1995. Decision Support Software: User Manual. Expert Choice, Pittsburgh, Pennsylvania, US.

Fedra, K., Reitsma, R.F., 1990. Decision support and geographic information systems. In: Scholten, J.J., Stillwell, J.C.H. (Eds.), Geographic Information Systems for Urban and Regional Planning. Kluwer, Dordrecht, Netherlands, pp. 177–188.

Ghosh, A., McLafferty, S.L., 1987. Location Strategies for Retail and Service Firms. Lexington Books, Massachusetts.

Grothe, M., Scholten, H.J., 1992. Modelling catchment areas: towards the development of spatial decision support systems for facility location problems. In: Harts, J.J., Ottens, H.F.L., Scholten, H.J. (Eds.), Proceedings of the Second European Conference on Geographical Information Systems 2. EGIS Foundation, Faculty of Geographical Sciences, Utrecht, Netherlands, pp. 978–987.

Jankowski, P., 1995. Integrating geographical information systems and multiple critera decision-making methods. International Journal of Geographical Information Systems 9, 251–273.

Jankowski, P., Ewart, G., 1996. Spatial decision support system for health practitioners: selecting a location for rural health. Geographical Systems 3, 279–299.

Janssen, R., 1991. Multiobjective decision support for environmental problems. Dissertation, Free University, Amsterdam, Netherlands.

Keenan, B., 1998. Spatial decision support systems for vehicle routing. Decision Support Systems 22, 65–71.

Kohsaka, H., 1993. A monitoring and locational decision support system for retail activity. Environment and planning A 25, 197–211.

Lin, H., Wan, Q., Li, X., Chan, J., Kong, Y., 1997. GIS-based multicriteria evaluation for investment environments. Environment and Planning B: Planning and Design 24, 403–414.

Macdonald, M., 1997. A spatial decision support system for collaborative solid waste planning. In: Craglia, M., Couclelis, H. (Eds.), Geographic Information Research: Bridging the Atlantic. Taylor & Francis, London, UK, pp. 510–522.

Malczewski, J., 1996. A GIS-based approach to multiple criteria group decision-making. International Journal of Geographical Information Systems 10, 955–997.

Maniezzo, I., Mendes, I., Paruccini, M., 1998. Decision support for siting problems. Decision Support Systems 23, 273–284.

Pereira, J.M.C., Duckstein, L., 1993. A multiple criteria decision-making approach to GIS-based land suitability evaluation. International Journal of Geographical Information Systems 7, 407–424.

Roy, J.R., Anderson, M., 1988. Assessing impacts of retail development and redevelopment. In: Taylor, M.A.P., Sharpe, R. (Eds.), Desktop Planning: Microcomputer Applications for Infrastructure & Services Planning & Management. Newton Hargreen Publishing Company, Melbourne, Australia, pp. 172–179.

Sprague, R.H., 1989. A framework for the development of decision support systems. In: Sprague, R.H., Watson, H.J. (Eds.), Decision Support Systems: Putting Theory Into Practice. Prentice-Hall, London, pp. 9–35.

Sprague Jr., R.H., Carlson, J.E.D., 1982. Building Effective Decision Support Systems. Prentice-Hall, New Jersey, US.

Thill, J.C., 1999. Spatial Multicriteria Decision-Making and Analysis: A Geographic Information Sciences Approach. Ashgate, Aldershot, UK.

PERGAMON

Transportation Research Part C 8 (2000) 381–408

TRANSPORTATION
RESEARCH
PART C

www.elsevier.com/locate/trc

Optimizing path query performance: graph clustering strategies ☆

Yun-Wu Huang [a,*,1], Ning Jing [b,2], Elke A. Rundensteiner [c,3]

[a] *IBM T.J. Watson Research Center, 30 Saw Mill River Road, Hawthrone, NY 10532, USA*
[b] *Changsha Institute of Technology, NY, USA*
[c] *Worcester Polytechnic Institute, NY, USA*

Abstract

Path queries over transportation networks are operations required by many Geographic Information Systems applications. Such networks, typically modeled as graphs composed of nodes and links and represented as link relations, can be very large and hence often need to be stored on secondary storage devices. Path query computation over such large persistent networks amounts to high I/O costs due to having to repeatedly bring in links from the link relation from secondary storage into the main memory buffer for processing. This paper is the first to present a comparative experimental evaluation of alternative graph clustering solutions in order to show their effectiveness in path query processing over transportation networks. Clustering optimization is attractive because it does not incur any run-time cost, requires no auxiliary data structures, and is complimentary to many of the existing solutions on path query processing. In this paper, we develop a novel clustering technique, called spatial partition clustering (SPC), that exploits unique properties of transportation networks such as spatial coordinates and high locality. We identify other promising candidates for clustering optimizations from the literature, such as two-way partitioning and approximate topological clustering. We fine-tune them to optimize their I/O behavior for path query processing. Our experimental evaluation of the performance of these graph clustering techniques using an actual city road network as well as randomly generated graphs considers variations in parameters such as memory buffer size, length of the paths, locality, and out-degree. Our experimental results are the foundation for establishing guidelines to select the best clustering technique based on the type of networks. We

☆ This work was supported in part by the University of Michigan ITS Research Center of Excellence grant (DTFH61-93-X-00017-Sub) sponsored by the US Department of Transportation and by the Michigan Department of Transportation. N. Jing was supported in part by the State Education Commission of People's Republic of China.

* Corresponding author. Tel.: +1-914-784-7523.

E-mail addresses: ywh@us.ibm.com (Y.-W. Huang), jning@eecs.umich.edu (N. Jing), rundenst@cs.wpi.edu (E.A. Rundensteiner).

[1] This work was performed while the author was Ph.D. student at the University of Michigan.

[2] This work was performed while the author was visiting the University of Michigan.

[3] This work was performed while the author was a faculty member of the University of Michigan.

PII: S0968-090X(00)00049-8

find that our SPC performs the best for the highly interconnected city map; the hybrid approach for random graphs with high locality; and the two-way partitioning based on link weights for random graphs with no locality.

Keywords: Path query processing; Transportation networks; Spatial clustering; Clustering optimization; Geographic information systems

---

1. Introduction

1.1. Background on path query processing

Transportation networks are essential components of many geographic information systems (GIS) applications. Such applications include navigation, route guidance, traveler information systems, fleet management, public transit, troop movement, urban planning, to name a few. Among the services provided by such GIS systems, path query processing is an important feature required by many of the above applications (Huang et al., 1998, 1997d,e, 1996,a; Shekhar and Liu, 1995; Zhao and Zaki, 1994). Examples of path queries are:

Q1: Find the most energy-efficient path from A to B that does not use toll roads.
Q2: Display all the garages reachable from A in 10 min.
Q3: Find the shortest path from A to B that does not pass through areas with altitude $>$ 1000 feet for more than 10 miles.

In addition to requiring path search support from the GIS system, queries such as the three above often also have embedded constraints that must be processed. For example, for Q1, the computed path contains no links of type *toll* road. This alpha-numeric filter can be applied to all links of the graph before path processing, thus effectively constructing a smaller subgraph on which to apply the path search. For Q2, the constraint is that the destination nodes are of *garage* type. Again prefiltering could be conducted. For Q3, we now deal with a spatial and hence much more expensive filter, namely that the computed path does not contain any link that traverses areas with *altitude* $>$ 1000 *feet for more than* 10 *miles*. In this case, we cannot a priori determine a valid subgraph of the complete network and worse yet, the spatial characteristics have to be stored with each link – thus significantly increasing the storage requirements and thus expected costs of path processing. Note that while the cost measurements used in path query computation for all examples may be different, such as based on fuel consumption in Q1, on travel time in Q2, and on distance in Q3, they can be all abstracted and then handled using the same path search techniques.

In order to process path queries such as the above, a GIS system must model the topological information of the transportation networks as well as maintain the attributes and cost measurements associated with each component of the network. Typically, a GIS system models the topological information of a transportation network by representing it as a graph composed of nodes and links. A node represents for example an intersection and a link represents a road segment which is one section of a road between two neighboring intersections where traffic flows in one direction. In this paper the topological information and other *attributes* associated with intersections and road segments are stored in two separate structures, called the *node table* and the *link table*. Each element in such a table is referred to as a tuple of the table. The attributes that describe a node tuple may include its *x*- and *y*-coordinates, the connecting road segments (incoming and outgoing), the traffic control configuration (traffic light, stop sign, etc.), points of interest, and so on. A link itself is identified by its *origin* and *destination* nodes. Additional attributes for describing each link for a road network include for example the number of lanes, maximum speed, length, up-to-date link travel speed, and so on. The sizes for each node and link therefore can be very large, up to hundreds of bytes in length.

Transportation networks are considered stable graphs for the purpose of this paper, since the addition and the removal of intersections or roads occurs only very infrequently in practice. The cost measurement data used for path query computation may however be either stable or unstable depending on the attributes. The up-to-date estimated link traversal time, for example, may depend on changing traffic conditions, and therefore is unstable because it needs to be updated as soon as the traffic changes occur. Link distance or the geographic coordinates of nodes on the network on the other hand are considered stable.

1.2. Clustering for path query processing: motivation

This paper investigates the optimization of path query processing based on graph clustering techniques. To compute paths for path queries such as those previously listed (Q1–Q3), we assume that popular graph-traversal search algorithms such as the *Dijkstra*, A^*, Breadth-First Search, and Depth-First Search algorithms or any of their variants are used. They search for paths by traversing from one node to another through their respective connecting link. Because path search computation is recursive in nature, searching a path means to recursively access links from the *link table*.

However, since the size of the *link table* is often larger than the capacity of the main memory buffer of a given GIS system, the *link table* may need to be stored on a secondary storage device, typically on disk. While state-of-the-art database engines may attempt to cache the link table into main memory during path evaluation, this will generally not be feasible due to size constraints. In this case, many tuples (links) in the *link table* may need to be retrieved over and over again from secondary storage and placed into the main memory buffer for evaluation. Given that such I/O operations on most modern computers are typically several 100-fold more expensive than CPU operations, the I/O costs are the dominant factor of path computation costs.

The high processing costs are thus incurred by the recursive nature of the graph traversal component of path query computation. Resolving embedded constraints may further increase I/O costs significantly. For example, in a related effort (Huang et al., 1998, 1997,a,b,d,e), we found that processing spatial constraints (see Q3 path query) is very I/O intensive. Thus such constraint resolution competes with the path finding component of the search process for computational resources such as the buffer space. This further motivates our research presented in this paper on optimizing the path computation process by reducing I/O activities.

Data is commonly not transferred between secondary storage and main memory one tuple at a time, but rather at the granularity of one or more buffer pages containing possibly many tuples each. Hence, one important performance consideration studied by the database community is how best to place tuples onto disk pages so to minimize the number of required I/O operations.

This is done by assuring that tuples brought into memory on one disk page are ideally all made use off whenever in the buffer. This optimization strategy of grouping data onto pages is commonly referred to as *clustering*.

The purpose of this paper is to demonstrate that clustering optimization for path query computation can be effective for many types of transportation networks. Clustering is attractive because it does not incur any run-time cost, nor does it require any auxiliary data structure that demands buffer space. Because transportation networks are stable graphs, clustering is a one-time a priori cost not affecting actual path processing. Most importantly, clustering is at a level lower than many other path query solutions that focus on auxiliary access structures or on algorithmic techniques, therefore results emerging from the comparative evaluation of our clustering research can be deployed by such other solutions that do not already employ specific link clustering (Agrawal and Jagadish, 1988, 1989; Bancilhon and Ramakrishnan, 1986; Zhao and Zaki, 1994). Our work thus is complimentary to much of the existing work on path finding and could be exploited to further optimize such techniques.

1.3. Contributions

The contributions of this paper can be summarized as follows:

1. We identify existing clustering techniques from the literature, namely topological clustering (adopted from Agrawal and Kiernan, 1993), two-way partition clustering (adopted from Cheng and Wei, 1991), and random clustering, and apply them for optimizing the I/O behavior for path query processing.
2. We develop a novel clustering technique, which we call spatial partition clustering (SPC), that exploits unique properties of transportation networks such as spatial coordinates or high locality of such networks. The basic idea of SPC is to achieve spatial partitioning by dividing the links in the networks into partitions such that the *origin* nodes of all links within a partition are bounded by an area that resembles a square on the map underlying the network.
3. In addition, we develop two extensions to the above techniques further tuned for path query processing for GIS systems. Namely, we combine our proposed spatial partition solution with the min-cut technique to create a hybrid approach. Second, we modify the two-way partition clustering to consider both connectivity and link weights in determining partitions.
4. We implement the six selected clustering techniques on a uniform testbed for fair comparison. To benchmark these six techniques, we also implemented the *Dijkstra* algorithm (Dijkstra, 1959), one of the more popular and effective path search algorithms, so that it can be applied as search technique over the network data once clustered.
5. We perform extensive experimental evaluation of both existing as well as our proposed clustering techniques to determine their relative effectiveness. These experiments are conducted both using real data, in particular, the road network of Ann Arbor, Michigan (5596 nodes, 14,033 links), as well as synthetically generated networks. We compare the performance results of running the *Dijkstra* algorithm on network data layed out on disk by the different clustering strategies.
6. Our experimental results are used to establish guidelines to select the best clustering technique based on the type of network found in a given application domain. Characteristics of the

network considered include parameters such as the size of the graph, the average out-degree or the locality.[4]

While a preliminary version of this work appears in an earlier conference paper (Huang et al., 1996b), this journal paper differs from it in many respects. First, we propose two additional new graph clustering techniques that are the extensions of the clustering techniques presented in (Huang et al., 1996b), namely the hybrid SPC technique and the link weight based partitioning technique. Second, we include the experimental results of the new extensions into the performance evaluation section in this paper. Third, this paper presents additional types of experiments that provide new insights into the behavior of proposed clustering techniques. Such experiments, not available in the previous report, are for example the path finding experiments based on paths of different lengths and networks of various average out-degrees. These new types of experiments have been instrumented for all clustering techniques, both the ones introduced in Huang et al. (1996b) as well as the new optimizations. Fourth, because the performance of the new extensions can be shown to lead to further improvement over the original techniques, the conclusions for this paper have been revised to reflect the new results. Lastly, this paper is written with more examples and illustrations for better understanding and accessibility to the material.

2. Related work

There are many recent research efforts reported in the literature that focus on minimizing the I/O costs of path computation in a database setting that assumes a fixed-size main memory I/O buffer. Most of such research has proposed solutions to solve recursive query problems for *general databases* that focused on pure transitive closure computation (Agrawal et al., 1998; Agrawal and Jagadish, 1990; Bancilhon, 1985; Ebert, 1981; Ioannidis, 1986; Ioannidis and Ramakrishna, 1988; Ioannidis et al., 1993; Schmitz, 1983). In our work, rather than aiming for generality, we now take an application-driven stance by proposing different disk page clustering algorithms for optimizing path query processing for GIS type of applications and then experimentally evaluating their relative advantages and disadvantages.

Two unresolved problems arise when applying transitive closure pre-computation techniques to path query processing for transportation networks. First, a single transitive closure computation cannot take different embedded constraints into account. For example, a transitive closure computed for path query Q1 cannot be used to answer path query Q3. To answer all path queries with a large set of different embedded constraints, we may need to compute numerous transitive closures, each based on a unique embedded constraint. Clearly, this is not feasible in practice.

Second, some link weights (cost measurements) may be unstable and can change very frequently. In order for the transitive closure computed based on such cost measurements to reflect the most up-to-date cost, re-computation may need to be conducted very frequently. However, performance results in Agrawal et al. (1998) and Ioannidis et al. (1993) have shown that their techniques are not efficient in computing the shortest path transitive closure for graphs with cycles

[4] We define that in a graph of high locality, the two end nodes of most links are located closely geographically, whereas for graphs of no locality, such restriction does not apply.

such as transportation networks in GIS applications. Re-computation of shortest path transitive closure using such techniques thus cannot be done frequently, under-cutting the correctness of the computed paths.

In the GIS community, there has been much work on data structures and representations for most efficiently being able to manage and access geograhical data sets (Goodchild, 1990; Goodchild and Shiren, 1992). Focus here typically was on the logical organization of data to support certain access patterns, whereas less effect has gone into the lower-level physical aspect of management of the data on disk (clustering). The later is the target of our current work, and is complimentary to data structure type of work. Similarly, there is also a body of literature on evaluating shortest path algorithms both in the database as well as in the GIS community. Focus here has been on a comparison of the performance of different types of shortest path algorithms itself (Zhan, 1998), on-line reordering (Huang et al., 1996), or on a priori materialization of best paths (Huang et al., 1997c).

In Agrawal and Jagadish (1988), a multi-level structure is proposed to prune the search space by conducting path pre-computations within each component data structure. Such a technique is less useful for processing path queries with embedded constraints because the pre-computation information is general and may not be effectively usable as heuristics in pruning the search space for path queries with embedded constraints.

In our previous work, we have explored the hierarchical path view approach which fragments a large graph into smaller subgraphs and pre-computes the path transitive closure for each subgraph (Huang et al., 1997c; Huang et al., 1997e; Huang et al., 1996a; Jing et al., 1998; Jing et al., 1996). The advantage of such a technique is more efficient computation in both transitive closures of the subgraphs and path search through the hierarchy.

In Zhao and Zaki (1994), a graph indexing technique is proposed to improve paging performance for graph traversal by building an auxiliary structure that predicts which nodes are to be accessed in the future. However, such an auxiliary structure itself requires storage that competes for buffer space. A further limiting factor is that the examples given in Zhao and Zaki (1994) have fanout of only 1 or 2. Typical transportation networks have a fanout of at least 2 to 5, meaning that the size of the auxiliary structure will be larger than indicated in Zhao and Zaki (1994). Thus more buffer space needs to be put aside to load such a structure. In Zhao and Zaki (1994), no path finding experiment on real networks was conducted. The degree of improvement for GIS road and other transportation networks therefore is unknown.

Clustering techniques, which are the focus of this paper, have been previously proposed to reduce the path query processing I/O costs by arranging the link relation in a certain order in secondary storage (Agrawal and Kiernan, 1993; Banerjee et al., 1988; Larson and Deshpande, 1998). However, pure topological clustering (Banerjee et al., 1988; Larson and Deshpande, 1998) is not effective for shortest path computation for cyclic graphs such as GIS transportation networks where the ancestor relation is mostly bidirectional. This is because pure topological clustering only preserves the ancestry relation in one direction; it does not guarantee good paging behavior when the computation searches the other direction when cycles exist. In Agrawal and Kiernan (1993), an approximate topological clustering was proposed that handles cyclic graphs using heuristics that first remove the acyclic subparts of the graph and next remove some links to break the remaining cyclic subparts. The goal of such clustering is to minimize the number of link tuples that trace backward topologically.

In this paper, a version of this clustering technique is implemented and benchmarked in order to apply it to networks and to compare against alternative strategies. Our experimental evaluation in Section 6 indicates that several alternative clustering strategies outperform such topological clustering technique for both GIS networks and for random ones.

The heuristic partitioning techniques (Cheng and Wei, 1991; Fiduccia and Mattheyses, 1982; Kernighan and Lin, 1970; Wei and Cheng, 1990) commonly deployed in very large scale integrated circuit (VLSI) design can also be adopted to cluster the link relation for a graph. Such techniques partition a graph in subgraphs based on certain partitioning objectives, of which the most common one is to minimize the total distance of inter-connection links between partitions. (Shekhar and Liu, 1995) adopted a two-way partitioning algorithm (Cheng and Wei, 1991) as a clustering mechanism for the proposed access structure for aggregate queries for transportation networks and found it effective. Their aggregate query experimentation given in Shekhar and Liu (1995) however is based on a linear path evaluation. Recursive path search such as the path queries discussed in this paper is not considered. In this paper, we also include the two-way partition algorithm (Cheng and Wei, 1991) in our evaluation in order to compare this technique to alternative clustering solutions.

3. Spatial partition clustering

In this section, we first present the transportation network characteristics exploited by our proposed SPC algorithm, followed by a description of the algorithm that creates the SPC.

3.1. Exploiting characteristics of transportation networks

We propose to cluster link tuples from the *link table* based on the spatial proximity of their *origin* nodes, that is, to group tuples of the link table into disk pages and then to transfer the link relation between secondary storage and main memory in the granularity of these pages. We call this the Spatial Partition Clustering, or short SPC. To understand why the SPC can be effective for transportation networks, we describe the unique characteristics of the (road) networks:

- Road networks are relatively sparse, have uniform fanout typically between 2 and 5.
- Road networks are strongly inter-connected, with each node typically reachable from near-by nodes by traversing only a few links.
- Road networks consist of mostly short links in comparison to the size of the underlying spatial region. In other words, most road links span a short distance from one intersection to the neighboring intersection.

Graph-traversal search algorithms conduct node expansions by traversing links. Because most road links are short, these algorithms therefore exhibit high expansion locality on transportation networks. Furthermore, page sizes in modern databases can be quite large. Therefore many link tuples in the *link table* can be stored within one page. Because road transportation networks are sparse with low fanout, multiple groups of links with the same *origin* can be stored within one page. We call them same-origin-link (SOL) groups. For example, with a 4 KB page size and link tuple size of 128 bytes, 32 links can be stored within one page. For a transportation network with average fanout of 3, roughly 11 SOL groups can be clustered in one page. This means that there are roughly 11 different nodes in each page that could potentially be expanded by the search algorithm.

If we cluster the *link table* so that every page contains links whose *origin* nodes are geographically closely located, we are grouping the expansion nodes based on their spatial proximity. Based on the fact that transportation networks are highly inter-connected and consist of mostly short links, the graph-traversal algorithms such as *Dijkstra* are likely to expand nodes within the same page by traversing the intra-page links before traversing cross-page links with such a clustering. Given a fixed-sized main memory buffer not large enough to hold the entire *link table*, such paging behavior would reduce page misses caused by cross-page link traversing. We now present the algorithm that creates the spatial partition clustering for a given network.

3.2. The spatial partition clustering algorithm

The algorithm that creates the SPC clustering is based on the *plane-sweep* techniques commonly found in multi-dimensional spatial data operations. The *plane-sweep* technique is for example used to implement the spatial intersect operation in Brinkhoff et al. (1994), Preparata and Shamos (1985) and Shamos and Hoey (1976). The basic idea of SPC is first to sort all links by the x-coordinate values of their *origin* nodes. The *plane-sweep* technique is then applied to sweep all x-sorted links along the x-coordinate from left to right. The sweeping process stops periodically to sort the links swept since last stoppage by the y-coordinate values of their *origin* nodes. Because the *origin* nodes of the links between two stoppage points span a short distance along the x-axis, sorting these links by the y-coordinate values of their *origin* nodes achieves a partial spatial ordering. After each y-sorting, the y-sorted links can be grouped into disk pages. We call the output of this clustering process the SPC-clustered *link table*, which as explained earlier corresponds to the layout of link tuples onto disk pages.

One critical decision to such a partition algorithm is to determine the proper stoppage points during plane sweeping when y-sorting takes place. Our goal is to achieve a balanced partitioning in which each resulting partition consists of links whose *origin* nodes are located within a bounding area that resembles a square block when the links are evenly distributed on the map. Below, we introduce a heuristic that dynamically computes the proper stoppage points in order to achieve such a balanced partitioning. To accommodate unevenly distributed maps, the heuristic we use will adjusts the bounding block for each partition by growing in the y-axis direction if the regional link distribution is sparse, and shrinking if otherwise. In either case, each partitioned page is maximally filled with links whose *origin* nodes are relatively closely located.

To present the algorithm that creates the SPC clustering, we use the following parameters:

- f refers to the number of link tuples that fit into a given page size, referred to as the link tuple blocking factor. We thus call every f consecutive link tuples an f-page.
- The *block table* is a temporary table that stores the links collected between two stoppage points during the sweeping process.
- dx_i is the difference between the minimum and maximum x-coordinate values of the *origin* nodes of the links in the first i f-pages in the *block table*.
- dy_i is the difference between the minimum and maximum y-coordinate values of *origin* nodes of the links in the first i f-pages in the *block table*.
- SPC-clustered *link table* is the resulting table.

Algorithm (*SPC*()).

Input: L: link table filled with all link tuples

Output: CL: link table clustered into pages

1 The unclustered *link table* L is sorted by the x-coordinate values of the *origin* nodes of its link tuples. The result is called the x-sorted *link table*.

2 Read the x-sorted *link table* sequentially one f-page at a time using the following process: read the next f-page and write the page to the end of the *block table* (*block table* is initially empty). Then check the following conditions:

- If all tuples in the x-sorted *link table* are read, go to step 3.
- If there is only one f-page in the *block table*, go to step 2 to read the next f-page and write it to the end of the *block table*.
- Otherwise, conduct the following evaluation:

  2.1 Let p be the number of f-pages in the *block table*. Compute the following:

$$d_p = |(\mathrm{d}y_p/p) - \mathrm{d}x_p|$$

$$d_{p-1} = |(\mathrm{d}y_{p-1}/(p-1)) - \mathrm{d}x_{p-1}|$$

  2.2 If $d_p > d_{p-1}$, this is a stoppage point. Perform the following:

    2.2.1 Sort the link tuples of the first $p-1$ f-pages by the y-coordinate values of their *origin* nodes, group them into pages, and sequentially append to the SPC-clustered *link table*.

    2.2.2 Move the pth f-page of link tuples in the *block table* to the first page in the *block table*. Set the number of pages in the *block table* to 1.

  2.3 Go to step 2 to read the next f-page.

3 Sort all remaining link tuples in the *block table* by the y-coordinate values of their *origin* nodes, group them into pages, and sequentially append to the SPC-clustered *link table* CL. Output CL.

The intuition behind the heuristic is that when the first few f-pages (e.g., 1 or 2) are written from the *link table* to the *block table*, p is small, and $d_p = |(\mathrm{d}y_p/p) - \mathrm{d}x_p|$ will likely be large, assuming a map with evenly distributed links. This is because in the *plane sweep* process, we are proceeding with small progress on the x-axis and with entire range on the y-axis. When more f-pages are added to the *block table*, p increases and d_p decreases. At some point, d_p will approach 0 and then starts picking up again when $\mathrm{d}x_p > (\mathrm{d}y_p/p)$. We capture this point by dynamically detecting $d_p > d_{p-1}$ and make it a stoppage point. At a stoppage point, links in the first $p-1$ f-pages in the *block table* are sorted by the y-coordinate values of their *origin* nodes. Because $d_{p-1} = |(\mathrm{d}y_{p-1}/(p-1)) - \mathrm{d}x_{p-1}|$ approaches 0, each partition will be bounded by an area that resembles a square box.

Fig. 1 illustrates the sweeping process and the reasoning behind the heuristics in determining the stoppage points. In Fig. 1(a), the link tuples are sorted by the x-coordinate values of their *origin* nodes. Next, f-pages of link tuples are written to the *block table* sequentially. In Fig. 1(b), when the fourth f-page is written to the *block table*, $d_4 > d_3$. This is a stoppage point. In Fig. 1(c), links in the first 3 f-pages in the *block table* are then sorted by the y-coordinate values of their

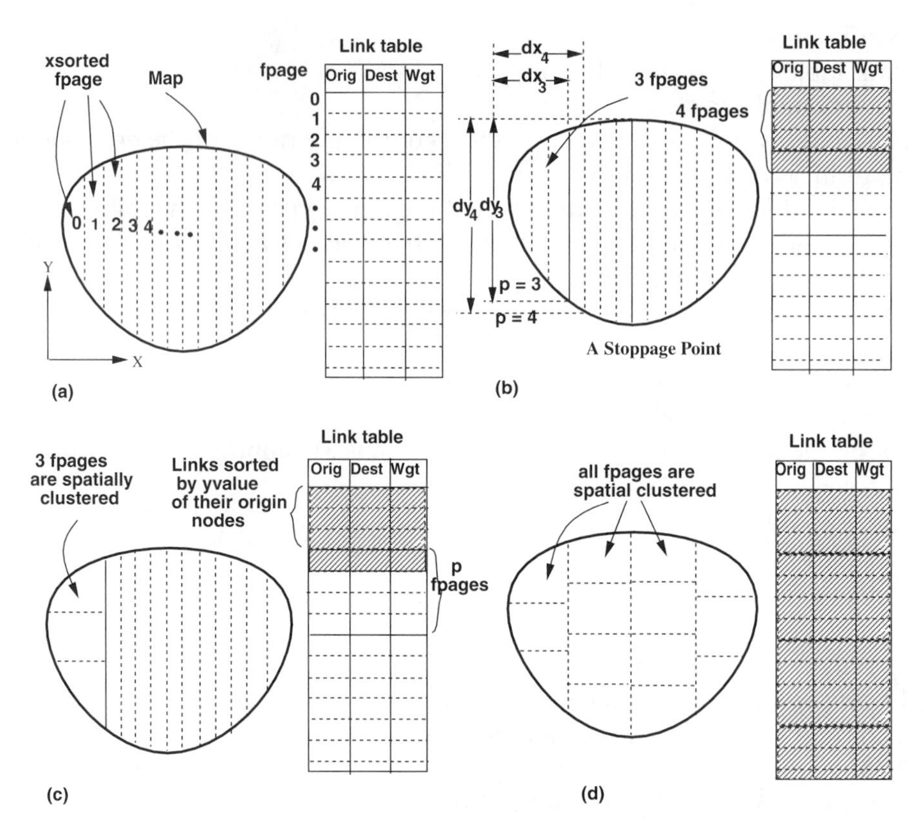

Fig. 1. Spatial partition clustering: (a) sort links by x-value of their origin nodes, (b) 4 f-pages loaded in block table when $d_4 > d_3$ where $d_4 = |(\mathrm{d}y_4/4) - \mathrm{d}x_4|$ and $d_3 = |(\mathrm{d}y_3/3) - \mathrm{d}x_3|$, (c) sort 3 f-pages links by y-value of their origin nodes, (d) spatial partition clustered links.

origin nodes and the y-sorted links are grouped in pages and written to the SPC-clustered *link table*. Note that at this point, the first 3 f-pages in the *link table* are properly clustered. When the sweeping process is complete, all f-pages in the *link table* are properly clustered as shown in Fig. 1(d).

4. Alternative graph clustering strategies

In this section, we present three alternative clustering strategies implemented for comparative studies. They are the two-way partition clustering (TWPC) (Cheng and Wei, 1991), the "approximately" Topological Clustering (TopoC) (Agrawal and Kiernan, 1993), and the Random Clustering (RandC). We assume that for each clustering technique, links of the same SOL group are always clustered together in the *link table*. Such a clustering is important because the graph-traversal path search algorithms typically expand a node by traversing all its outgoing links to the connecting nodes. Grouping links by their *origin* nodes makes sure such expansions exhibit good I/O behavior.

4.1. Two-way partition clustering

Partitioning algorithms have been widely deployed in the design and fabrication of very large scale integrated circuit (VLSI) chips. Most such algorithms partition a network into two subnetworks (Cheng and Wei, 1991; Fiduccia and Mattheyses, 1982; Kernighan and Lin, 1970), and through a *divide-and-conquer* process, reduce a complex problem into smaller and hence more manageable subproblems. The common objective of such partitioning is to shorten the total interconnection distance between all subnetworks in achieving a reduced layout cost and better system performance. We now propose that these partitioning algorithms could also be applied to our problem of transportation network clustering, namely to cluster the *link table* by storing each partition within a single page. In our context, the size of each partition therefore is bounded by the size of a buffer page. Our goal of such partitioning is to reduce the page misses that occur during path query computation to a minimum. Because each cross-page traversal in path computation may potentially incur a page miss, our partition objective is then to *minimize the number of inter-partition (cross-page) links*.

4.1.1. The two-way partitioning algorithm

We implement a partition clustering based on the two-way algorithm proposed in Cheng and Wei (1991). Because the partition problem with specified size constraints belongs to the class of NP-complete problems (Carey et al., 1976), all partition algorithms focus on finding heuristics in providing solutions in polynomial time. The most common heuristic used in two-way partitioning is based on a two-stage process (Kernighan and Lin, 1970). First, an initial cut that separates a network into two is derived. Next, nodes are swapped from one partition to another as long as such swapping results in a better cut. For example, in Fig. 2, swapping node x from partition A to B creates a cut that reduces the connecting links from 3 to 2 between the two partitions. Swapping can also be conducted between two nodes from different partitions. For example, in Fig. 3, swapping node x and node y to their opposite partition reduces the cross-partition links from 4 to 2. During each swapping run, priority is always given to the swap that yields the best cut. Swapping can continue until it no longer creates a better cut. To avoid cyclic swapping that results in an infinitive loop, a restriction is imposed that allows one node to be swapped only once during each swapping run. To remedy such a restriction, multiple iterations of swapping runs may be necessary to achieve an acceptable result.

The two-way partitioning algorithm we implement is also based on the two-stage heuristics (Cheng and Wei, 1991). We add a contraction stage that has been shown to be an improvement over the traditional two-stage approach (Cheng and Wei, 1991). Because our partitioning objective is to reduce the inter-connection links, we abbreviate this clustering technique as TWPC_con. We now give an overview of the algorithm, while a detailed presentation of the algorithm can be found in the original paper (Cheng and Wei, 1991).

Algorithm (*Two-Way Partitioning TWPC_con*()).

Input:

- G: network with all link tuples.
- p: integer denoting maximal size of a partition.

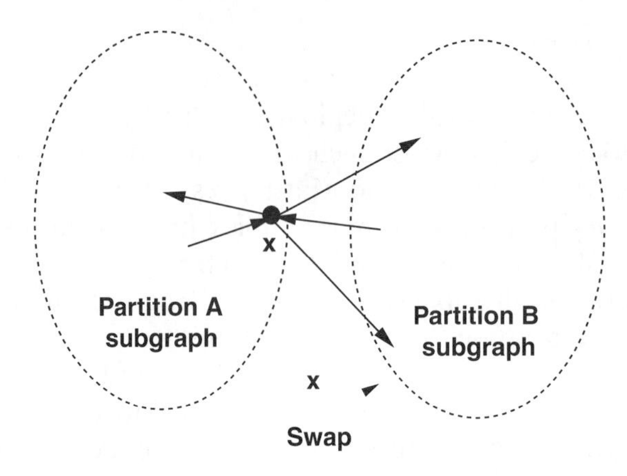

Fig. 2. An example of single-node swapping.

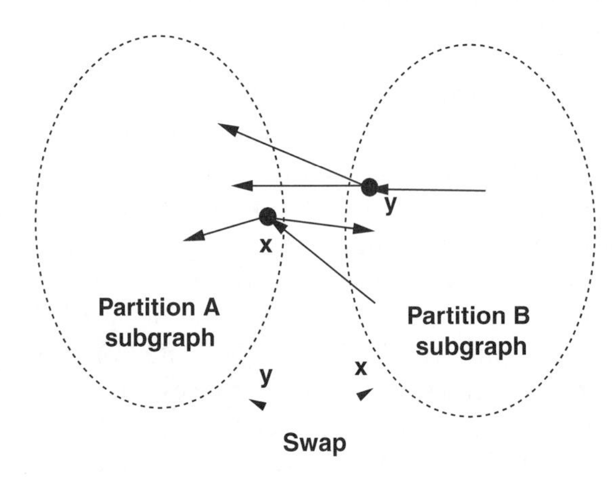

Fig. 3. An example of pair-wise swapping.

- i: integer denoting number of swaps allowed.
- s1,s2: integers denoting size constraints on partitions.

Output:

- G': network G partitioned into two groups.

 1 **Contracting stage**:

 1.1 Initially, the network G has only one partition.

 1.2 Based on *divide-and-conquer*, recursively apply the ratio-cut routine in (Wei and Cheng, 1990) to the partitions whose sizes are greater than a specified value p.

 1.3 Based on the resulting partitions, contract G to a condensed graph G' such that each partition in G is a node in G' and each interconnection link between two partitions is a link between the two corresponding nodes in G'.

 2 **Swapping stage**:

 2.1 Randomly select a cut that partitions G'. into two groups.

 2.2 Iteratively, apply the Fiduccia–Mattheyses algorithm (Fiduccia and Mattheyses, 1982) to the partitioned G' i times for better swapping result, with the size constraints of the two

resulting partitions set to $s1$ and $s2$. The i, $s1$, $s2$ are pre-specified input parameters.
2.3 The result of step 2.2 is a two-way cut of G'.

3 **Restoring stage**:
3.1 Restore the two partitions in G' created in step 2 by replacing each condensed node in each partition by its original nodes in the correspondent partition created in step 1. The result is two-way cut in G.
3.2 Apply the Fiduccia–Mattheyses algorithm on the two restored partitions in G one time, and the ending two partitions are the final result.

The ratio-cut routine (Wei and Cheng, 1990) in step 1.2 and the Fiduccia–Mattheyses algorithm (Fiduccia and Mattheyses, 1982) in step 2.2 are two partitioning algorithms based on the two-stage heuristics described earlier. The former relies on the ratio-cut property to achieve a more balanced split while the latter deploys a data structure that reduces the computational complexity and a specific size restriction to achieve a desired partitioning.

The intuition behind the contraction approach is that nodes that are more strongly connected are identified by the ratio-cut routine in step 1.2 and treated as one node in the swapping stage. This way the chance of them being split into different partitions by a bad split is reduced. For example, in Fig. 4, the ratio-cut routine may group the nodes forming a circular versus a triangular configuration into two different subgraphs A and B. Because partitions A and B are subsequently contracted into two inseparable units, if there is a cut that goes through A and B, it has to go through the link between node x and node y. Note that this is an optimal cut between A and B and no further swapping will change this cut-link. If no contraction is performed, an initial cut of all the nodes in A and B may look like the cut in Fig. 5. Note that subsequent swapping will not alter this cut because any single-node or pair-wise swapping of nodes a, b, c, d does not yield a cut with less inter-partition links. Therefore the optimal cut that goes through the link between nodes x and y would be lost in the case if we were to not utilize the contraction heuristic.

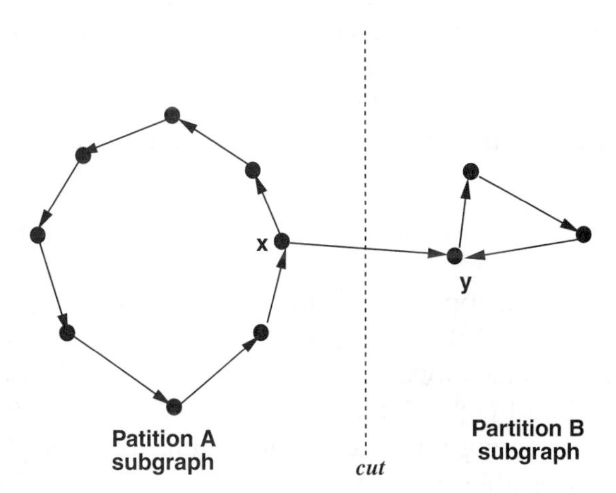

Fig. 4. An example of partitioning with contraction.

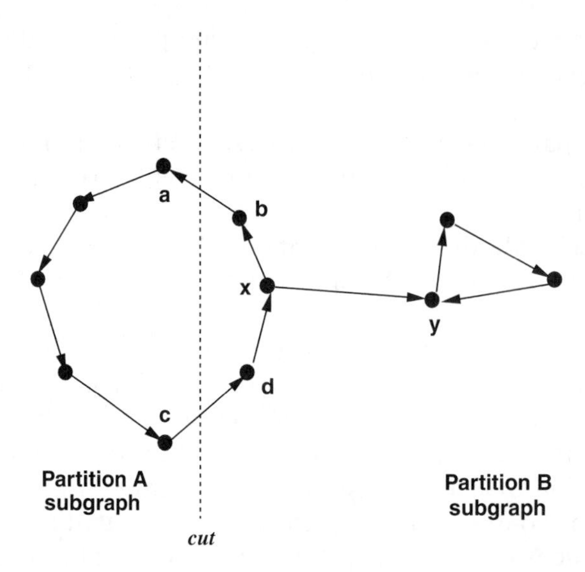

Fig. 5. An example of partitioning without contraction.

4.1.2. Our adaptation of the two-way partitioning algorithm

The above two-way partition algorithm cuts a network into two partitions based on ratio-cut heuristics. To adapt it to our page clustering, we recursively apply it until each partition fits into one page. Because the tremendous potential I/O required to process a path query, we desire to load as much graph information into one page as possible. Consequently, the occupancy rate of each page is set to be very high. As a result, we allow for an uneven two-way partitioning as long as the size of one partition approximates that of a page. To achieve this, we set a relatively high minimum occupancy rate for each partition. To avoid a "local minima" trap, we allow the swapping process to overflow a partition to make it larger than a page with the hope that further swapping will decrease its size to be within a page. Thus it is possible that after several iterations of swapping, one partition may have a size that is slightly greater than that of a page. To avoid partitions with such an unsatisfactory occupancy rate in our context of paging, we instead introduce a heuristic during the final Fiduccia and Mattheyses algorithm run in step 3.2 to favor swapping nodes out of such partitions whose sizes are slightly over that of a page. We find that this adaptation achieves a uniformly high occupancy rate among all final partitions.

4.2. Approximately topological graph clustering

For an acyclic directed graph, topological clustering consists of arranging all its links in a topological order. The advantage of this topological clustering of links is that a path query over it can be processed by accessing links in one pass. Thus, if the link table does fit into main memory, the path query will exhibit good I/O performance by avoiding to access any page more than once. However, this topological clustering approach is not applicable to cyclic graphs as there is no topological order that can be established due to cycles. Agrawal and Kiernan (1993) proposed an approach which extended the topological clustering to cyclic graphs by recursively breaking cycles and removing acyclic portions of the cyclic graph. By this approach, the cyclic graph can be "approximately" topologically sorted.

In this paper, we implement and evaluate the approximately topological clustering algorithm proposed in Agrawal and Kiernan (1993) and call it TopoC (for Topological Clustering). The following is a description of the main steps of the TopoC.

Algorithm (*TopoC*()).

Input: L: link table representing transportation network

Output: CL: link table clustered into pages

1. Move a root-link [5] of the link table L into the clustered link table CL. Repeat this process until no more root-links can be found in the remaining table If the remaining link table is empty, go to step 4.
2. Move a sink-link[6] to a temporary link table. Repeat this process until no more sink-links remain in the link table L.
3. Randomly pick a node in the remaining link table L. Move all its outgoing links to the temporary link table. Go to step 1.
4. Append the links in the temporary link table in reverse order to the clustered link table CL.

The TopoC achieves approximately topological clustering in three phases: Steps 1 and 2 remove the acyclic portions of the graph. Step 3 breaks cycles by removing all out-going links of one selected node. When the remaining graph is empty, step 4 appends the links in the temporary link table to the resulting clustered link table.

As an example, Fig. 6(a) shows a cyclic directed graph and its unclustered link table. We use this graph to illustrate how the clustered link table is built step by step by the TopoC algorithm. Fig. 6(b) depicts the graph after TopoC moves root-links L_{fc} and L_{fg} into the clustered link table. Since Fig. 6(b) has no sink-links, TopoC skips step 2. In Fig. 6(c), TopoC breaks a cycle by moving link L_{ab} into the temporary link table. Repeating step 1, Fig. 6(d) depicts the graph by moving root-links L_{bd} and L_{be} to the clustered link table. Then step 2 removes sink-link L_{ca} from the graph into the temporary link table (Fig. 6(e)). By breaking another cycle in Fig. 6 (f), link L_{de} is moved over to the temporary link table. Since the remaining graph at this time is an acyclic graph, its links are topologically sorted by repeating step 1 to move root-links to the clustered link table (Fig. 6(g)). Finally, Fig. 6(h) depicts the topologically clustered link table by appending the links of the temporary link table in reverse order.

4.3. Random clustering

Random Clustering (RandC) corresponds to a clustering of the *link table* in which the link tuples are in random order with the exception that links of the same *origin* node are clustered together. Random Clustering is included in our experimental evaluation as the straw-man to

[5] If the *origin* node of a link has no incoming link, this link is referred to as root-link.

[6] If the *destination* node of a link has no out-going link, and the *origin* node of this link does not belong to any of the cycles, this link is referred to as sink-link.

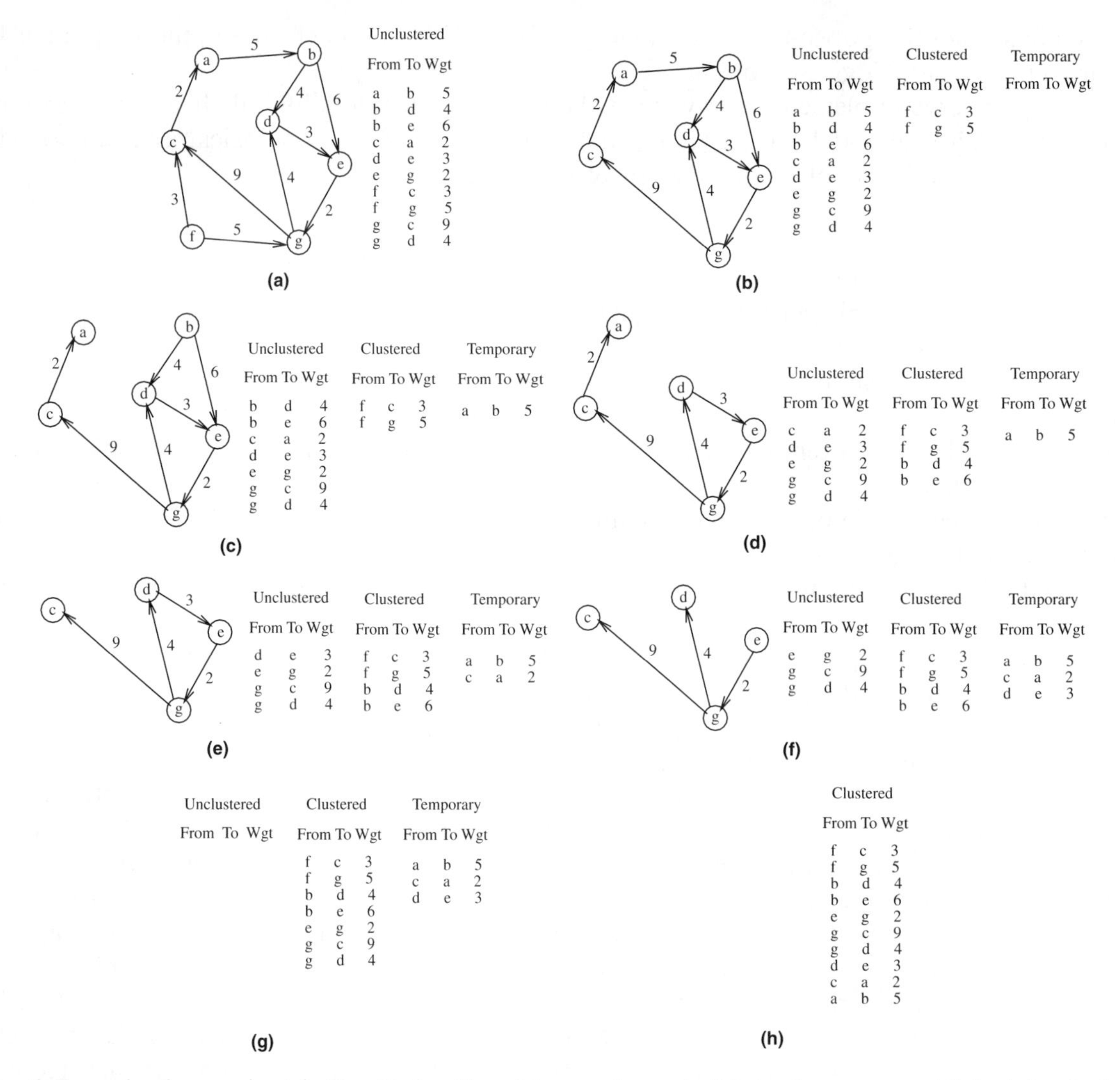

Fig. 6. Example of approximately TopoC algorithm: (a) example graph, (b) step 1: remove root-links (of node f), (c) step 3: break cycle (of node a), (d) step 1: remove root-links (of node b), (e) step 2: remove sink-link (of node c), (f) step 3: break cycle (of node d), (g) step 1: remove root-links, (h) step 4: attend temporary link.

determine the path query processing cost when no clustering strategy is deployed. In this paper, we compare its performance in path query processing with those of all other clustering strategies.

5. Extended clustering strategies

5.1. Combining spatial partitioning with swapping (hybrid)

One of the techniques commonly used in min-cut graph partitioning algorithms is node swapping (see Section 4.1.1 for a detailed discussion). We develop a hybrid clustering approach that combines the spatial partition clustering (Section 3.2) with the swapping technique (Section 4.1.1). We call it the hybrid approach. This hybrid approach starts by performing the spatial partition clustering on the graph, striving as before to reach high occupancy rate for each partition. In contrast to the SPC approach where the page occupancy rate is approximately 100%, we allow the minimum occupancy rate to be as low as 90% for the hybrid approach.[7] This relaxation on the occupancy rate leaves some space on each page that can be used for more effective single-node swapping.

When the spatial partitioning process is complete, we perform single-node swapping with the maximum occupancy rate set to 100%. This means that swapping cannot fill a partition so as to exceed the size of a page. We conduct such swapping for several iterations with the partitioning objective based on minimum number of inter-partition links. Next, we continue by performing pair-wise swapping for several more iterations. This is designed to give the partitions that are full a chance to swap nodes with other partitions to reach a better cut. In contrast to the swapping in the two-way partitioning approach in which each node has only one other partition to swap to (see Section 4.1), the swapping in our hybrid approach has to consider all other partitions a node can potentially swap to. This is because the spatial partitioning process creates many initial partitions. It is to be expected (and our experiments confirm) that the resulting clustered *link table* has a few more pages than the SPC-clustered *link table* because of its slightly lowered occupancy rate, but the number of inter-partition links could potentially be reduced by the swapping process.

5.2. Two-way partitioning based on link weight (TWPC_wgt)

In applying the two-way partitioning algorithm to graph clustering, our objective is to reduce the number of cross-page links (TWPC_con in Section 4.1). We believe such an approach will lead graph-traversal path computation to traverse more intra-page links and less inter-page links since the numbers of the latter are reduced. As a result, the page misses happening during path search are likely to be reduced. We now extend the partitioning objective to include link weights also.

As before, we set our objective to first also minimize the number of total cross-page links. However, if two or more possible cuts have the same number of reductions in terms of inter-partition links, we break the tie by giving the swapping priority to the cut that results in the maximum sum of weights of all cross-page links.[8] This objective is based on the fact that our path search algorithm, the *Dijkstra* algorithm, is a priority search that gives priority to the node with the minimum traversed weight so far. With everything else being equal, a link with a larger weight is less likely to be favored by the *Dijkstra* algorithm to be traversed next than a link with a smaller weight. Therefore, given an equal number of cross-page links, it can be expected that a partition that has a larger total weight of all cross-page links could potentially further improve the I/O efficiency of path query computation based on the *Dijkstra* algorithm. Note that to reduce the number of cross-page links is still the first priority because if we put the maximum sum of all

[7] Our experiments showed that 90% is a good compromise between high occupancy rate and sufficient room for subsequent node swapping.

[8] Note that for stable graph clustering that we consider in this paper, the link weight used in this partitioning objective should not be an unstable link attribute that changes frequently.

weights of all cross-page links as the first priority, the resulting cut will possibly have a large number of cross-page links, making the partitioning ineffective.

6. Testbed environment

In this section, we first discuss our experimental testbed setup, followed by the graph representation, and data sets, and then experimental parameters and measurements.

6.1. Experimental testbed setup

Our experimental testbed is implemented on a SUN Sparc-20 workstation running the Unix operating system. It includes the clustering algorithms presented in this paper, a heap-based *Dijkstra* algorithm, an I/O buffer manager, and many other supporting data structures. All programs are written in C++.

6.2. Graph representation

We use the *link table* to model the topology of the graph. Each link tuple in the *link table* models a link in the graph. The path queries discussed in this paper are assumed to be path queries with embedded constraints (see examples in Section 1). Because to resolve such constraints may require the retrieval of link attributes in order to evaluate the validity of each link traversed during path finding, we must store relevant link attributes in their corresponding link tuples. In this paper, the link tuple adopted in our experiments is set to 128 bytes. The reader can find a listing of possible attributes that could be associated with such links. As discussed in Section 1, depending on the type of query, the link attributes must be kept with the link itself in order to allow for the filtering of links from the candidate paths during path processing, such as for the spatial constraints in query Q3 from Section 1.

6.3. Experimental transportation networks

We test two kinds of graphs: randomly generated graphs and a real (fine-granularity) network representing the streets of Ann Arbor City that has 5596 nodes and 14,033 links. We experiment with random graphs with 5000 nodes and vary the average out-degree from 2 to 8. To create a random graph with average out-degree d, we randomly select, for each node, from 1 to $2 \times d - 1$ outgoing links and, for each link, we randomly select a *destination*. The *destination* must be different from the *origin* of the same link, and the *destinations* of two different outgoing links from the same *origin* must be different. The *weight* for each link is chosen to be a random integer between 1 and 100. We also create two sets of such random graphs, one with high locality, and the other with no locality. To control the locality of a random graph, we associate each node with an x-coordinate and a y-coordinate value. For graphs of high locality, we allow a link to exist only when its *origin* and *destination* are within a relatively close vicinity as compared to the total area. For graphs of no locality, we set no such limitation.

The reason we experiment with random networks of both high and no locality is because more advanced GIS applications such as Intelligent Transportation Systems need to model graphs beyond the road transportation networks (such as the Ann Arbor city network). For example, the graphs of airline flight routes exhibit no planarity and locality therefore can be better modeled by random graphs with no locality. An inter-modal network of both subway train and bus routes, however, exhibits high locality without planarity, therefore can be modeled by random graphs with high locality.

6.4. Clustering and path search algorithms

In our experiments, we first prepare the network data using the various clustering techniques proposed in this paper, namely SPC, Hybrid (the hybrid approach that combines SPC and node swapping), two-way partition clustering based on connectivity (TWPC_con), two-way partition clustering based on stable link weights (TWPC_wgt), topological clustering (TopoC), and random clustering (RandC). The data then is layed out on disk based on this preprocessing stage.

Thereafter, for each experiment, we apply the Dijkstra algorithm to conduct a single-source shortest path search for randomly selected source nodes i to all other nodes in the network. Such computation corresponds to the graph-traversal search for the shortest-path from node i to the one node j that is the farthest away from i. Hence this set of experiments tests the worst-case scenario in searching a shortest path from i.

6.5. Settings of parameters and measurements

In this paper, the size of pages on disk and in the main memory buffe is set to be 4 KB each. The experiments are based on a buffer containing up to 240 pages, i.e., varying the size of the buffer from 64 KB up to 960 KB. The size of the entire *link table* is about 2 Mb, with a small difference between the various clustering techniques.

Although the experiments presented in this paper are based on buffer sizes up to 960 KB, the adequate buffer size for path query processing is proportional to the size of the underlying network. The experimental networks in this paper (i.e., the Ann Arbor city network) are of medium sizes (Section 6.3). The sizes of larger cities can be many times larger. For example, the Detroit road network we are using for related research in this project has more than 50,000 links which is about three times the size of the Ann Arbor map. Consequently, the buffer requirement should increase for larger maps. Second, resolving constraints embedded in the path queries may incur heavy I/O activities which takes away the buffer space from the path search process. Lastly, in a multi-user and multi-tasking database system, one cannot assume that the entire resources such as the buffer space are available to one single query process. We therefore are motivated to find the clustering strategy that can process path queries efficiently while using as small a portion of the buffer as possible. Therefore, in reality, the buffer requirement for processing constrained path finding on a large network in a multi-user database environment can be many times larger than the various buffer sizes depicted in our experimental evaluation in Section 7.

Given that we use the same search algorithm Dijkstra in our experiments, the CPU processing costs are all fairly comparable whereas the number of disk pages that must be transferred between the slower secondary storage device to the faster main memory system for processing, referred to number of page input/output operations or in short I/Os, varies significantly based on the placement of link tuples on pages. In most systems, I/Os are on the order of 100-fold more expensive than CPU operations, and hence the critical factor for determining system performance are the I/O and not the CPU costs. Hence, we measure performance in our work in terms of number of I/Os, with the actual total processing time being a multiplicative of this I/O count based on the average I/O cost in the given system.

7. Experimental evaluation

This section presents experimental results and performance evaluation measured using the simulated number of disk pages transferred between the slower secondary storage device to the faster main memory system for processing, referred to number of page input/output operations or in short I/Os. In most systems, I/Os are on the order of 100-fold more expensive than CPU operations, and hence the critical factor for determining system performance. The actual performance of course varies with each system.

7.1. Experiments on Ann Arbor road network

In the first set of experiments, we use the Ann Arbor road network and conduct single-source shortest path search for randomly selected nodes using the clustering techniques proposed in this paper. The results in Fig. 7 show that Random clustering performs much worse than any other clustering, confirming our claim in this paper that proper graph clustering can be a key to efficient path query processing. Because the cost of the Random clustering is very high, making it hard to see the difference in performance between the other five clustering approaches, we plotted Fig. 8 without showing the Random clustering results. In Fig. 8, it is clear that SPC performs the best, followed by, in exact order, TWPC_wgt, TWPC_con, Hybrid, and TopoC. It is surprising to see that although TopoC has the worst performance among the five clustering optimizations, it is still much more effective than RandC. This is contradictory to the suggestion in Shekhar and Liu (1995) and Zhao and Zaki (1994) that topological clustering is not effective for highly cyclic graphs such as our road network.

It is also interesting to note that the Hybrid approach performs worst than the SPC. This indicates that the partitioning objective we set to minimize the cross-page links during the swapping process may not be as relevant to highly interconnected near-planar graphs like the Ann Arbor network as spatial proximity which SPC is based upon. This also helps to explain why SPC performs better than TWPC_wgt and TWPC_con. The fact that TWPC_wgt performs better than TWPC_con indicates that by incorporating link weights, the partitioning objective of TWPC_wgt catches the expansion behavior of the *Dijkstra* algorithm better than that of TWPC_con. Note that when the buffer size is greater than 512 KB, all five clustering strategies perform the same. This is because the size of the buffer is large enough to contain the expansion locality of the *Dijkstra* algorithm captured by all five clustering optimizations. Therefore, roughly one pass for such a large buffer would be sufficient to compute the single-source shortest paths.

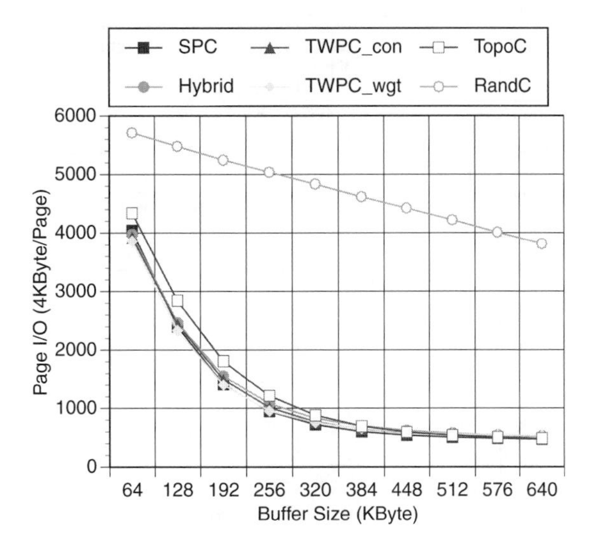

Fig. 7. I/O cost of searching the longest path on Ann Arbor map.

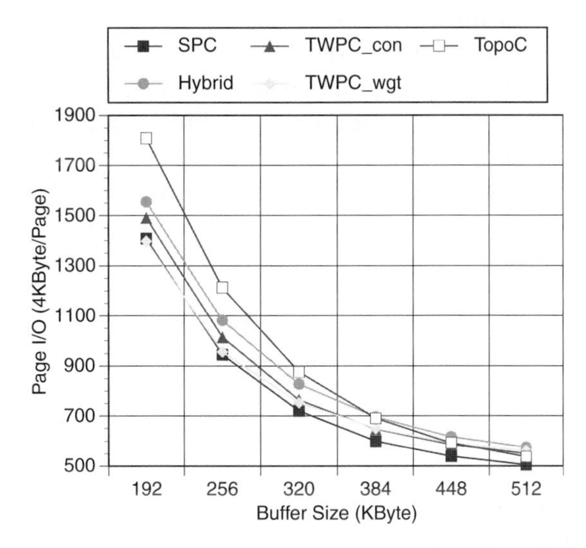

Fig. 8. I/O cost of searching the longest path (excluding random clustering) on Ann Arbor map.

7.2. Experiments on the high-locality random graphs

The second set of experiments is based on a randomly generated graph with 5000 nodes, average out-degree of 3, and high locality. We conduct the same single-source path search experiments described above. While the Ann Arbor network is very planar and interconnected, the high-locality random graph does not guarantee planarity and high interconnection. The results in Fig. 9 show that RandC remains the distant worst, with the TopoC significantly worse than the other four clustering approaches.

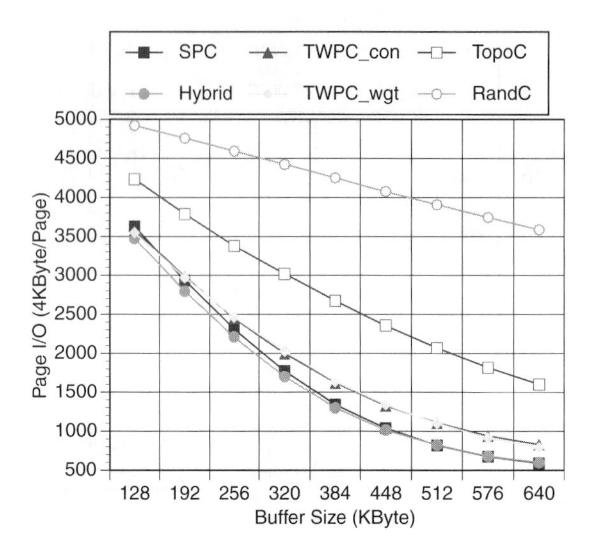

Fig. 9. I/O cost of searching the longest path on the high-locality random graph.

The close-up results in Fig. 10 show that the Hybrid and SPC perform better than TWPC_con and TWPC_wgt, with the Hybrid having a slight edge. This is different from the experimental results on the Ann Arbor network where the Hybrid is worse than the other three. This indicates that the partitioning objective of minimizing cross-page links does help in bringing down the I/O cost incurred by the *Dijkstra* path search for high-locality graphs without high interconnection and planarity. The superior performance of both the Hybrid and SPC over the TWPC_con and TWPC_wgt indicates that each partition created by our proposed SPC partitioning algorithm is tailored to fit perfectly into a buffer page. Therefore the resulting graph partitions are better, in

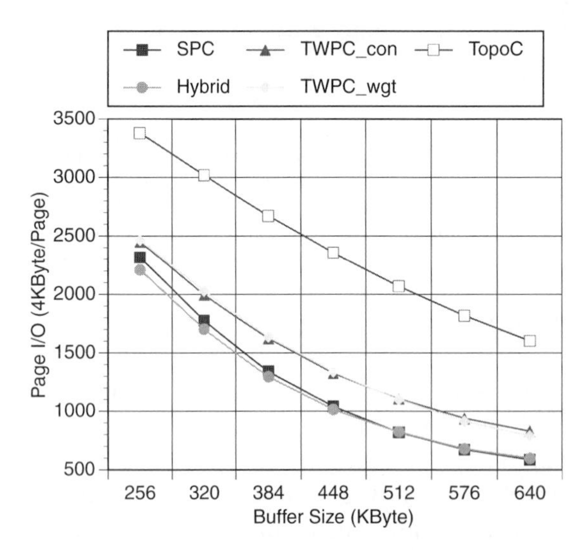

Fig. 10. I/O cost of searching the longest path (excluding random clustering) on high-locality random graph.

terms of both page occupancy rate and expansion locality exhibited by the *Dijkstra* algorithm, than the partitions created by the TWPC approaches that use *divide-and-conquer* to distribute partitions into pages.

7.3. Experiments on the low-locality random graphs

In the third set of experiments, we test a randomly generated graph with 5000 nodes, average out-degree of 3, and no locality. Interestingly, the results in Fig. 11 show that RandC and SPC are equally the worst. This can be explained by the fact that without locality, the spatial proximity is irrelevant for proper partitioning. Consequently, the SPC performs the same as the RandC on graphs with no locality. The swapping process in the Hybrid approach tries to correct the irrelevant spatial partitioning, but its performance is still worse than the TWPC approaches, and even worse than the TopoC for some buffer sizes. The results also show that TWPC_wgt has the best performance, followed by TWPC_con. This indicates that the two TWPC approaches are not locality-dependent, therefore have better performance than the SPC and the Hybrid approaches on graphs with no locality. The link-weights based partitioning objective in TWPC_wgt that we have proposed is an effective optimization over the pure connectivity based objective in TWPC_con. We note however that the TWPC_wgt has the limitation that the link weight used as the partitioning objective must be stable.

7.4. Experiments on paths of different length

So far, our experiments focus on the worst-case scenario, i.e., the cost of computing the longest among all shortest-paths to all possible destinations. We now explore the path search performance on paths of different length, namely short paths, medium paths, and long paths. We define the direct distance between the farthest node-pair as d_{max}, and the direct distance between the two end nodes of a path as d_{short} for short paths, d_{medium} for medium paths, and d_{long} for long paths. Then the following relations hold:

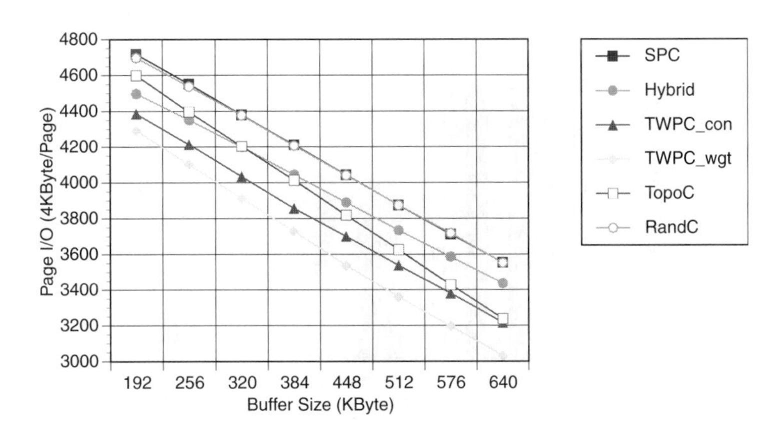

Fig. 11. I/O cost of searching the longest path on the low-locality random graph.

$$d_{\text{short}} < (d_{\max}/3),$$
$$(d_{\max}/3) \leqslant d_{\text{medium}} < (d_{\max} \times 2/3),$$
$$d_{\text{long}} \geqslant (d_{\max} \times 2/3).$$

Because such a direct-distance based classification is only meaningful if the graph has locality and is highly inter-connected, we conduct this set of experiments on the real Ann Arbor city map only.

We randomly select a number of node-pairs from each category, conduct path search, and collect the average results. The buffer size is set to 256 KB. Because the performance for RandC is much worse than any other clustering approach, we use the log scale in Fig. 12. Fig. 13 shows the results in linear scale without RandC. In Fig. 13, we can see more clearly that SPC consistently has the best performance for all three kinds of paths, while TopoC generally has the worst performance.

7.5. Experiments on average out-degree

In this last set of experiments, we randomly generated graphs with 5000 nodes, varying the average out-degrees from 2 to 8. We set the buffer size to 960 KB in order to accommodate graphs with large average out-degree. The purpose of the experiments is to find out which clustering strategies work better for graphs of various average out-degrees. Because graphs of high out-degrees automatically lose locality with evenly distributed nodes, we only generate graphs with no locality for this set of experiments.

The experimental results in Fig. 14 show that both TWPC_wgt and TWPC_con perform better as the average out-degree increases. Although TWPC_wgt has a better performance than TWPC_con, the performance curve of TWPC_con actually becomes (i.e., drops) more favorably

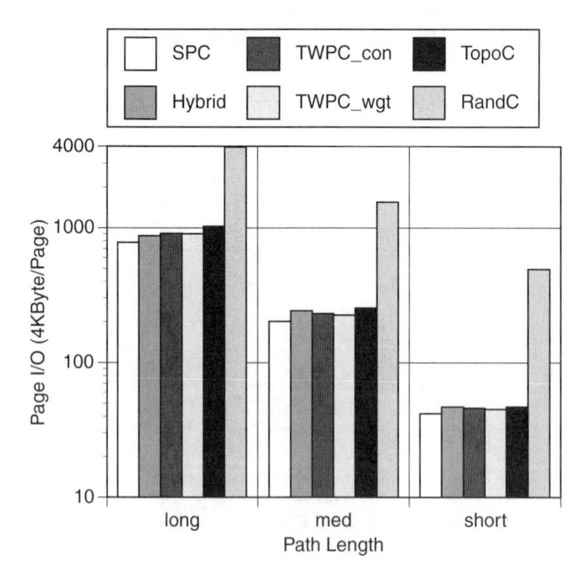

Fig. 12. I/O cost by length of paths on the Ann Arbor network.

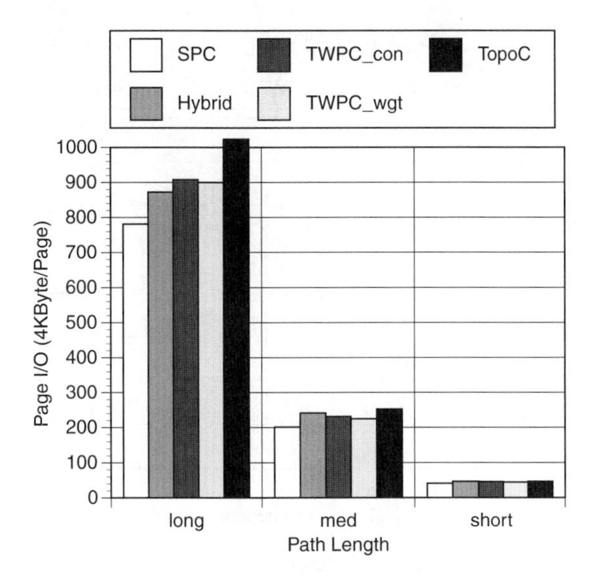

Fig. 13. I/O cost by length of paths (excluding random clustering) on the Ann Arbor network.

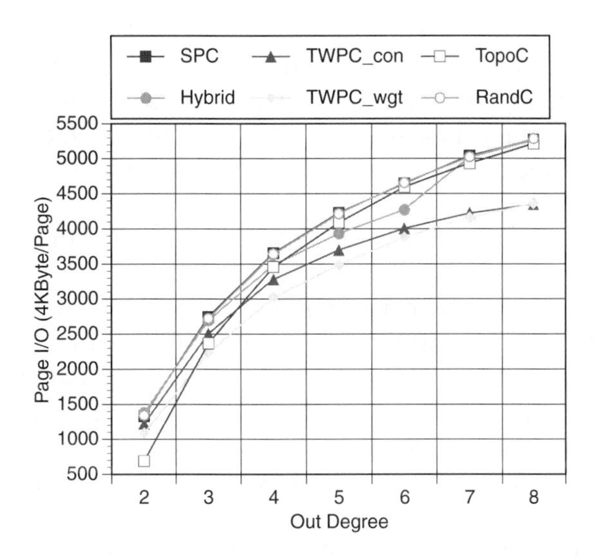

Fig. 14. I/O cost by average out-degree on the random graphs.

as the average out-degree increases. The other four clustering approaches all fair poorly as the average out-degree goes up. Fig. 15 is a close-up of Fig. 14 with average out-degrees from 2 to 5. In Fig. 15, the performance of TopoC is much better than the others when the average out-degree is 2. This is because when the average out-degree is small, there are primarily acyclic components in the graph. The TopoC strategy is extremely efficient for the acyclic graphs and therefore performs much better when there are many acyclic subgraphs in the graph.

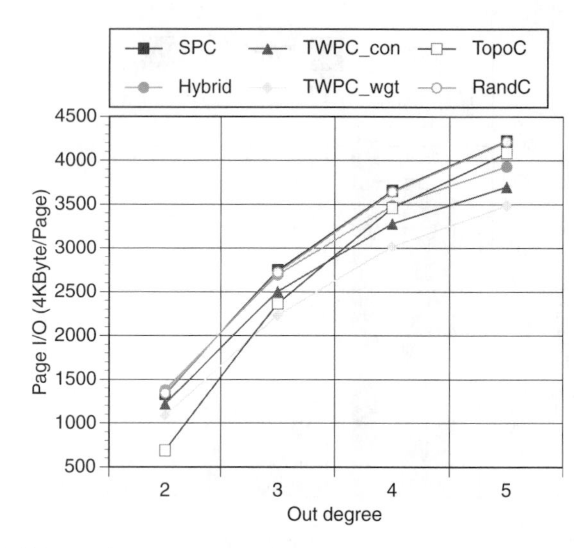

Fig. 15. I/O cost by average out-degree on the random graphs (close-up).

8. Conclusions

Efficient path query processing over transportation networks is important for many GIS applications. In this paper, we consider the optimization of path query processing using graph clustering techniques. Clustering optimization is attractive because it does not incur any run-time cost, nor does it require auxiliary data structures that demand memory. More importantly, it is complimentary to many of the existing path query solutions typically at the data structure or at the algorithmic level.

In this paper, we first propose a new clustering technique, called the spatial partition clustering (SPC), for path query optimization for transportation networks. Next, three other graph clustering techniques, namely the two-way partitioning clustering, topological clustering, and random clustering, are identified from the literature, fine-tuned for GIS path query optimization, and then implemented in our uniform testbed. In addition, based on the spatial partition clustering and the two-way partitioning clustering, we develop two extensions, called the hybrid approach and the two-way partitioning based on link weight, respectively.

This paper presents an extensive experimental evaluation of the comparative performance of the above six graph clustering techniques. Our experiments are based on three kinds of networks. They are the Ann Arbor city road map; randomly generated graphs with high locality modeling GIS maps such as inter-modal bus and subway routes; randomly generated graphs with no locality modeling GIS networks such as airline routes. The experiments are conducted by varying testing parameters such as memory buffer size, path length, locality, and average out-degree. The experimental results show that our spatial partition clustering performs the best for the real network; the hybrid approach has the best performance for random graphs with high locality; whereas the two-way partitioning based on link weights works the best for random graphs with no locality. Such experimental results are important, representation a foundation for establishing guidelines in the selection of the best clustering technique based on the type of network prevalent in a given application.

For future work, effective graph clustering techniques can also be extended to solve more general path problems such as recursive query processing. Results from this paper can also be exploited for further optimization of complex path query processing with embedded constraints. For instance, our exploration of a framework of spatial path queries (Huang et al., 1997) is based on the SPC solution first introduced in this paper. Lastly, clustering techniques could be explored to take into account knowledge about in which s spatial location paths with certain properties can be found. Such knowledge could be utilized to constrain the search and also to adjust the clustering of links for origin-destination pairs that meet this particular class of queries.

References

Agrawal, R., Dar, S., Jagadish, H.V., 1998. Direct transitive closure algorithms: design and performance evaluation. ACM Transactions on Database Systems 15 (3), 427–458.

Agrawal, R., Jagadish, H.V., 1988. Efficient search in very large databases. In: Proceedings of the 14th VLDB Conference. Los Angeles, CA, pp. 407–418.

Agrawal, R., Jagadish, H.V., 1989. Materialization and incremental update of path information. In: IEEE Fifth International Conference on Data Engineering, pp. 374–383.

Agrawal, R., Jagadish, H.V., 1990. Hybrid transitive closure algorithms. In: Proceedings of the 16th VLDB Conference. Brisbane, Australia, pp. 326–334.

Agrawal, R., Kiernan, J., 1993. An access structure for generalized transitive closure queries. In: IEEE Ninth International Conference on Data Engineering, pp. 429–438.

Banerjee, J., Kim, W., Kim, S.J., Garza, J.F., 1988. Clustering a DAG for CAD databases. IEEE Transactions on Software Engineering 14 (11).

Bancilhon, F., 1985. Naive evaluation of recursively defined relations, 1985. In: Brodie, M., Mylopoulos, J. (Eds.), On Knowledge Base Management Systems – Integrating Database and AI systems. Springer, New York.

Bancilhon, F., Ramakrishnan, R., 1986. An Amateur's introduction to recursive query processing strategies. In: Proceedings of the 1986 ACM SIGMOD International Conference on Management of Data.

Brinkhoff, T., Kriegel, H., Schneider, R., Seeger, B., 1994. Multi-step processing of spatial joints. In: Proceedings of the 1994 ACM SIGMOD International Conference on Management of Data, pp. 197–208.

Carey, M.R., Johnson, D.S., Stockmeyer, L., 1976. Some simplified np-complete graph problems. Theoretical Computer Science 237–267.

Cheng, C.K., Wei, T.C., 1991. An improved two-way partitioning algorithm with stable performance. IEEE Transactions on Computer-Aided Design 10 (12), 1502–1511.

Dijkstra, E.W. 1959. A note on two problems in connection with graphs. Numer. 269–271.

Ebert, J., 1981. A sensitive transitive closure algorithm. Information Processing Letters 12, 255–258.

Fiduccia, C.M., Mattheyses, R.M., 1982. A linear time heuristic for improving network partitions. In: Proceedings of ACM/IEEE 19th Design Automatic Conference, pp. 175–181.

Goodchild, M.F., 1990. Tiling large geograhical databases. In: Buchmann, A., Gnther, O., Smith, T.R., Wang, Y.-F. (Eds.), Design and Implementation of Large Spatial Databases. Springer, New York, pp. 137–146.

Goodchild, M.F., Shiren, Y., 1992. A hierarchical spatial data structure for global geographic information systems. Computer Vision Graphics and Image Processing: Graphical Models and Image Processing 54 (1), 31–44.

Huang, Y.W., Jones, M.C., Rundensteiner, E.A., 1998. Symbolic intersect detection: a method for improving spatial intersect joints. Journal of GeoInformatica Special issue on Spatial Database Systems 2 (2), 149–174.

Huang, Y.W., Jing, N., Rundensteiner, E., 1997. Integrated query processing strategies for spatial path queries. In: IEEE International Conference on Data Engineering, pp. 477–486.

Huang, Y.W., Jing, N., Rundensteiner, E., Huang, Yun-Wu., Jones, Matthew C., Rundensteiner, Elke A., 1997a. Improving spatial intersect joints using symbolic intersect detection. In: SSD Conference, pp. 165–177.

Huang, Y.W., Jing, N., Rundensteiner, E., 1997b. A cost model for estimating the performance of spatial joints using R-trees. SSDBM (1997) 30–38.

Huang, Y.W., Jing, N., Rundensteiner, E.A., 1997c. A hierarchical path view model for path finding in intelligent transportation systems. Journal of GeoInformatica 1 (2), 125–159.

Huang, Y.W., Jing, N., Rundensteiner, E., 1997d. Spatial joints using R-trees: breadth-first traversal with global optimizations. VLDB (1997) 396–405.

Huang, Y.W., Jing, N., Rundensteiner, E.A., 1997e. Query processing strategies for spatial path queries. In: IEEE International Conference on Data Engineering, ICDE-13, England.

Huang, Y.W., Jing, N., Rundensteiner, E.A., 1996. Path view algorithm for transportation networks: the dynamic reordering approach. In: ACM Workshop on Geographic Information Systems, ACM GIS'96, Washington, DC.

Huang, Y.W., Jing, N., Rundensteiner, E.A., 1996. Evaluation of hierarchical path finding techniques for ITS route guidance. In: Proceedings of ITS-America, Houston, April.

Huang, Y.W., Jing, N., Rundensteiner, E., 1996. Effective graph clustering for path queries in digital map databases. In: Proceedings of the Fifth International Conference on CIKM 1996, Washington, DC, pp. 215–222.

Ioannidis, Y.E., 1986. On the computation of the transitive closure of relational operators. In: Proceedings of the 12th International Conference on VLDB, pp. 403–411.

Ioannidis, Y.E., Ramakrishnan, R., 1988. An efficient transitive closure algorithm. In: Proceedings of the 14th International Conference on VLDB, pp. 382–394.

Ioannidis, Y.E., Ramakrishnan, R., Winger, L., 1993. Transitive closure algorithms based on graph traversal. ACM Transactions on Database Systems 18 (3), 512–576.

Jing, N., Huang, Y.W., Rundensteiner, E.A., 1998. Hierarchical encoded path views for path query processing: an optimal model and its performance evaluation. IEEE Transactions of Knowledge and Data Eng. 10 (3), 409–432.

Jing, N., Huang, Y. W., Rundensteiner, E., 1996. Hierarchical optimization of optimal path finding for transportation applications. In: Proceedings of the Fifth International Conference on CIKM, pp. 261–268.

Larson, P.A., Deshpande, V., 1998. A file structure supporting traversal recursion. In: Proceedings of the 1989 ACM SIGMOD International Conference on Management of Data, pp. 243–252.

Kernighan, B.W., Lin, S., 1970. An efficient heuristic procedure for partitioning graphs. Bell System Technical Journal 49 (2), 291–307.

Preparata, F.P., Shamos, M.I., 1985. Computational Geometry. Springer, NewYork.

Schmitz, I., 1983. An improved transitive closure algorithm. Computing 30, 359–371.

Shamos, M.I., Hoey, D.J., 1976. Geometric intersection problems. In: Proceedings of the 17th Annual Conference on Foundations of Computer Science, pp. 208–215.

Shekhar, S., Liu, D.R., 1995. CCAM: a connectivity-clustered access method for aggregate queries on transportation networks: a summary of results. In: IEEE 11th International Conference on Data Engineering, pp. 410–419.

Wei, Y.-C., Cheng, C.-K, 1990. Ratio cut partitioning for hierarchical designs. Technical Report CS90-164, University of California, San Diego, January.

Zhan, Noon, 1998. Shortest path algorithms: an evaluation using real road networks. Transportation Science 32, 65–73.

Zhao, J.L., Zaki, A., 1994. Spatial data traversal in road map databases: a graph indexing approach. In: Proceedings of the Third International Conference on CIKM, pp. 355 –362.

PERGAMON

Transportation Research Part C 8 (2000) 409–425

TRANSPORTATION
RESEARCH
PART C

www.elsevier.com/locate/trc

Design and development of interactive trip planning for web-based transit information systems

Zhong-Ren Peng [a,*], Ruihong Huang [b]

[a] *Department of Urban Planning, School of Architecture and Urban Planning, University of Wisconsin – Milwaukee, P.O. Box 413, Milwaukee, WI 53201, USA*

[b] *Department of Geography, University of Wisconsin – Milwaukee, Milwaukee, USA*

Abstract

This article presents a Web-based transit information system design that uses Internet Geographic Information Systems (GIS) technologies to integrate Web serving, GIS processing, network analysis and database management. A path finding algorithm for transit network is proposed to handle the special characteristics of transit networks, e.g., time-dependent services, common bus lines on the same street, and non-symmetric routing with respect to an origin/destination pair. The algorithm takes into account the overall level of services and service schedule on a route to determine the shortest path and transfer points. A framework is created to categorize the development of transit information systems on the basis of content and functionality, from simple static schedule display to more sophisticated real time transit information systems. A unique feature of the reported Web-based transit information system is the Internet-GIS based system with an interactive map interface. This enables the user to interact with information on transit routes, schedules, and trip itinerary planning. Some map rendering, querying, and network analysis functions are also provided. © 2000 Elsevier Science Ltd. All rights reserved.

Keywords: Internet GIS; Transit networks; Shortest path; Intelligent transportation systems

1. Introduction

This paper describes and discusses a way of designing a Web-based transit information system that allows transit users to plan a trip itinerary and to query service-related information, such as schedules and routes using Internet Geographic Information Systems (GIS) technologies. Historically, transit agencies have relied on printed schedules to provide customers with information

* Corresponding author. Tel.: +1-414-229-5887; fax: +1-414-229-6976.
E-mail address: zpeng@uwm.edu (Z.-R. Peng).

PII: S0968-090X(00)00016-4

about transit routing and schedules. Transit users have had to select proper routes and transfer points based on the information printed on the schedule. Use of the schedule is complex and can be confusing for many people. In addition, since schedules are infrequently updated, many service changes cannot be reflected in the brochure in a timely manner. Most transit agencies also staff customer service agents to provide telephone assistance in answering customer inquiries about schedules and directions. Customer service agents can suggest itinerary plans for customers based on printed route maps, published schedules, and service updates not yet made on the printed schedules. This manual itinerary planning process is tedious, time-consuming, redundant, and often error-prone; information can be inconsistent from one service agent to another (Salters, 1996). Recently, computer assisted trip planning systems were developed to automate the process (Salters, 1996; Peng, 1997; Casey et al., 1998). However, this early computer-aided trip planning system is mainly designed to assist customer service agents and requires proprietary software installed in the users' local computers. Transit users have limited or no direct access to it and have to call in to get updated service information and an itinerary.

The Internet and the World Wide Web are revolutionizing the process of information dissemination, communications, and transactions, which have brought some important changes to traditional functions of transit services. For example, most of the traditional customer service functions (e.g., schedules, routing, itinerary planning) can be enhanced or even substituted by Web-based information systems. The beauty of the Web-based information is that it could tie together and make more intelligible routing and scheduling information that traditional brochure designers struggled with for years in the pre-Web days. Many transit agencies are now in the process of creating and upgrading their transit information on the Web. With the rapid development of the Internet technology and the proliferation of online information, the number and use of transit information Web sites are increasing rapidly. For instance, the UK Public Transport Information Web site had 1000 visits a month at the end of 1996; it received over 13,000 per month by July 1998 (http://www.ul.ie/~infopolis/).

There are many transit information systems on the Web. These include simple static schedule display to more sophisticated real time bus location systems. This paper provides a taxonomy to review the state-of-the-art of the existing and future development of transit information systems on the Web to form a framework for Web-based transit information systems.

2. Review of transit information on the Internet

The purpose of transit information on the Web, like any other online information systems, is evolving from information dissemination to interactive communications and online transactions. Transit information dissemination serves the purpose of transit information announcement and display, such as information about published schedules and routing, as well as service changes. Users receive the information passively. Interactive communication provides user interactivity and feedback channels. Users can actively manipulate and search for specific information based on their own needs and give feedback to the system providers. Transactions offer instant interactions between system providers and users, for example, online ticketing and reservations.

Based on these evolving purposes, online transit information systems can vary significantly in terms of content and function. Table 1 provides a framework for online transit information

Table 1
Taxonomy of online transit information systems

	Content level	Functions and interface				
		Web browsing (HTML, PDF)	Text search, static graphic links (map images)	Interactive map-based search, query and analysis (Internet GIS)	Customization and information delivery	Online transaction
Function level		0	1	2	3	4
Contents						
General information	A	A0	A1	–	–	–
Static information (routes, schedule and fare)	B	B0	B1	B2	B3	B4
Trip itinerary planning	C	C0	C1	C2	C3	C4
Real time information (bus locations and delays)	D	D0	D1	D2	D3	D4

systems and categorizes the quality of service provided on different Web sites based on information content and functionality.

The rows in Table 1 represent the contents of information on an online transit information site. Information provided may range from static information to dynamic real time information. A transit Web site can simply provide the basic information about the agency and its services (Level A). While useful, it is not very informative, and the site may not attract users.

The next level of content (Level B) may include static information about transit routes, networks, service schedules, and fares. This content is the minimum that is needed for transit users to access basic transit service information. At this level, the content may also include multi-modal information, such as rail, air, and highway traffic information. This would help the travelers make a more informed decision on mode choices. Multi-modal travel information would be especially valuable for Internet-enabled kiosks on airports or other trip generators. These can be used by travelers to make a more informed decision on whether to ride a bus, take a taxi, or rent a car.

The third level of content (Level C) may include trip planning or trip itinerary information. The trip-planning program can provide the user the optimal path between the user's trip origin and destination, as well as the time of travel based on transit schedule information.

The fourth level of content (Level D) may include real time information about bus locations and its expected arrival time, possible delays, and incident and/or weather information. This real time information can be derived from automatic vehicle locator (AVL) systems (Casey et al., 1998).

The columns in Table 1 represent the level of function that the Web interface supports, and the rows represent the content level. The entries in each cell represent their row and column locations. The function of a Web interface ranges from information dissemination (browsing) to interactive communications (search, query, analysis, and customization) and, finally, to online transactions. The simplest Web interface (Level 0) can provide text browsing only. It provides no clickable maps or search capability. This is implemented by using a simple two-tier architecture (Web browser to server) where all information is provided as Hyper Text Markup Language (HTML) or Portable Document Format (PDF) and/or static image maps. This is the minimal function the online transit information system should support.

The next level (Level 1) of interface support can provide graphic browsing using graphic links embedded in map images (clickable maps). A transit network map can be provided to link schedule data on each time point on the map. The user is able to select a transit line or time point on the transit network map. Each transit line or time point is linked with the schedule data. This is useful for users who know the location of their trip ends and are already familiar with the transit routes and services. This is still a two-tier architecture (Web browser to Web server) implementation, but the system at this level does not provide data search and query capability.

The next level of functional support (Level 2) can provide spatial search and attribute data search. Furthermore, it can also provide graphic interface to allow users to directly interact with transit network maps. For example, if the user enters an address or points to a specific location on the map, the system can find all the bus routes and stops within walking distance of that location. The user can render the transit network and street network maps by zooming in and out, panning, or by conducting a spatial search. The network location data and schedule data, or real time bus locations data, are linked with a relational database in a database management system (DBMS) on the server. Typically, this level of service requires a three-tier network architecture that handles client-side user interaction, server-side network analysis, and database management.

There is a significant difference between functions and architectures at Level 1 and Level 2. For the Level 1 function support, the system architecture is still the two-tier Web browser to server structure. Static documents are linked with the portions of the map images by the underlying Universal Resource Locators (URLs). When the user clicks a location on a map image, a linked document page is displayed. The linked document pages, like bus schedules, are static and pre-prepared. Any changes in the bus schedules have to be made manually on those pre-prepared document pages. For Level 2 functionality, the system architecture is a three-tier structure. That is, the user interface on the Web browser is linked with the Web server, which is further linked with a GIS application server and/or database server. The spatial features (e.g., routes, stops, time points) and their attributes (e.g., schedules, real time bus locations) on the map are connected with a map server and/or DBMS. Any changes in the schedules in the database will be automatically updated and available instantly. When the user makes a request, that request is transferred to the server; the server then searches the database and returns the query results to the user. The Extensible Markup Language (XML) can also be used to facilitate spatial and attribute data search.

Level 3 of function support is capable of providing customization and information delivery. The system can store the user's personal profile, such as customer's frequent origins and destinations, and the usual time of travel. When the user logs on, the system can retrieve the information based on the user profile, such as bus arrival information on the user's frequently patronized routes and stops. Furthermore, this customized information can also be delivered to the customer via pagers or hand-held devices (Peng and Jan, 1999). Wireless information delivery is clearly a future trend of online information systems. Customization facilitates the formation and timely delivery of personalized information. Use of real time information is especially important for customized services. The desirability of user profiles is somewhat controversial. On the one hand, the storage of personal profiles makes it easier to retrieve and deliver personalized information. On the other hand, some users may resent someone else watching their travel behavior. Storing personal profiles is common in e-commerce, but the extent to which such surveillance technologies infringe on personal liberties has rarely been addressed in the literature. However, technologies exist to allow personal information to be available only to users and cannot be retrieved even by Web site providers. Similar technologies have been implemented in online tax filing services. The taxpayers can enter his/her tax information online with his/her own password for protection. The password is encrypted so that the online tax filing service has no knowledge of it and cannot retrieve it. This could be a secure means to ensure that the transit user's personal profile and privacy are protected.

Finally, at the highest level of function support (Level 4), online transactions can be provided, such as on-line ticket sales (B4), online reservation in the case of para-transit (C4), and online matching for car-poolers (C4). The difference between B4 and D4 is that B4 is a conventional physical ticketing and D4 is an electronic ticketing. At level B4, transit tickets and passes can be purchased on line using a credit card, but real tickets and passes have to be delivered via mail. At level D4, no physical ticket delivery is necessary. Customers can pay for the trip on the Web. The user does not even need to print out a ticket; the machine on the bus can automatically validate the electronic ticket or the payment from a smart card. In terms of online registration, such as online ride-share matching and online para-transit scheduling, at the B4 level, users can enter personal information about home and work addresses, travel time to work, and so on, but they cannot obtain matching information on the fly. Whereas at the D4 level, matching information can be immediately available on the Web. Furthermore, at the D4 level, registration for para-transit services can be done in real or near real time based on actual bus location.

These levels of services reflect the current and foreseeable future development of Internet technology. The technology may evolve as Internet technology advances, but the basic concept may remain relatively stable over the near term. The level of service of a transit information Web site ranges from the lowest at the upper left corner (A0) to the highest at the lower right corner of Table 1.

Most current Internet-based transit information services are at levels A0, A1, B0, B1, C0 and C1. For example, the Toronto Transit Commission (Canada) Web site (http://www.city.toronto.on.ca/ttc/schedules/index.htm) is at level B0, where both schedules and route maps are in PDF. The Metropolitan Atlanta Rapid Transit Authority's (USA) Web site (http://www.itsmarta.com/) and the SunTran System of the Tucson Metropolitan area, Arizona, USA (http://www.suntran.com/index.htm) are also at level B0, with schedules in HTML or PDF and route sketch maps (unscaled transit route illustrations) in PDF or GIF format. Washington Metro in Washington, DC, USA, has implemented a clickable map for its rail system (http://www.wmata.com/). Each rail station in a map image is linked with a description about the station and its related schedule and bus transfer information, a B1 service. It also has an itinerary planning system for its rail system (http://www.wmata.com/), a C1 service. The Denver, Colorado, Regional Transportation District (RDT) (http://www.rtd-denver.com/index.html) has implemented a text-based searchable schedule database at the B1 level. The San Francisco Bay Area Transit Information System (USA) (http://www.transitinfo.org/) is at level B2, where transit schedules can be directly queried from the transit maps, and interactive mapping functions are also provided. The trip planning system at the Road Management System for Europe (ROMANSE) in Southampton, England (http://www.romanse.org.uk/) is at level C1, which provides pull-down list input boxes and text output. Another trip planning system at the Los Angeles County Metropolitan Transportation

Authority, (California, USA) (http://www.mta.net/) is also at level C1. A map is shown to illustrate the trip origins, destinations, and routes, but there is no map rendering function. Transit schedules are presented as separate PDF documents. Users cannot directly find service schedules from transit route maps. Similarly, C1 service is also available through the Ann Arbor Transportation Authority (Michigan, USA) (http://theride.org/home.html) and train service at the New York–New Jersey–Connecticut region, USA (http://www.itravel.scag.ca.gov/itravel/). A few Web sites have service at level D1 such as the busview_X program (http://www.its.washington.edu/projects/busviewX.html) and Superoute 66 (http://travel.labs.bt.com/route66/). Chicago Transit Authority (USA) has implemented an online ticketing pass system at its Web site (http://www.transitchicago.com/welcome/index.html), a service at level B4. Customers can purchase transit passes using a credit card and the pass will arrive through the mail within 7–10 days. Higher level services usually, but not always, include the lower level services. For example, if a Web site provides C2 function, it usually provides functions at levels A0, B0 and B1. This paper describes an Internet GIS approach to design a higher level of transit information services at levels C2, C3, and expandable to levels D2, D3.

The remainder of this paper is organized as follows. It starts with a system architecture design of a distributed transit information system using Internet GIS. The system architecture has three main components: a Web/Map server component, a database management component, and a network analysis component. A path finding algorithm for the transit network is introduced as part of the network analysis component. An example design of a transit information system on the Web is provided using data from a large city in the United States. Finally, the paper concludes by offering an outline for further improvements.

3. An architecture of a distributed transit information system

One of the recent developments in GIS technology is to deliver GIS data and analysis functions on the Web through the Internet (Batty, 1999; Colman, 1999; Plewe, 1997). Internet GIS, an emerging technology to serve GIS data and provide GIS functionality on the Web (Plewe, 1997; Peng and Beimborn, 1998; Peng, 1999), is designed to integrate the Web and GIS in order to manipulate, visualize, and analyze GIS data on the Web. Internet GIS has been used in many applications (Sarjakoski, 1998; Doyle et al., 1998; Peng and Beimborn, 1998; Muro-Medrano et al., 1999). The system architecture of Internet GIS is evolving already in its short existence. Early development looked at the Internet as a way to disseminate spatial data (Coleman and McLaughlin, 1997; Peng and Nebert, 1997). But accessing spatial data on the Internet did not provide any GIS analysis functionality and was thus a very limited application. Later developments linked existing GIS programs with the Web server to provide users some limited GIS functionality on the Web (Colman, 1999; Conquest and Speer, 1996). This approach takes advantage of existing GIS programs and their functions and delivers them to users through Web browsers. Recent advances explore distributed components and three-tier system architecture (Ran et al., 1999). The distributed component approach adopts the client server model to distribute data and GIS processing components from the server to the Web client and is more efficient and scalable.

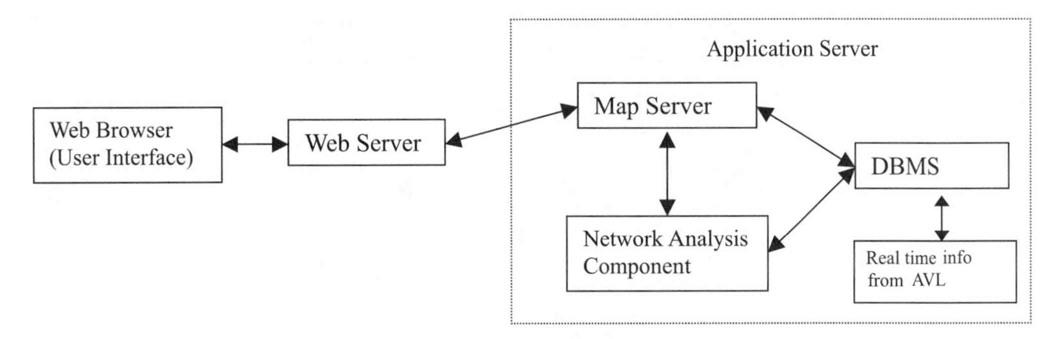

Fig. 1. A three-tier architecture for Advanced Transit Information Systems.

Three-tier system architecture is used to design a transit information system at Levels C2, C3, D2 and D3 as shown in Fig. 1. The three-tier architecture is composed of the Web browser (client tier), Web sever (server tier), and one or more application servers (application tier). The Web browser is a user interface used to gather user input. The Web server acts as middleware to handle users' requests and transfer the requests to an application server. The application server is used to process user requests. It is composed of three components: a map server, a network analysis server, and a database server. The map server is designed for map rendering and spatial analysis; the network analysis server is used to provide network analysis functions; and the database server is used to handle data management via DBMSs.

The architecture shown in Fig. 1 is a server-based information system. That is, users make queries at the Web browser, but the process is conducted at the application server. User queries from a Web browser are transferred to the Web server, which sends the user's request to the map server. Based on the user's request, the map server either processes the query itself or sends the task to a network analysis component and/or a DBMS for processing. The output is then delivered to the Web server and ultimately to the user at the Web browser.

4. System components and implementation

Although this three-tier architecture can be used in all Internet GIS programs in general and online transit information systems in particular, different implementation may result in totally different systems. This study implemented the system inside MapObjects by Environmental Systems Research Institute (ESRI) using Visual Basic. We used MapObjects to display maps, acquire input (e.g., trip origins and destinations, time of travel) and display graphic output to take advantage of its capability of feature rendering and map manipulation. The network analysis component is a modified NetEngine library (ESRI, 1998), a network analysis program by ESRI. The MapObjects Internet Map Server (IMS) is used to serve the application on the Internet.

4.1. User interface design

The system implementation process is shown in Fig. 2. It starts with user input on trip origin, destination, and travel date and time. An interface needs to be developed where users interact with

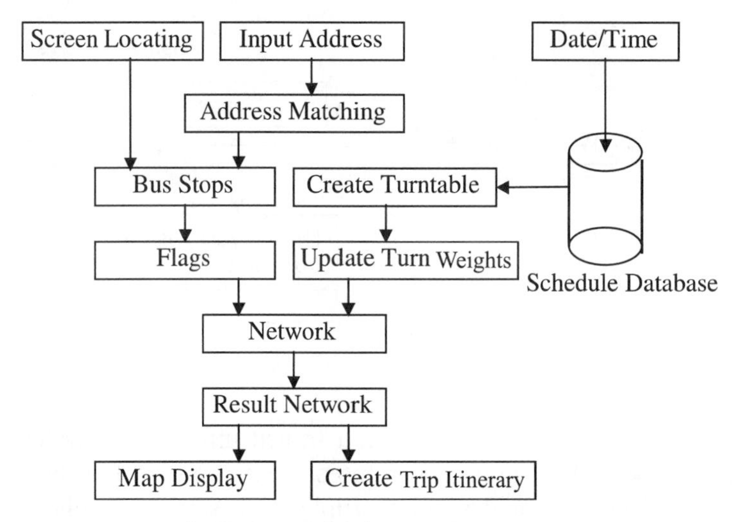

Fig. 2. System implementation process.

the application. Its flexibility and ease of use is critical to the use of the application (Howard and MacEachren, 1996). There are several options that can be used to design the user interface as discussed in Table 1, i.e., text, sketch map images, and interactive maps. The problem of the text-only interface, besides the lack of vivid visual effect that a map can provide, is that the program may not find the location of the trip origin and destinations if inexact addresses are provided. This is quite often the case when the user is not familiar with the area or does not know the exact address. To avoid this situation, pull-down list boxes containing street intersections and landmarks are sometimes used (e.g., http://www.romanse.org.uk/). Sketch route maps offer little or no references to surrounding streets, nor do they provide proper scale.

A unique feature of this user interface (Fig. 3) is that it is a GIS based system with an interactive map interface, which provides users a map interface to select locations directly on the map. Some map rendering functions, such as zoom and pane, are also provided. Users can also enter their trip origins and destinations on an input box or select intersections or landmarks from a pull-down list.

Furthermore, the map-based user interface also provides spatial query and searching functions to obtain street information (e.g., street names and locations) from the map, as well as search locations of streets by typing street names. The user can find a particular street from the database and the map will automatically center at that location. The street maps use selective labeling. That is, as the user zooms into the detail of the map, the street name will be shown on the map. This makes it easy to browse around the map to find more information about the neighborhood, other bus lines and local attractions.

To enhance the interactivity between the user and the map, Web client-side applications, such as plug-ins, ActiveX controls and Java applets, could be developed. But these client-side applications are perceived to be too technical for transit users by the transit agency for which this system was designed. Therefore, a server-side approach is used to build the interface to interact with users. This is the thin-client approach, the simplest, yet the most user-friendly. It has no limitation on a user's computer platform and local resources. Basic map rendering functions like

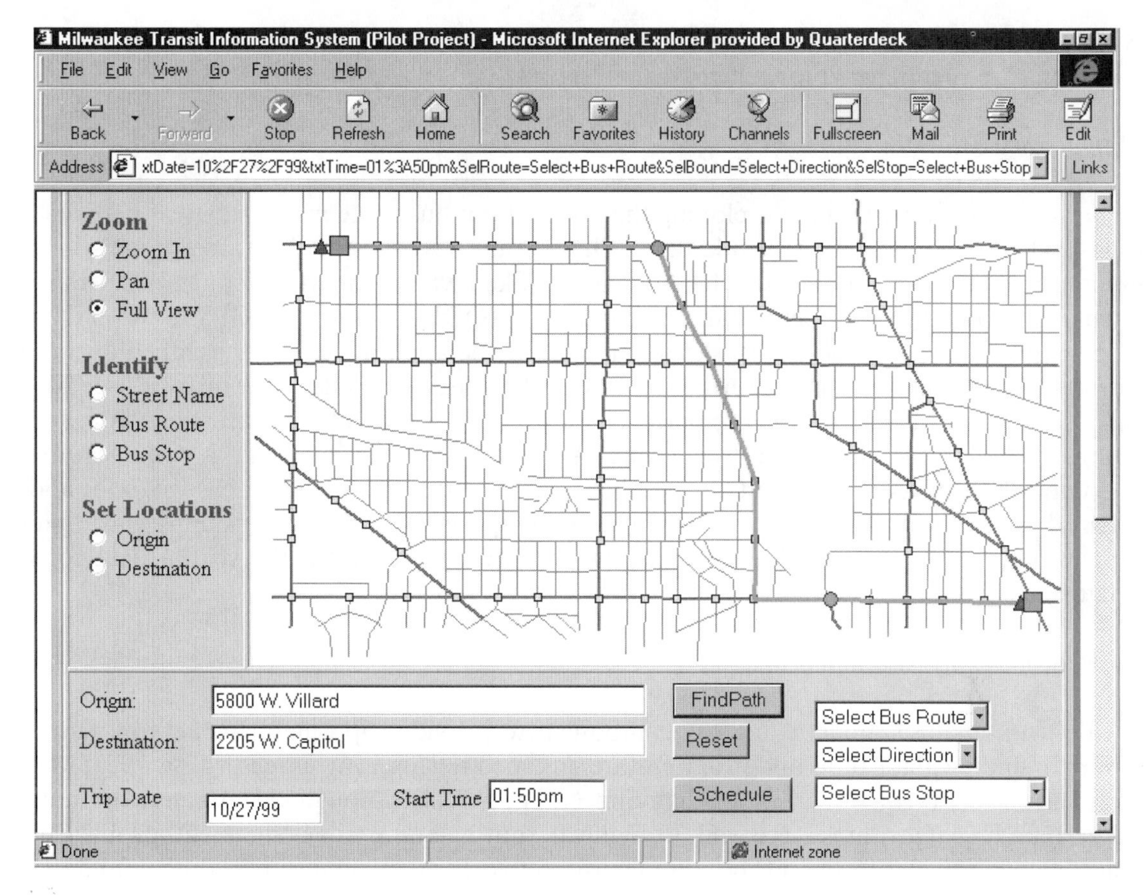

Fig. 3. The graphic user interface.

zoom, query and search are provided in HTML form. Users are able to select features directly from the map. However, they are not able to draw a box or a circle directly from a map image because of the limitations of the HTML. XML can be used for further improvement.

4.2. Map server functions design

The system is intended to offer users interactivity with the map by allowing users to browse service and other information directly on the map. Therefore, the system has to be able to offer map-rendering and address matching capability. This is handled by the Map server (e.g., MapObjects and MapObjects IMS by ESRI).

Trip origins and destinations are not necessarily on the transit network. Consequently, the system needs to find all bus stops that are within walking distance (a quarter mile or 0.4 km) of the user's trip origin and destination. The reason that all stops within walking distance are searched rather than the one that is the closest is that the closest transit stop may not be on the shortest route path (Peng, 1997). Sometimes a little longer walking time may result in a shorter overall travel time. If there is no stop within walking distance of trip origin and destination, a longer walking distance should be used. After those stops from the origin and destination are found, these stops are then flagged for network analysis.

4.3. Data and DBMSs

Four data files were used in developing the application: bus route network, street network, bus stops, and time points with schedule data. These data files are stored in a relational database system using Access by Microsoft. The database is linked with the map server and network server (discussed below) through open database connectivity (ODBC). Real time GPS location data are not available at the time of project development, but the application was developed as an open system to incorporate real time information when it becomes available in the near future. The bus route file is derived from the street centerline file with the addition of some attribute data, such as the street length, speed limit, and travel time. A line feature street map and a point feature bus stop map were used as background layers. The street map was also used as a base map for address matching of trip origins and destinations. The bus stop map was used for defining start and end stops of a trip. The bus schedule database is separately stored from the spatial data for easy update and management.

4.4. Network analysis component

A network analysis model is the key component to provide trip itinerary planning. However, most path finding algorithms and programs are designed for highway usage (Moor, 1957; Martin, 1963; Dial, 1971; Ikeda et al., 1994; Zhan and Noon, 1998). Although existing network analysis and path finding algorithms serve well for highway routing and traffic assignment, problems arise when they are applied to transit, because transit networks have significantly different characteristics from highway networks (Spear, 1994; Peng, 1997). Many researchers have pointed out the inadequacy of applying the path finding algorithms of highway networks to solve the minimal path finding problems for transit networks (Le Clercq, 1972; Chriqui and Robillard, 1975; Last and Leak, 1976; Tong and Richardson, 1984; De Cea and Fernandez, 1989; Spiess and Florian, 1989; Wong and Tong, 1998). Because transit service is time dependent, different times of the day or different days of the week have different levels of transit service. Some services are available only at the peak time period. Second, one street segment may serve different bus routes and many routes may stop at the same bus stop. This is the so-called "common bus lines problem" (Chriqui and Robillard, 1975). Third, unlike the highway routing problem, where the computation of shortest path is symmetric with respect to an origin/destination pair, the routing on transit networks from origins to destinations is not symmetric with that from destinations to origins. Fourth, transit transfers depend on the arrival time of another bus. Hence the best path between an origin and destination can change depending upon the timing of services available. Furthermore, many routes have loop routes and layover time. These unique characteristics make the minimal path finding application for transit networks much more challenging.

In the case of highway network analysis, one street segment and one intersection has one unique value of travel time and turn weight (or turn penalty). However, in a transit system, one street segment may have several bus routes, and each has its own headway. Some are regular buses and some are express buses. It is even more difficult to determine the turn weight (wait time) at each intersection in a transit system. At the same intersection, different buses may have different turn weights. Take the intersection A in Fig. 4 as an example. For riders taking bus B-3 north, the left turn weight time is very small because the rider does not need to transfer. But for the same turn, if the rider has to transfer from B-3 to B-2 west, the turn weight could be very large because the rider has to transfer. However, conventional path finding programs based on the street network require a single turn weight for each turn at every intersection. This makes the actual turn weight extremely difficult to determine.

Adding more complexity to the situation is that the turn weight changes over time because the bus headway changes; some buses even stop serving at certain times of the day. This problem is similar to constructing a shortest path based on the congestion level on roadways, sometimes referred to time dependent constrained shortest path or TCSP (Frank et al., 2000). Even when the turn weight is calculated for every second, it is still difficult to determine which time period to use. If the time of trip origin is used, when the bus gets to that transfer point the transfer route may not be in service. When the expected arrival time to that intersection is used, how do you know the expected arrival time if you have not determined the path? Because of the complexity of the transit route network, conventional highway network topology and analysis methods are difficult to apply in transit networks.

Several transit network models were proposed in the literature. Most of these prior models for the shortest path finding in transit networks can be categorized into two groups: headway-based and schedule-based. The headway-based path finding algorithms assign passengers to the first arriving vehicle based on the combined frequencies of common bus lines (Spiess and Florian, 1989). Constant (average) headway during a time period on a segment is usually assumed. The shortest path finding algorithms are usually variants of traffic assignment procedures used for highway networks that are modified to reflect the waiting time inherent to transit networks (Dial, 1967; Last and Leak, 1976). The schedule-based transit network models used a branch-and-bound type algorithm to determine the time dependent least cost paths between all origin/destination pairs in the transit network. The passengers are assumed to board the first vehicle to arrive as specified in a pre-determined schedule (Tong and Richardson, 1984; Wong and Tong, 1998). The path assignment for the headway-based approach is stochastic and heuristic in nature in the sense that a passenger always takes the first vehicle that arrives. But the algorithm itself does not identify which routes the passenger should take if there is more than one bus line going to the same destination. The results could be more than one path for any origin/destination pair. On the contrary, the schedule-based transit models deterministically identify the arrival of the first vehicle

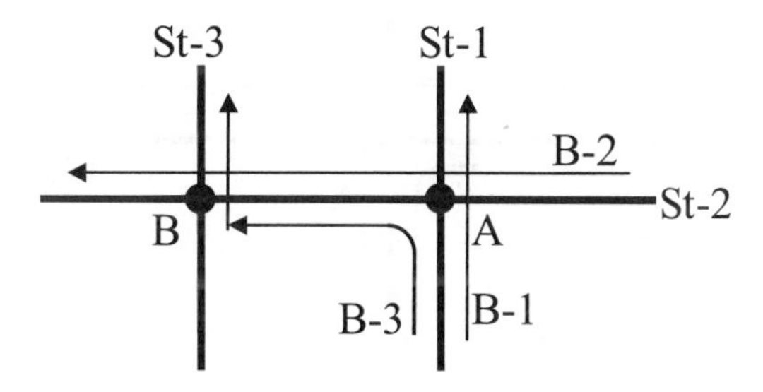

Fig. 4. Illustration of turn weight determination.

based on a fixed schedule. Therefore, it yields one and only one optimal solution for any given origin/destination pair.

This study combines the headway-based and schedule-based approach in a two-stage path finding process. In the first stage, average headway is calculated for each segment on the transit network. The average headway is used to calculate travel time on each link and the turn weight at each intersection. A vine-building type shortest path algorithm (Kirby and Potts, 1969) is then used to estimate the shortest path for each origin/destination pair. Since transit service frequency varies at different times of the day, the average headway needs to be calculated for each time period (every half-hour in this case). In other words, the shortest path problem has been transformed into many time dependent shortest path problems over many discrete time intervals. This is similar to the time independent solution method in Frank et al. (2000) and Handler and Zang (1980).

An example is shown in Fig. 5. Assume a user takes bus B-1 from the street St-1 and needs to change to the west on Street St-2 at intersection A. The user needs to transfer at point A. The waiting time at point A depends on the number of bus routes on the street St-2 running to the west and the headway of each bus route on that street at a specific time.

The value of the turn weight is derived statistically from the possible wait time at the intersection that may involve bus transfers. One-half of the average headway of all buses in one direction at the intersection is used to determine the turn weight at a specific time. One-half of average headway for each bus line from the possible next three consecutive buses' headway is used as an aggregated measure of turn weight. For example, if the trip start time is 4:00 p.m., the average of three consecutive headways of route B-2 and route B-3 at intersection A after 4:00 p.m. is 600 and 900 seconds, respectively. Thus the weight value for that turn can be calculated by $(1/2) * 3600/(3600/600 + 3600/900) = 180$ s; the average waiting time is calculated to be 180 s. Intersections that have more transit routes will have smaller headway and smaller turn weight values, and hence less impedance. The larger the number of alternative bus routes and the shorter the headway, the smaller the turn weight value and therefore the more likely the intersection is to be selected as a turn point (or transfer point). Because the bus headway may change by time of day, the turn weight value varies accordingly.

This method of calculating the value of turn weight is only a proxy for potential wait time. The turn weight is only used in the process of finding the shortest path. It is updated at the time of travel. The travel time for each link on the transit network is determined by the average travel time of bus lines. Except for occasional express buses, the travel time on each link for every bus is

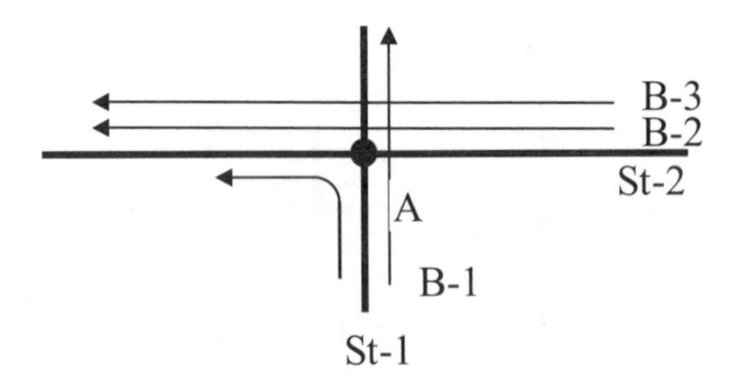

Fig. 5. Turn weight definition for transit networks.

very similar. The link travel time and turn weight are then used to search for the shortest path using a vine-building algorithm.

Once a shortest path is defined, actual wait time and transfer points are retrieved from the schedule database, and the user is assigned to specific bus lines based on the actual schedule. Although the shortest path finding program identifies transfer points at intersection, some actual transfer points may be at non-intersections. For example, buses B-1 and B-4 run on the same street, St-1, and a passenger needs to transfer from B-4 to B-1 as shown in Fig. 6. The shortest path program will identify the transfer point at intersection A. The program will then check for the bus stop location database to identify the shortest distance between stops on routes B-1 and B-4. Usually a shared bus stop (in this case, stop Z) will be chosen as the actual transfer point.

In the case of multiple routes going to the same destination that could be transferred to, the next arriving bus is chosen as the first choice for the transfer. However, the program also lists other available bus lines so that the passenger has backup options in case the current bus and/or the transfer bus is not on schedule. If there are express bus lines, the first regular bus that arrives at the transfer point may not be the best choice. The passenger is better off to wait a little longer for the next express bus than to take the first arriving regular bus. To take this into consideration, the program compares the arrival times of all possible buses that go through the trip origin or current transfer point at the next transfer point or destination. The one that gets the next transfer point or destination first would be the fastest one. The program lists that choice as the first path choice. As expected, the shortest path that goes through major streets that have more bus routes is usually selected, and transfer points are usually chosen at intersections that have more alternative buses on the same street.

This hybrid method, relying on the average headway as turn weight, may not produce *the* shortest path. To produce one single best route, one has to calculate the actual wait time at each potential transfer point based on the service schedule. For example, there would be two alternative turn weights at point A in Fig. 5 for a given time period. One is the wait time from bus B-1 transferring to bus B-2 and the other is the wait time from bus B-1 to bus B-3. But all vine-building-based path finding algorithms can handle only one turn weight for each intersection; they cannot handle two or more turn weights at one intersection. To solve this problem, additional pseudo-links and pseudo-nodes need to be added to the transit network to represent each individual bus route. Theoretically, this method could produce the shortest path if the bus is always 100% on time. However, buses are not always on time. Therefore, there are at least two problems associated with this approach. First, since the bus is not always on time, we have to give enough

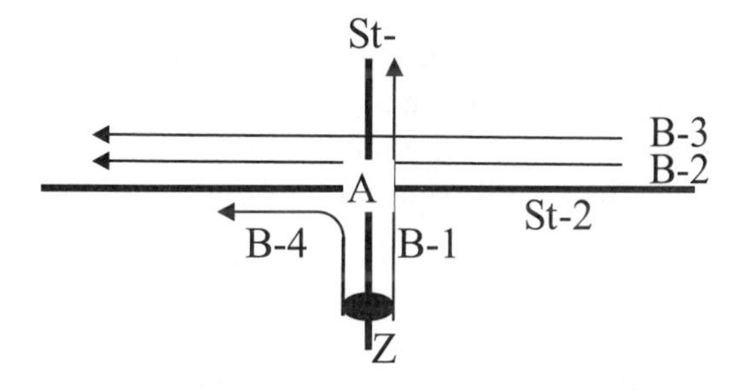

Fig. 6. Finding transfer point.

cushion time for passengers to transfer. If the cushion time is too long, the resulting shortest path may not be the shortest. If the cushion time is too short, it is possible that the passenger may miss the transfer bus at the transfer point. Second, the headway-based turn weight usually selects a transfer point with more transit routes. This gives passengers more options in case they miss the first bus. But the schedule-only-based algorithm depends solely on the schedule of individual routes; it does not give preference to streets with multiple bus routes. Therefore, it offers fewer route options. For example, based on the schedule match, the schedule-only approach may result in the shortest path that goes through a link with a single bus route with a headway of 30 min. The hybrid approach yields a path that goes through a link with multiple bus routes. In the former case, the wait time is only 5 min; in the second case, the weight time is 7 min, but there are two bus lines each with 15 min headway. If for some reason the passenger misses the first bus, the passenger then has to wait up to 30 min for the next bus in the first case, but he/she needs only to wait for another 15 min or less in the second case. Therefore, given the not-so-reliable transit services, the hybrid approach will produce a better solution.

The performance of the path finding program is very good. The shortest path has been pre-constructed from every node to all other nodes for every half-hour time period and is stored in the server. Therefore, there is no need to estimate the shortest path on the fly. Depending on network traffic, it usually takes a couple of seconds to retrieve the information from the server.

Once the shortest path has been determined, the system creates path directions to report to the user the names of starting and ending bus stops, the bus route(s) to take, and transfer stops and transfer routes, as well as bus arrival and departure information. The system also conducts a separate shortest path search for walk directions from trip origin to the starting transit stop and from ending transit stops to trip destination. Since the street distance is the only link cost, Dijkstra's (1959) shortest path algorithm is used to determine the walking path. The whole path direction (including walking and transit) is reported to the user in both the text format and maps.

4.5. Construct personalized travel information

A user profile is used to store customers' personal information, including home address, work address, locations of other common destinations, and usual time to work and to home. This information is stored in a user profile database. Each user is assigned a user identifier and a password. When the user enters the site, his/her personalized information will be automatically retrieved. The personalized information includes trip itinerary and bus schedules. A timer is set up to check the current time against the usual time to work and time to home. If the current time is closer to the time to home, a work-to-home trip itinerary will be presented by default. The user can always change his/her profile information, and the database will be updated accordingly. The personal profile is not mandatory. But creating a personal profile allows the system to inform the user via e-mail or other means of any service changes. In the future, as real time information is available, the user can get more timely information on delays and service changes.

4.6. Future improvements

The system design allows for displaying real time bus location once the real time AVL data are available. The bus locations on the map can be updated in a pre-defined time interval such as

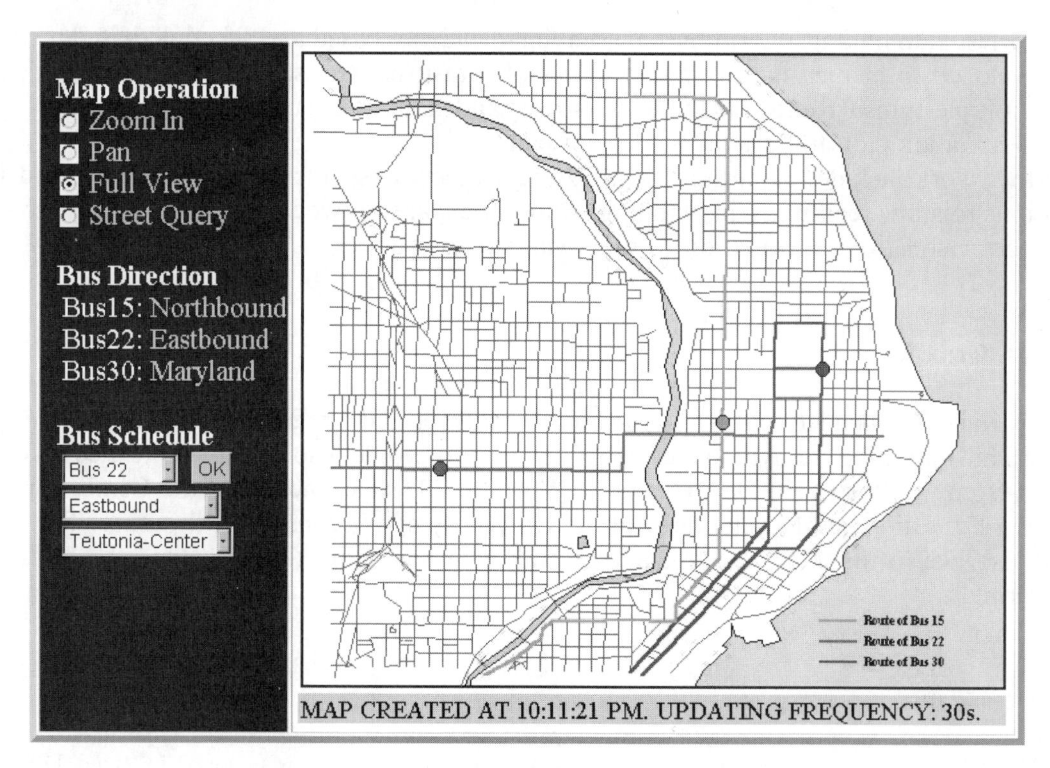

Fig. 7. An example of displaying real time bus locations.

every 30 seconds or every minute. An example is shown in Fig. 7 to display real time bus locations and bus movement animation using Global Positioning System (GPS) data. One important improvement to be made in the future would be using real time bus location based on the AVL data in the path finding process, raising the level of service to D2 and D3. Since AVL have been installed on buses in many transit systems, serving real time data is a matter of technical and institutional coordination. The current online transit information system is designed flexibly enough to accommodate future real time GPS data.

5. Conclusion

This paper presents a distributed Web-based transit information system. A unique feature of the system is that it integrates Internet GIS into the system design so that the user interface is map-based. The user can interact with the transit network and street maps, conducting query, search, and map rendering. The interactive map-based user interface also allows the system to incorporate other information, such as shops, theaters, parks, and other local attractions. This is very important for visitors who may want to explore these sites around their destinations. Internet GIS has been proven to be a powerful tool to develop flexible and versatile functions and to deliver rich information content to the users through the Internet and the World Wide Web.

A shortest path finding algorithm that combines headway-based and schedule-based methods is also developed to fit the unique characteristics of transit networks, namely time-dependent services, multiple transit routes on the same street, and non-symmetry of shortest paths between origin/destination pair and destination/origin pair.

Further work includes expansion to utilize real time transit location information and traffic conditions to allow real time trip planning. Another expansion would be developing mechanisms to deliver personalized information to users via wireless devices.

Acknowledgements

This article has benefited greatly from the comments and suggestions from Professor Jean-Claude Thill and three anonymous referees. The authors would also like to acknowledge financial support from the Wisconsin Department of Workforce Development and the Center for Transportation Education and Development at the University of Wisconsin – Milwaukee. The authors would also like to thank Professor Nancy Frank who has painstakingly edited the draft of the manuscript.

References

Batty, M., 1999. New technology and GIS. In: Longley, P.A., Goodchild, M.F., Maguire, D.J., Rhind, D.W. (Eds.), Geographic Information Systems. Wiley, Chichester, pp. 309–316.

Casey, R.F., Labell, L.N., Prensky, S.P., Schweiger, C.L., 1998. Advanced Public Transportation Systems: The State of the Art Update'98. Federal Transit Administration, Washington.

Chriqui, C., Robillard, P., 1975. Common bus lines. Transportation Science 9, 115–121.

Colman, D.J., 1999. Geographic information systems in networked environments. In: Longley, P.A., Goodchild, M.F., Maguire, D.J., Rhind, D.W. (Eds.), Information Systems. Wiley, Chichester, pp. 317–329.

Coleman, D.J., McLaughlin, J.D., 1997. Information access and usage in a spatial information marketplace. Journal of Urban and Regional Information Systems 9 (1), 8–19.

Conquest, J., Speer, E., 1996. Disseminating ARC/INFO dataset documentation in a distributed computing environment. In: Proceedings of 1996 ESRI User Conference, Redlands, CA (ESRI URL: http://www.esri.com/resources/userconf/proc96/TO200/PAP166/P165.m).

De Cea, J., Fernandez, J.E., 1989. Transit assignment to minimal routes: an efficient new algorithm. Traffic Engineering and Control 30, 491–494.

Dial, R.B., 1967. Transit pathfinder algorithm. Highway Research Records 205, 67–85.

Dial, R.B., 1971. A probabilistic multipath assignment model which obviates path enumeration. Transportation Research 5, 83–111.

Dijkstra, E.W., 1959. A note on two problems in connection with graphs. Numerische Mathematik 1, 269–271.

Doyle, S., Dodge, M., Smith, A., 1998. The potential of web-based mapping and virtual reality technologies for modeling urban environments. Computers Environment and Urban Systems 22 (2), 137–155.

Environmental Systems Research Institute, Inc., 1998. NetEngine: A Programmer's Library for Network Analysis. Environmental Systems Research Institute, Inc., Redland, CA.

Frank, W.C., Thill, J.-C., Batta, R., 2000. Spatial decision support system for hazardous material truck routing. Transportation Research C 8 (1–6), 337–359.

Handler, G.Y., Zang, I., 1980. A dual algorithm for the constrained shortest path problem. Networks 10, 293–310.

Howard, D., MacEachren, A.M., 1996. Interface design for geographic visualization: tools for representing reliability. Cartography and Geographic Information Systems 23 (2), 59–77.

Ikeda, T., Hsu, M.Y., Imai, H., Nishimura, S., Shimoura, H., Hashimoto, T., Temmoku, K., Mitoh, K., 1994. A fast algorithm for finding better routes by AI search techniques. IEEE Vehicle Navigation & Information Systems Conference Proceedings B 3–6, 291–296.

Kirby, R., Potts, R.B., 1969. The minimal route problem for networks with turn penalties and prohibition. Transportation Research 3.3, 397–408.

Last, A., Leak, S.E., 1976. Transept: a bus model. Traffic Engineering and Control 17, 14–20.

Le Clercq, F., 1972. A public transport assignment method. Traffic Engineering and Control 13, 91–96.

Moor, E.F., 1957. The shortest path through a maze. In: Proceedings of the International Symposium on the Theory of Switching, Harvard University.

Martin, B.V., 1963. Minimum path algorithms for transportation planning. Research Report R63-52. Department of Civil Engineering, Massachusetts Institute of Technology.

Muro-Medrano, P.R., Infante, D., Guillo, J., Zarazaga, J., Banares, J.A., 1999. A CORBA infrastructure to provide distributed GPS data in real time to GIS applications. Computers Environment and Urban Systems 23 (4), 271–285.

Peng, Z.R., 1997. A methodology for design of GIS-based automatic transit traveler information systems. Computers Environment and Urban Systems 21 (5), 359–372.

Peng, Z.R., 1999. An assessment framework of the development strategies of internet GIS. Environment and Planning B: Planning and Design 26 (1), 117–132.

Peng, Z.R., Beimborn, E., 1998. Internet GIS: applications in transportation. TR News, Number 195, March/April, pp. 22–26.

Peng, Z.R., Jan, O., 1999. An assessment of means of transit information delivery. Transportation Research Record, forthcoming.

Peng, Z.R., Nebert, D., 1997. An internet-based GIS data access system. Journal of Urban and Regional Information Systems 9 (1), 20–30.

Plewe, B., 1997. GIS Online: information retrieval, mapping, and the internet. OnWorld Press, Santa Fe, NM.

Ran, B., Chang, B.P., Chen, J., 1999. Architecture development for web-based GIS applications in transportation. In: Paper presented at the 78th Transportation Research Board Annual meeting, Washington, 10–14 January.

Sarjakoski, T., 1998. Networked GIS for public participation – emphasis on utilizing image data. Computers Environment and Urban Systems 22 (4), 381–392.

Salters, T., 1996. DART on target. ITS World, May/June.

Spear, B. D., 1994. GIS and spatial data needs for urban transportation applications. In: David Moyer, D., Ries, T. (Eds.), Proceedings of the 1994 Geographic Information Systems for Transportation (GIS-T) Symposium, Norfolk, Virginia, pp. 31–41.

Spiess, H., Florian, M., 1989. Optimal strategies: a new assignment model for transit networks. Transportation Research B 23, 83–102.

Tong, C.O., Richardson, A.J., 1984. A computer model for finding the time-dependent minimum path in a transit system with fixed schedules. Journal of Advanced Transportation 18 (2), 145–161.

Wong, S.C., Tong, C.O., 1998. Estimation of time-dependent origin–destination matrices for transit networks. Transportation Research B 32 (1), 35–48.

Zhan, F.B., Noon, C.E., 1998. Shortest path algorithms: an evaluation using real road networks. Transportation Science 32 (1), 65–73.

PERGAMON

Transportation Research Part C 8 (2000) 427–444

TRANSPORTATION RESEARCH PART C

www.elsevier.com/locate/trc

An Internet-based geographic information system that integrates data, models and users for transportation applications

Athanasios K. Ziliaskopoulos [a,*], S. Travis Waller [b]

[a] *Department of Civil Engineering, Northwestern University, Evanston, IL 60208, USA*
[b] *Department of Industrial and Systems Engineering, Northwestern University, Evanston, IL 60208, USA*

Abstract

This paper is concerned with the development of an Internet-based geographic information system (GIS) that brings together spatio-temporal data, models and users in a single efficient framework to be used for a wide range of transportation applications – planning, engineering and operational. The functional requirements of the system are outlined taking into consideration the various enabling technologies, such as Internet tools, large-scale databases and distributed computing systems. Implementation issues as well as the necessary models needed to support the system are briefly discussed. © 2000 Elsevier Science Ltd. All rights reserved.

Keywords: Geographic information systems; Intelligent Transportation Systems; Distributed systems; Dynamic traffic assignment

1. Introduction

Transportation systems are complex entities that require substantial data to be monitored, controlled, maintained and improved, as well as various elaborate models to help a diverse group of agencies to operate the system. Data currently collected by transportation agencies can be loosely classified into the following types: planning, engineering and operational; a similar classification can be made for the models used. These data have certain similarities: (i) they are all concerned with the same network, traffic demand and control devices, albeit at different levels of aggregation and precision of representation, (ii) they are all spatio-temporal, i.e., they can be

* Corresponding author. Tel.: +1-847-467-4540; fax: +1-847-491-5264.
E-mail addresses: a-z@nwu.edu (A.K. Ziliaskopoulos), s-waller@nwu.edu (S.T. Waller).

PII: S0968-090X(00)00027-9

associated with spatial and temporal coordinates, (iii) they are often used by more than one models for different applications. Great benefits could be realized, if these data are integrated into a single database or linked distributed databases and become accessible by all necessary models. Data and models could, in turn, be available to all involved entities: planners, engineers, operators, as well as various stakeholders, such as researchers, consultants, trucking companies, special interest groups and even the traveling public.

Presently, transportation professionals have to cope with fragmented databases, multiple and incompatible models, redundant and often conflicting data acquisition efforts, lack of coordination between various agencies and private companies operating on the same transportation facilities. This results in serious inefficiencies and waste of resources. For example, a transportation planner uses the network of a metropolitan area for air quality analysis with a certain format that is usually incompatible with a signal optimization or simulation software that the traffic engineer of the same urban area uses to optimize the operations on the urban network. Typically, both professionals independently collect and code the data for the same network and demand (albeit at different aggregation levels) without taking advantage of the existing resources at each other's agency.

Enabling technologies developed over the past few years have created unprecedented opportunities to overcome some of the problems above. These technologies include the explosion of the Internet and Internet support tools, terabyte size databases, distributed computing architectures, client server technologies as well as a new generation of transportation tools resulted from the evolution of Intelligent Transportation Systems. Many of these technologies have already been adopted by corporations and have led into the development of new business models. In fact, from supply chain dynamics to customer relationship management to the enabling of an e-commerce infrastructure, the opportunity and need have never been greater to link business processes and people throughout an organization (Chambers, 1999). Internet-enabled geographic information system (GIS) has also attracted a lot of attention in the last few years: Jankowski and Stasik (1997) introduced an Internet-based GIS to make possible collaborative spatial decision making via public participation. Keisler and Sundell (1997) presented an integrated geographic multi-attribute utility system with application to park planning. An extensive survey of applications and research issues for geographic information technologies applications in business is provided in Mennecke (1997).

This paper introduces a prototype Internet-based GIS that aims to integrate spatio-temporal data and models for a wide range of transport applications: planning, engineering and operational. The GIS graphic user interface (GUI) is built in JAVA, so that it can be used over the Internet (or any other large network). The database efficiently stores and retrieves spatio-temporal data, by associating geographic coordinates and time stamps. The database is designed to efficiently manage a wide range of transport data: from off-line planning and engineering to streams of real-time, such as those coming in from street sensors and newer vehicle based devices. Furthermore, the control and encapsulation of the data that becomes possible with such a system, help deal with problems arising from agency specific requirements.

A number of models have been coded in the same framework and can by accessed via the GIS' GUI in a client server setting. The models can be remotely accessed via the GIS and run in a distributed environment based on the common object request broker architecture (CORBA). The implemented models include traditional signal control and analysis tools, planning models as well as newer dynamic traffic assignment (DTA) and routing algorithms. This paper discusses the interactions between the models, the potential efficiencies achieved by the integration of data and models, implementation difficulties as well as the implications to planning, engineering and operational practices. It should be noted that the primary focus is on the technical aspects of this integration and not on the institutional/policy issues even though the latter would obviously constrain the final system. Such issues, however, deserve attention, which is beyond the scope of this paper. Furthermore, due to the wide variety of policies and agencies, the introduced system is not currently presented in the context of any particular state agency or entity. Instead, the underlying characteristics of transportation data and models are examined in order to develop a system, which can later be placed within the institutional framework of a particular entity.

In the next section, we identify the needs and the functional requirements of the system. The overall model architecture is presented in Section 3. The models currently implemented or that need to be part of the framework are described in Section 4. Implementation details are provided in Section 5, including a justification for the Internet-based GUI and the distributed computing environment. Section 6 discusses the potential benefits that can be realized by the integration of the data, model and users. Section 7 concludes this paper and identifies directions for future research.

2. Needs and functional requirements

The transportation practice involves data, models and users. Data currently collected by transportation agencies can be loosely classified into the following types: planning, engineering and operational; a similar classification can be made for the models used. *Planning* data range from trip surveys, socioeconomic data by region and network infrastructure data, to aggregate daily link traffic flows. These data are used for long-range planning purposes, such as estimating the impact of infrastructure improvements, mode split, and environmental impact. Models used involve the traditional four-step planning models as well as other forecasting and discrete choice models. *Engineering* applications typically require more detailed network, control and traffic data, obtained by direct observation and traffic engineering studies. Network data involve detail representation of intersection geometry, turning movements and prohibitions; traffic data are often aggregated every 15 min, while signal or sign control information, such as controller type and precise timing plan are needed. These data are used for shorter-term engineering applications, such as warrant analysis, signal timing plan development, and freeway management. Models used by engineers are typically comprised of signal optimization, simulation and capacity analysis. Finally, traffic operators collect real-time data from online sensors, based on which real-time *operational* decisions are being made. The network infrastructure data are also continuously updated based on construction zone plan changes, incidents and traffic control. Urban traffic control system (UTCS) type models are often used by control centers and occasionally ramp metering and variable message algorithms by freeway operators. A rather crude list of the various types of the data, models and users is included in Table 1.

As mentioned earlier, the state of the transportation practice is characterized by fragmented databases, redundant and often ill-conceived data acquisition efforts, legacy database software,

Table 1
Types of data, models and users for transport applications

USERS	DATA	MODELS	APPLICATIONS
Operators	Real-Time	Traffic Simulators	Planning
Engineers	Sensors	Micro-	Evaluation
DOT	AVL	Meso-	Infrastructure Imp
Cities	I-PASS	Macro-scopic	Evaluation
County	Transit/Trucks	DTA	CM/Air Quality
Planners	Anecdotal	Simulation Based	Transit Routes
MPO	Historical/Offline	Analytical	Engineering
DOT	Aggregated RT	FAT/UE/SO	Design
Policy Makers	Census	Multiuser	Control
Public	Special Studies	Routing	Maintenance
Stakeholders	Surveys	Control	Safety
Truckers	Offline	Isolated/Arterial/Network/Ramp	VMS Loc, ITMS
AAA	Online	Metering	Policy Making
Consultants	Special	Online/Offline	Project Monitoring
Transit	Workzone	HCS	Economic Impact
	Crashes	UE/SUE	Legislation
	Warrant Analysis	Discrete Choice	Public
		Logit/Multi-/Nested	Trip Planning
			ATIS

lack of professional software support, incompatible and incomplete data sets. The models are a rather peculiar mix of government developed and maintained tools (such as TRANSYT, NETSIM and TRANSIMS) that compete with models developed and maintained by the private sector (Synchro, TransCad and Integration). Most models are typically incompatible and the only way to convert datasets from one model to another is via a painstaking process of converting ASCII files. Finally, the number of users of the data and models tends to be large, typically remotely located with different needs, objectives and levels of sophistication. The lack of coordination among professionals inside an agency as well as between various agencies and private companies operating on the same transportation facilities is also well established.

The first issue we were confronted with during the development phase of our framework was the identification of the needs of transportation systems in the light of the opportunities presented by the various enabling technologies. These needs were grouped in three sets:

1. Interfaces and user connectivity needs.
2. Database integration needs.
3. Modeling needs.

In order to meet the above needs, the following functional requirements are identified.

2.1. Internet-based interfaces, common to all users accessing the system

This functional requirement identifies the Internet as the means of accessing the system, though an Intranet or any other large scale corporate network structure could also satisfy this requirement. Given, however, that one class of users could ultimately be the public, the Internet seems to be the logical and least expensive approach to adopt. In addition, the existing vast infrastructure of the Internet guaranties continuous update, new tools and familiarity of the users with these tools. If the interface of the GIS is accessible via the Internet, many users can access the system simultaneously and perform a plethora of operations.

2.2. An integrated (and possibly distributed) Internet-enabled database that can manage large sets of off-line and online data

Database support and reporting functionality can be obtained from off-the-shelf commercial packages such as ORACLE 8i that also supports Internet access. An issue that needs special consideration is the ability to deal with real-time data coming from infrastructure sensors, their filtering, fusing and archiving. The advantage of having a single database (even distributed or a number of linked databases) accessed via a common interface is that all users have access to similar consistent datasets. In addition, if the database is updated by one user, the changes are transparent to all other users. Design issues related to a GIS database are extensively discussed in Healey (1991).

2.3. Distributed models so that the embedded algorithms with large needs in computational power can be executed in a reasonable time

Simply put, distributed systems allow simultaneous execution of many objects on many computers spatially distributed and interconnected via the Internet or another network. Distributed systems posses intrinsic abilities for both synchronous and asynchronous communication among objects. This allows the interface to be executed remotely from the implementation of the algorithms, and multiple processors to be used for the execution of the algorithm modules if needed.

2.4. An efficient reporting system that can query the database and produce intuitive reports that can address the needs of all users

This functionality will facilitate communication among professionals. As mentioned above, this functionality typically comes as a support tool to commercial databases, such as ORACLE 8i. It should be emphasized, however, that there is a trade-off between a reporting system that is easy to use and allowing for the user to customize the query and reporting system, which increases the need for training and the difficulty of using the system.

2.5. An efficient administrative support system for allowing users to manage their datasets and reports

During the course of using the system, many reports and secondary datasets will be produced. The users should have the ability to manage their files, remove temporary data sets and permanently store and publish the final ones.

2.6. *An effective security system that protects data integrity and protects authorized data access*

Different users will have access to the database at various authorization levels. For example, a construction engineer can update the construction plans for a workzone area that changes the available capacity on the particular facility. These changes will be immediately known to the operator of that facility, who may not have, however, the authority to change them. The opposite may happen for the ramp metering strategy of the facility under construction: it may be known to the construction or resident engineer but cannot be altered by them.

2.7. *Scalability of the system so that data and models can be expanded as user needs evolve*

The system will continuously evolve as more users, data and models are included. Thus, the system should be scalable to accommodate future growth and changes.

2.8. *Communication among users allowing the creation of virtual professional communities*

The Internet and the integrated database provide for a transparent system, where besides data and models, secondary data, reports, analysis and actions are available to other users operating on the system. This will allow for not only information but also for information exchange; if researchers are included, this could be a convenient medium for technology transfer.

3. Overall approach and model architecture

The overall approach introduced in this paper was inspired of and is implemented according to practices adopted by the InfoTech (IT) Industry. The objective is to meet the functional requirements identified in the previous section and bring together data, models, users and applications into a seamless efficient, interwoven system. The introduced framework is called visual interactive system for transportation algorithms (VISTA). VISTA is a CORBA compliant distributed system accessible over a network (including the Internet[1]), the client is machine-independent (JAVA technology), user-friendly and accessible based on various authorization levels. Data and models can be accessed by all users at different capacities, for retrieval, maintenance and analysis. CORBA is employed since application modules are written in a variety of languages (C, C++, FORTRAN, etc.) and may run on various Windows or UNIX platforms. While other distribution technologies (RMI, DCOM) have developed towards this more open object-oriented framework structure, it has been a relatively recent move and not found sufficient to meet the functional requirements at the time of development.

Specifically, the system implemented so far includes:

- A data warehouse module (similar to ORACLE 8i) accessed over a network (Intranet/Internet).
- Existing and new transportation models and tools (planning, engineering, control, monitoring, evaluation and operational).

[1] http://its.civil.nwu.edu/vista-beta/

- User interfaces for the various stakeholders (planners, engineers, policy makers, and operators) that enable access to the data and models from any computer hardware, at any location, at any time, by any user.
- The system is a CORBA compliant distributed system, allowing for CPU intensive models to be executed on many computers.
- Support capabilities for all relevant transportation applications.
- Functionality for interaction among users.
- Some reporting capability.
- Security features for many users accessing the same database at various authorization levels.
- Some basic administrative capabilities.

The system is intended to be deployed at a State level, where the State Department of Transportation (DOT), the Metropolitan Planning Organizations (MPOs), County Engineers, City Engineers, Transit Agencies, Freight Agencies, and other stakeholders will have access to the system at various authorization levels to obtain/maintain data, run models and perform analysis. Policy makers at the Federal, State and Local governments will be able to monitor projects, obtain data, evaluate impacts of policies and make decisions. The overall structure of VISTA is outlined in Fig. 1.

The user interface is written entirely in JAVA so that it can easily be used on multiple platforms and across the Internet. It communicates with the Management Module through the standard Java 1.2 object request broker (ORB), allowing any user with a recent Netscape or Internet Explorer browser to access the system by using the Java 1.2 plug-in. The interface works as a client exclusively, and communicates only with the Management and Database modules. If the client is functioning as an Applet, it is constrained by tight security restrictions with regard to network communication. Essentially, the Applet is allowed to communicate only with the machine from which it originated. Therefore, it is important that the Management Module be run on the same machine which services the VISTA web page.

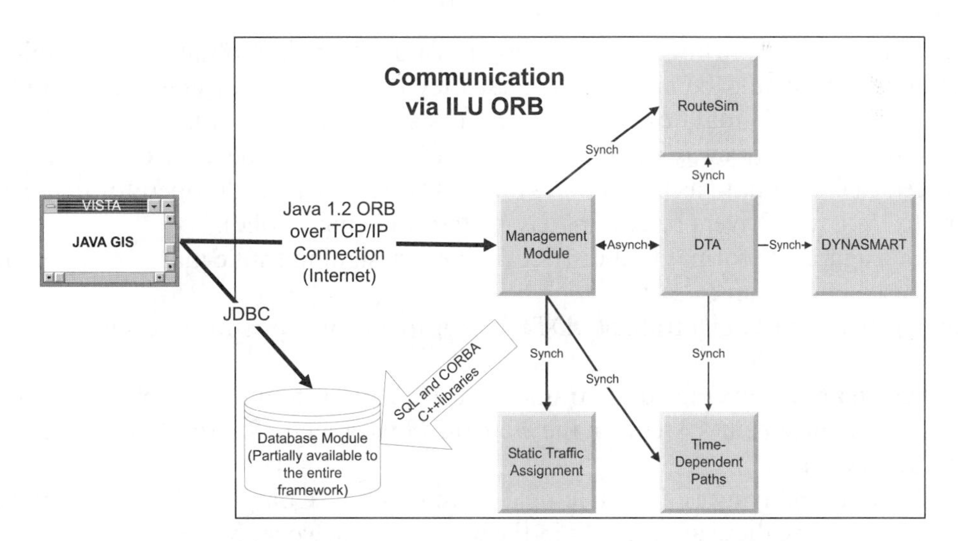

Fig. 1. VISTA structure.

When the VISTA HTML page is loaded, the interoperable object reference (IOR) string is read in as a parameter. This IOR string consists of a sequence of characters which uniquely identifies the Management Module and allows the interface to look it up through the CORBA function *ORB.string_to_object(ior)*. This function returns a reference that can be used to get a JAVA object of class Management. Once this reference is obtained, the Management object can be manipulated as if it were any other local object.

The current database module is based on a combination of the PostgresSQL database management system and specialized file handling routines. The specialized routines are implemented in order to store intermediate model data in an optimal manner. Such data include the travel costs as reported by the simulator within each iteration of the DTA algorithm. However, once the algorithm is complete, the cost data is transferred into the structured query language (SQL) database in both desegregate and filtered aggregate form in order to support a unified reporting model. The original intermediate binary data can still be accessed through C/C++ and CORBA library functions. The database module can be accessed through CORBA, C/C++, open database connectivity (ODBC), or java database connectivity (JDBC) libraries. Since the database module supports the ODBC interface, widely available tools such as Microsoft Access can be used in order to manipulate VISTA data. Finally, by implementing a standard database protocol, the underlying data management tool is compatible with, or can easily be changed to other compliant database systems, such as ORACLE 8i.

A typical concern when dealing with unified transportation systems involving GIS tools is the representation of geographical information. A great deal of work has been done within this area such as the linear referencing system (NHCRP 20-27, 1997) presented work does not attempt to expand upon this specific field nor describe a system which is dependant on any one representation methodology. Instead, by using the object-oriented nature of the framework, internal data representations can be maintained that are independent of specific standards. This can be accomplished through the encapsulation of data within the database, and access provided only through proxy data values.

Another concern is data fusing and archiving transportation data obtained on a real-time basis from sensing devices should be archived so that they can be used for engineering and planning applications. For example, data obtained by sensors should be available:

- at the 6-s format to a freeway operator, who can invoke real-time control strategies;
- at a "processed 6-s format" as travel times on the freeway to a freight operator that can issue an advisory to his truck drivers (or ultimately to the traveling public);
- at the 15-min aggregate format to a district engineer designing traffic maintenance plans for upcoming construction plans;
- at an aggregate average daily traffic (ADT) format to a transit operator deciding on new transit routes;
- at an aggregate annual average daily traffic (AADT) format to an MPO planner who develops a congestion management system or the Federal Government for the Highway Performance Monitoring System.

The area of data archiving is an active field of research in Computer Science. Little has been done with incorporating these approaches in the current framework, but there is a need for further development in this area.

Another example of the improved efficiencies by maintaining the data in a single database is maintaining crash data and performing safety analysis: Currently, accident data are maintained by multiple jurisdictions, involve many people employed by the agencies in coding and cleaning the data. If one needs to access the data for year 1999, in most States, she has to wait at least a year for the data to become available (if at all). With the proposed system, the officer will ultimately have the capability to directly input the data via the Internet and a handheld device (in an interface form resembling the paper forms he fills up). These data will immediately become available to all agencies with minimum additional processing. Having the data available in a single database will allow safety experts to accurately correlate it to prevailing infrastructure, weather, traffic and control conditions (at the time of the accident) and develop appropriate countermeasures on time.

The presence of abundant data does not itself guarantee useful information for potential users. Furthermore, specific information requirements are often not known in advance when dealing with complex systems. These are the specific issues which have motivated the SQL, ODBC, and JDBC protocols. By supporting all of these protocols with complete network and Internet availability, the database module allows a multitude of tools complete (or limited for security reasons) access to framework data. This well-established design model allows for the straightforward implementation of complex querying and reporting through the support of JDBC from the GIS. Typical features that users expect can easily be incorporated into the framework as simple forms and buttons within a GUI. Furthermore, this open-architecture design ensures that if a desired functionality is omitted, a user can acquire the desired information through third part tools without the need for changes within the framework.

In addition to the benefits of SQL/ODBC to the requirement of useful information reporting, benefits are experienced with regard to the administration of user datasets. Data tables can be manipulated and moved through typical SQL commands (which can easily be automated through a GUI interface). In addition to the SQL functionality, users can manipulate the intermediate binary module data through a Java explorer interface, which references CORBA calls to the Management Module.

The data can be accessed via three methods over a network: CORBA, JDBC, and ODBC. For CORBA access, a user must first login with an accepted account. Only then access can be gained to the intermediate binary data. Access to this form of data can easily be restricted as desired by server administrators. JDBC and ODBC support includes all typical security features (which far exceed the scope of this brief description, such as encryption). Typical SQL user account and security information is maintained such as user privileges and data ownership.

Due to the distributed nature of the framework, one method available in order to scale the functional ability of the system includes the addition of workstations. While this is obviously a benefit when multiple users are running various algorithms, or a single user is running multiple ones, many of the models can take direct advantage of additional CPUs through inherent parallel properties. Additional servers are also an option for the management of data since many SQL servers support data replication to other ODBC compliant databases as well as various local caching strategies.

Next, we describe the model structure, while the implementation details of the distributed system are discussed in Section 5.

4. Model structure

The primary modules currently implemented in the VISTA framework include a traffic simulator (RouteSim), traditional (static) planning models, DTA models, network routing algorithms, signal optimization models, ramp metering and incident management models. The interactions among models are coordinated by the central Management Module. Although each of these models may have different data type and structure requirements, the format for this data is kept uniform. The way in which the VISTA modules interact is represented in Fig. 1. Each interaction is specified as either a synchronous or asynchronous invocation.

4.1. Management Module

The Management Module is the central component of the VISTA framework, and one of the only modules the user interface directly communicates with. It continuously runs on the server, handles incoming requests from remote interface modules, and executes the algorithm modules. A remote CORBA object can be described by its IDL file (Object Management Group, 1995).

As discussed in Section 3, when the Management Module first runs, it creates an HTML file, which contains the IOR string as an HTML parameter. This string uniquely identifies the Management Module as a CORBA object. By knowing this string, any CORBA enabled object present on the Internet has the ability to lookup and communicate with the Management object. Alternatively, a remote module could contact various system modules by accessing the CORBA naming service available on the central server.

When the Management Module receives a remote call to execute an algorithm, it does so through various means. The simplest method is to use the ANSI system( ) function to execute a separate program. The input parameters are specified as command line arguments, and the resulting output read from a temporary file, then returned to the invoking interface object. This method is best for relatively small algorithms which need little computational time and data manipulation.

For larger, more complex systems such as DTA, the algorithm appears as a different CORBA module, which contains its own remote methods. For these cases, the remote methods are specified to be *asynchronous* or *oneway* (which is interpreted as *asynchronous* within the employed ORB implementations). For very complex system where distribution is a possibility, this option also allows further distribution within the algorithm to take place using the framework services.

4.2. RouteSim

RouteSim is a mesoscopic simulator based on an extension of Daganzo's (1994) cell transmission model introduced by Ziliaskoupoulos and Lee (1997). RouteSim is one of the fundamental modules, since it is used for simulation, DTA and evaluation. The main enhancements over the basic cell transmission model are: (i) the concept of adjustable size cells that improves the flexibility, accuracy and computational requirements of the model, and (ii) a modeling approach to represent signalized intersections. The basic cell transmission model along with the enhancements yield a model that can simulate integrated freeway/surface street networks with varying degree of detail. RouteSim requires as inputs network geometry and path flow data. The path flow data can be generated from time-dependent or static origin-destination matrices or input directly by the user. RouteSim assigns every generated vehicle to a path, similar to the DYNASMART model introduced by Mahmassani et al. (1993). An advantage of RouteSim is that the simulation step and the representational detail are adjustable to the geometry of the network. Lengthy freeway segments that do not need to be modeled in detail are simulated as aggregate long cells and their state is updated infrequently – for example, a two-mile freeway segment without on- and off-ramps could be modeled as a single cell and be updated every 2 min. On the other hand, close to intersections or problematic points where the evolution of queues, spatio-temporal traffic dynamics and signalization phases need to be captured in detail, the simulation step can be as small as 2 s. Simulation steps of this magnitude allow detail representation of signalized intersections – that is, signal control strategies, phasing, start-up/lost times and gap acceptance behavior. Note that while detail data (e.g., geometry, timing plans, turning movements) are required for accurately simulating a network with signalized intersections, RouteSim will run even if no such data are provided, by assuming (and prompting the user), geometry, control and traffic data.

4.3. Planning models

System optimum and user equilibrium static assignment algorithms have been implemented and can be invoked through VISTA. The algorithms are deterministic approaches based on Frank–Wolfe's convex combinations method (Sheffi, 1985); a stochastic user equilibrium model is currently under development using a paired combinatorial logit model (Gliebe et al., 1999). The demand tables are part of the input data, since no trip generation, distribution and mode split modules are currently implemented. VISTA, however, provides a convenient framework for embedding such models, as well as using them in conjunction with DTA models.

In addition, highway capacity analysis modules are currently being implemented so that the level of service for intersections and street segments can be computed for the equilibrium flows. The computational procedures are done according to the Highway Capacity Manual suggestions. Existing software, such as the HCS, could also be interfaced.

4.4. Signal control models

Signal timing plans can be computed for isolated intersections based on simple delay functions and offsets for intersections along an arterial (McShane and Roess, 1990). Network-wide signal optimization models are currently under development, although any of the already existing models (e.g., TRANSYT) can be easily interfaced. A user-friendly graphic interface for viewing (or modifying) the intersection signal time plans has also been developed.

4.5. Dynamic traffic assignment

Various DTA models have been implemented within VISTA:

1. A departure-based and fixed arrival time version of simulation-based user equilibrium (UE) DTA approaches using RouteSim to propagate traffic and satisfy capacity constraints (Ziliaskopoulos and Rao, 1998).

2. A modified version of DYNASMART-X (Mahmassani et al., 1992) that is capable of modeling multiple user classes including user equilibrium and system optimum (SO) users. This version is departure time-based only, and uses DYNASMART to simulate traffic (Hawas et al., 1997).
3. Two analytical DTA models: a departure time and a arrival time approach; both approaches are linear programming and are solved with CPLEX (Ziliaskopoulos, 2000).
4. A combined departure and arrival time-based analytical model that is also solved using CPLEX (Yue et al., 1999).

All these DTA models use the same geometry, control and demand data inputs; the demand tables need to be arrival and/or departure time based, depending on the model invoked. The simulation-based DTA models access one of the simulator modules (RouteSim or DYNASMART), time-dependent least time and cost path modules, as well as various other modules. Since these systems work in an iterative scheme, the computational time of sub-modules becomes very important. DTA models are the most time consuming models, but many of these modules have operations that can be run in parallel. For instance, the time-dependent shortest path algorithms have the ability to be distributed over multiple processors (Ziliaskopoulos et al., 1998). Furthermore, all modules related to the routing of the various classes in DYNASMART-X can be run on separate processors. Since VISTA is based on the CORBA specification, it can handle the communication and invocation of these modules on separate processors.

Some computational performance information for the modified DYNASMART-X system is included in Table 2.

This data is for one iteration of the DTA algorithm on a 444 link network, with a demand of 8000 vehicles during a 50-min time period. The demand was broken evenly into two classes consisting of UE and SO behavior. There are three modules within this DTA system: UE consisting primarily of the time-dependent shortest path (TDSP) algorithm, SO consisting primarily of the time-dependent least cost (TDLC) algorithm, and the simulator module consisting of DYNASMART. As mentioned before, the two classes can be run in parallel. Therefore, the time incurred by the UE module is essentially negligible and the total time consists only of the SO and Simulator modules. This test was performed on a 167 MHz dual processor UltraSparc II.

4.6. Routing algorithms

Various routing algorithms can be invoked through VISTA: static and dynamic shortest path algorithms based on time or cost on the links. Versions of the dynamic algorithms that simultaneously optimize route and departure time are also being developed. The algorithms are im-

Table 2
DTA CPU time per iteration

Module	CPU time (s)
UE (TDSP algorithm)	72
SO (TDLC algorithm)	177
Simulator (DYNASMART)	265

plemented in C++. Implementation details can be found in Ziliaskopoulos and Mahmassani (1994). The routing algorithms require as input the link travel times and/or costs, the network topology and have the capability to account for intersection movement delays. The output is typically a tree, rooted at the origin or destination.

5. Implementation of the distributed system

The client module has access to two primary framework resources: the Management and Database modules. Although a primary design principle for the framework is the centralization of all data within the database, certain temporary data can efficiently be communicated directly between modules via remote method processing. Since the Management Module must coordinate the execution of numerous algorithm modules and handle certain data elements, the robust CORBA protocol as specified by the OMG is implemented.

The OMG does not develop products; it focuses on creating specifications which can be implemented by vendors. For this design, the inter-language unification (ILU) (Janssen et al., 1998) package originally developed by Xerox is used. This package was developed independently from CORBA, but has moved consistently towards the OMG standards. It is freely available in source code form, and has support for C, C++, Java, Modula-2, Python, and FOTRAN by wrapping it with C code. Since ILU is available in source code form, it can be compiled and used on a wide variety of platforms including Windows NT, Solaris, SGI, HP, and Linux. Although ILU has some support for Java, it lacks strong support for Applets. However, since ILU follows the OMG standards, the Java 1.2 ORB, as well as many others, are completely compatible with modules written using ILU.

5.1. Benefits of CORBA

The foundation for the introduced framework is CORBA, which is a part of the object management architecture (OMA) (Soley, 1993). The goal of the OMA is to facilitate the construction of systems from objects distributed across multiple computers. Some of the benefits from using this system include:

- *Object orientation*. CORBA is an inherently object-oriented system. Therefore, each component is recognized as a distinct object with an internal state. This aids code reuse, framework structure and expansion.
- *Distributed capabilities*. CORBA posses intrinsic abilities for both synchronous and asynchronous communication between objects. This allows the interface to be executed remotely from the implementation of the algorithms, and multiple processors to be used for the execution of the algorithm modules if needed.
- *The OMG interface definition language* (*IDL*). This is a basic scripting language used to specify the interface for each module. Due to this, each module needs only to know the interface of other modules in order to communicate with them. This essentially separates the published interface of an object from its implementation.
- *The object request broker* (*ORB*). This handles the underlying network communication issues and allows remote objects to be invoked as if they were local. Furthermore, the ORB handles

data transmission (data marshaling), execution of object implementations (activation of objects), and exception reporting to the invoking object.

Through the combination of CORBA and the open database structure, the framework allows for the addition and modification of modules in a straightforward manner. Each module within the framework, which can act as a server, is represented via an IDL interface. By abstracting the link between interface and implementation, a module can simply publish its' interface within the framework and then any other component, including third-part additions, can access the models public functions and data. This allows for the interaction of both proprietary and open models within the framework and lessens the need for specific programming details.

5.2. The IDL standard

IDL is a universally applicable notation for application program interfaces. The international standards organization (ISO) has extracted the IDL section from CORBA to be drafted as an international standard (ISO DIS 14750). Basically, IDL is a simple scripting language which allows programmers to define an opaque boundary between client code and object implementation. It includes the names of framework modules, and the method prototypes for each of these modules. Once an IDL file is written, a parser, or stubber, is used to process the IDL file and generate program source code for a specific language. Since IDL is language-independent, each programming language must have a separate stubber program. The resulting code is referred to as skeleton code and is then compiled with the user written implementation code to create the CORBA program. In the VISTA framework a C, C++, and Java stubber are required since all three languages are used for various modules.

IDL also supports the definition of new data types in a fashion similar to C. By defining a new data type in the IDL file, each language present in the framework will understand this data type, and be capable of passing and retrieving it. The IDL standard also supports attribute, and exception specification.

The GIS maintains connections for two primary external protocols: the JDBC and CORBA systems. From the CORBA system, access is provided for complex module interactions, while the JDBC protocol provides robust data functionality. The GIS itself is designed with the Swing model-view-controller (MVC) architecture according to Sun Microsystems' JDK1.2 specification. The GIS–GUI is broken into five primary families of classes:

View. Maintains the specific screen coordinate data and drawing functionality for updating the actual viewable screen such as simple querying and zooming. Interacts with the *Data* classes to obtain drawable data and complex geographic queries, *User* classes for user-specific information such as preferences and permissions, and *Module* classes for application output and functionality.

Data. Handles the connection to the JDBC protocol and encapsulates all network data within the GIS. Has access to complex SQL querying on the remote database through the JDBC connection. Returns Cartesian coordinate data to the *View* family of classes and executes queries based on the remote database or from its own internal representation depending on the nature of the request.

User. Maintains user-specific information such as preferences, permissions, and execution logs. Possesses a limited capacity for communication with the CORBA Management Module with regard to user-specific information.

Module. Coordinates the control and interaction with remote applications via the CORBA Management Module. Encapsulates all interaction with the modules within the GIS.

Reporting. Handles the representation of non-graphic reporting within the GIS. Interacts primarily with the *Data* classes for data querying, and *User* classes for preferences, etc.

A key set of structures, "*Draw* classes", maintain an internal representation for each data entity, such as vehicle, signal, node, link, etc. Furthermore, the individual drawing rules are delegated to the entity classes themselves, which are capable of being used as Java *Shape* objects. For instance, a *link* class is responsible for requesting the necessary information for a specific use from the remote SQL server, converting (if necessary) this information into its' internal representation, and passing the required drawing capability to the view if requested.

The above structure is intended to abstract and isolate those issues most likely to represent problem-specific issues. The network data, for example, are encapsulated within the GIS *Data* family of classes as well as the SQL database itself. This was found to be necessary, since geographic representations as well as many of the various applications require the network data to be represented in a vastly different manner. Furthermore, the numerous applications are capable of being treated uniformly from within the GIS by deferring implementation specific requirements to the Management Module and beyond.

5.3. Specialized data types

When specific data needs to be returned from a remote module, one method is to define a specialized data type within the framework. This allows any necessary data to be returned by remote procedures using the typical function return capability. For example, the traffic simulator function RouteSim returns information such as total system travel time, and number of nodes, links, cells, and vehicles. To create this specialized data type it is first specified in the frameworks IDL file as:

```
struct s_RSimData {
   double ttime;
   int nodes,links,cell,vehicles;
   };
typedef s\_RSimData RSimData;
```

Other specialized data types include such things as summary data from the static assignment and DTA module, administrative user messages and warnings, module execution history, and path information from network routing algorithms. An alternative method would be to use the central SQL database. The database offers increased robustness, whereas direct CORBA transfers typically yield significantly quicker communication and would prove most useful for real-time requirements.

5.4. CORBA data transfer

As has been mentioned previously, the CORBA standard allows remote objects to be called as if they were typical local objects. This allows a straightforward means for data communication between modules. A typical remote execution and retrieval of data for a synchronous method would appear as:

```
try {
   RSimData rsimdata = Management.RunRouteSim(start, stop, step);
```

```
} catch(Exception e) { ... }
```

In this example, the traffic simulator RouteSim is run with the specified start, stop and step times. The RunRouteSim() method lacks the ONEWAY specification in the IDL file which implies it is to be a synchronous function. Therefore, the interface will wait until the simulator is complete before continuing with operation. When completed, the remote call will return a variable of type RSimData. This type was defined from within the IDL file, so all CORBA modules in the VISTA framework understand this data type. This is the simplest kind of data communication: data are passed to remote modules as parameters in the remote procedure calls, and data are returned back to the client as a specialized data type.

6. Challenges and expected benefits

No major technical challenges appear to exist for implementing and deploying the proposed framework, but quite a few institutional. The system requires close collaboration among many agencies and other entities that traditionally do not interact. It assumes also that certain organizations will relinquish control of data and will be asked to change certain practices. We recognize that this may be a formidable obstacle. However, the expected benefits are so obvious and substantial that as these technologies are routinely accepted and used by corporations, the pressure on public agencies will mountain, and ultimately some version of the proposed system will be adopted.

Next, some of the expected benefits from deploying the proposed system are briefly identified and discussed:

Data consistency. The system ensures the use of the same data by all participating agencies and professionals.

Efficiency. Data duplication and redundant data collection activities will be eliminated.

Economies of scale. Resources could be pooled by various agencies to obtain additional data or functionality that will benefit all. In addition, the value of collecting additional data or developing an additional model can be easily identified; this can help ensure equity for all participating entities.

New data collection and other approaches. The acceptance of this new technology can act as a catalyst for new approaches to be considered by agencies. For example, the fact that the system is accessible over the Internet, makes it open to the general public. Travelers could use it to obtain traffic and transit information, either pre-route or en-route. At the same time, though, the system could obtain data from the travelers, such as origin-destination node, time of travel and possibly other type of information that traditionally is obtained through surveys.

Data availability. Anytime, anywhere, anyone (authorized) can access the system and functionality.

Productivity improvement. It will enable transportation professionals to better do their job by freeing them from tedious tasks of data conversion and manipulation as well as help them better interact with other professionals since they are using consistent data and models. In addition, having all tools and data at hand at any place and time helps them easily perform look-up tasks and presenting their results. Information on any component of the transportation system will be available on-line, enabling time-sensitive decisions to be taken within a much shorter time frame. Implementation of the models will be conducted seamlessly and on time.

Creation of a professional community. Communication among users could provide for the creation of virtual professional communities. Secondary data, reports, analysis and actions can be available to all responsible parties via the database. This will allow for not only information but also for knowledge exchange. If a county engineer has difficulties performing a task, he can access expertise at another county or state office with a click of a button.

New product development. The open architecture of the system will enable developers to validate and test new models based on the same data that all other transportation professionals use, providing credibility to their work and reducing the time to enter the market. Furthermore, the proposed system will enable new users to develop transportation software based on a common platform, reducing software development and implementation costs.

Transparency and accountability. By using electronic signatures, agencies will know who did, what, when and for what purpose.

7. Conclusions

This paper introduced an integrated framework for transportation analysis, control and optimization algorithms that unifies several transportation tools so that they share a common data specification and user interface. The main focus is on interactions between data and models particular to transportation systems rather than specific issues of geographic representation. A mesoscopic traffic simulator (RouteSim), static traffic assignment, DTA, control, and routing algorithms have been implemented and embedded in the framework. The framework is based on the common object request brokerage architecture (CORBA) specification that allows the modules to be written in separate programming languages, and to be run on different machines over a network. A Java-based geographic information system (GIS) is used as a user interface with zooming, panning, and query capabilities and can be executed over the World Wide Web. The GIS runs on the users' computer, and submits requests remotely to the server for execution. The design of the CORBA framework, data and model integration, algorithm development, and network communication issues were discussed.

References

Chambers, J.T., 1999. Virtual close-virtual manufacturing. Fortune 139 (10), 104.

Daganzo, C.F., 1994. The cell transmission model: a simple dynamic representation of highway traffic consistent with the hydrodynamic theory. Transportation Research B 28 (4), 269–287.

Gliebe, J., Ziliaskopoulos, A.K., Koppelman, F., 1999. Stochastic network loading using a paired combinatorial logit model. Transportation Research B (forthcoming).

Hawas, Y., Mahmassani, H.S., Peeta, S., Taylor, R., Ziliaskopoulos, A.K., Chang, G., Peeta, S., 1997. Development of DYNASMART-X software for real-time dynamic traffic assignment. Technical Report ST0067-85-TASK E, Center for Transportation Research, The University of Texas at Austin.

Healey, R.G., 1991. Database management systems. In: Maguire, D.J., Goodchild, M.F., Rhind, D.W. (Eds.), Geographic Information Systems: Principles and Applications, vol. 1. Longman Scientific and Technical, London, pp. 251–267.

Jankowski, P., Stasik, M., 1997. Design considerations for space and time distributed collaborative spatial decision making. Journal of Geographic Information and Decision Analysis 1 (1), 1–8.

Janssen, B., Spreitzer, M., Larner, D., Jacobi, C., 1998. ILU 2.0alpha12 Reference Manual. ftp://ftp.parc.xerox.com/ilu/ilu.html.

Keisler, J.M., Sundell, R.C., 1997. Combining multi-attribute utility and geographic information for boundary decisions: an application to park planning. Journal of Geographic Information and Decision Analysis 1 (2), 101–118.

Mahmassani, H.S., Peeta, S., Hu, T.Y., Ziliaskopoulos, A.K., 1992. Dynamic traffic assignment and simulation procedures for ATIS/ATMS applications. Technical Report DTFH61-90-R-00074-2, Center for Transportation Research, The University of Texas at Austin.

Mahmassani, H.S., Peeta, S., Hu, T.Y., Ziliaskopoulos, A.K., 1993. Dynamic traffic assignment with multiple user classes for real-time ATIS/ATMS applications. In: Proceedings of the Advanced Traffic management Conference, St. Petersburg, Florida, pp. 91–117.

McShane, W.R., Roess, R.P., 1990. Traffic Engineering. Prentice-Hall, Englewood Cliffs, NJ.

Mennecke, B., 1997. Understanding the role of GIS in business: application and research directions. Journal of Geographic Information and Decision Analysis 3 (1), 44–68.

NCHRP 20-27, 1997. A generic data model for linear referencing systems. National Cooperative Highway Research Program of the Transportation Research Board (National Research Council), September 1997.

Object Management Group, 1995. The Common Object Request Broker Architecture and Specification, Revision 2.0. Framingham, M.A., July 1995.

Sheffi, Y., 1985. Urban Transportation Networks: Equilibrium Analysis with Mathematical Programming Methods. Prentice-Hall, Englewood Cliffs, NJ.

Soley, R.M. (Ed.), 1993. Object Management Architecture Guide, second ed. Object Management Group.

Yue, L., Ziliaskopoulos, A.K., Waller, S.T., 1999. Linear programming formulations for system optimum DTA with arrival time based and departure time based demands. Transportation Research Board (submitted).

Ziliaskopoulos, A.K., 2000. A linear programming model for the single destination system optimum dynamic traffic assignment problem. Transportation Science 34, 1–14.

Ziliaskopoulos, A.K., Lee, S., 1997. A cell transmission based assignment-simulation model for integrated freeway/surface street systems. Transportation Research Record 1701, 12–23.

Ziliaskopoulos, A.K., Mahmassani, H.S., 1994. A time-dependent shortest path algorithm for real-time intelligent vehicle/highway systems. Transportation Research Record 1408, 94–104.

Ziliaskopoulos, A.K., Rao, L., 1998. A simultaneous route and departure time choice equilibrium model on dynamic networks. International Transactions of Operational Research 6 (1), 21–37.

Ziliaskopoulos, A.K., Kotzinos, D., Mahmassani, H.S., 1998. Design and implementation of parallel time-dependent shortest path algorithms for real-time intelligent transportation systems. Transportation Research C 5 (2), 95–107.

Subject Index

PII: S096-8090X(00)00051-6